D1203778

# Computational Intelligence

# Computational Intelligence

An Introduction

Second Edition

**Andries P. Engelbrecht**
*University of Pretoria*
*South Africa*

John Wiley & Sons, Ltd

### Other Wiley Editorial Offices

John Wiley & Sons Inc., 111 River Street, Hoboken, NJ 07030, USA

Jossey-Bass, 989 Market Street, San Francisco, CA 94103-1741, USA

Wiley-VCH Verlag GmbH, Boschstr. 12, D-69469 Weinheim, Germany

John Wiley & Sons Australia Ltd, 42 McDougall Street, Milton, Queensland 4064, Australia

John Wiley & Sons (Asia) Pte Ltd, 2 Clementi Loop #02-01, Jin Xing Distripark, Singapore 129809

John Wiley & Sons Canada Ltd, 6045 Freemont Blvd, Mississauga, Ontario, L5R 4J3, Canada

### Library of Congress Cataloging-in-Publication Data

Engelbrecht, Andries P.
    Computational intelligence : an introduction / Andries P. Engelbrecht. – 2nd ed.
        p. cm.
    Includes bibliographical references.
    ISBN 978-0-470-03561-0 (cloth)
    1. Computational intelligence. 2. Neural networks (Computer science) 3. Evolutionary programming
(Computer science) I. Title.
    Q342.E54 2007
    006.3–dc22

                                                            2007021101

### British Library Cataloguing in Publication Data

A catalogue record for this book is available from the British Library

ISBN   978-0-470-03561-0 (HB)

Typeset by the author

*To my parents, Jan and Magriet Engelbrecht,
without whose loving support
this would not have happened.*

# Contents

# Part VI   FUZZY SYSTEMS          451

# List of Figures

# List of Tables

# List of Algorithms

# Preface to the Second Edition

Man has learned much from studies of natural systems, using what has been learned to develop new algorithmic models to solve complex problems. This book presents an introduction to some of these technological paradigms, under the umbrella of computational intelligence (CI). In this context, the book includes artificial neural networks, evolutionary computation, swarm intelligence, artificial immune systems, and fuzzy systems, which are respectively models of the following natural systems: biological neural networks, evolution, swarm behavior of social organisms, natural immune systems, and human thinking processes.

Why this book on computational intelligence? Need arose from a graduate course, where students did not have a deep background of artificial intelligence and mathematics. Therefore the introductory perspective is essential, both in terms of the CI paradigms and mathematical depth. While the material is introductory in nature, it does not shy away from details, and does present the mathematical foundations to the interested reader. The intention of the book is not to provide thorough attention to all computational intelligence paradigms and algorithms, but to give an overview of the most popular and frequently used models. For these models, detailed overviews of different implementations are given. As such, the book is appropriate for beginners in the CI field. The book is also applicable as prescribed material for a third year undergraduate course.

In addition to providing an overview of CI paradigms, the book provides insights into many new developments on the CI research front to tempt the interested reader. As such, the material is useful to graduate students and researchers who want a broader view of the different CI paradigms, also researchers from other fields who have no knowledge of the power of CI techniques, e.g. bioinformaticians, biochemists, mechanical and chemical engineers, economists, musicians and medical practitioners.

The book is organized in six parts. Part I provides a short introduction to the different CI paradigms and a historical overview. Parts II to VI cover the different paradigms, and can be reviewed in any order.

Part II deals with artificial neural networks (NN), including the following topics: Chapter 2 introduces the artificial neuron as the fundamental part of a neural network, including discussions on different activation functions, neuron geometry and learning rules. Chapter 3 covers supervised learning, with an introduction to different types of supervised networks. These include feedforward NNs, functional link NNs, product unit NNs, cascade NNs, and recurrent NNs. Different supervised learning algorithms are discussed, including gradient descent, conjugate gradient methods, LeapFrog and particle swarm optimization. Chapter 4 covers unsupervised learning. Different unsupervised NN models are discussed, including the learning vector quantizer and self-organizing feature maps. Chapter 5 discusses radial basis function NNs. Reinforcement learning is dealt with in Chapter 6. Much attention is given to performance issues of supervised networks in Chapter 7. The focus of the chapter is on accuracy measures and ways to improve performance.

Part III introduces several evolutionary computation models. Topics covered include: an overview of the computational evolution process and basic operators in Chapter 8. Chapter 9 covers genetic algorithms, Chapter 10 genetic programming, Chapter 11 evolutionary programming, Chapter 12 evolution strategies, Chapter 13 differential evolution, Chapter 14 cultural algorithms, and Chapter 15 covers coevolution, introducing both competitive and symbiotic coevolution.

Part IV presents an introduction to two types of swarm-based models: Chapter 16 discusses particle swarm optimization, while ant algorithms are discussed in Chapter 17.

Artificial immune systems are covered in Part V, with the natural immune system being discussed in Chapter 18 and a number of artificial immune models in Chapter 19.

Part VI deals with fuzzy systems. Chapter 20 presents an introduction to fuzzy logic with a discussion of membership functions. Fuzzy inferencing systems are explained in Chapter 21, while fuzzy controllers are discussed in Chapter 22. An overview of rough sets is given in Chapter 23.

Throughout the book, assignments are given to highlight certain aspects of the covered material and to stimulate thought. Some example applications are given where they seemed appropriate to better illustrate the theoretical concepts.

The accompanying website of this book, which can be located at http://ci.cs.up.ac.za, provides algorithms to implement many of the CI models discussed in this book. These algorithms are implemented in Java, and form part of an opensource library, CIlib, developed by the Computational Intelligence Research Group in the Department of Computer Science, University of Pretoria. CIlib (http://cilib.sourceforge.net) is a generic framework for easy implementation of new CI algoithms, and currently contains frameworks for particle swarm optimization, neural networks, and evolutionary computation. Lists with acronyms and symbols used in the book can also be downloaded from the book's website.

As a final remark, it is necessary to thank a number of people who have helped to produce this book. First of all, thanks to my mother, Magriet Engelbrecht, who has helped with typing and proofreading most of the text. Also, thanks to Anri Henning who spent a number of nights proofreading the material. The part on artificial immune systems was written by one of my PhD students, Attie Graaff. Without his help, this book would not have been so complete. Lastly, I thank all of my postgraduate students who have helped with the development of CIlib.

Pretoria, South Africa

# Part I

# INTRODUCTION

# Chapter 1

# Introduction to Computational Intelligence

A major thrust in algorithmic development is the design of algorithmic models to solve increasingly complex problems. Enormous successes have been achieved through the modeling of biological and natural intelligence, resulting in so-called "intelligent systems". These intelligent algorithms include artificial neural networks, evolutionary computation, swarm intelligence, artificial immune systems, and fuzzy systems. Together with logic, deductive reasoning, expert systems, case-based reasoning and symbolic machine learning systems, these intelligent algorithms form part of the field of *Artificial Intelligence* (AI). Just looking at this wide variety of AI techniques, AI can be seen as a combination of several research disciplines, for example, computer science, physiology, philosophy, sociology and biology.

*But what is intelligence?* Attempts to find definitions of intelligence still provoke heavy debate. Dictionaries define intelligence as the ability to comprehend, to understand and profit from experience, to interpret intelligence, having the capacity for thought and reason (especially to a high degree). Other keywords that describe aspects of intelligence include creativity, skill, consciousness, emotion and intuition.

*Can computers be intelligent?* This is a question that to this day causes more debate than the definitions of intelligence. In the mid-1900s, Alan Turing gave much thought to this question. He believed that machines could be created that would mimic the processes of the human brain. Turing strongly believed that there was nothing the brain could do that a well-designed computer could not. More than fifty years later his statements are still visionary. While successes have been achieved in modeling small parts of biological neural systems, there are still no solutions to the complex problem of modeling intuition, consciousness and emotion – which form integral parts of human intelligence.

In 1950 Turing published his test of computer intelligence, referred to as the *Turing test* [858]. The test consisted of a person asking questions via a keyboard to both a person and a computer. If the interrogator could not tell the computer apart from the human, the computer could be perceived as being intelligent. Turing believed that it would be possible for a computer with $10^9$ bits of storage space to pass a 5-minute version of the test with 70% probability by the year 2000. Has his belief come true? The answer to this question is left to the reader, in fear of running head first into

*Computational Intelligence: An Introduction*, Second Edition A.P. Engelbrecht
©2007 John Wiley & Sons, Ltd

another debate! However, the contents of this book may help to shed some light on the answer to this question.

A more recent definition of artificial intelligence came from the IEEE Neural Networks Council of 1996: the study of how to make computers do things at which people are doing better. A definition that is flawed, but this is left to the reader to explore in one of the assignments at the end of this chapter.

This book concentrates on a sub-branch of AI, namely Computational Intelligence (CI) – the study of adaptive mechanisms to enable or facilitate intelligent behavior in complex and changing environments. These mechanisms include those AI paradigms that exhibit an ability to learn or adapt to new situations, to generalize, abstract, discover and associate. The following CI paradigms are covered: artificial neural networks, evolutionary computation, swarm intelligence, artificial immune systems, and fuzzy systems. While individual techniques from these CI paradigms have been applied successfully to solve real-world problems, the current trend is to develop hybrids of paradigms, since no one paradigm is superior to the others in all situations. In doing so, we capitalize on the respective strengths of the components of the hybrid CI system, and eliminate weaknesses of individual components.

The rest of this chapter is organized as follows: Section 1.1 of this chapter presents a short overview of the different CI paradigms, also discussing the biological motivation for each paradigm. A short history of AI is presented in Section 1.2.

At this point it is necessary to state that there are different definitions of what constitutes CI. This book reflects the opinion of the author, and may well cause some debate. For example, swarm intelligence (SI) and artificial immune systems (AIS) are classified as CI paradigms, while many researchers consider these paradigms to belong only under Artificial Life. However, both particle swarm optimization (PSO) and ant colony optimization (ACO), as treated under SI, satisfy the definition of CI given above, and are therefore included in this book as being CI techniques. The same applies to AISs.

## 1.1   Computational Intelligence Paradigms

This book considers five main paradigms of Computation Intelligence (CI), namely artificial neural networks (NN), evolutionary computation (EC), swarm intelligence (SI), artificial immune systems (AIS), and fuzzy systems (FS). Figure 1.1 gives a summary of the aim of the book. In addition to CI paradigms, probabilistic methods are frequently used together with CI techniques, which is also shown in the figure. Soft computing, a term coined by Lotfi Zadeh, is a different grouping of paradigms, which usually refers to the collective set of CI paradigms and probabilistic methods. The arrows indicate that techniques from different paradigms can be combined to form hybrid systems.

Each of the CI paradigms has its origins in biological systems. NNs model biological

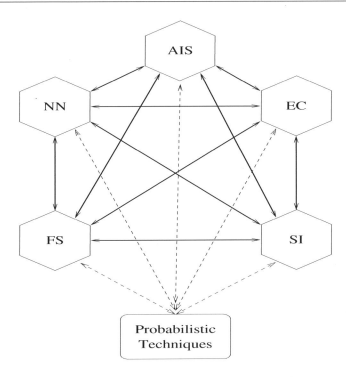

**Figure 1.1** Computational Intelligence Paradigms

neural systems, EC models natural evolution (including genetic and behavioral evolution), SI models the social behavior of organisms living in swarms or colonies, AIS models the human immune system, and FS originated from studies of how organisms interact with their environment.

## 1.1.1   Artificial Neural Networks

The brain is a complex, nonlinear and parallel computer. It has the ability to perform tasks such as pattern recognition, perception and motor control much faster than any computer – even though events occur in the nanosecond range for silicon gates, and milliseconds for neural systems. In addition to these characteristics, others such as the ability to learn, memorize and still generalize, prompted research in algorithmic modeling of biological neural systems – referred to as *artificial neural networks* (NN).

It is estimated that there is in the order of 10-500 billion neurons in the human cortex, with 60 trillion synapses. The neurons are arranged in approximately 1000 main modules, each having about 500 neural networks. *Will it then be possible to truly model the human brain?* Not now. Current successes in neural modeling are for small artificial NNs aimed at solving a specific task. Problems with a single objective can be solved quite easily with moderate-sized NNs as constrained by the capabilities of modern computing power and storage space. The brain has, however, the ability to solve several problems simultaneously using distributed parts of the brain. We still have a long way to go ...

The basic building blocks of biological neural systems are nerve cells, referred to as neurons. As illustrated in Figure 1.2, a neuron consists of a cell body, dendrites and an axon. Neurons are massively interconnected, where an interconnection is between the axon of one neuron and a dendrite of another neuron. This connection is referred to as a *synapse*. Signals propagate from the dendrites, through the cell body to the axon; from where the signals are propagated to all connected dendrites. A signal is transmitted to the axon of a neuron only when the cell "fires". A neuron can either inhibit or excite a signal.

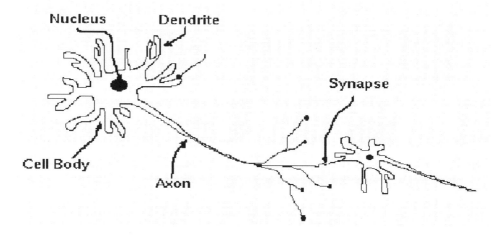

**Figure 1.2** A Biological Neuron

An artificial neuron (AN) is a model of a biological neuron (BN). Each AN receives signals from the environment, or other ANs, gathers these signals, and when fired, transmits a signal to all connected ANs. Figure 1.3 is a representation of an artificial neuron. Input signals are inhibited or excited through negative and positive numerical weights associated with each connection to the AN. The firing of an AN and the strength of the exiting signal are controlled via a function, referred to as the activation function. The AN collects all incoming signals, and computes a net input signal as a function of the respective weights. The net input signal serves as input to the activation function which calculates the output signal of the AN.

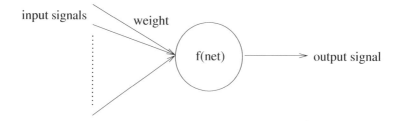

**Figure 1.3** An Artificial Neuron

An artificial neural network (NN) is a layered network of ANs. An NN may consist of an input layer, hidden layers and an output layer. ANs in one layer are connected, fully or partially, to the ANs in the next layer. Feedback connections to previous layers are also possible. A typical NN structure is depicted in Figure 1.4.

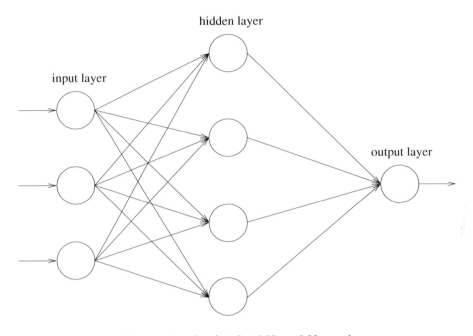

**Figure 1.4** An Artificial Neural Network

Several different NN types have been developed, for example (the reader should note that the list below is by no means complete):

- single-layer NNs, such as the Hopfield network;
- multilayer feedforward NNs, including, for example, standard backpropagation, functional link and product unit networks;
- temporal NNs, such as the Elman and Jordan simple recurrent networks as well as time-delay neural networks;
- self-organizing NNs, such as the Kohonen self-organizing feature maps and the learning vector quantizer;
- combined supervised and unsupervised NNs, e.g. some radial basis function networks.

These NN types have been used for a wide range of applications, including diagnosis of diseases, speech recognition, data mining, composing music, image processing, forecasting, robot control, credit approval, classification, pattern recognition, planning game strategies, compression, and many others.

## 1.1.2   Evolutionary Computation

Evolutionary computation (EC) has as its objective to mimic processes from natural evolution, where the main concept is survival of the fittest: the weak must die. In natural evolution, survival is achieved through reproduction. Offspring, reproduced from two parents (sometimes more than two), contain genetic material of both (or all) parents – hopefully the best characteristics of each parent. Those individuals that inherit bad characteristics are weak and lose the battle to survive. This is nicely illustrated in some bird species where one hatchling manages to get more food, gets stronger, and at the end kicks out all its siblings from the nest to die.

Evolutionary algorithms use a *population* of individuals, where an individual is referred to as a *chromosome*. A chromosome defines the characteristics of individuals in the population. Each characteristic is referred to as a *gene*. The value of a gene is referred to as an *allele*. For each generation, individuals compete to reproduce offspring. Those individuals with the best survival capabilities have the best chance to reproduce. Offspring are generated by combining parts of the parents, a process referred to as *crossover*. Each individual in the population can also undergo mutation which alters some of the allele of the chromosome. The survival strength of an individual is measured using a *fitness* function which reflects the objectives and constraints of the problem to be solved. After each generation, individuals may undergo *culling*, or individuals may survive to the next generation (referred to as *elitism*). Additionally, behavioral characteristics (as encapsulated in phenotypes) can be used to influence the evolutionary process in two ways: phenotypes may influence genetic changes, and/or behavioral characteristics evolve separately.

Different classes of evolutionary algorithms (EA) have been developed:

- **Genetic algorithms** which model genetic evolution.
- **Genetic programming** which is based on genetic algorithms, but individuals are programs (represented as trees).
- **Evolutionary programming** which is derived from the simulation of adaptive behavior in evolution (*phenotypic* evolution).
- **Evolution strategies** which are geared toward modeling the strategy parameters that control variation in evolution, i.e. the evolution of evolution.
- **Differential evolution**, which is similar to genetic algorithms, differing in the reproduction mechanism used.
- **Cultural evolution** which models the evolution of culture of a population and how the culture influences the genetic and phenotypic evolution of individuals.
- **Coevolution** where initially "dumb" individuals evolve through cooperation, or in competition with one another, acquiring the necessary characteristics to survive.

Other aspects of natural evolution have also been modeled. For example, mass extinction, and distributed (island) genetic algorithms, where different populations are maintained with genetic evolution taking place in each population. In addition, aspects such as migration among populations are modeled. The modeling of parasitic behavior has also contributed to improved evolutionary techniques. In this case parasites infect individuals. Those individuals that are too weak die. On the other hand, immunology has been used to study the evolution of viruses and how antibodies should evolve to kill virus infections.

Evolutionary computation has been used successfully in real-world applications, for example, data mining, combinatorial optimization, fault diagnosis, classification, clustering, scheduling, and time series approximation.

### 1.1.3 Swarm Intelligence

Swarm intelligence (SI) originated from the study of colonies, or swarms of social organisms. Studies of the social behavior of organisms (individuals) in swarms prompted the design of very efficient optimization and clustering algorithms. For example, simulation studies of the graceful, but unpredictable, choreography of bird flocks led to the design of the particle swarm optimization algorithm, and studies of the foraging behavior of ants resulted in ant colony optimization algorithms.

Particle swarm optimization (PSO) is a stochastic optimization approach, modeled on the social behavior of bird flocks. PSO is a population-based search procedure where the individuals, referred to as particles, are grouped into a swarm. Each particle in the swarm represents a candidate solution to the optimization problem. In a PSO system, each particle is "flown" through the multidimensional search space, adjusting its position in search space according to its own experience and that of neighboring particles. A particle therefore makes use of the best position encountered by itself and the best position of its neighbors to position itself toward an optimum solution. The effect is that particles "fly" toward an optimum, while still searching a wide area around the current best solution. The performance of each particle (i.e. the "closeness" of a particle to the global minimum) is measured according to a predefined fitness function which is related to the problem being solved. Applications of PSO include function approximation, clustering, optimization of mechanical structures, and solving systems of equations.

Studies of ant colonies have contributed in abundance to the set of intelligent algorithms. The modeling of pheromone depositing by ants in their search for the shortest paths to food sources resulted in the development of shortest path optimization algorithms. Other applications of ant colony optimization include routing optimization in telecommunications networks, graph coloring, scheduling and solving the quadratic assignment problem. Studies of the nest building of ants and bees resulted in the development of clustering and structural optimization algorithms.

### 1.1.4 Artificial Immune Systems

The natural immune system (NIS) has an amazing pattern matching ability, used to distinguish between foreign cells entering the body (referred to as *non-self*, or *antigen*) and the cells belonging to the body (referred to as *self*). As the NIS encounters antigen, the adaptive nature of the NIS is exhibited, with the NIS memorizing the structure of these antigen for faster future response the antigen.

In NIS research, four models of the NIS can be found:

- The **classical view** of the immune system is that the immune system distinguishes between self and non-self, using lymphocytes produced in the lymphoid organs. These lymphocytes "learn" to bind to antigen.

- **Clonal selection theory**, where an active B-Cell produces antibodies through a cloning process. The produced clones are also mutated.

- **Danger theory**, where the immune system has the ability to distinguish between dangerous and non-dangerous antigen.

- **Network theory**, where it is assumed that B-Cells form a network. When a B-Cell responds to an antigen, that B-Cell becomes activated and stimulates all other B-Cells to which it is connected in the network.

An artificial immune system (AIS) models some of the aspects of a NIS, and is mainly applied to solve pattern recognition problems, to perform classification tasks, and to cluster data. One of the main application areas of AISs is in anomaly detection, such as fraud detection, and computer virus detection.

## 1.1.5  Fuzzy Systems

Traditional set theory requires elements to be either part of a set or not. Similarly, binary-valued logic requires the values of parameters to be either 0 or 1, with similar constraints on the outcome of an inferencing process. Human reasoning is, however, almost always not this exact. Our observations and reasoning usually include a measure of uncertainty. For example, humans are capable of understanding the sentence: "Some Computer Science students can program in most languages". But how can a computer represent and reason with this fact?

Fuzzy sets and fuzzy logic allow what is referred to as *approximate reasoning*. With fuzzy sets, an element belongs to a set to a certain degree of certainty. Fuzzy logic allows reasoning with these uncertain facts to infer new facts, with a degree of certainty associated with each fact. In a sense, fuzzy sets and logic allow the modeling of common sense.

The uncertainty in fuzzy systems is referred to as *nonstatistical uncertainty*, and should not be confused with *statistical uncertainty*. Statistical uncertainty is based on the laws of probability, whereas nonstatistical uncertainty is based on vagueness, imprecision and/or ambiguity. Statistical uncertainty is resolved through observations. For example, when a coin is tossed we are certain what the outcome is, while before tossing the coin, we know that the probability of each outcome is 50%. Nonstatistical uncertainty, or fuzziness, is an inherent property of a system and cannot be altered or resolved by observations.

Fuzzy systems have been applied successfully to control systems, gear transmission and braking systems in vehicles, controlling lifts, home appliances, controlling traffic signals, and many others.

## 1.2   Short History

Aristotle (384–322 bc) was possibly the first to move toward the concept of artificial intelligence. His aim was to explain and codify styles of deductive reasoning, which he referred to as *syllogisms*. Ramon Llull (1235–1316) developed the *Ars Magna*: an optimistic attempt to build a machine, consisting of a set of wheels, which was supposed to be able to answer all questions. Today this is still just a dream – or rather, an illusion. The mathematician Gottfried Leibniz (1646–1716) reasoned about the existence of a *calculus philosophicus*, a universal algebra that can be used to represent all knowledge (including moral truths) in a deductive system.

The first major contribution was by George Boole in 1854, with his development of the foundations of propositional logic. In 1879, Gottlieb Frege developed the foundations of predicate calculus. Both propositional and predicate calculus formed part of the first AI tools.

It was only in the 1950s that the first definition of artificial intelligence was established by Alan Turing. Turing studied how machinery could be used to mimic processes of the human brain. His studies resulted in one of the first publications of AI, entitled *Intelligent Machinery*. In addition to his interest in intelligent machines, he had an interest in how and why organisms developed particular shapes. In 1952 he published a paper, entitled *The Chemical Basis of Morphogenesis* – possibly the first studies in what is now known as *artificial life*.

The term *artificial intelligence* was first coined in 1956 at the Dartmouth conference, organized by John MacCarthy – now regarded as the father of AI. From 1956 to 1969 much research was done in modeling biological neurons. Most notable was the work on perceptrons by Rosenblatt, and the *adaline* by Widrow and Hoff. In 1969, Minsky and Papert caused a major setback to artificial neural network research. With their book, called *Perceptrons*, they concluded that, in their "intuitive judgment", the extension of simple perceptrons to multilayer perceptrons "is sterile". This caused research in NNs to go into hibernation until the mid-1980s. During this period of hibernation a few researchers, most notably Grossberg, Carpenter, Amari, Kohonen and Fukushima, continued their research efforts.

The resurrection of NN research came with landmark publications from Hopfield, Hinton, and Rumelhart and McLelland in the early and mid-1980s. From the late 1980s research in NNs started to explode, and is today one of the largest research areas in Computer Science.

The development of evolutionary computation (EC) started with genetic algorithms in the 1950s with the work of Fraser, Bremermann and Reed. However, it is John Holland who is generally viewed as the father of EC, most specifically of genetic algorithms. In these works, elements of Darwin's theory of evolution [173] were modeled algorithmically. In the 1960s, Rechenberg developed evolutionary strategies (ES). Independently from this work, Lawrence Fogel developed evolutionary programming as an approach to evolve behavioral models. Other important contributions that shaped the field were by De Jong, Schaffer, Goldberg, Koza, Schwefel, Storn, and Price.

Many people believe that the history of fuzzy logic started with Gautama Buddha (563 bc) and Buddhism, which often described things in shades of gray. However, the Western community considers the work of Aristotle on two-valued logic as the birth of fuzzy logic. In 1920 Lukasiewicz published the first deviation from two-valued logic in his work on three-valued logic – later expanded to an arbitrary number of values. The quantum philosopher Max Black was the first to introduce quasi-fuzzy sets, wherein degrees of membership to sets were assigned to elements. It was Lotfi Zadeh who contributed most to the field of fuzzy logic, being the developer of fuzzy sets [944]. From then, until the 1980s fuzzy systems was an active field, producing names such as Mamdani, Sugeno, Takagi and Bezdek. Then, fuzzy systems also experienced a dark age in the 1980s, but was revived by Japanese researchers in the late 1980s. Today it is a very active field with many successful applications, especially in control systems. In 1991, Pawlak introduced rough set theory, where the fundamental concept is that of finding a lower and upper approximation to input space. All elements within the lower approximation have full membership, while the boundary elements (those elements between the upper and lower approximation) belong to the set to a certain degree.

Interestingly enough, it was an unacknowledged South African poet, Eugene N Marais (1871-1936), who produced some of the first and most significant contributions to swarm intelligence in his studies of the social behavior of both apes and ants. Two books on his findings were published more than 30 years after his death, namely *The Soul of the White Ant* [560] and *The Soul of the Ape* [559]. The algorithmic modeling of swarms only gained momentum in the early 1990s with the work of Marco Dorigo on the modeling of ant colonies. In 1995, Eberhart and Kennedy [224, 449] developed the particle swarm optimization algorithm as a model of bird flocks. Swarm intelligence is in its infancy, and is a promising field resulting in interesting applications.

The different theories in the science of immunology inspired different artificial immune models (AISs), which are either based on a specific theory on immunology or a combination of the different theories. The initial *classical view* and theory of *clonal selection* in the natural immune system was defined by Burnet [96] as B-Cells and Killer-T-Cells with antigen-specific receptors. This view was enhanced by the definition of Bretscher and Cohn [87] by introducing the concept of a helper T-Cell. Lafferty and Cunningham [497] added a co-stimulatory signal to the helper T-Cell model of Bretscher and Cohn [87].

The first work in AIS on the modeling of the discrimination between *self* and *non-self* with mature T-Cells was introduced by Forrest *et al.* [281]. Forrest *et al.* introduced a training technique known as the *negative selection* of T-Cells [281]. The model of Mori *et al* [606] was the first to implement the *clonal selection* theory, which was applied to optimization problems. The *network theory* of the natural immune system was introduced and formulated by Jerne [416] and further developed by Perelson [677]. The theory of Jerne is that the B-Cells are interconnected to form a network of cells

[416, 677]. The first mathematical model on the theory of Jerne was proposed by Farmer *et al.* [255]. The *network theory* has been modeled into artificial immune systems (AISs) for data mining and data analysis tasks. The earliest AIS research based on the mathematical model of the *network theory* [255], was published by Hunt and Cooke [398]. The model of Hunt and Cooke was applied to the recognition of DNA sequences. The *danger theory* was introduced by Matzinger [567, 568] and is based on the co-stimulated model of Lafferty and Cunningham [497]. The main idea of the *danger theory* is that the immune system distinguishes between what is dangerous and non-dangerous in the body. The first work on danger theory inspired AISs was published by Aickelin and Cayzer [14].

## 1.3 Assignments

1. Comment on the eligibility of Turing's test for computer intelligence, and his belief that computers with $10^9$ bits of storage would pass a 5-minute version of his test with 70% probability.

2. Comment on the eligibility of the definition of artificial intelligence as given by the 1996 IEEE Neural Networks Council.

3. Based on the definition of CI given in this chapter, show that each of the paradigms (NN, EC, SI, AIS, and FS) does satisfy the definition.

# Part II

# ARTIFICIAL NEURAL NETWORKS

Artificial neural networks (NN) were inspired from brain modeling studies. Chapter 1 illustrated the relationship between biological and artificial neural networks. But why invest so much effort in modeling biological neural networks? Implementations in a number of application fields have presented ample rewards in terms of efficiency and ability to solve complex problems. Some of the classes of applications to which artificial NNs have been applied include:

- *classification*, where the aim is to predict the class of an input vector;
- *pattern matching*, where the aim is to produce a pattern best associated with a given input vector;
- *pattern completion*, where the aim is to complete the missing parts of a given input vector;
- *optimization*, where the aim is to find the optimal values of parameters in an optimization problem;
- *control*, where, given an input vector, an appropriate action is suggested;
- *function approximation/times series modeling*, indexfunction approximation where the aim is to learn the functional relationships between input and desired output vectors;
- *data mining*, with the aim of discovering hidden patterns from data – also referred to as knowledge discovery.

A neural network is basically a realization of a nonlinear mapping from $\mathbb{R}^I$ to $\mathbb{R}^K$, i.e.

$$f_{NN} : \mathbb{R}^I \rightarrow \mathbb{R}^K \tag{1.1}$$

where $I$ and $K$ are respectively the dimension of the input and target (desired output) space. The function $f_{NN}$ is usually a complex function of a set of nonlinear functions, one for each neuron in the network.

Neurons form the basic building blocks of NNs. Chapter 2 discusses the single neuron, also referred to as the *perceptron*, in detail. Chapter 3 discusses NNs under the supervised learning regime, while Chapter 4 covers unsupervised learning NNs. Hybrid supervised and unsupervised learning paradigms are discussed in Chapter 5. Reinforcement learning is covered in Chapter 6. Part II is concluded by Chapter 7 which discusses NN performance issues, with reference to supervised learning.

# Chapter 2

# The Artificial Neuron

An artificial neuron (AN), or neuron, implements a nonlinear mapping from $\mathbb{R}^I$ usually to $[0, 1]$ or $[-1, 1]$, depending on the activation function used. That is,

$$f_{AN} : \mathbb{R}^I \rightarrow [0, 1] \tag{2.1}$$

or

$$f_{AN} : \mathbb{R}^I \rightarrow [-1, 1] \tag{2.2}$$

where $I$ is the number of input signals to the AN. Figure 2.1 presents an illustration of an AN with notational conventions that will be used throughout this text. An AN receives a vector of $I$ input signals,

$$\mathbf{z} = (z_1, z_2, \cdots, z_I) \tag{2.3}$$

either from the environment or from other ANs. To each input signal, $z_i$, is associated a weight, $v_i$, to strengthen or deplete the input signal. The AN computes the net input signal, and uses an activation function $f_{AN}$ to compute the output signal, $o$, given the net input. The strength of the output signal is further influenced by a threshold value, $\theta$, also referred to as the *bias*.

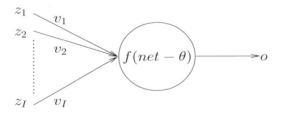

**Figure 2.1** An Artificial Neuron

## 2.1 Calculating the Net Input Signal

The net input signal to an AN is usually computed as the weighted sum of all input signals,

$$net = \sum_{i=1}^{I} z_i v_i \tag{2.4}$$

*Computational Intelligence: An Introduction*, Second Edition A.P. Engelbrecht
©2007 John Wiley & Sons, Ltd

Artificial neurons that compute the net input signal as the weighted sum of input signals are referred to as *summation units* (SU). An alternative to compute the net input signal is to use *product units* (PU) [222], where

$$net = \prod_{i=1}^{I} z_i^{v_i} \tag{2.5}$$

Product units allow higher-order combinations of inputs, having the advantage of increased information capacity.

## 2.2   Activation Functions

The function $f_{AN}$ receives the net input signal and bias, and determines the output (or firing strength) of the neuron. This function is referred to as the *activation function*. Different types of activation functions can be used. In general, activation functions are monotonically increasing mappings, where (excluding the linear function)

$$f_{AN}(-\infty) = 0 \quad \text{or} \quad f_{AN}(-\infty) = -1 \tag{2.6}$$

and

$$f_{AN}(\infty) = 1 \tag{2.7}$$

Frequently used activation functions are enumerated below:

1. **Linear function** (see Figure 2.2(a) for $\theta = 0$):

$$f_{AN}(net - \theta) = \lambda(net - \theta) \tag{2.8}$$

   where $\lambda$ is the slope of the function. The linear function produces a linearly modulated output, where $\lambda$ is a constant.

2. **Step function** (see Figure 2.2(b) for $\theta > 0$):

$$f_{AN}(net - \theta) = \begin{cases} \gamma_1 & \text{if } net \geq \theta \\ \gamma_2 & \text{if } net < \theta \end{cases} \tag{2.9}$$

   The step function produces one of two scalar output values, depending on the value of the threshold $\theta$. Usually, a binary output is produced for which $\gamma_1 = 1$ and $\gamma_2 = 0$; a bipolar output is also sometimes used where $\gamma_1 = 1$ and $\gamma_2 = -1$.

3. **Ramp function** (see Figure 2.2(c) for $\theta > 0$):

$$f_{AN}(net - \theta) = \begin{cases} \gamma & \text{if } net - \theta \geq \epsilon \\ net - \theta & \text{if } -\epsilon < net - \theta < \epsilon \\ -\gamma & \text{if } net - \theta \leq -\epsilon \end{cases} \tag{2.10}$$

   The ramp function is a combination of the linear and step functions.

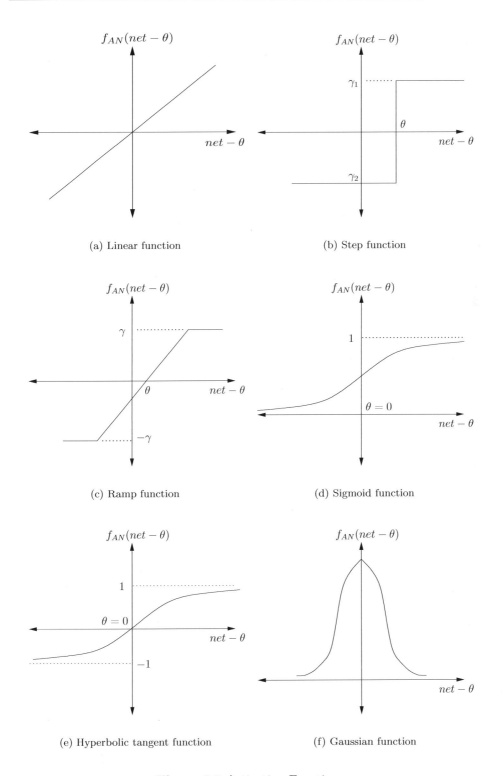

(a) Linear function

(b) Step function

(c) Ramp function

(d) Sigmoid function

(e) Hyperbolic tangent function

(f) Gaussian function

**Figure 2.2** Activation Functions

4. **Sigmoid function** (see Figure 2.2(d) for $\theta = 0$):

$$f_{AN}(net - \theta) = \frac{1}{1 + e^{-\lambda(net-\theta)}} \qquad (2.11)$$

The sigmoid function is a continuous version of the ramp function, with $f_{AN}(net) \in (0, 1)$. The parameter $\lambda$ controls the steepness of the function. Usually, $\lambda = 1$.

5. **Hyperbolic tangent** (see Figure 2.2(e) for $\theta = 0$):

$$f_{AN}(net - \theta) = \frac{e^{\lambda(net-\theta)} - e^{-\lambda(net-\theta)}}{e^{\lambda(net-\theta)} + e^{-\lambda(net-\theta)}} \qquad (2.12)$$

or also approximated as

$$f_{AN}(net - \theta) = \frac{2}{1 + e^{-\lambda(net-\theta)}} - 1 \qquad (2.13)$$

The output of the hyperbolic tangent is in the range $(-1, 1)$.

6. **Gaussian function** (see Figure 2.2(f) for $\theta = 0$):

$$f_{AN}(net - \theta) = e^{-(net-\theta)^2/\sigma^2} \qquad (2.14)$$

where $net - \theta$ is the mean and $\sigma$ the standard deviation of the Gaussian distribution.

## 2.3 Artificial Neuron Geometry

Single neurons can be used to realize linearly separable functions without any error. Linear separability means that the neuron can separate the space of $I$-dimensional input vectors yielding an above-threshold response from those having a below-threshold response by an $I$-dimensional hyperplane. The hyperplane forms the boundary between the input vectors associated with the two output values. Figure 2.3 illustrates the decision boundary for a neuron with the ramp activation function. The hyperplane separates the input vectors for which $\sum_i z_i v_i - \theta > 0$ from the input vectors for which $\sum_i z_i v_i - \theta < 0$.

Figure 2.4 shows how two Boolean functions, AND and OR, can be implemented using a single perceptron. These are examples of linearly separable functions. For such simple functions, it is easy to manually determine values for the bias and the weights. Alternatively, given the input signals and a value for $\theta$, the weight values $v_i$, can easily be calculated by solving

$$\mathbf{v}\mathbf{Z} = \theta \qquad (2.15)$$

where $\mathbf{Z}$ is the matrix of input patterns as given in the truth tables.

An example of a Boolean function that is not linearly separable is the XOR as illustrated in Figure 2.5. A single perceptron can not implement this function. If a single

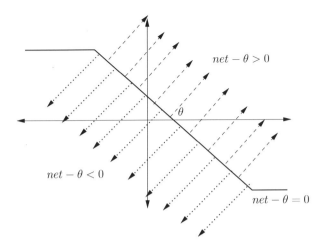

**Figure 2.3** Artificial Neuron Boundary

perceptron is used, then the best accuracy that can be obtained is 75%. To be able to learn functions that are not linearly separable, a layered NN of several neurons is required. For example, the XOR function requires two input units, two hidden units and one output unit.

# 2.4 Artificial Neuron Learning

The question that now remains to be answered is whether an automated approach exists for determining the values of the weights $v_i$ and the threshold $\theta$? As illustrated in the previous section, it is easy to calculate these values for simple problems. But suppose that no prior knowledge exists about the function – except for data – how can the $v_i$ and $\theta$ values be computed? The answer is through learning. The AN learns the best values for the $v_i$ and $\theta$ from the given data. Learning consists of adjusting weight and threshold values until a certain criterion (or several criteria) is (are) satisfied.

There are three main types of learning:

- **Supervised learning**, where the neuron (or NN) is provided with a data set consisting of input vectors and a target (desired output) associated with each input vector. This data set is referred to as the training set. The aim of supervised training is then to adjust the weight values such that the error between the real output, $o = f(net - \theta)$, of the neuron and the target output, $t$, is minimized.

- **Unsupervised learning**, where the aim is to discover patterns or features in the input data with no assistance from an external source. Many unsupervised learning algorithms basically perform a clustering of the training patterns.

- **Reinforcement learning**, where the aim is to reward the neuron (or parts of a NN) for good performance, and to penalize the neuron for bad performance.

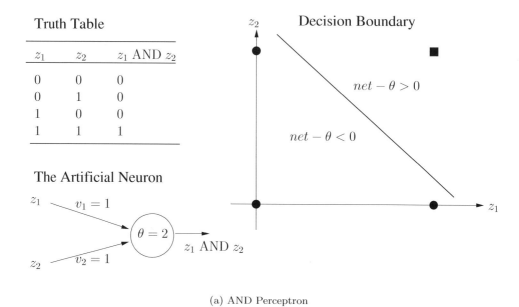

(a) AND Perceptron

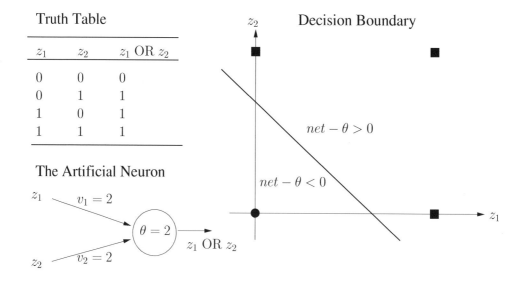

(b) OR Perceptron

**Figure 2.4** Linear Separable Boolean Perceptrons

Truth Table

| $z_1$ | $z_2$ | $z_1$ XOR $z_2$ |
|-------|-------|-----------------|
| 0 | 0 | 0 |
| 0 | 1 | 1 |
| 1 | 0 | 1 |
| 1 | 1 | 0 |

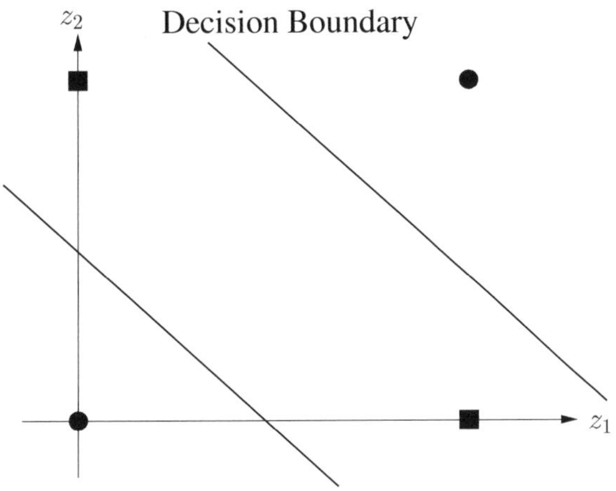

**Figure 2.5** XOR Decision Boundaries

Several learning rules have been developed for the different learning types. Before continuing with these learning rules, we simplify our AN model by introducing augmented vectors.

## 2.4.1 Augmented Vectors

An artificial neuron is characterized by its weight vector $\mathbf{v}$, threshold $\theta$ and activation function. During learning, both the weights and the threshold are adapted. To simplify learning equations, the input vector is augmented to include an additional input unit, $z_{I+1}$, referred to as the *bias unit*. The value of $z_{I+1}$ is always -1, and the weight $v_{I+1}$ serves as the value of the threshold. The net input signal to the AN (assuming SUs) is then calculated as

$$
\begin{aligned}
net &= \sum_{i=1}^{I} z_i v_i - \theta \\
&= \sum_{i=1}^{I} z_i v_i + z_{I+1} v_{I+1} \\
&= \sum_{i=1}^{I+1} z_i v_i
\end{aligned}
\tag{2.16}
$$

where $\theta = z_{I+1} v_{I+1} = -v_{I+1}$.

In the case of the step function, an input vector yields an output of 1 when $\sum_{i=1}^{I+1} z_i v_i \geq 0$, and 0 when $\sum_{i=1}^{I+1} z_i v_i < 0$.

The rest of this chapter considers training rules for single neurons.

### 2.4.2   Gradient Descent Learning Rule

While gradient descent (GD) is not the first training rule for ANs, it is possibly the approach that is used most to train neurons (and NNs for that matter). GD requires the definition of an error (or objective) function to measure the neuron's error in approximating the target. The sum of squared errors

$$\mathcal{E} = \sum_{p=1}^{P_T}(t_p - o_p)^2 \tag{2.17}$$

is usually used, where $t_p$ and $o_p$ are respectively the target and actual output for the $p$-th pattern, and $P_T$ is the total number of input-target vector pairs (*patterns*) in the training set.

The aim of GD is to find the weight values that minimize $\mathcal{E}$. This is achieved by calculating the gradient of $\mathcal{E}$ in weight space, and to move the weight vector along the negative gradient (as illustrated for a single weight in Figure 2.6).

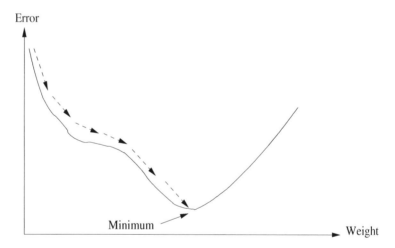

**Figure 2.6** Gradient Descent Illustrated

Given a single training pattern, weights are updated using

$$v_i(t) = v_i(t-1) + \Delta v_i(t) \tag{2.18}$$

with

$$\Delta v_i(t) = \eta\left(-\frac{\partial \mathcal{E}}{\partial v_i}\right) \tag{2.19}$$

where

$$\frac{\partial \mathcal{E}}{\partial v_i} = -2(t_p - o_p)\frac{\partial f}{\partial net_p}z_{i,p} \tag{2.20}$$

and $\eta$ is the learning rate (i.e. the size of the steps taken in the negative direction of the gradient). The calculation of the partial derivative of $f$ with respect to $net_p$ (the net input for pattern $p$) presents a problem for all discontinuous activation functions, such as the step and ramp functions; $z_{i,p}$ is the $i$-th input signal corresponding to pattern $p$. The Widrow-Hoff learning rule presents a solution for the step and ramp functions, while the generalized delta learning rule assumes continuous functions that are at least once differentiable.

### 2.4.3 Widrow-Hoff Learning Rule

For the Widrow-Hoff learning rule [907], assume that $f = net_p$. Then $\frac{\partial f}{\partial net_p} = 1$, giving

$$\frac{\partial \mathcal{E}}{\partial v_i} = -2(t_p - o_p)z_{i,p} \tag{2.21}$$

Weights are then updated using

$$v_i(t) = v_i(t-1) + 2\eta(t_p - o_p)z_{i,p} \tag{2.22}$$

The Widrow-Hoff learning rule, also referred to as the least-means-square (LMS) algorithm, was one of the first algorithms used to train layered neural networks with multiple adaptive linear neurons. This network was commonly referred to as the Madaline [907, 908].

### 2.4.4 Generalized Delta Learning Rule

The generalized delta learning rule is a generalization of the Widrow-Hoff learning rule that assumes differentiable activation functions. Assume that the sigmoid function (from equation (2.11)) is used. Then,

$$\frac{\partial f}{\partial net_p} = o_p(1 - o_p) \tag{2.23}$$

giving

$$\frac{\partial \mathcal{E}}{\partial v_i} = -2(t_p - o_p)o_p(1 - o_p)z_{i,p} \tag{2.24}$$

### 2.4.5 Error-Correction Learning Rule

For the error-correction learning rule it is assumed that binary-valued activation functions are used, for example, the step function. Weights are only adjusted when the neuron responds in error. That is, only when $(t_p - o_p) = 1$ or $(t_p - o_p) = -1$, are weights adjusted using equation (2.22).

## 2.5 Assignments

1. Explain why the threshold $\theta$ is necessary. What is the effect of $\theta$, and what will the consequences be of not having a threshold?

2. Explain what the effects of weight changes are on the separating hyperplane.

3. Explain the effect of changing $\theta$ on the hyperplane that forms the decision boundary.

4. Which of the following Boolean functions can be realized with a single neuron that implements a SU? Justify your answer by giving weight and threshold values.

    (a) $z_1 z_2 \bar{z}_3$

    (b) $z_1 \bar{z}_2 + \bar{z}_1 z_2$

    (c) $z_1 + z_2$

    where $z_1 z_2$ denotes $(z_1 \ AND \ z_2)$; $z_1 + z_2$ denotes $(z_1 \ OR \ z_2)$; $\bar{z}_1$ denotes $(NOT \ z_1)$.

5. Is it possible to use a single PU to learn problems that are not linearly separable?

6. In the calculation of error, why is the error per pattern squared?

7. Can errors be calculated as $|t_p - o_p|$ instead of $(t_p - o_p)^2$ if gradient descent is used to adjust weights?

8. Is the following statement true or false: *A single neuron can be used to approximate the function $f(z) = z^2$?* Justify your answer.

9. What are the advantages of using the hyperbolic tangent activation function instead of the sigmoid activation function?

# Chapter 3

# Supervised Learning Neural Networks

Single neurons have limitations in the type of functions they can learn. A single neuron (implementing a SU) can be used to realize linearly separable functions only. As soon as functions that are not linearly separable need to be learned, a layered network of neurons is required. Training these layered networks is more complex than training a single neuron, and training can be supervised, unsupervised or through reinforcement. This chapter deals with supervised training.

Supervised learning requires a training set that consists of input vectors and a target vector associated with each input vector. The NN learner uses the target vector to determine how well it has learned, and to guide adjustments to weight values to reduce its overall error. This chapter considers different NN types that learn under supervision. These network types include standard multilayer NNs, functional link NNs, simple recurrent NNs, time-delay NNs, product unit NNs, and cascade networks. These different architectures are first described in Section 3.1. Different learning rules for supervised training are then discussed in Section 3.2. The chapter ends with a short discussion on ensemble NNs in Section 3.4.

## 3.1 Neural Network Types

Various multilayer NN types have been developed. Feedforward NNs such as the standard multilayer NN, functional link NN and product unit NN receive external signals and simply propagate these signals through all the layers to obtain the result (output) of the NN. There are no feedback connections to previous layers. Recurrent NNs, on the other hand, have such feedback connections to model the temporal characteristics of the problem being learned. Time-delay NNs, on the other hand, memorize a window of previously observed patterns.

*Computational Intelligence: An Introduction*, Second Edition A.P. Engelbrecht
©2007 John Wiley & Sons, Ltd

### 3.1.1  Feedforward Neural Networks

Figure 3.1 illustrates a standard feedforward neural network (FFNN), consisting of three layers: an input layer (note that some literature on NNs do not count the input layer as a layer), a hidden layer and an output layer. While this figure illustrates only one hidden layer, a FFNN can have more than one hidden layer. However, it has been proved that FFNNs with monotonically increasing differentiable functions can approximate any continuous function with one hidden layer, provided that the hidden layer has enough hidden neurons [383]. A FFNN can also have direct (linear) connections between the input layer and the output layer.

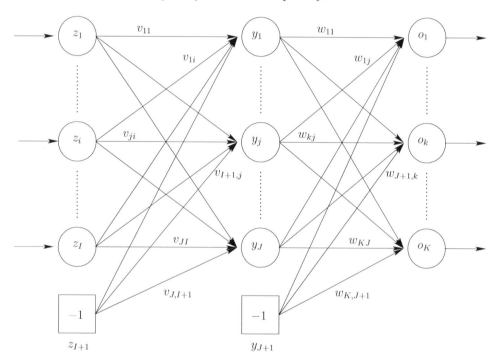

**Figure 3.1** Feedforward Neural Network

The output of a FFNN for any given input pattern $\mathbf{z}_p$ is calculated with a single forward pass through the network. For each output unit $o_k$, we have (assuming no direct connections between the input and output layers),

$$
\begin{aligned}
o_{k,p} &= f_{o_k}(net_{o_k,p}) \\
&= f_{o_k}\left(\sum_{j=1}^{J+1} w_{kj} f_{y_j}(net_{y_j,p})\right) \\
&= f_{o_k}\left(\sum_{j=1}^{J+1} w_{kj} f_{y_j}\left(\sum_{i=1}^{I+1} v_{ji} z_{i,p}\right)\right)
\end{aligned}
\tag{3.1}
$$

where $f_{o_k}$ and $f_{y_j}$ are respectively the activation function for output unit $o_k$ and hidden unit $y_j$; $w_{kj}$ is the weight between output unit $o_k$ and hidden unit $y_j$; $z_{i,p}$ is the value of input unit $z_i$ of input pattern $\mathbf{z}_p$; the $(I+1)$-th input unit and the $(J+1)$-th hidden unit are bias units representing the threshold values of neurons in the next layer.

Note that each activation function can be a different function. It is not necessary that all activation functions be the same. Also, each input unit can implement an activation function. It is usually assumed that input units have linear activation functions.

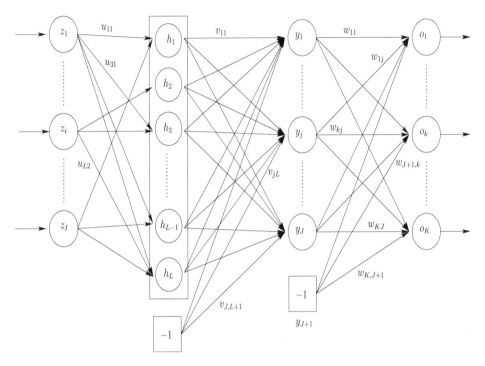

**Figure 3.2** Functional Link Neural Network

## 3.1.2  Functional Link Neural Networks

In functional link neural networks (FLNN) input units do implement activation functions (or rather, transformation functions). A FLNN is simply a FFNN with the input layer expanded into a layer of functional higher-order units [314, 401]. The input layer, with dimension $I$, is therefore expanded to functional units $h_1, h_2, \cdots, h_L$, where $L$ is the total number of functional units, and each functional unit $h_l$ is a function of the input parameter vector $(z_1, \cdots, z_I)$, i.e. $h_l(z_1, \cdots, z_I)$ (see Figure 3.2). The weight matrix $U$ between the input layer and the layer of functional units is defined as

$$u_{li} = \begin{cases} 1 & \text{if functional unit } h_l \text{ is dependent of } z_i \\ 0 & \text{otherwise} \end{cases} \tag{3.2}$$

For FLNNs, $v_{jl}$ is the weight between hidden unit $y_j$ and functional link $h_l$.

Calculation of the activation of each output unit $o_k$ occurs in the same manner as for FFNNs, except that the additional layer of functional units is taken into account:

$$o_{k,p} = f_{o_k} \left( \sum_{j=1}^{J+1} w_{kj} f_{y_j} \left( \sum_{l=1}^{L} v_{jl} h_l(\mathbf{z}_p) \right) \right) \tag{3.3}$$

The use of higher-order combinations of input units may result in faster training times and improved accuracy (see, for example, [314, 401]).

### 3.1.3   Product Unit Neural Networks

Product unit neural networks (PUNN) have neurons that compute the weighted product of input signals, instead of a weighted sum [222, 412, 509]. For product units, the net input is computed as given in equation (2.5).

Different PUNNs have been suggested. In one type each input unit is connected to SUs, and to a dedicated group of PUs. Another PUNN type has alternating layers of product and summation units. Due to the mathematical complexity of having PUs in more than one hidden layer, this section only illustrates the case for which just the hidden layer has PUs, and no SUs. The output layer has only SUs, and linear activation functions are assumed for all neurons in the network. Then, for each hidden unit $y_j$, the net input to that hidden unit is (note that no bias is included)

$$
\begin{aligned}
net_{y_{j,p}} &= \prod_{i=1}^{I} z_{i,p}^{v_{ji}} \\
&= \prod_{i=1}^{I} e^{v_{ji} \ln(z_{i,p})} \\
&= e^{\sum_i v_{ji} \ln(z_{i,p})}
\end{aligned}
\tag{3.4}
$$

where $z_{i,p}$ is the activation value of input unit $z_i$, and $v_{ji}$ is the weight between input $z_i$ and hidden unit $y_j$.

An alternative to the above formulation of the net input signal for PUs is to include a "distortion" factor within the product [406], such as

$$net_{y_{j,p}} = \prod_{i=1}^{I+1} z_{i,p}^{v_{ji}} \tag{3.5}$$

where $z_{I+1,p} = -1$ for all patterns; $v_{j,I+1}$ represents the distortion factor. The purpose of the distortion factor is to dynamically shape the activation function during training to more closely fit the shape of the true function represented by the training data.

If $z_{i,p} < 0$, then $z_{i,p}$ can be written as the complex number $z_{i,p} = \imath^2 |z_{i,p}|$ $(\imath = \sqrt{-1})$ that, substituted in (3.4), yields

$$net_{y_{j,p}} = e^{\sum_i v_{ji} \ln |z_{i,p}|} e^{\sum_i v_{ji} \ln \imath^2} \tag{3.6}$$

Let $c = 0 + \imath = a + b\imath$ be a complex number representing $\imath$. Then,

$$\ln c = \ln re^{\imath\theta} = \ln r + \imath\theta + 2\pi k\imath \tag{3.7}$$

where $r = \sqrt{a^2 + b^2} = 1$.

Considering only the main argument, $arg(c)$, $k = 0$ which implies that $2\pi k\imath = 0$. Furthermore, $\theta = \frac{\pi}{2}$ for $\imath = (0, 1)$. Therefore, $\imath\theta = \imath\frac{\pi}{2}$, which simplifies equation (3.10) to $\ln c = \imath\frac{\pi}{2}$, and consequently,

$$\ln \imath^2 = \imath\pi \tag{3.8}$$

Substitution of (3.8) in (3.6) gives

$$\begin{aligned}
net_{y_{j,p}} &= e^{\sum_i v_{ji} \ln |z_{i,p}|} e^{\sum_i v_{ji}\pi\imath} \\
&= e^{\sum_i v_{ji} \ln |z_{i,p}|} \left[ \cos\left(\sum_{i=1}^{I} v_{ji}\pi\right) + \imath \sin\left(\sum_{i=1}^{I} v_{ji}\pi\right) \right] \tag{3.9}
\end{aligned}$$

Leaving out the imaginary part ([222] show that the added complexity of including the imaginary part does not help with increasing performance),

$$net_{y_{j,p}} = e^{\sum_i v_{ji} \ln |z_{i,p}|} \cos\left(\pi \sum_{i=1}^{I} v_{ji}\right) \tag{3.10}$$

Now, let

$$\rho_{j,p} = \sum_{i=1}^{I} v_{ji} \ln |z_{i,p}| \tag{3.11}$$

$$\phi_{j,p} = \sum_{i=1}^{I} v_{ji}\mathcal{I}_i \tag{3.12}$$

with

$$\mathcal{I}_i = \begin{cases} 0 & \text{if } z_{i,p} > 0 \\ 1 & \text{if } z_{i,p} < 0 \end{cases} \tag{3.13}$$

and $z_{i,p} \neq 0$.

Then,

$$net_{y_{j,p}} = e^{\rho_{j,p}} \cos(\pi\phi_{j,p}) \tag{3.14}$$

The output value for each output unit is then calculated as

$$o_{k,p} = f_{o_k} \left( \sum_{j=1}^{J+1} w_{kj} f_{y_j} \left( e^{\rho_{j,p}} \cos(\pi\phi_{j,p}) \right) \right) \tag{3.15}$$

Note that a bias is now included for each output unit.

### 3.1.4   Simple Recurrent Neural Networks

Simple recurrent neural networks (SRNN) have feedback connections which add the ability to also learn the temporal characteristics of the data set. Several different types of SRNNs have been developed, of which the Elman and Jordan SRNNs are simple extensions of FFNNs.

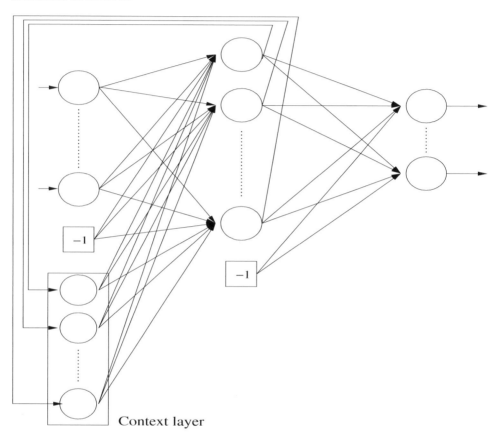

**Figure 3.3** Elman Simple Recurrent Neural Network

The Elman SRNN [236], as illustrated in Figure 3.3, makes a copy of the hidden layer, which is referred to as the *context layer*. The purpose of the context layer is to store the previous state of the hidden layer, i.e. the state of the hidden layer at the previous pattern presentation. The context layer serves as an extension of the input layer, feeding signals representing previous network states, to the hidden layer. The input vector is therefore

$$\mathbf{z} = (\underbrace{z_1, \cdots, z_{I+1}}_{actual\ inputs}, \underbrace{z_{I+2}, \cdots, z_{I+1+J}}_{context\ units}) \tag{3.16}$$

Context units $z_{I+2}, \cdots, z_{I+1+J}$ are fully interconnected with all hidden units. The connections from each hidden unit $y_j$ (for $j = 1, \cdots, J$) to its corresponding context unit $z_{I+1+j}$ have a weight of 1. Hence, the activation value $y_j$ is simply copied to $z_{I+1+j}$. It is, however, possible to have weights not equal to 1, in which case the influence of previous states is weighted. Determining such weights adds additional complexity to the training step.

Each output unit's activation is then calculated as

$$o_{k,p} = f_{o_k} \left( \sum_{j=1}^{J+1} w_{kj} f_{y_j} ( \sum_{i=1}^{I+1+J} v_{ji} z_{i,p} ) \right) \quad (3.17)$$

where $(z_{I+2,p}, \cdots, z_{I+1+J,p}) = (y_{1,p}(t-1), \cdots, y_{J,p}(t-1))$.

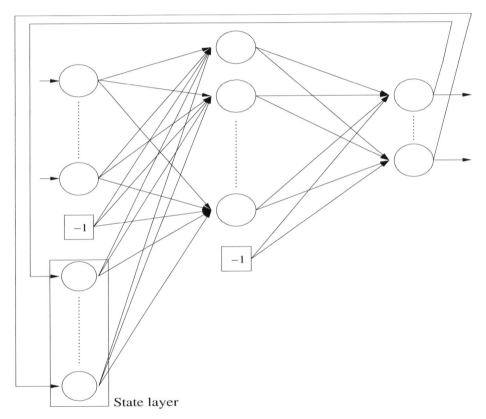

**Figure 3.4** Jordan Simple Recurrent Neural Network

Jordan SRNNs [428], on the other hand, make a copy of the output layer instead of the hidden layer. The copy of the output layer, referred to as the *state layer*, extends the input layer to

$$\mathbf{z} = (\underbrace{z_1, \cdots, z_{I+1}}_{actual\ inputs}, \underbrace{z_{I+2}, \cdots, z_{I+1+K}}_{state\ units}) \quad (3.18)$$

The previous state of the output layer then also serves as input to the network. For each output unit,

$$o_{k,p} = f_{o_k}\left(\sum_{j=1}^{J+1} w_{kj} f_{y_j}\left(\sum_{i=1}^{I+1+K} v_{ji} z_{i,p}\right)\right) \qquad (3.19)$$

where $(z_{I+2,p}, \cdots, z_{I+1+K,p}) = (o_{1,p}(t-1), \cdots, o_{K,p}(t-1))$.

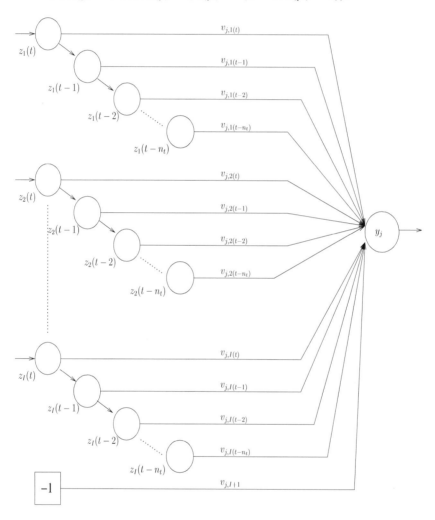

**Figure 3.5** A Single Time-Delay Neuron

## 3.1.5   Time-Delay Neural Networks

A time-delay neural network (TDNN) [501], also referred to as backpropagation-through-time, is a temporal network with its input patterns successively delayed in time. A single neuron with $n_t$ time delays for each input unit is illustrated in Figure 3.5. This type of neuron is then used as a building block to construct a complete feedforward TDNN.

Initially, only $z_{i,p}(t)$, with $t = 0$, has a value and $z_{i,p}(t - t')$ is zero for all $i = 1, \cdots, I$ with time steps $t' = 1, \cdots, n_t$; $n_t$ is the total number of time steps, or number of delayed patterns. Immediately after the first pattern is presented, and before presentation of the second pattern,

$$z_{i,p}(t - 1) = z_{i,p}(t) \qquad (3.20)$$

After presentation of $t'$ patterns and before the presentation of pattern $t' + 1$, for all $t = 1, \cdots, t'$,

$$z_{i,p}(t - t') = z_{i,p}(t - t' + 1) \qquad (3.21)$$

This causes a total of $n_t$ patterns to influence the updates of weight values, thus allowing the temporal characteristics to drive the shaping of the learned function. Each connection between $z_{i,p}(t - t')$ and $z_{i,p}(t - t' + 1)$ has a value of 1.

The output of a TDNN is calculated as

$$o_{k,p} = f_{o_k} \left( \sum_{j=1}^{J+1} w_{kj} f_{y_j} \left( \sum_{i=1}^{I} \sum_{t=0}^{n_t} v_{j,i(t)} z_{i,p}(t) + z_{I+1} v_{j,I+1} \right) \right) \qquad (3.22)$$

### 3.1.6 Cascade Networks

A cascade NN (CNN) [252, 688] is a multilayer FFNN where all input units have direct connections to all hidden units and to all output units. Furthermore, the hidden units are cascaded. That is, each hidden unit's output serves as an input to all succeeding hidden units and all output units. Figure 3.6 illustrates a CNN.

The output of a CNN is calculated is

$$o_{k,p} = f_{o_k} \left( \sum_{i=1}^{I+1} u_{ki} z_i + \sum_{j=1}^{J} w_{kj} f_{y_j} \left( \sum_{i=1}^{I+1} v_{ji} z_i + \sum_{l=1}^{j-1} s_{jl} y_l \right) \right) \qquad (3.23)$$

where $u_{ki}$ represents a weight between output unit $k$ and input unit $i$, $s_{jl}$ is a weight between hidden units $j$ and $l$, and $y_l$ is the activation of hidden unit $l$.

At this point it is important to note that training of a CNN consists of finding weight values and the size of the NN. Training starts with the simplest architecture containing only the $(I+1)K$ direct weights between input and output units (indicated by a solid square in Figure 3.6). If the accuracy of the CNN is unacceptable one hidden unit is added, which adds another $(I + 1)J + (J - 1) + JK$ weights to the network. If $J = 1$, the network includes the weights indicated by the filled squares and circles in Figure 3.6. When $J = 2$, the weights marked by filled triangles are added.

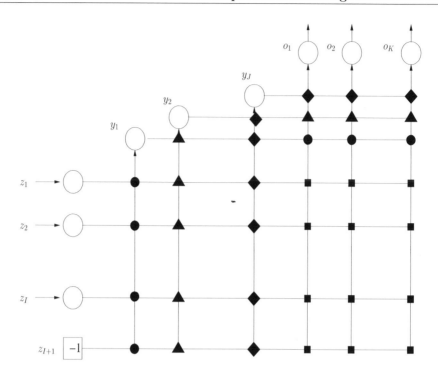

**Figure 3.6** Cascade Neural Network

## 3.2   Supervised Learning Rules

Up to this point it was shown how NNs can be used to calculate an output value given an input pattern. This section explains approaches to train the NN such that the output of the network is an accurate approximation of the target values. First, the learning problem is explained, and then different training algorithms are described.

### 3.2.1   The Supervised Learning Problem

Consider a finite set of input-target pairs $D = \{d_p = (\mathbf{z}_p, \mathbf{t}_p)|p = 1, \cdots, P\}$ sampled from a stationary density $\Omega(D)$, with $z_{i,p}, t_{k,p} \in \mathbb{R}$ for $i = 1, \cdots, I$ and $k = 1, \cdots, K$; $z_{i,p}$ is the value of input unit $z_i$ and $t_{k,p}$ is the target value of output unit $o_k$ for pattern $p$. According to the signal-plus-noise model,

$$\mathbf{t}_p = \mu(\mathbf{z}_p) + \zeta_p \tag{3.24}$$

where $\mu(\mathbf{z})$ is the unknown function. The input values $z_{i,p}$ are sampled with probability density $\omega(\mathbf{z})$, and the $\zeta_{k,p}$ are independent, identically distributed noise sampled with density $\phi(\zeta)$, having zero mean. The objective of learning is then to approximate the unknown function $\mu(\mathbf{z})$ using the information contained in the finite data set $D$. For NN learning this is achieved by dividing the set $D$ randomly into a training set $D_T$, a validation set $D_V$, and a test set $D_G$ (all being dependent from one another). The approximation to $\mu(\mathbf{z})$ is found from the training set $D_T$, memorization is determined from $D_V$ (more about this later), and the generalization accuracy is estimated from the test set $D_G$ (more about this later).

Since prior knowledge about $\Omega(D)$ is usually not known, a nonparametric regression approach is used by the NN learner to search through its hypothesis space $\mathcal{H}$ for a function $f_{NN}(D_T, \mathbf{W})$ which gives a good estimation of the unknown function $\mu(\mathbf{z})$, where $f_{NN}(D_T, \mathbf{W}) \in \mathcal{H}$. For multilayer NNs, the hypothesis space consists of all functions realizable from the given network architecture as described by the weight vector $W$.

During learning, the function $f_{NN} : \mathbb{R}^I \longrightarrow \mathbb{R}^K$ is found which minimizes the empirical error

$$\mathcal{E}_T(D_T; \mathbf{W}) = \frac{1}{P_T} \sum_{p=1}^{P_T} (F_{NN}(\mathbf{z}_p, \mathbf{W}) - \mathbf{t}_p)^2 \qquad (3.25)$$

where $P_T$ is the total number of training patterns. The hope is that a small empirical (training) error will also give a small true error, or generalization error, defined as

$$\mathcal{E}_G(\Omega; \mathbf{W}) = \int (f_{NN}(\mathbf{z}, \mathbf{W}) - \mathbf{t})^2 d\Omega(\mathbf{z}, \mathbf{t}) \qquad (3.26)$$

For the purpose of NN learning, the empirical error in equation (3.25) is referred to as the objective function to be optimized by the optimization method. Several optimization algorithms for training NNs have been developed [51, 57, 221]. These algorithms are grouped into two classes:

- **Local optimization**, where the algorithm may get stuck in a local optimum without finding a global optimum. Gradient descent and scaled conjugate gradient are examples of local optimizers.

- **Global optimization**, where the algorithm searches for the global optimum by employing mechanisms to search larger parts of the search space. Global optimizers include LeapFrog, simulated annealing, evolutionary algorithms and swarm optimization.

Local and global optimization techniques can be combined to form hybrid training algorithms.

Learning consists of adjusting weights until an acceptable empirical error has been reached. Two types of supervised learning algorithms exist, based on when weights are updated:

- **Stochastic/online learning**, where weights are adjusted after each pattern presentation. In this case the next input pattern is selected randomly from the training set, to prevent any bias that may occur due to the order in which patterns occur in the training set.

- **Batch/offline learning**, where weight changes are accumulated and used to adjust weights only after all training patterns have been presented.

## 3.2.2    Gradient Descent Optimization

Gradient descent (GD) optimization has led to one of the most popular learning algorithms, namely backpropagation, popularized by Werbos [897]. Learning iterations (one learning iteration is referred to as an *epoch*) consists of two phases:

1. **Feedforward pass**, which simply calculates the output value(s) of the NN for each training pattern (as discussed in Section 3.1).

2. **Backward propagation**, which propagates an error signal back from the output layer toward the input layer. Weights are adjusted as functions of the backpropagated error signal.

### Feedforward Neural Networks

Assume that the sum squared error (SSE) is used as the objective function. Then, for each pattern, $\mathbf{z}_p$,

$$\mathcal{E}_p = \frac{1}{2} \left( \frac{\sum_{k=1}^{K}(t_{k,p} - o_{k,p})^2}{K} \right) \tag{3.27}$$

where $K$ is the number of output units, and $t_{k,p}$ and $o_{k,p}$ are respectively the target and actual output values of the $k$-th output unit.

The rest of the derivations refer to an individual pattern. The pattern subscript, $p$, is therefore omitted for notational convenience. Also assume sigmoid activation functions in the hidden and output layers with augmented vectors. All hidden and output units use SUs. Then,

$$o_k = f_{o_k}(net_{o_k}) = \frac{1}{1 + e^{-net_{o_k}}} \tag{3.28}$$

and

$$y_j = f_{y_j}(net_{y_j}) = \frac{1}{1 + e^{-net_{y_j}}} \tag{3.29}$$

Weights are updated, in the case of stochastic learning, according to the following equations:

$$w_{kj}(t) \;\; += \;\; \Delta w_{kj}(t) + \alpha \Delta w_{kj}(t-1) \tag{3.30}$$
$$v_{ji}(t) \;\; += \;\; \Delta v_{ji}(t) + \alpha \Delta v_{ji}(t-1) \tag{3.31}$$

where $\alpha$ is the momentum (discussed later).

In the rest of this section the equations for calculating $\Delta w_{kj}(t)$ and $\Delta v_{ji}(t)$ are derived. The reference to time, $t$, is omitted for notational convenience.

From (3.28),

$$\frac{\partial o_k}{\partial net_{o_k}} = \frac{\partial f_{o_k}}{\partial net_{o_k}} = (1 - o_k)o_k = f'_{o_k} \tag{3.32}$$

and

$$\frac{\partial net_{o_k}}{\partial w_{kj}} = \frac{\partial}{\partial w_{kj}} \left( \sum_{j=1}^{J+1} w_{kj} y_j \right) = y_j \tag{3.33}$$

where $f'_{o_k}$ is the derivative of the corresponding activation function. From equations (3.32) and (3.33),

$$
\begin{aligned}
\frac{\partial o_k}{\partial w_{kj}} &= \frac{\partial o_k}{\partial net_{o_k}} \frac{\partial net_{o_k}}{\partial w_{kj}} \\
&= (1 - o_k) o_k y_j \\
&= f'_{o_k} y_j
\end{aligned}
\tag{3.34}
$$

From equation (3.27),

$$\frac{\partial E}{\partial o_k} = \frac{\partial}{\partial o_k} \left( \frac{1}{2} \sum_{k=1}^{K} (t_k - o_k)^2 \right) = -(t_k - o_k) \tag{3.35}$$

Define the output error that needs to be back-propagated as $\delta_{o_k} = \frac{\partial E}{\partial net_{o_k}}$. Then, from equation (3.35) and (3.32),

$$
\begin{aligned}
\delta_{o_k} &= \frac{\partial E}{\partial net_{o_k}} \\
&= \frac{\partial E}{\partial o_k} \frac{\partial o_k}{\partial net_{o_k}} \\
&= -(t_k - o_k)(1 - o_k) o_k = -(t_k - o_k) f'_{o_k}
\end{aligned}
\tag{3.36}
$$

Then, the changes in the hidden-to-output weights are computed from equations (3.35), (3.34) and (3.36) as

$$
\begin{aligned}
\Delta w_{kj} &= \eta \left( -\frac{\partial E}{\partial w_{kj}} \right) \\
&= -\eta \frac{\partial E}{\partial o_k} \frac{\partial o_k}{\partial w_{kj}} \\
&= -\eta \delta_{o_k} y_j
\end{aligned}
\tag{3.37}
$$

Continuing with the input-to-hidden weights,

$$\frac{\partial y_j}{\partial net_{y_j}} = \frac{\partial f_{y_j}}{\partial net_{y_j}} = (1 - y_j) y_j = f'_{y_j} \tag{3.38}$$

and

$$\frac{\partial net_{y_j}}{\partial v_{ji}} = \frac{\partial}{\partial v_{ji}} \left( \sum_{i=1}^{I+1} v_{ji} z_i \right) = z_i \tag{3.39}$$

From equations (3.38) and (3.39),

$$\frac{\partial y_j}{\partial v_{ji}} = \frac{\partial y_j}{\partial net_{y_j}} \frac{\partial net_{y_j}}{\partial v_{ji}}$$

$$= (1 - y_j)y_j z_i = f'_{y_j} z_i \qquad (3.40)$$

and

$$\frac{\partial net_{o_k}}{\partial y_j} = \frac{\partial}{\partial y_j}\left(\sum_{j=1}^{J+1} w_{kj} y_j\right) = w_{kj} \qquad (3.41)$$

From equations (3.36) and (3.41),

$$\frac{\partial E}{\partial y_j} = \frac{\partial}{\partial y_j}\left(\frac{1}{2}\sum_{k=1}^{K}(t_k - o_k)^2\right)$$

$$= \sum_{k=1}^{K} \frac{\partial E}{\partial o_k} \frac{\partial o_k}{\partial net_{o_k}} \frac{\partial net_{o_k}}{\partial y_j}$$

$$= \sum_{k=1}^{K} \frac{\partial E}{\partial net_{o_k}} \frac{\partial net_{o_k}}{\partial y_j}$$

$$= \sum_{k=1}^{K} \delta_{o_k} w_{kj} \qquad (3.42)$$

Define the hidden layer error, which needs to be back-propagated, from equations (3.42) and (3.38) as,

$$\delta_{y_j} = \frac{\partial E}{\partial net_{y_j}}$$

$$= \frac{\partial E}{\partial y_j} \frac{\partial y_j}{\partial net_{y_j}}$$

$$= \sum_{k=1}^{K} \delta_{o_k} w_{kj} f'_{y_j} \qquad (3.43)$$

Finally, the changes to input-to-hidden weights are calculated from equations (3.42), (3.40) and (3.43) as

$$\Delta v_{ji} = \eta\left(-\frac{\partial E}{\partial v_{ji}}\right)$$

$$= -\eta\frac{\partial E}{\partial y_j} \frac{\partial y_j}{\partial v_{ji}}$$

$$= -\eta\delta_{y_j} z_i \qquad (3.44)$$

If direct weights from the input to the output layer are included, the following additional weight updates are needed:

$$\Delta u_{ki} = \eta\left(-\frac{\partial E}{\partial u_{ki}}\right)$$

$$\begin{aligned} &= -\eta \frac{\partial E}{\partial o_k} \frac{\partial o_k}{\partial u_{ki}} \\ &= -\eta \delta_{o_k} z_i \end{aligned} \qquad (3.45)$$

where $u_{ki}$ is a weight from the $i$-th input unit to the $k$-th output unit.

In the case of batch learning, weights are updated as given in equations (3.30) and (3.31), but with

$$\Delta w_{kj}(t) = \sum_{p=1}^{P_T} \Delta w_{kj,p}(t) \qquad (3.46)$$

$$\Delta v_{ji}(t) = \sum_{p=1}^{P_T} \Delta v_{ji,p}(t) \qquad (3.47)$$

where $\Delta w_{kj,p}(t)$ and $\Delta v_{ji,p}(t)$ are weight changes for individual patterns $p$, and $P_T$ is the total number of patterns in the training set.

Stochastic learning is summarized in Algorithm 3.1.

---

**Algorithm 3.1** Stochastic Gradient Descent Learning Algorithm

---

Initialize weights, $\eta$, $\alpha$, and the number of epochs $t = 0$;
**while** *stopping condition(s) not true* **do**
    Let $\mathcal{E}_T = 0$;
    **for** *each training pattern p* **do**
        Do the feedforward phase to calculate $y_{j,p}$ ($\forall\ j = 1, \cdots, J$) and $o_{k,p}$
        ($\forall\ k = 1, \cdots, K$);
        Compute output error signals $\delta_{o_{k,p}}$ and hidden layer error signals $\delta_{y_{j,p}}$;
        Adjust weights $w_{kj}$ and $v_{ji}$ (backpropagation of errors);
        $\mathcal{E}_T + = [\mathcal{E}_p = \sum_{k=1}^{K}(t_{k,p} - o_{k,p})^2]$;
    **end**
    $t = t + 1$;
**end**

---

Stopping criteria usually includes:

- Stop when a maximum number of epochs has been exceeded.
- Stop when the mean squared error (MSE) on the training set,

$$\mathcal{E}_T = \frac{\sum_{p=1}^{P_T} \sum_{k=1}^{K}(t_{k,p} - o_{k,p})^2}{P_T K} \qquad (3.48)$$

  is small enough (other error measures such as the root mean squared error can also be used).

- Stop when overfitting is observed, i.e. when training data is being memorized. An indication of overfitting is when $\mathcal{E}_V > \overline{\mathcal{E}}_V + \sigma_{\mathcal{E}_V}$, where $\overline{\mathcal{E}}_V$ is the average validation error over the previous epochs, and $\sigma_{\mathcal{E}_V}$ is the standard deviation in validation error.

It is straightforward to apply GD optimization to the training of FLNNs, SRNNs and TDNNs, so derivations of the weight update equations are left to the reader. GD learning for PUNNs is given in the next section.

**Product Unit Neural Networks**

This section derives learning equations for PUs used in the hidden layer only, assuming GD optimization and linear activation functions. Since only the equations for the input-to-hidden weights change, only the derivations of these weight update equations are given. The change $\Delta v_{ji}$ in weight $v_{ji}$ is

$$\begin{aligned}
\Delta v_{ji} &= \eta \left( -\frac{\partial E}{\partial v_{ji}} \right) \\
&= -\eta \frac{\partial E}{\partial net_{y_{j,p}}} \frac{net_{y_{j,p}}}{\partial v_{ji}} \\
&= -\eta \delta_{y_{j,p}} \frac{\partial net_{y_{j,p}}}{\partial v_{ji}}
\end{aligned} \tag{3.49}$$

where $\delta_{y_{j,p}}$ is the error signal, computed in the same way as for SUs, and

$$\begin{aligned}
\frac{net_{y_{j,p}}}{\partial v_{ji}} &= \frac{\partial}{\partial v_{ji}} \left( \prod_{i=1}^{I} z_{i,p}^{v_{ji}} \right) \\
&= \frac{\partial}{\partial v_{ji}} \left( e^{\rho_{j,p}} \cos(\pi \phi_{j,p}) \right) \\
&= e^{\rho_{j,p}} [\ln|z_{i,p}| \cos(\pi \phi_{j,p}) - \mathcal{I}_i \pi \sin(\pi \phi_{j,p})]
\end{aligned} \tag{3.50}$$

A major advantage of product units is an increased information capacity compared to summation units [222, 509]. Durbin and Rumelhart showed that the information capacity of a single PU (as measured by its capacity for learning random Boolean patterns) is approximately $3I$, compared to $2I$ for a single SU ($I$ is the number of inputs to the unit) [222]. The larger capacity means that functions approximated using PUs will require less processing elements than required if SUs were used. This point can be illustrated further by considering the minimum number of processing units required for learning the simple polynomial functions in Table 3.1. The minimal number of SUs were determined using a sensitivity analysis variance analysis pruning algorithm [238, 246], while the minimal number of PUs is simply the number of different powers in the expression (provided a polynomial expression).

While PUNNs provide the advantage of having smaller network architectures, a major drawback of PUs is an increased number of local minima, deep ravines and valleys. The search space for PUs is usually extremely convoluted. Gradient descent, which works best when the search space is relatively smooth, therefore frequently gets trapped in local minima or becomes paralyzed (which occurs when the gradient of the error with

**Table 3.1** SUs and PUs Needed for Simple Functions

| Function | SUs | PUs |
|---|---|---|
| $f(z) = z^2$ | 2 | 1 |
| $f(z) = z^6$ | 3 | 1 |
| $f(z) = z^2 + z^5$ | 3 | 2 |
| $f(z_1, z_2) = z_1^3 z_2^7 - 0.5z_1^6$ | 8 | 2 |

respect to the current weight is close to zero). Leerink *et al.* [509] illustrated that the 6-bit parity problem could not be trained using GD and PUs. Two reasons were identified to explain why GD failed: (1) weight initialization and (2) the presence of local minima. The initial weights of a network are usually computed as small random numbers. Leerink *et al.* argued that this is the worst possible choice of initial weights, and suggested that larger initial weights be used instead. But, large weights lead to large weight updates due to the exponential term in the weight update equation (see equation (3.50)), which consequently cause the network to overshoot the minimum. Experience has shown that GD only manages to train PUNNs when the weights are initialized in close proximity of the optimal weight values – the optimal weight values are, however, usually not available.

As an example to illustrate the complexity of the search space for PUs, consider the approximation of the function $f(z) = z^3$, with $z \in [-1, 1]$. Only one PU is needed, resulting in a 1-1-1 NN architecture (that is, one input, one hidden and one output unit). In this case the optimal weight values are $v = 3$ (the input-to-hidden weight) and $w = 1$ (the hidden-to-output weight). Figures 3.7(a)-(b) present the search space for $v \in [-1, 4]$ and $w \in [-1, 1.5]$. The error is computed as the mean squared error over 500 randomly generated patterns. Figure 3.7(b) clearly illustrates 3 minima, with the global minimum at $v = 3, w = 1$. These minima are better illustrated in Figure 3.7(c) where $w$ is kept constant at its optimum value of 1. Initial small random weights will cause the network to be trapped in one of the local minima (having very large MSE). Large initial weights may also be a bad choice. Assume an initial weight $v \geq 4$. The derivative of the error with respect to $v$ is extremely large due to the steep gradient of the error surface. Consequently, a large weight update will be made which may cause jumping over the global minimum. The neural network either becomes trapped in a local minimum, or oscillates between the extreme points of the error surface.

A global stochastic optimization algorithm is needed to allow searching of larger parts of the search space. The optimization algorithm should also not rely heavily on the calculation of gradient information. Simulated annealing [509], genetic algorithms [247, 412], particle swarm optimization [247, 866] and LeapFrog [247] have been used successfully to train PUNNs.

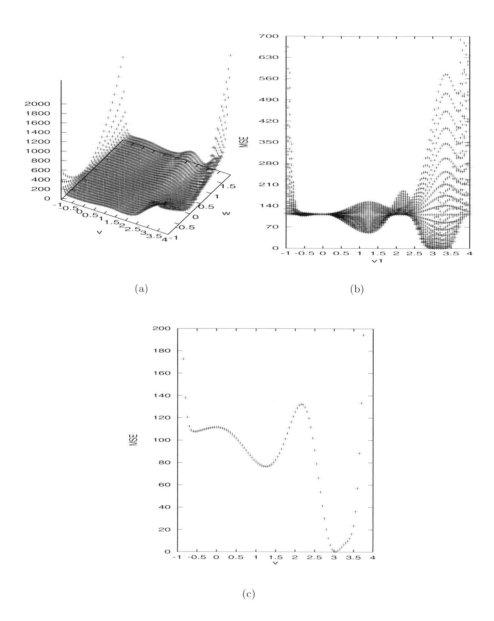

**Figure 3.7** Product Unit Neural Network Search Space for $f(z) = z^3$

### 3.2.3 Scaled Conjugate Gradient

Conjugate gradient optimization trades off the simplicity of GD and the fast quadratic convergence of Newton's method. Several conjugate gradient learning algorithms have been developed (look at the survey in [51]), most of which are based on the assumption that the error function of all weights in the region of the solution can be accurately approximated by

$$\mathcal{E}_T(D_T, \mathbf{w}) = \frac{1}{2}\mathbf{w}^T\mathbf{H}\mathbf{w} - \theta^T\mathbf{w}$$

where $\mathbf{H}$ is the Hessian matrix. Since the dimension of the Hessian matrix is the total number of weights in the network, the calculation of conjugate directions on the error surface becomes computationally infeasible. Computationally feasible conjugate gradient algorithms compute conjugate gradient directions without explicitly computing the Hessian matrix, and perform weight updates along these directions.

---

**Algorithm 3.2** Conjugate Gradient Neural Network Training Algorithm

---

Initialize the weight vector, $\mathbf{w}(0)$;
Define the initial direction vector as

$$\mathbf{p}(0) = -\mathcal{E}'(\mathbf{w}(0)) = \theta - \mathbf{H}\mathbf{w}(0) \tag{3.51}$$

**for** $t = 1, \ldots, n_w$ **do**
    Calculate the step size,

$$\eta(t) = -\frac{\mathcal{E}'(\mathbf{w}(t))^T\mathbf{p}(t)}{\mathbf{p}(t)^T\mathbf{H}\mathbf{p}(t)} \tag{3.52}$$

    Calculate a new weight vector,

$$\mathbf{w}(t+1) = \mathbf{w}(t) + \eta(t)\mathbf{p}(t) \tag{3.53}$$

    Calculate scale factors,

$$\beta(t) = \frac{\mathcal{E}'(\mathbf{w}(t+1))^T\mathcal{E}'(\mathbf{w}(t+1))}{\mathcal{E}'(\mathbf{w}(t))^T\mathcal{E}'(\mathbf{w}(t))} \tag{3.54}$$

    Calculate a new direction vector,

$$\mathbf{p}(t+1) = -\mathcal{E}(\mathbf{w}(t+1)) + \beta(t)\mathbf{p}(t) \tag{3.55}$$

**end**
Return weight vector, $\mathbf{w}(t+1)$;

---

An important aspect in conjugate gradient methods is that of direction vectors, $\{\mathbf{p}(0), \mathbf{p}(1), \ldots, \mathbf{p}(t-1)\}$. These vectors are created to be conjugate with the weight vector, $\mathbf{w}$. That is, $\mathbf{p}^T(t_1)\mathbf{w}\mathbf{p}(t_2) = 0$ for $t_1 \neq t_2$. A new conjugate direction vector is generated at each iteration by adding to the calculated current negative gradient vector of the error function a linear combination of the previous direction vectors. The standard conjugate gradient algorithm is summarized in Algorithm 3.2. Note that this algorithm assumes a quadratic error function, in which case the algorithm converges in no more than $n_w$ steps, where $n_w$ is the total number of weights and biases.

---

**Algorithm 3.3** Fletcher-Reeves Conjugate Gradient Algorithm

---

Initialize the weight vector, $\mathbf{w}(0)$;
Calculate the gradient, $\mathcal{E}'(\mathbf{w}(0))$;
Compute the first direction vector as $\mathbf{p}(0) = -\mathcal{E}'(\mathbf{w}(0))$;
**while** *stopping conditions(s) not true* **do**
    **for** $t = 0, \ldots, n_w - 1$ **do**
        Calculate the step size,

$$\eta(t) = \min_{\eta \geq 0} \mathcal{E}(\mathbf{w}(t) + \eta\mathbf{p}(t)) \tag{3.56}$$

        Calculate a new weight vector,

$$\mathbf{w}(t + 1) = \mathbf{w}(t) + \eta(t)\mathbf{p}(t) \tag{3.57}$$

        Calculate scale factors,

$$\beta(t) = \frac{\mathcal{E}'(\mathbf{w}(t+1))^T \mathcal{E}'(\mathbf{w}(t+1))}{\mathcal{E}'(\mathbf{w}(t))^T \mathcal{E}'(\mathbf{w}(t))} \tag{3.58}$$

        Calculate a new direction vector,

$$\mathbf{p}(t + 1) = -\mathcal{E}'(\mathbf{w}(t + 1)) + \beta(t)\mathbf{p}(t) \tag{3.59}$$

    **end**
    **if** *stopping condition(s) not true* **then**

$$\mathbf{w}(0) = \mathbf{w}(n_w) \tag{3.60}$$

    **end**
**end**
Return $\mathbf{w}(n_w)$ as the solution;

---

The Fletcher-Reeves conjugate gradient algorithm does not assume a quadratic error function. The algorithm restarts after $n_w$ iterations if a solution has not yet been found. The Fletcher-Reeves conjugate gradient algorithm is summarized in Algorithm 3.3.

The scale factors in Algorithms 3.2 and 3.3 can also be calculated in the following ways:

- **Polak-Ribiere** method:

$$\beta(t) = \frac{(\mathcal{E}'(\mathbf{w}(t + 1)) - \mathcal{E}'(\mathbf{w}(t)))^T \mathcal{E}'(\mathbf{w}(t + 1))}{\mathcal{E}'(\mathbf{w}(t))^T \mathcal{E}'(\mathbf{w}(t))} \tag{3.61}$$

- **Hestenes-Stiefer** method:

$$\beta(t) = \frac{(\mathcal{E}'(\mathbf{w}(t+1)) - \mathcal{E}'(\mathbf{w}(t)))^T \mathcal{E}'(\mathbf{w}(t+1))}{\mathbf{p}(t)^T (\mathcal{E}'(\mathbf{w}(t+1)) - \mathcal{E}'(\mathbf{w}(t)))} \tag{3.62}$$

---

**Algorithm 3.4** Scaled Conjugate Gradient Algorithm

---

Initialize the weight vector $\mathbf{w}(1)$ and the scalars $\sigma > 0$, $\lambda_1 > 0$ and $\bar{\lambda} = 0$;
Let $\mathbf{p}(1) = \mathbf{r}(1) = -\mathcal{E}'(\mathbf{w}(1))$, $t = 1$ and *success* = *true*;
Label A: **if** *success* = *true* **then**
    Calculate the second-order information;
**end**
Scale $\mathbf{s}(t)$;
**if** $\delta(t) \le 0$ **then**
    Make the Hessian matrix positive definite;
**end**
Calculate the step size;
Calculate the comparison parameter;
**if** $\Delta(t) \ge 0$ **then**
    A successful reduction in error can be made, so adjust the weights;
    $\bar{\lambda}(t) = 0$;
    success = true;
    **if** $t \bmod n_w = 0$ **then**
        Restart the algorithm, with $\mathbf{p}(t+1) = \mathbf{r}(t+1)$ and go to label A;
    **end**
    **else**
        Create a new conjugate direction;
    **end**
    **if** $\Delta(t) \ge 0.75$ **then**
        Reduce the scale parameter with $\lambda(t) = \frac{1}{2}\lambda(t)$;
    **end**
**end**
**else**
    A reduction in error is not possible, so let $\bar{\lambda}(t) = \lambda(t)$ and *success* = *false*;
**end**
**if** $\Delta(t) < 0.25$ **then**
    Increase the scale parameter to $\lambda(t) = 4\lambda(t)$;
**end**
**if** *the steepest descent direction* $\mathbf{r}(t) \ne 0$ **then**
    Set $t = t + 1$ and go to label A;
**end**
**else**
    Terminate and return $\mathbf{w}(t+1)$ as the desired minimum;
**end**

---

Møller [533] proposed the scaled conjugate gradient (SCG) algorithm as a batch learning algorithm. Step sizes are automatically determined, and the algorithm is restarted after $n_w$ iterations if a good solution was not found. The SCG is summarized in Algorithm 3.4. With reference to the different steps of this algorithm, find detail below:

- Calculation of second-order information:

$$\sigma(t) = \frac{\sigma}{||\mathbf{p}(t)||} \tag{3.63}$$

$$\mathbf{s}(t) = \frac{\mathcal{E}'(\mathbf{w}(t) + \sigma(t)\mathbf{p}(t)) - \mathcal{E}'(\mathbf{w}(t))}{\sigma(t)} \tag{3.64}$$

$$\delta(t) = \mathbf{p}(t)^T\mathbf{s}(t) \tag{3.65}$$

where $\mathbf{p}(t)^T$ is the transpose of vector $\mathbf{p}(t)$, and $||\mathbf{p}(t)||$ is the Euclidean norm.

- Perform scaling:

$$\mathbf{s}(t) \; += \; (\lambda(t) - \overline{\lambda}(t))\mathbf{p}(t) \tag{3.66}$$

$$\delta(t) \; += \; (\lambda(t) - \overline{\lambda}(t))||\mathbf{p}(t)||^2 \tag{3.67}$$

- Make the Hessian matrix positive definite:

$$\mathbf{s}(t) = \mathbf{s}(t) + \left(\lambda(t) - 2\frac{\delta(t)}{||\mathbf{p}(t)||^2}\right)\mathbf{p}(t) \tag{3.68}$$

$$\overline{\lambda}(t) = 2\left(\lambda(t) - 2\frac{\delta(t)}{||\mathbf{p}(t)||^2}\right) \tag{3.69}$$

$$\delta(t) = -\delta(t) + \lambda(t)||\mathbf{p}(t)||^2 \tag{3.70}$$

$$\lambda(t) = \overline{\lambda}(t) \tag{3.71}$$

- Calculate the step size:

$$\mu(t) = \mathbf{p}(t)^T\mathbf{r}(t) \tag{3.72}$$

$$\eta(t) = \frac{\mu(t)}{\delta(t)} \tag{3.73}$$

- Calculate the comparison parameter:

$$\Delta(t) = \frac{2\delta(t)[\mathcal{E}(\mathbf{w}(t)) - \mathcal{E}(\mathbf{w}(t) + \eta(t)\mathbf{p}(t))]}{\mu(t)^2} \tag{3.74}$$

- Adjust the weights:

$$\mathbf{w}(t+1) = \mathbf{w}(t) + \eta(t)\mathbf{p}(t) \tag{3.75}$$

$$\mathbf{r}(t+1) = -\mathcal{E}'(\mathbf{w}(t+1)) \tag{3.76}$$

- Create a new conjugate direction:

$$\beta(t) = \frac{||\mathbf{r}(t+1)||^2 - \mathbf{r}(t+1)^T\mathbf{r}(t)}{\mu(t)} \tag{3.77}$$

$$\mathbf{p}(t+1) = \mathbf{r}(t+1) + \beta(t)\mathbf{p}(t) \tag{3.78}$$

The algorithm restarts each $n_w$ consecutive epochs for which no reduction in error could be achieved, at which point the algorithm finds a new direction to search. The function to calculate the derivative, $\mathcal{E}'(\mathbf{w}) = \frac{\partial \mathcal{E}}{\partial \mathbf{w}}$, computes the derivative of $\mathcal{E}$ with respect to each weight for each of the patterns. The derivatives over all the patterns are then summed, i.e.

$$\frac{\partial \mathcal{E}}{\partial w_i} = \sum_{p=1}^{P_T} \frac{\partial \mathcal{E}}{\partial w_{i,p}} \tag{3.79}$$

where $w_i$ is a single weight.

### 3.2.4 LeapFrog Optimization

LeapFrog is an optimization approach based on the physical problem of the motion of a particle of unit mass in an $n$-dimensional conservative force field [799, 800]. The potential energy of the particle in the force field is represented by the function to be minimized – in the case of NNs, the potential energy is the MSE. The objective is to conserve the total energy of the particle within the force field, where the total energy consists of the particle's potential and kinetic energy. The optimization method simulates the motion of the particle, and by monitoring the kinetic energy, an interfering strategy is adapted to appropriately reduce the potential energy. The LeapFrog NN training algorithm is given in Algorithm 3.5. The reader is referred to [799, 800] for more information on this approach.

### 3.2.5 Particle Swarm Optimization

Particle swarm optimization (PSO), which is a stochastic population-based search method (refer to Chapter 16), can be used to train a NN. In this case, each particle represents a weight vector, and fitness is evaluated using the MSE function (refer to Section 16.7 for more detail on NN training using PSO). What should be noted is that weights and biases are adjusted without using any error signals, or any gradient information. Weights are also not adjusted per training pattern. The PSO velocity and position update equations are used to adjust weights and biases, after which the training set is used to calculate the fitness of a particle (or NN) in $P_T$ feedforward passes.

Evolutionary algorithms can also be used in a similar way to train NNs.

## 3.3 Functioning of Hidden Units

Section 2.3 illustrated the geometry and functioning of a single perceptron. This section illustrates the tasks of the hidden units in supervised NNs. For this purpose, consider a standard FFNN consisting of one hidden layer employing SUs. To simplify visual illustrations, consider the case of two-dimensional input for classification and one-dimensional input for function approximation.

**Algorithm 3.5** LeapFrog Algorithm

Create a random initial solution $\mathbf{w}(0)$, and let $t = -1$;
Let $\Delta t = 0.5, \delta = 1, m = 3, \delta_1 = 0.001, \epsilon = 10^{-5}, i = 0, j = 2, s = 0, p = 1$;
Compute the initial acceleration $\mathbf{a}(0) = -\nabla\mathcal{E}(\mathbf{w}(0))$ and velocity $\mathbf{v}(0) = \frac{1}{2}\mathbf{a}(0)\Delta t$;
**repeat**
     $t = t + 1$;
     Compute $\|\Delta\mathbf{w}(t)\| = \|\mathbf{v}(t)\|\Delta t$;
     **if** $\|\Delta\mathbf{w}(t)\| < \delta$ **then**
         $p = p + \delta_1$, $\Delta t = p\Delta t$;
     **end**
     **else**
         $\mathbf{v}(t) = \delta\mathbf{v}(t)/(\Delta t\|\mathbf{v}(t)\|)$;
     **end**
     **if** $s \geq m$ **then**
         $\Delta t = \Delta t/2$, $s = 0$;
         $\mathbf{w}(t) = (\mathbf{w}(t) + \mathbf{w}(t-1))/2$;
         $\mathbf{v}(t) = (\mathbf{v}(t) + \mathbf{v}(t-1))/4$;
     **end**
     $\mathbf{w}(t+1) = \mathbf{w}(t) + \mathbf{v}(t)\Delta t$;
     **repeat**
         $\mathbf{a}(t+1) = -\nabla\mathcal{E}(\mathbf{w}(t+1))$;
         $\mathbf{v}(t+1) = \mathbf{v}(t) + \mathbf{a}(t+1)\Delta t$;
         **if** $\mathbf{a}^T(t+1)\mathbf{a}(t) > 0$ **then**
             $s = 0$;
         **end**
         **else**
             $s = s + 1, p = 1$;
         **end**
         **if** $\|\mathbf{a}(t+1)\| > \epsilon$ **then**
             **if** $\|\mathbf{v}(t+1)\| > \|\mathbf{v}(t)\|$ **then**
                 $i = 0$;
             **end**
             **else**
                 $\mathbf{w}(t+2) = (\mathbf{w}(t+1) + \mathbf{w}(t))/2$;
                 $i = i + 1$;
                 Perform a restart: **if** $i \leq j$ **then**
                     $\mathbf{v}(t+1) = (\mathbf{v}(t+1) + \mathbf{v}(t))/4$;
                     $t = t + 1$;
                 **end**
                 **else**
                     $\mathbf{v}(t+1) = 0$, $j = 1, t = t + 1$;
                 **end**
             **end**
         **end**
     **until** $\|\mathbf{v}(t+1)\| > \|\mathbf{v}(t)\|$;
**until** $\|\mathbf{a}(t+1)\| \leq \epsilon$;
Return $\mathbf{w}(t)$ as the solution;

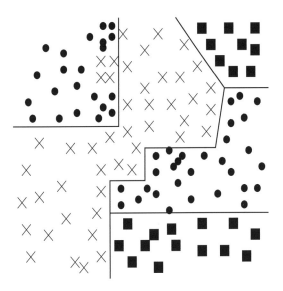

**Figure 3.8** Feedforward Neural Network Classification Boundary Illustration

For classification problems, the task of hidden units is to form the decision boundaries to separate different classes. Figure 3.8 illustrates the boundaries for a three-class problem. Solid lines represent boundaries. For this artificial problem ten boundaries exist. Since each hidden unit implements one boundary, ten hidden units are required to perform the classification as illustrated in the figure. Less hidden units can be used, but at the cost of an increase in classification error. Also note that in the top left corner there are misclassifications of class ×, being part of the space for class •. This problem can be solved by using three additional hidden units to form these boundaries. How can the number of hidden units be determined without using any prior knowledge about the input space? This very important issue is dealt with in Chapter 7, where the relationship between the number of hidden units and performance is investigated.

In the case of function approximation, assuming a one-dimensional function as depicted in Figure 3.9, five hidden units with sigmoid activation functions are required to learn the function. A sigmoid function is then fitted for each inflection point of the target function. The number of hidden units is therefore the number of turning points plus one. In the case of linear activation functions, the hidden units perform the same task. However, more linear activation functions may be required to learn the function to the same accuracy as obtained using sigmoid functions.

## 3.4   Ensemble Neural Networks

Training of NNs starts on randomly selected initial weights. This means that each time a network is retrained on the same data set, different results can be expected, since learning starts at different points in the search space; different NNs may disagree, and make different errors. This problem in NN training prompted the development of

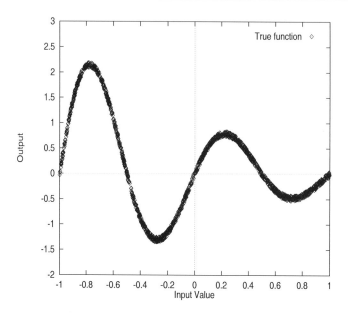

**Figure 3.9** Hidden Unit Functioning for Function Approximation

ensemble networks, where the aim is to optimize results through the combination of a number of individual networks, trained on the same task.

In its most basic form, an ensemble network – as illustrated in Figure 3.10 – consists of a number of NNs all trained on the same data set, using the same architecture and learning algorithm. At convergence of the individual NN members, the results of the different NNs need to be combined to form one, final result. The final result of an ensemble can be calculated in several ways, of which the following are simple and efficient approaches:

- Select the NN within the ensemble that provides the best generalization performance.

- Take the average over the outputs of all the members of the ensemble.

- Form a linear combination of the outputs of each of the NNs within the ensemble. In this case a weight, $w_n$, is assigned to each network as an indication of the credibility of that network. The final output of the ensemble is therefore a weighted sum of the outputs of the individual networks.

The combination of inputs as discussed above is sensible only when there is disagreement among the ensemble members, or if members make their errors on different parts of the search space.

Several adaptations of the basic ensemble model are of course possible. For example, instead of having each NN train on the same data set, different data sets can be used. One such approach is bagging, which is a bootstrap ensemble method that creates individuals for its ensemble by training each member network on a random redistribution of the original training set [84]. If the original training set contained

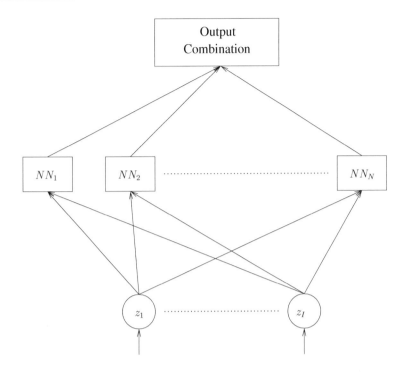

**Figure 3.10** Ensemble Neural Network

$P_T$ patterns, then a data set of $P_T$ patterns is randomly sampled from the original training set for each of the ensemble members. This means that patterns may be duplicated in the member training sets. Also, not all of the patterns in the original training set will necessarily occur in the member training sets.

Alternatively, the architectures of the different NNs may differ. Even different NN types can be used. It is also not necessary that each of the members be trained using the same optimization algorithm.

The above approaches to ensemble networks train individual NNs in parallel, independent of one another. Much more can be gained under a cooperative ensemble strategy, where individual NNs (referred to as agents) exchange their experience and knowledge during the training process. Research in such cooperative agents is now very active, and the reader is recommended to read more about these.

One kind of cooperative strategy for ensembles is referred to as boosting [220, 290]. With boosting, members of the ensemble are not trained in parallel. They are trained sequentially, where already trained members filter patterns into easy and hard patterns. New, untrained members of the ensemble then focus more on the hard patterns as identified by previously trained networks.

# 3.5    Assignments

1. Give an expression for $o_{k,p}$ for a FFNN with direct connections between the input and output layer.

2. Why is the term $(-1)^{v_{j,I+1}}$ possible in equation (3.5)?

3. Explain what is meant by the terms overfitting and underfitting. Why is $\mathcal{E}_V > \overline{\mathcal{E}}_V + \sigma_{\mathcal{E}_V}$ a valid indication of overfitting?

4. Investigate the following aspects:

    (a) Are direct connections between the input and output layers advantageous? Give experimental results to illustrate.

    (b) Compare a FFNN and an Elman RNN trained using GD. Use the following function as benchmark: $z_t = 1 + 0.3z_{t-2} - 1.4z_{t-1}^2$, with $z_1, z_2 \sim U(-1,1)$, sampled from a uniform distribution in the range $(-1,1)$.

    (c) Compare stochastic learning and batch learning using GD for the function $o_t = z_t$ where $z_t = 0.3z_{t-6} - 0.6z_{t-4} + 0.5z_{t-1} + 0.3z_{t-6}^2 - 0.2z_{t-4}^2 + \zeta_t$, and $z_t \sim U(-1,1)$ for $t = 1, \cdots, 10$, and $\zeta_t \sim N(0, 0.05)$.

    (d) Compare GD and SCG on any classification problem from the UCI machine learning repository at http://www.ics.uci.edu/~mlearn/MLRepository.html.

    (e) Show if PSO performs better than GD in training a FFNN.

5. Assume that gradient descent is used as the optimization algorithm, and derive learning equations for the Elman SRNN, the Jordan SRNN, TDNN and FLNN.

6. Explain how a SRNN learns the temporal characteristics of data.

7. Show how a FLNN can be used to fit a polynomial through data points given in a training set.

8. Explain why bias for only the output units of a PUNN, as discussed in this chapter, is sufficient. In other words, the PUs do not have a bias. What will be the effect if a bias is included in the PUs?

9. Explain why the function $f(z_1, z_2) = z_1^3 z_2^7 - 0.5z_1^6$ requires only two PUs, if it is assumed that PUs are only used in the hidden layer, with linear activations in both the hidden and output layers.

10. Assume that a PUNN with PUs in the hidden layer, SUs in that output layer, and linear activation functions in all layers, is used to approximate a polynomial. Explain why the minimal number of hidden units is simply the total number of non-constant, unique terms in the polynomial.

11. What is the main requirement for activation and error functions if gradient descent is used to train supervised neural networks?

12. What is the main advantage of using recurrent neural networks instead of feedforward neural networks?

13. What is the main advantage in using PUs instead of SUs?

14. Propose a way in which a NN can learn a functional mapping and its derivative.

15. Show that the PUNN as given in Section 3.1.3 implements a polynomial approximation.

# Chapter 4

# Unsupervised Learning
# Neural Networks

An important feature of NNs is their ability to learn from their environment. Chapter 3 covered NN types that learned under the guidance of a supervisor or teacher. The supervisor presents the NN learner with an input pattern and a desired response. Supervised learning NNs then try to learn the functional mapping between the input and desired response vectors. In contrast to supervised learning, the objective of unsupervised learning is to discover patterns or features in the input data with no help from a teacher. This chapter deals with the unsupervised learning paradigm.

Section 4.1 presents a short background on unsupervised learning. Hebbian learning is presented in Section 4.2, while Section 4.3 covers principal component learning, Section 4.4 covers the learning vector quantizer version I, and Section 4.5 discusses self-organizing feature maps.

## 4.1    Background

Aristotle observed that human memory has the ability to connect items (e.g. objects, feelings and ideas) that are similar, contradictory, that occur in close proximity, or in succession [473]. The patterns that we associate may be of the same or different types. For example, a photo of the sea may bring associated thoughts of happiness, or smelling a specific fragrance may be associated with a certain feeling, memory or visual image. Also, the ability to reproduce the pitch corresponding to a note, irrespective of the form of the note, is an example of the pattern association behavior of the human brain.

Artificial neural networks have been developed to model the pattern association ability of the human brain. These networks are referred to as associative memory NNs. Associative memory NNs are usually two-layer NNs, where the objective is to adjust the weights such that the network can store a set of pattern associations – without any external help from a teacher. The development of these associative memory NNs is mainly inspired from studies of the visual and auditory cortex of mammalian organisms, such as the bat. These artificial NNs are based on the fact that parts of the

*Computational Intelligence: An Introduction*, Second Edition A.P. Engelbrecht
©2007 John Wiley & Sons, Ltd

brain are organized such that different sensory inputs are represented by topologically ordered computational maps. The networks form a topographic map of the input patterns, where the coordinates of the neurons correspond to intrinsic features of the input patterns.

An additional feature modeled with associative memory NNs is to preserve old information as new information becomes available. In contrast, supervised learning NNs have to retrain on all the information when new data becomes available; if not, supervised networks tend to focus on the new information, forgetting what the network has already learned.

Unsupervised learning NNs are functions that map an input pattern to an associated target pattern, i.e.

$$f_{NN} : \mathbb{R}^I \rightarrow \mathbb{R}^K \tag{4.1}$$

as illustrated in Figure 4.1. The single weight matrix determines the mapping from the input vector $\mathbf{z}$ to the output vector $\mathbf{o}$.

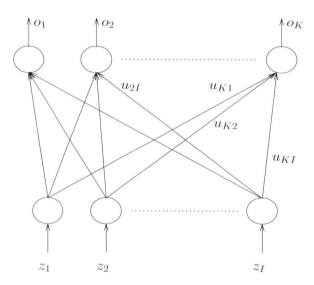

**Figure 4.1** Unsupervised Neural Network

## 4.2 Hebbian Learning Rule

The Hebbian learning rule, named after the neuropsychologist Hebb, is the oldest and simplest learning rule. With Hebbian learning [203], weight values are adjusted based on the correlation of neuron activation values. The motivation of this approach is from Hebb's hypothesis that the ability of a neuron to fire is based on that neuron's ability to cause other neurons connected to it to fire. In such cases the weight between the two correlated neurons is strengthened (or increased). Using the notation from

Figure 4.1, the change in weight at time step $t$ is given as

$$\Delta u_{ki}(t) = \eta o_{k,p} z_{i,p} \tag{4.2}$$

Weights are then updated using

$$u_{ki}(t) = u_{ki}(t-1) + \Delta u_{ki}(t) \tag{4.3}$$

where $\eta$ is the learning rate.

From equation (4.2), the adjustment of weight values is larger for those input-output pairs for which the input value has a greater effect on the output values.

The Hebbian learning rule is summarized in Algorithm 4.1. The algorithm terminates when there is no significant change in weight values, or when a specified number of epochs has been exceeded.

---

**Algorithm 4.1** Hebbian Learning Algorithm

---

Initialize all weights such that $u_{ki} = 0$, $\forall i = 1, \cdots, I$ and $\forall k = 1, \cdots, K$;
**while** *stopping condition(s) not true* **do**
    **for** *each input pattern* $\mathbf{z}_p$ **do**
        Compute the corresponding output vector $\mathbf{o}_p$;
    **end**
    Adjust the weights using equation (4.3);
**end**

---

A problem with Hebbian learning is that repeated presentation of input patterns leads to an exponential growth in weight values, driving the weights into saturation. To prevent saturation, a limit is posed on the increase in weight values. One type of limit is to introduce a nonlinear *forgetting factor*:

$$\Delta u_{ki}(t) = \eta o_{k,p} z_{i,p} - \gamma o_{k,p} u_{ki}(t-1) \tag{4.4}$$

where $\gamma$ is a positive constant, or equivalently,

$$\Delta u_{ki}(t) = \gamma o_{k,p}[\beta z_{i,p} - u_{ki}(t-1)] \tag{4.5}$$

with $\beta = \eta/\gamma$. Equation (4.5) implies that inputs for which $z_{i,p} < u_{ki}(t-1)/\beta$ have their corresponding weights $u_{ki}$ decreased by a value proportional to the output value $o_{k,p}$. When $z_{i,p} > u_{ki}(t-1)/\beta$, weight $u_{ki}$ is increased proportional to $o_{k,p}$.

Sejnowski proposed another way to formulate Hebb's postulate, using the covariance correlation of the neuron activation values [773]:

$$\Delta u_{ki}(t) = \eta[(z_{i,p} - \bar{z}_i)(o_{k,p} - \bar{o}_k)] \tag{4.6}$$

with

$$\bar{z}_i = \sum_{p=1}^{P_T} z_{i,p}/P \tag{4.7}$$

$$\overline{o}_k = \sum_{p=1}^{P_T} o_{k,p}/P \tag{4.8}$$

Another variant of the Hebbian learning rule uses the correlation in the changes in activation values over consecutive time steps. For this learning rule, referred to as *differential Hebbian learning*,

$$\Delta u_{ki}(t) = \eta \Delta z_i(t) \Delta o_k(t-1) \tag{4.9}$$

where

$$\Delta z_i(t) = z_{i,p}(t) - z_{i,p}(t-1) \tag{4.10}$$

and

$$\Delta o_k(t-1) = o_{k,p}(t-1) - o_{k,p}(t-2) \tag{4.11}$$

## 4.3   Principal Component Learning Rule

Principal component analysis (PCA) [426] is a statistical technique used to transform a data space into a smaller space of the most relevant features. The aim is to project the original $I$-dimensional space onto an $I'$-dimensional linear subspace, where $I' < I$, such that the variance in the data is maximally explained within the smaller $I'$-dimensional space. Features (or inputs) that have little variance are thereby removed. The principal components of a data set are found by calculating the covariance (or correlation) matrix of the data patterns, and by getting the minimal set of orthogonal vectors (the eigenvectors) that span the space of the covariance matrix. Given the set of orthogonal vectors, any vector in the space can be constructed with a linear combination of the eigenvectors.

Oja developed the first principal components learning rule, with the aim of extracting the principal components from the input data [635]. Oja's principal components learning rule is an extension of the Hebbian learning rule, referred to as normalized Hebbian learning, to include a feedback term to constrain weights. In doing so, principal components could be extracted from the data. The weight change is given as

$$
\begin{aligned}
\Delta u_{ki}(t) &= u_k i(t) - u_{ki}(t-1) \\
&= \eta o_{k,p}[z_{i,p} - o_{k,p} u_{ki}(t-1)] \\
&= \underbrace{\eta o_{k,p} z_{i,p}}_{Hebbian} - \underbrace{\eta o_{k,p}^2 u_{ki}(t-1)}_{forgetting\ factor}
\end{aligned}
\tag{4.12}
$$

The first term corresponds to standard Hebbian learning (refer to equation (4.2)), while the second term is a forgetting factor to prevent weight values from becoming unbounded.

The value of the learning rate, $\eta$, above is important to ensure convergence to a stable state. If $\eta$ is too large, the algorithm will not converge due to numerical unstability. If $\eta$ is too small, convergence is extremely slow. Usually, the learning rate is time dependent, starting with a large value that decays gradually as training progresses. To ensure numerical stability of the algorithm, the learning rate $\eta_k(t)$ for output unit $o_k$ must satisfy the inequality:

$$0 < \eta_k(t) < \frac{1}{1.2\lambda_k} \tag{4.13}$$

where $\lambda_k$ is the largest eigenvalue of the covariance matrix of the inputs to the unit [636]. A good initial value is given as $\eta_k(0) = 1/[2\mathcal{Z}^T\mathcal{Z}]$, where $\mathcal{Z}$ is the input matrix.

Cichocki and Unbehauen [130] provided an adaptive learning rate that utilizes a forgetting factor, $\gamma$, as follows:

$$\eta_k(t) = \frac{1}{\frac{\gamma}{\eta_k(t-1)} + o_k^2(t)} \tag{4.14}$$

with

$$\eta_k(0) = \frac{1}{o_k^2(0)} \tag{4.15}$$

Usually, $0.9 \leq \gamma \leq 1$.

The above can be adapted to allow the same learning rate for all the weights in the following way:

$$\eta_k(t) = \frac{1}{\frac{\gamma}{\eta_k(t-1)} + \|\mathbf{o}(t)\|_2^2} \tag{4.16}$$

with

$$\eta_k(0) = \frac{1}{\|\mathbf{o}(0)\|_2^2} \tag{4.17}$$

Sanger [756] developed another principal components learning algorithm, similar to that of Oja, referred to as generalized Hebbian learning. The only difference is the inclusion of more feedback information and a decaying learning rate $\eta(t)$:

$$\Delta u_{ki}(t) = \eta(t)[\underbrace{z_{i,p}o_{k,p}}_{Hebbian} - o_{k,p}\sum_{j=0}^{k} u_{ji}(t-1)o_{j,p}] \tag{4.18}$$

For more information on principal component learning, the reader is referred to the summary in [356].

## 4.4 Learning Vector Quantizer-I

One of the most frequently used unsupervised clustering algorithms is the learning vector quantizer (LVQ) developed by Kohonen [472, 474]. While several versions of LVQ exist, this section considers the unsupervised version, LVQ-I.

Ripley [731] defined clustering algorithms as those algorithms where the purpose is to divide a set on $n$ observations into $m$ groups such that members of the same group are more alike than members of different groups. The aim of a clustering algorithm is therefore to construct clusters of similar input vectors (patterns), where similarity is usually measured in terms of Euclidean distance. LVQ-I performs such clustering.

The training process of LVQ-I to construct clusters is based on competition. Referring to Figure 4.1, each output unit $o_k$ represents a single cluster. The competition is among the cluster output units. During training, the cluster unit whose weight vector is the "closest" to the current input pattern is declared as the winner. The corresponding weight vector and that of neighboring units are then adjusted to better resemble the input pattern. The "closeness" of an input pattern to a weight vector is usually measured using the Euclidean distance. The weight update is given as

$$\Delta u_{ki}(t) = \begin{cases} \eta(t)[z_{i,p} - u_{ki}(t-1)] & \text{if } k \in \kappa_{k,p}(t) \\ 0 & \text{otherwise} \end{cases} \qquad (4.19)$$

where $\eta(t)$ is a decaying learning rate, and $\kappa_{k,p}(t)$ is the set of neighbors of the winning cluster unit $o_k$ for pattern $p$. It is, of course, not strictly necessary that LVQ-I makes use of a neighborhood function, thereby updating only the weights of the winning output unit.

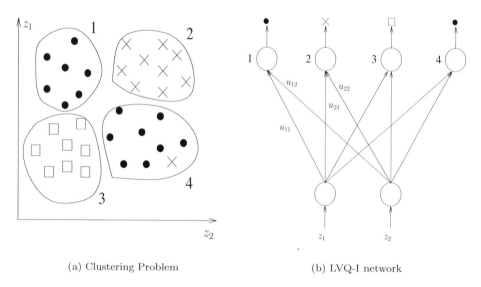

(a) Clustering Problem                    (b) LVQ-I network

**Figure 4.2** Learning Vector Quantizer to Illustrate Clustering

An illustration of clustering, as done by LVQ-I, is given in Figure 4.2. The input space, defined by two input units $z_1$ and $z_2$, is represented in Figure 4.2(a), while Figure 4.2(b) illustrates the LVQ-I network architecture required to form the clusters. Note that although only three classes exist, four output units are necessary – one for each cluster. Less output units will lead to errors since patterns of different classes will be grouped in the same cluster, while too many clusters may cause overfitting. For the problem illustrated in Figure 4.2(a), an additional cluster unit may cause a separate cluster to learn the single × in cluster 4.

The Kohonen LVQ-I algorithm is summarized in Algorithm 4.2. For the LVQ-I, weights are either initialized to random values, sampled from a uniform distribution, or by

---

**Algorithm 4.2** Learning Vector Quantizer-I Training Algorithm

---

Initialize the network weights, the learning rate, and the neighborhood radius;
**while** *stopping condition(s) not true* **do**
    **for** *each pattern p* **do**
        Compute the Euclidean distance, $d_{k,p}$, between input vector $\mathbf{z}_p$ and each
        weight vector $\mathbf{u}_k = (u_{k1}, u_{k2}, \cdots, u_{KI})$ as

$$d_{k,p}(\mathbf{z}_p, \mathbf{u}_k) = \sqrt{\sum_{i=1}^{I} (z_{i,p} - u_{ki})^2} \qquad (4.20)$$

        Find the output unit $o_k$ for which the distance $d_{k,p}$ is the smallest;
        Update all the weights for the neighborhood $\kappa_{k,p}$ using equation (4.19);
    **end**
    Update the learning rate;
    Reduce the neighborhood radius at specified learning iterations;
**end**

---

taking the first input patterns as the initial weight vectors. For the example in Figure 4.2(b), the latter will result in the weights $u_{11} = z_{1,1}, u_{12} = z_{2,1}, u_{21} = z_{1,2}, u_{22} = z_{2,2}$, etc.

Stopping conditions may be

- a maximum number of epochs is reached,

- stop when weight adjustments are sufficiently small,

- a small enough quantization error has been reached, where the quantization error is defined as

$$Q_T = \frac{\sum_{p=1}^{P_T} ||\mathbf{z}_p - \mathbf{u}_k||_2^2}{P_T} \qquad (4.21)$$

One problem that may occur in LVQ networks is that one cluster unit may dominate as the winning cluster unit. The danger of such a scenario is that most patterns will be in one cluster. To prevent one output unit from dominating, a "conscience" factor is incorporated in a function to determine the winning output unit. The conscience factor penalizes an output for winning too many times. The activation value of output units is calculated using

$$o_{k,p} = \begin{cases} 1 & \text{for } \min_{\forall k} \{ d_{k,p}(\mathbf{z}_p, \mathbf{u}_k) - b_k(t) \} \\ 0 & \text{otherwise} \end{cases} \qquad (4.22)$$

where

$$b_k(t) = \gamma \left( \frac{1}{I} - g_k(t) \right) \qquad (4.23)$$

and

$$g_k(t) = g_k(t-1) + \beta(o_{k,p} - g_k(t-1)) \qquad (4.24)$$

In the above, $d_{k,p}$ is the Euclidean distance as defined in equation (4.20), $I$ is the total number of input units, and $g_k(0) = 0$. Thus, $b_k(0) = \frac{1}{I}$, which initially gives each output unit an equal chance to be the winner; $b_k(t)$ is the conscience factor defined for each output unit. The more an output unit wins, the larger the value of $g_k(t)$ becomes, and $b_k(t)$ becomes larger negative. Consequently, a factor $|b_k(t)|$ is added to the distance $d_{k,p}$. Usually, for normalized inputs, $\beta = 0.0001$ and $\gamma = 10$.

## 4.5    Self-Organizing Feature Maps

Kohonen developed the self-organizing feature map (SOM) [474, 475, 476], as motivated by the self-organization characteristics of the human cerebral cortex. Studies of the cerebral cortex showed that the motor cortex, somatosensory cortex, visual cortex and auditory cortex are represented by topologically ordered maps. These topological maps form to represent the structures sensed in the sensory input signals.

The self-organizing feature map is a multidimensional scaling method to project an $I$-dimensional input space to a discrete output space, effectively performing a compression of input space onto a set of codebook vectors. The output space is usually a two-dimensional grid. The SOM uses the grid to approximate the probability density function of the input space, while still maintaining the topological structure of input space. That is, if two vectors are close to one another in input space, so is the case for the map representation.

The SOM closely resembles the learning vector quantizer discussed in the previous section. The difference between the two unsupervised algorithms is that neurons are usually organized on a rectangular grid for SOM, and neighbors are updated to also perform an ordering of the neurons. In the process, SOMs effectively cluster the input vectors through a competitive learning process, while maintaining the topological structure of the input space.

Section 4.5.1 explains the standard stochastic SOM training rule, while a batch version is discussed in Section 4.5.2. A growing approach to SOM is given in Section 4.5.3. Different approaches to speed up the training of SOMs are overviewed in Section 4.5.4. Section 4.5.5 explains the formation of clusters for visualization purposes. Section 4.5.6 discusses in brief different ways how the SOM can be used after training.

### 4.5.1    Stochastic Training Rule

SOM training is based on a competitive learning strategy. Assume $I$-dimensional input vectors $\mathbf{z}_p$, where the subscript $p$ denotes a single training pattern. The first step of the training process is to define a map structure, usually a two-dimensional grid (refer to Figure 4.3). The map is usually square, but can be of any rectangular shape. The number of elements (neurons) in the map is less than the number of training patterns. Ideally, the number of neurons should be equal to the number of independent training patterns.

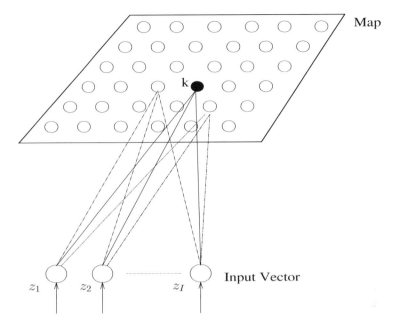

**Figure 4.3** Self-organizing Map

Each neuron on the map is associated with and $I$-dimensional weight vector that forms the centroid of one cluster. Larger cluster groupings are formed by grouping together "similar" neighboring neurons.

Initialization of the codebook vectors can occur in various ways:

- Assign random values to each weight $\mathbf{w}_{kj} = (w_{kj1}, w_{kj2}, \cdots, w_{KJI})$, with $K$ the number of rows and $J$ the number of columns of the map. The initial values are bounded by the range of the corresponding input parameter. While random initialization of weight vectors is simple to implement, this form of initialization introduces large variance components into the map which increases training time.

- Assign to the codebook vectors randomly selected input patterns. That is,

$$\mathbf{w}_{kj} = \mathbf{z}_p \qquad (4.25)$$

with $p \sim U(1, P_T)$.

This approach may lead to premature convergence, unless weights are perturbed with small random values.

- Find the principal components of the input space, and initialize the codebook vectors to reflect these principal components.

- A different technique of weight initialization is due to Su *et al.* [818], where the objective is to define a large enough hyper cube to cover all the training patterns [818]. The algorithm starts by finding the four extreme points of the map by determining the four extreme training patterns. Firstly, two patterns are found with the largest inter-pattern Euclidean distance. A third pattern is located at the furthest point from these two patterns, and the fourth pattern with largest Euclidean distance from these three patterns. These four patterns form the corners of the map. Weight values of the remaining neurons are found through interpolation of the four selected patterns, in the following way:

— Weights of boundary neurons are initialized as

$$\mathbf{w}_{1j} = \frac{\mathbf{w}_{1J} - \mathbf{w}_{11}}{J - 1}(j - 1) + \mathbf{w}_{11} \tag{4.26}$$

$$\mathbf{w}_{Kj} = \frac{\mathbf{w}_{KJ} - \mathbf{w}_{K1}}{J - 1}(j - 1) + \mathbf{w}_{K1} \tag{4.27}$$

$$\mathbf{w}_{k1} = \frac{\mathbf{w}_{K1} - \mathbf{w}_{11}}{K - 1}(k - 1) + \mathbf{w}_{11} \tag{4.28}$$

$$\mathbf{w}_{kJ} = \frac{\mathbf{w}_{KJ} - \mathbf{w}_{1J}}{K - 1}(k - 1) + \mathbf{w}_{1J} \tag{4.29}$$

for all $j = 2, \cdots, J - 1$ and $k = 2, \cdots, K - 1$.

— The remaining codebook vectors are initialized as

$$\mathbf{w}_{kj} = \frac{\mathbf{w}_{kJ} - \mathbf{w}_{k1}}{J - 1}(j - 1) + \mathbf{w}_{k1} \tag{4.30}$$

for all $j = 2, \cdots, J - 1$ and $k = 2, \cdots, K - 1$.

The standard training algorithm for SOMs is stochastic, where codebook vectors are updated after each pattern is presented to the network. For each neuron, the associated codebook vector is updated as

$$\mathbf{w}_{kj}(t + 1) = \mathbf{w}_{kj}(t) + h_{mn,kj}(t)[\mathbf{z}_p - \mathbf{w}_{kj}(t)] \tag{4.31}$$

where $mn$ is the row and column index of the winning neuron. The winning neuron is found by computing the Euclidean distance from each codebook vector to the input vector, and selecting the neuron closest to the input vector. That is,

$$||\mathbf{w}_{mn} - \mathbf{z}_p||_2 = \min_{\forall kj}\{||\mathbf{w}_{kj} - \mathbf{z}_p||_2^2\} \tag{4.32}$$

The function $h_{mn,kj}(t)$ in equation (4.31) is referred to as the neighborhood function. Thus, only those neurons within the neighborhood of the winning neuron $mn$ have their codebook vectors updated. For convergence, it is necessary that $h_{mn,kj}(t) \to 0$ when $t \to \infty$.

The neighborhood function is usually a function of the distance between the coordinates of the neurons as represented on the map, i.e.

$$h_{mn,kj}(t) = h(||c_{mn} - c_{kj}||_2^2, t) \tag{4.33}$$

with the coordinates $c_{mn}, c_{kj} \in \mathbb{R}^2$. With increasing value of $||c_{mn} - c_{kj}||_2^2$ (that is, neuron $kj$ is further away from the winning neuron $mn$), $h_{mn,kj} \to 0$. The neighborhood can be defined as a square or hexagon. However, the smooth Gaussian kernel is mostly used:

$$h_{mn,kj}(t) = \eta(t)e^{-\frac{||c_{mn} - c_{kj}||_2^2}{2\sigma^2(t)}} \tag{4.34}$$

where $\eta(t)$ is the learning rate and $\sigma(t)$ is the width of the kernel. Both $\eta(t)$ and $\sigma(t)$ are monotonically decreasing functions.

The learning process is iterative, continuing until a "good" enough map has been found. The quantization error is usually used as an indication of map accuracy, defined as the sum of Euclidean distances of all patterns to the codebook vector of the winning neuron, i.e.

$$\mathcal{E}_T = \sum_{p=1}^{P_T} ||\mathbf{z}_p - \mathbf{w}_{mn}(t)||_2^2 \tag{4.35}$$

Training stops when $\mathcal{E}_T$ is sufficiently small.

## 4.5.2  Batch Map

The stochastic SOM training algorithm is slow due to the updates of weights after each pattern presentation: all the weights are updated. Batch versions of the SOM training rule have been developed that update weight values only after all patterns have been presented. The first batch SOM training algorithm was developed by Kohonen [475], and is summarized in Algorithm 4.3.

---

**Algorithm 4.3** Batch Self-Organizing Map

---

Initialize the codebook vectors by assigning the first $KJ$ training patterns to them, where $KJ$ is the total number of neurons in the map;
**while** *stopping condition(s) not true* **do**
    **for** *each neuron, kj* **do**
        Collect a list of copies of all patterns $\mathbf{z}_p$ whose nearest codebook vector belongs to the topological neighborhood of that neuron;
    **end**
    **for** *each codebook vector* **do**
        Compute the codebook vector as the mean over the corresponding list of patterns;
    **end**
**end**

---

Based on the batch learning approach above, Kaski *et al.* [442] developed a faster version, as summarized in Algorithm 4.4.

## 4.5.3  Growing SOM

One of the design problems when using a SOM is deciding on the size of the map. Too many neurons may cause overfitting of the training patterns, with each training pattern assigned to a different neuron. Alternatively, the final SOM may have succeeded in forming good clusters of similar patterns, but with many neurons with a zero or close to zero frequency. The frequency of a neuron refers to the number of patterns for

---

**Algorithm 4.4** Fast Batch Self-Organizing Map

---

Initialize the codebook vectors, $\mathbf{w}_{kj}$, using any initialization approach;
**while** *stopping condition(s) not true* **do**
    **for** *each neuron, $kj$* **do**
        Compute the mean over all patterns for which that neuron is the winner;
        Denote the average by $\overline{\mathbf{w}}_{kj}$;
    **end**
    Adapt the weight values for each codebook vector using

$$\mathbf{w}_{kj} = \frac{\sum_{nm} N_{nm} h_{nm,kj} \overline{\mathbf{w}}_{nm}}{\sum_{nm} N_{nm} h_{nm,kj}} \qquad (4.36)$$

    where $nm$ iterates over all neurons, $N_{nm}$ is the number of patterns for which
    neuron $nm$ is the winner, and $h_{nm,kj}$ is the neighborhood function which
    indicates if neuron $nm$ is in the neighborhood of neuron $kj$, and to what degree.
**end**

---

which that neuron is the winner, referred to as the best matching neuron (BMN). Too many neurons also cause a substantial increase in computational complexity. Too few neurons, on the other hand, will result in clusters with a high variance among the cluster members.

An approach to find near optimal SOM architectures is to start training with a small architecture, and to grow the map when more neurons are needed. One such SOM growing algorithm is given in Algorithm 4.5, assuming a square map structure. Note that the map-growing process coexists with the training process.

Growing of the map is stopped when any one of the following criteria is satisfied:

- the maximum map size has been reached;
- the largest neuron quantization error is less than a user specified threshold, $\epsilon$;
- the map has converged to the specified quantization error.

A few aspects of the growing algorithm above need some explanation. These are the constants $\epsilon, \gamma$, and the maximum map size as well as the different stopping conditions. A good choice for $\gamma$ is 0.5. The idea of the interpolation step is to assign a weight vector to the new neuron $ab$ such that it removes patterns from neuron $kj$ with the largest quantization erro in order to reduce the error of that neuron. A value less than 0.5 will position neuron $ab$ closer to $kj$, with the chance that more patterns will be removed from neuron $kj$. A value larger than 0.5 will have the opposite effect.

The quantization error threshold, $\epsilon$, is important to ensure that a sufficient map size is constructed. A small value for $\epsilon$ may result in a too large map architecture, while a too large $\epsilon$ may result in longer training times to reach a large enough architecture.

An upper bound on the size of the map is easy to determine: it is simply the number of training patterns, $P_T$. This is, however, undesirable. The maximum map size is

---

**Algorithm 4.5** Growing Self-Organizing Map Algorithm

---

Initialize the codebook vectors for a small, undersized SOM;
**while** *stopping condition(s) not true* **do**
    **while** *growing condition not triggered* **do**
        Train the SOM for $t'$ pattern presentations using any SOM training method;
    **end**
    **if** *grow condition is met* **then**
        Find the neuron $kj$ with the largest quantization error;
        Find the furthest immediate neighbor $mn$ in the row-dimension of the map,
        and the furthest neuron $rs$ in the column-dimension;
        Insert a column between neurons $kj$ and $rs$ and a row between neurons $kj$
        and $mn$ (this step preserves the square structure of the map);
        For each neuron $ab$ in the new column, initialize the corresponding codebook
        vectors $\mathbf{w}_{ab}$ using

$$\mathbf{w}_{ab} = \gamma(\mathbf{w}_{a,b-1} + \mathbf{w}_{a,b+1}) \qquad (4.37)$$

        and for each neuron in the new row,

$$\mathbf{w}_{ab} = \gamma(\mathbf{w}_{a-1,b} + \mathbf{w}_{a+1,b}) \qquad (4.38)$$

        where $\gamma \in (0,1)$
    **end**
**end**
Refine the weights of the final SOM architecture with additional training steps until
convergence has been reached.

---

rather expressed as $\beta P_T$, with $\beta \in (0,1)$. Ultimately, the map size should be at least equal to the number of independent variables in the training set. The optimal value of $\beta$ is problem dependent, and care should be taken to ensure that $\beta$ is not too small if a growing SOM is not used. If this is the case, the final map may not converge to the required quantization error since the map size will be too small.

## 4.5.4 Improving Convergence Speed

Training of SOMs is slow, due to the large number of weight updates involved (all the weights are updated for standard SOM training). Several mechanisms have been developed to reduce the number of training calculations, thereby improving speed of convergence. BatchMap is one such mechanism. Other approaches include the following:

**Optimizing the neighborhood**

If the Gaussian neighborhood function as given in equation (4.34) is used, all neurons will be in the neighborhood of the BMN, but to different degrees, due to the asymptotic characteristics of the function. Thus, all codebook vectors are updated even if they are far from the BMN. This is strictly not necessary, since neurons far away from the BMN are dissimilar to the presented pattern, and will have negligible weight changes. Many calculations can therefore be saved by clipping the Gaussian neighborhood at a certain threshold – without degrading the performance of the SOM.

Additionally, the width of the neighborhood function can change dynamically during training. The initial width is large, with a gradual decrease in the variance of the Gaussian, which controls the neighborhood. For example,

$$\sigma(t) = \sigma(0)e^{-t/\tau_1} \tag{4.39}$$

where $\tau_1$ is a positive constant, and $\sigma(0)$ is the initial, large variance.

If the growing SOM (refer to Section 4.5.3) is used, the width of the Gaussian neighborhood function should increase with each increase in map size.

## Learning Rate

A time-decaying learning rate may be used, where training starts with a large learning rate which gradually decreases. That is,

$$\eta(t) = \eta(0)e^{-t/\tau_2} \tag{4.40}$$

where $\tau_2$ is a positive constant and $\eta(0)$ is the initial, large learning rate.

## Shortcut Winner Search

The shortcut winner search decreases the computational complexity by using a more efficient search for the BMN. The search is based on the premise that the BMN of a pattern is in the vicinity of the BMN for the previous epoch. The search for a BMN is therefore constrained to the current BMN and its neighborhood. In short, the search for a BMN for each pattern is summarized in Algorithm 4.6.

---

**Algorithm 4.6** Shortcut Winner Search

---

Retrieve the previous BMN;
Calculate the distance of the pattern to the codebook vector of the previous BMN;
Calculate the distance of the pattern to all direct neighbors of the previous BMN;
**if** *the previous BMN is still the best* **then**
    Terminate the search;
**end**
**else**
    Let the new BMN be the neuron (within the neighborhood) closest to that
    pattern;
**end**

---

Shortcut winner search does not perform a search for the BMN over the entire map, but just within the neighborhood of the previous BMN, thereby substantially reducing computational complexity.

### 4.5.5 Clustering and Visualization

The effect of the SOM training process is to cluster together similar patterns, while preserving the topology of input space. After training, all that is given is the set of trained weights with no explicit cluster boundaries. An additional step is required to find these cluster boundaries.

One way to determine and visualize these cluster boundaries is to calculate the unified distance matrix (U-matrix) [403], which contains a geometrical approximation of the codebook vector distribution in the map. The U-matrix expresses for each neuron, the distance to the neighboring codebook vectors. Large values within the U-matrix indicate the position of cluster boundaries. Using a gray-scale scheme, Figure 4.4(a) visualizes the U-matrix for the iris classification problem.

For the same problem, Figure 4.4(b) visualizes the clusters on the actual map. Boundaries are usually found by using Ward clustering [23] of the codebook vectors. Ward clustering follows a bottom-up approach where each neuron initially forms its own cluster. At consecutive iterations, two clusters that are closest to one another are merged, until the optimal or specified number of clusters has been constructed. The end result of Ward clustering is a set of clusters with a small variance over its members, and a large variance between separate clusters.

The Ward distance measure is used to decide which clusters should be merged. The distance measure is defined as

$$d_{rs} = \frac{n_r n_s}{n_r + n_s} \|\mathbf{w}_r - \mathbf{w}_s\|_2^2 \qquad (4.41)$$

where $r$ and $s$ are cluster indices, $n_r$ and $n_s$ are the number of patterns within the clusters, and $\mathbf{w}_r$ and $\mathbf{w}_s$ are the centroid vectors of these clusters (i.e. the average of all the codebook vectors within the cluster). The two clusters are merged if their distance, $d_{rs}$, is the smallest. For the newly formed cluster, $q$,

$$\mathbf{w}_q = \frac{1}{n_r + n_s}(n_r \mathbf{w}_r + n_s \mathbf{w}_s) \qquad (4.42)$$

and

$$n_q = n_r + n_s \qquad (4.43)$$

Note that, in order to preserve topological structure, two clusters can only be merged if they are adjacent. Furthermore, only clusters that have a nonzero number of patterns associated with them are merged.

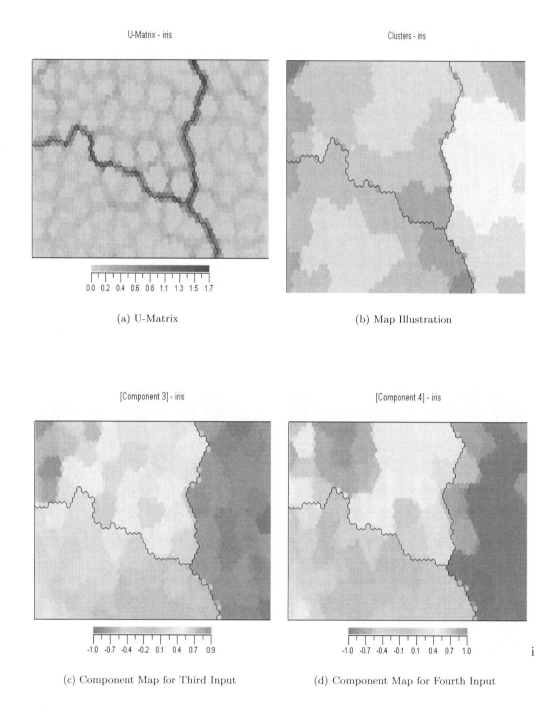

(a) U-Matrix

(b) Map Illustration

(c) Component Map for Third Input

(d) Component Map for Fourth Input

**Figure 4.4** Visualization of SOM Clusters for Iris Classification

## 4.5.6 Using SOM

The SOM has been applied to a variety of real-world problems, including image analysis, speech recognition, music pattern analysis, signal processing, robotics, telecommunications, electronic-circuit design, knowledge discovery, and time series analysis. The main advantage of SOMs comes from the easy visualization and interpretation of clusters formed by the map.

In addition to visualizing the complete map as illustrated in Figure 4.4(b), the relative component values in the codebook vectors can be visualized as illustrated in the same figure. Here a component refers to an input attribute. That is, a component plane can be constructed for each input parameter (component) to visualize the distribution of the corresponding weight (using some color scale representation). The map and component planes can be used for exploratory data analysis. For example, a marked region on the visualized map can be projected onto the component planes to find the values of the input parameters for that region.

A trained SOM can also be used as a classifier. However, since no target information is available during training, the clusters formed by the map should be manually inspected and labeled. A data vector is then presented to the map, and the winning neuron determined. The corresponding cluster label is then used as the class.

Used in recall mode, the SOM can be used to interpolate missing values within a pattern. Given such a pattern, the BMN is determined, ignoring the inputs with missing values. A value is then found by either replacing the missing value with the corresponding weight of the BMN, or through interpolation among a neighborhood of neurons (e.g. take the average of the weight values of all neurons in the neighborhood of the BMN).

## 4.6 Assignments

1. Implement and test a LVQ-I network to distinguish between different alphabetical characters of different fonts.

2. Explain why it is necessary to retrain a supervised NN on all the training data, including any new data that becomes available at a later stage. Why is this not such an issue with unsupervised NNs?

3. Discuss an approach to optimize the LVQ-I network architecture.

4. How can PSO be used for unsupervised learning?

5. What is the main difference between the LVQ-I and SOM as an approach to cluster multi-dimensional data?

6. For a SOM, if the training set contains $P_T$ patterns, what is the upper bound on the number of neurons necessary to fit the data? Justify your answer.

7. Explain the purpose of the neighborhood function of SOMs.

8. Assuming a Gaussian neighborhood function for SOMs, what can be done to reduce the number of weight updates in a sensible way?

9. Explain how a SOM can be used to distinguish among different hand gestures.

10. Discuss a number of ways in which the SOM can be adapted to reduce its computational complexity.

11. Explain how a SOM can be used as a classifier.

12. Explain how it is possible for the SOM to train on data with missing values.

13. How can a trained SOM be used to determine an appropriate value if for a given input pattern an attribute does not have a value.

# Chapter 5

# Radial Basis Function Networks

Several neural networks have been developed for both the supervised and the unsupervised learning paradigms. While these NNs were seen to perform very well in their respective application fields, improvements have been developed by combining supervised and unsupervised learning. This chapter discusses two such learning algorithms, namely the learning vector quantizer-II in Section 5.1 and radial basis function NNs in Section 5.2.

## 5.1 Learning Vector Quantizer-II

The learning vector quantizer (LVQ-II), developed by Kohonen, uses information from a supervisor to implement a reward and punish scheme. The LVQ-II assumes that the classifications of all input patterns are known. If the winning cluster unit correctly classifies the pattern, the weights to that unit are rewarded by moving the weights to better match the input pattern. On the other hand, if the winning unit misclassified the input pattern, the weights are penalized by moving them away from the input vector.

For the LVQ-II, the weight updates for the winning output unit $o_k$ are given as

$$\Delta u_{ki} = \begin{cases} \eta(t)[z_{i,p} - u_{ki}(t-1)] & \text{if } o_{k,p} = t_{k,p} \\ -\eta(t)[z_{i,p} - u_{ki}(t-1)] & \text{if } o_{k,p} \neq t_{k,p} \end{cases} \tag{5.1}$$

Similarly to the LVQ-I, a conscience factor can be incorporated to penalize frequent winners.

## 5.2 Radial Basis Function Neural Networks

A radial basis function (RBF) neural network (RBFNN) is a FFNN where hidden units do not implement an activation function, but represents a radial basis function. An RBFNN approximates a desired function by superposition of nonorthogonal, radially

*Computational Intelligence: An Introduction*, Second Edition A.P. Engelbrecht
©2007 John Wiley & Sons, Ltd

symmetric functions. RBFNNs have been independently proposed by Broomhead and Lowe [92], Lee and Kill [506], Niranjan and Fallside [630], and Moody and Darken [605] as an approach to improve accuracy and to decrease training time complexity.

The RBFNN architecture is overviewed in Section 5.2.1, while different radial basis functions are discussed in Section 5.2.2. Different training algorithms are given in Section 5.2.3. Variations of RBFNNs are discussed in Section 5.2.4.

## 5.2.1   Radial Basis Function Network Architecture

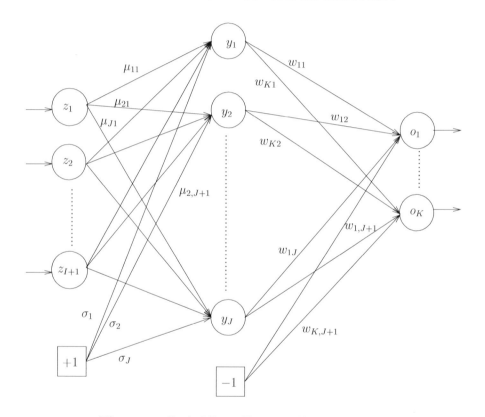

**Figure 5.1** Radial Basis Function Neural Network

Figure 5.1 illustrates a general architecture of the RBFNN. The architecture is very similar to that of a standard FFNN, with the following differences:

- Hidden units implement a radial basis function, $\Phi$. The output of each hidden unit is calculated as

$$y_{j,p}(\mathbf{z}_p) = \Phi(||\mathbf{z}_p - \mu_j||_2) \qquad (5.2)$$

  where $\mu_j$ represents the center of the basis function, and $|| \bullet ||_2$ is the Euclidean norm.

- Weights from the input units to a hidden unit, referred to as $\mu_{ij}$, represent the center of the radial basis function of hidden unit $j$.

- Some radial basis functions are characterized by a width, $\sigma_j$. For such basis functions, the weight from the basis unit in the input layer to each hidden unit represents the width of the basis function. Note that input unit $z_{I+1}$ has an input signal of $+1$.

The output of an RBFNN is calculated as

$$o_{k,p} = \sum_{j=1}^{J+1} w_{kj} y_{j,p} \tag{5.3}$$

Note that the output units of an RBFNN implement linear activation functions. The output is therefore just a linear combination of basis functions.

As with FFNNs, it has been shown that RBFNNs are universal approximators [47, 349, 682].

## 5.2.2 Radial Basis Functions

Each hidden unit implements a radial basis function. These functions, also referred to as kernel functions, are strictly positive, radially symmetric functions. A radial basis function (RBF) has a unique maximum at its center, $\mu_j$, and the function usually drops off to zero rapidly further away from the center. The output of a hidden unit indicates the closeness of the input vector, $\mathbf{z}_p$, to the center of the basis function.

In addition to the center of the function, some RBFs are characterized by a width, $\sigma_j$, which specifies the width of the receptive field of the RBF in the input space for hidden unit $j$.

A number of RBFs have been proposed [123, 130]:

- **Linear** function, where

$$\Phi(||\mathbf{z}_p - \mu_j||_2) = ||\mathbf{z}_p - \mu_j||_2 \tag{5.4}$$

- **Cubic** function, where

$$\Phi(||\mathbf{z}_p - \mu_j||_2) = ||\mathbf{z}_p - \mu_j||_2^3 \tag{5.5}$$

- **Thin-plate-spline** function, where

$$\Phi(||\mathbf{z}_p - \mu_j||_2) = ||\mathbf{z}_p - \mu_j||_2^2 \ln ||\mathbf{z}_p - \mu_j||_2 \tag{5.6}$$

- **Multiquadratic** function, where

$$\Phi(||\mathbf{z}_p - \mu_j||_2, \sigma_j) = \sqrt{||\mathbf{z}_p - \mu_j||_2^2 + \sigma_j^2} \tag{5.7}$$

- **Inverse multiquadratic** function, where

$$\Phi(||\mathbf{z}_p - \mu_j||_2, \sigma_j) = \frac{1}{\sqrt{||\mathbf{z}_p - \mu_j||_2^2 + \sigma_j^2}} \tag{5.8}$$

- **Gaussian** function, where

$$\Phi(||\mathbf{z}_p - \mu_j||_2, \sigma_j) = e^{-||\mathbf{z}_p - \mu_j||_2^2/(2\sigma_j^2)} \tag{5.9}$$

- **Logistic** function, where

$$\Phi(||\mathbf{z}_p - \mu_j||_2, \sigma_j) = \frac{1}{1 + e^{||\mathbf{z}_p - \mu_j||_2^2/\sigma_j^2 - \theta_j}} \tag{5.10}$$

where $\theta_j$ is an adjusted bias.

Considering the above functions, the accuracy of an RBFNN is influenced by:

- The **number of basis functions** used. The more basis functions that are used, the better the approximation of the target function will be. However, unnecessary basis functions increase computational complexity.

- The **location of the basis functions** as defined by the center vector, $\mu_j$, for each basis function. Basis functions should be evenly distributed to cover the entire input space.

- For some functions, the **width of the receptive field**, $\sigma_j$. The larger $\sigma_j$ is, the more of the input space is represented by that basis function.

Training of an RBFNN should therefore consider methods to find the best values for these parameters.

## 5.2.3   Training Algorithms

A number of methods have been developed to train RBFNNs. These methods differ mainly in the number of parameters that are learned. The fixed centers algorithm adapts only the weights between the hidden and output layers. Adaptive centers training algorithms adapt both weights, centers, and deviations. This section reviews some of these training algorithms.

### Training RBFNNs with Fixed Centers

Broomhead and Lowe [92] proposed a training method where it is assumed that RBF centers are fixed. Centers are randomly selected from the training set. Provided that a sufficient number of centers are uniformly selected from the training set, an adequate sampling of the input space will be obtained. Common practice is to select a large number of centers, and then to prune, after training, redundant basis functions. This is usually done in a systematic manner, removing only those RBFs that do not cause a significant degradation in accuracy.

The fixed centers training algorithm is summarized in Algorithm 5.1. With reference to this algorithm, Gaussian RBFs are used, with widths calculated as

$$\sigma_j = \sigma = \frac{d_{max}}{\sqrt{J}}, \ j = 1, \ldots, J \tag{5.11}$$

where $J$ is the number of centers (or hidden units), and $d_{max}$ is the maximum Euclidean distance between centers.

Weight values of connections between the hidden and output layers are found by solving for $\mathbf{w}_k$ in

$$\mathbf{w}_k = (\Phi^T \Phi)^{-1} \Phi^T \mathbf{t}_k \qquad (5.12)$$

where $\mathbf{w}_k$ is the weight vector of output unit $k$, $\mathbf{t}_k$ is the vector of target outputs, and $\Phi \in \mathbb{R}^{P_T \times J}$ is the matrix of RBF nonlinear mappings performed by the hidden layer.

---

**Algorithm 5.1** Training an RBFNN with Fixed Centers

---

Set $J$ to indicate the number of centers;
Choose the centers, $\mu_j$, $j = 1, \ldots, J$, as

$$\mu_j = \mathbf{z}_p, \ p \sim U(1, P_T) \qquad (5.13)$$

Calculate the width, $\sigma_j$, using equation (5.11);
Initialize all $w_{kj}$, $k = 1, \ldots, K$ and $j = 1, \ldots, J$ to small random values;
Calculate the output for each output unit using equation (5.3) with Gaussian radial basis functions;
Solve for the network weights using equation (5.12) for each $k = 1, \ldots, K$;

---

### Training an RBFNN using Gradient Descent

Moody and Darken [605] and Poggio and Girosi [682] used gradient descent to adjust weights, centers, and widths. The algorithm is summarized in Algorithm 5.2.

In Algorithm 5.2, $\eta_w, \eta_\mu$, and $\eta_\sigma$ respectively indicate the learning rate for weights, centers, and widths. In this algorithm, centers are initialized by sampling from the training set. The next subsection shows that these centers can be obtained in an unsupervised training step, prior to training the weights between hidden units (radial basis) and output units.

### Two-Phase RBFNN Training

The training algorithms discussed thus far have shown slow convergence times [899]. In order to increase training time, RBFNN training can be done in two phases [605, 881]: (1) unsupervised learning of the centers, $\mu_j$, and then, (2) supervised training of the $\mathbf{w}_k$ weights between the hidden and output layers using gradient descent. Algorithm 5.3 summarizes a training algorithm where the first phase utilizes an LVQ-I to cluster input patterns [881].

---

**Algorithm 5.2** Gradient Descent Training of RBFNN

---

Select the number of centers, $J$;
**for** $j = 1, \ldots, J$ **do**
    $p \sim U(1, P_T)$;
    $\mu_j(t) = \mathbf{z}_p$;
    $\sigma_j(t) = \frac{d_{max}}{\sqrt{J}}$;
**end**
**for** $k = 1, \ldots, K$ **do**
    **for** $j = 1, \ldots, J$ **do**
        $w_{kj} \sim U(w_{min}, w_{max})$;
    **end**
**end**
**while** *stopping condition(s) not true* **do**
    Select an input pattern, $d_p = (\mathbf{z}_p, \mathbf{t}_p)$;
    **for** $k = 1, \ldots, K$ **do**
        Compute $o_{k,p}$ using equation (5.3);
        **for** $j = 1, \ldots, J$ **do**
            Compute weight adjustment step size,

$$\Delta w_{kj}(t) = -\eta_w \frac{\partial E}{\partial w_{kj}}(t) \tag{5.14}$$

            Adjust weights using

$$w_{kj}(t+1) = w_{kj}(t) + \Delta w_{kj}(t) \tag{5.15}$$

        **end**
    **end**
    **for** $j = 1, \ldots, J$ **do**
        **for** $i = 1, \ldots, I$ **do**
            Compute center step size,

$$\Delta\mu_{ji}(t) = -\eta_\mu \frac{\partial E}{\partial \mu_{ji}}(t) \tag{5.16}$$

            Adjust centers using

$$\mu_{ji}(t+1) = \mu_{ji}(t) + \Delta\mu_{ji}(t) \tag{5.17}$$

        **end**
        Compute width step size,

$$\Delta\sigma_j(t) = -\eta_\sigma \frac{\partial E}{\partial \sigma_j}(t) \tag{5.18}$$

        Adjust widths using

$$\sigma_j(t+1) = \sigma_j(t) + \Delta\sigma_j(t) \tag{5.19}$$

    **end**
**end**

---

---

**Algorithm 5.3** Two-Phase RBFNN Training

---

Initialize $w_{kj}$, $k = 1, \ldots, K$ and $j = 1, \ldots, J$;
Initialize $\mu_{ji}$, $j = 1, \ldots, J$ and $i = 1, \ldots, I$;
Initialize $\sigma_j$, $j = 1, \ldots, J$;
**while** *LVQ-I has not converged* **do**
    Apply one epoch of LVQ-I to adjust $\mu_j$, $j = 1, \ldots, J$;
    Adjust $\sigma_j$, $j = 1, \ldots, J$;
**end**
$t = 0$;
**while** *gradient descent has not converged* **do**
    Select an input pattern, $(\mathbf{z}_p, \mathbf{t}_p)$;
    Compute the weight step sizes,

$$\Delta w_{kj}(t) = \eta \sum_{k=1}^{K} (t_{k,p} - o_{k,p}) y_{j,p} \qquad (5.20)$$

    Adjust the weights,
$$w_{kj}(t+1) = w_{kj}(t) + \Delta w_{kj}(t) \qquad (5.21)$$

**end**

---

Before the LVQ-I training phase, the RBFNN is initialized as follows:

- The centers are initialized by setting all the $\mu_{ji}$ weights to the average value of all inputs in the training set.

- The weights are initialized by setting all $\sigma_j$ to the standard deviation of all input values over the training set.

- The hidden-to-output weights, $w_{kj}$, are initialized to small random values.

At the end of each LVQ-I iteration, basis function widths are recalculated as follows: For each hidden unit, find the average of the Euclidean distances between $\mu_j$ and the input patterns for which the hidden unit was selected as the winner. The width, $\sigma_j$, is set to this average.

Instead of using LVQ-I, Moody and Darken [605] uses K-means clustering in the first phase. The K-means algorithm is initialized by setting each $\mu_j$ to a randomly selected input pattern. Training patterns are assigned to their closest center, after which each center is recomputed as

$$\mu_j = \frac{\sum_{p \in C_j} \mathbf{z}_p}{|C_j|} \qquad (5.22)$$

where $C_j$ is the set of patterns closest to center $\mu_j$. Training patterns are again reassigned to their closest center, after which the centers are recalculated. This process continues until there is no significant change in the centers.

After the K-means clustering, the widths are determined as follow:

$$\sigma_j = \tau ||\mu_l - \mu_j|| \tag{5.23}$$

where $\mu_l$ is the nearest neighbor of $\mu_j$, and $\tau \in [1, 1.5]$.

The second-phase is then executed to learn the weight values, $w_{kj}$ using gradient descent, or by solving for $\mathbf{w}_k$ as in equation (5.12).

## 5.2.4 Radial Basis Function Network Variations

Two variations of the standard RBFNN are discussed in this section. These variations were developed as an attempt to improve the performance of RBFNNs.

### Normalized Hidden Unit Activations

Moody and Darken [605] proposed that hidden unit activations must be normalized using,

$$y_{j,p}(\mathbf{z}_p) = \frac{\Phi(||\mathbf{z}_p - \mu_j||_2, \sigma_j)}{\sum_{l=1}^{J} \Phi(||\mathbf{z}_p - \mu_l||_2, \sigma_l)} \tag{5.24}$$

This introduces the property that

$$\sum_{j=1}^{J} y_{j,p}(\mathbf{z}_p) = 1, \forall p = 1, \ldots, P_T \tag{5.25}$$

which means that the above normalization represents the conditional probability of hidden unit $j$ generating $\mathbf{z}_p$. This probability is given as

$$P(j|\mathbf{z}_p) = \frac{P_j(\mathbf{z}_p)}{\sum_{l=1}^{J} P_l(\mathbf{z}_p)} = \frac{y_{j,p}(\mathbf{z}_p)}{\sum_{l=1}^{J} y_{l,p}(\mathbf{z}_p)} \tag{5.26}$$

### Soft-Competition

The K-means clustering approach proposed by Moody and Darken can be considered as a hard competition winner-takes-all action. An input pattern is assigned to the cluster of patterns of the $\mu_j$ to which the input pattern is closest. Adjustment of $\mu_j$ is then based only on those patterns for which it was selected as the winner.

In soft-competition [632], all input vectors have an influence on the adjustment of all centers. For each hidden unit,

$$\mu_j = \frac{\sum_{p=1}^{P_T} P(j|\mathbf{z}_p)\mathbf{z}_p}{\sum_{p=1}^{P_T} P(j|\mathbf{z}_p)} \tag{5.27}$$

where $P(j|\mathbf{z}_p)$ is defined in equation (5.26).

## 5.3   Assignments

1. Compare the performance of an RBFNN and a FFNN on a classification problem from the UCI machine learning repository (http://www.ics.uci.edu/~mlearn/MLRepository.html).

2. Compare the performance of the Gaussian and logistic basis functions.

3. Suggest an alternative to compute the hidden-to-output weights instead of using GD.

4. Suggest an alternative to compute the input-to-hidden weights instead of using LVQ-I.

5. Investigate alternative methods to initialize an RBF NN.

6. Is it crucial that all $w_{kj}$ be initialized to small random values? Motivate your answer.

7. Develop a PSO, DE, and EP algorithm to train an RBFNN.

# Chapter 6

# Reinforcement Learning

The last learning paradigm to be discussed is that of reinforcement learning (RL) [823], with its origins in the psychology of animal learning. The basic idea is that of awarding the learner (agent) for correct actions, and punishing wrong actions. Intuitively, RL is a process of trial and error, combined with learning. The agent decides on actions based on the current environmental state, and through feedback in terms of the desirability of the action, learns which action is best associated with which state. The agent learns from interaction with the environment.

While RL is a general learning paradigm in AI, this chapter focuses on the role that NNs play in RL. The LVQ-II serves as one example where RL is used to train a NN to perform data clustering (refer to Section 5.1).

Section 6.1 provides an overview of RL. Model-free learning methods are given in Section 6.2. Connectionist approaches to RL are described in Section 6.3.

## 6.1 Learning through Awards

Formally defined, reinforcement learning is the learning of a mapping from situations to actions with the main objective to maximize the scalar reward or reinforcement signal [824]. Informally, reinforcement learning is defined as learning by trial-and-error from performance feedback from the environment or an external evaluator. The agent has absolutely no prior knowledge of what action to take, and has to discover (or explore) which actions yield the highest reward.

A typical reinforcement learning problem is illustrated in Figure 6.1. The agent receives sensory inputs from its environment, as a description of the current state of the perceived environment. An action is executed, upon which the agent receives the reinforcement signal or reward. This reward can be a positive or negative signal, depending on the correctness of the action. A negative reward has the effect of punishing the agent for a bad action.

The action may cause a change in the agent's environment, thereby affecting the future options and actions of the agent. The effects of actions on the environment and future states can not always be predicted. It is therefore necessary that the agent frequently

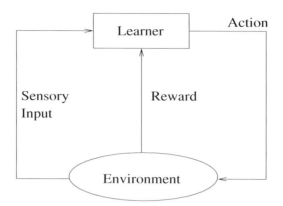

**Figure 6.1** Reinforcement Learning Problem

monitors its environment.

One of the important issues in RL (which occurs in most search methods) is that of the exploration–exploitation trade-off. As already indicated, RL has two important components:

- A trial and error search to find good actions, which forms the exploration component of RL.

- A memory of which actions worked well in which situations. This is the exploitation component of RL.

It is important that the agent exploits what it has already learned, such that a reward can be obtained. However, via the trial and error search, the agent must also explore to improve action selections in the future.

A reinforcement learning agent has the following components:

- A **policy**, which is the decision making function of the agent. This function is used to specify which action to execute in each of the situations that the agent may encounter. The policy is basically a set of associations between actions and situations, or alternatively, a set of stimulus-response rules.

- A **reward** function, which defines the goal of the agent. The reward function defines what are good and bad actions for the agent for specific situations. The reward is immediate, and represents only the current environment state. The goal of the agent is to maximize the total reward that it receives over the long run.

- A **value** function, which specifies the goal in the long run. The value function is used to predict future reward, and is used to indicate what is good in the long run.

- Optionally, an RL agent may also have a **model** of the environment. The environmental model mimics the behavior of the environment. This can be done by transition functions that describe transitions between different states.

For the value function, an important aspect is how the future should be taken into account. A number of models have been proposed [432]:

- The **finite-horizon model**, in which the agent optimizes its expected reward for the next $n_t$ steps, i.e.

$$E\left[\sum_{t=1}^{n_t} r(t)\right] \tag{6.1}$$

  where $r(t)$ is the reward for time-step $t$.

- The **infinite-horizon discounted model**, which takes the entire long-run reward of the agent into consideration. However, each reward received in future is geometrically discounted according to a discount factor, $\gamma \in [0, 1)$:

$$E\left[\sum_{t=0}^{\infty} \gamma^t r(t)\right] \tag{6.2}$$

  The discount factor enforces a bound on the infinite sum.

- The **average reward model**, which prefers actions that optimize the agent's long-run average reward:

$$\lim_{n_t \to \infty} E\left[\frac{1}{n_t} \sum_{t=0}^{n_t} r(t)\right] \tag{6.3}$$

  A problem with this model is that it is not possible to distinguish between a policy that gains a large amount of reward in the initial phases, and a policy where the largest gain is obtained in the later phases.

In order to find an optimal policy, $\pi^*$, it is necessary to find an optimal value function. A candidate optimal value function is [432],

$$V^*(s) = \max_{a \in \mathcal{A}} \left\{ R(s, a) + \gamma \sum_{s' \in \mathcal{S}} T(s, a, s') V^*(s') \right\}, \quad s \in \mathcal{S} \tag{6.4}$$

where $\mathcal{A}$ is the set of all possible actions, $\mathcal{S}$ is the set of environmental states, $R(s, a)$ is the reward function, and $T(s, a, s')$ is the transition function. Equation (6.4) states that the value of a state, $s$, is the expected instantaneous reward, $R(s, a)$, for action $a$ plus the expected discounted value of the next state, using the best possible action.

From the above, a clear definition of the model in terms of the transition function, $T$, and the reward function, $R$, is required. A number of algorithms have been developed for such RL problems. The reader is referred to [432, 824] for a summary of these methods. Of more interest to this chapter are model-free learning methods, as described in the next section.

# 6.2   Model-Free Reinforcement Learning Model

This section considers model-free RL methods, where the objective is to obtain an optimal policy without a model of the environment. This section reviews two approaches, namely temporal difference (TD) learning (in Section 6.2.1) and Q-learning (in Section 6.2.2).

## 6.2.1   Temporal Difference Learning

Temporal difference (TD) learning [824] learns the value policy using the update rule,

$$V(s) = V(s) + \eta(r + \gamma V(s^{'}) - V(s)) \tag{6.5}$$

where $\eta$ is a learning rate, $r$ is the immediate reward, $\gamma$ is the discount factor, $s$ is the current state, and $s^{'}$ is a future state. Based on equation (6.5), whenever a state, $s$, is visited, its estimated value is updated to be closer to $r + \eta V(s^{'})$.

The above model is referred to as TD(0), where only one future step is considered. The TD method has been generalized to TD($\lambda$) strategies [825], where $\lambda \in [0,1]$ is a weighting on the relevance of recent temporal differences of previous predictions. For TD($\lambda$), the value function is learned using

$$V(u) = V(u) + \eta(r + \gamma V(s^{'}) - V(s))e(u) \tag{6.6}$$

where $e(u)$ is the eligibility of state $u$. The eligibility of a state is the degree to which the state has been visited in the recent past, computed as

$$e(s) = \sum_{t^{'}=1}^{t} (\lambda\gamma)^{t-t^{'}} \delta_{s,s_t} \tag{6.7}$$

where

$$\delta_{s,s_t} = \begin{cases} 1 & s = s_t \\ 0 & \text{otherwise} \end{cases} \tag{6.8}$$

The update in equation (6.6) is applied to every state, according to its eligibility, and not just the previous state as for TD(0).

## 6.2.2   Q-Learning

In Q-learning [891], the task is to learn the expected discounted reinforcement values, $Q(s,a)$, of taking action $a$ in state $s$, then continuing by always choosing actions optimally. To relate Q-values to the value function, note that

$$V^{*}(s) = \max_a Q^{*}(s,a) \tag{6.9}$$

where $V^{*}(s)$ is the value of $s$ assuming that the best action is taken initially.

The Q-learning rule is given as

$$Q(s,a) = Q(s,a) + \eta(r + \gamma \max_{a' \in \mathcal{A}} Q(s',a') - Q(s,a)) \tag{6.10}$$

The agent then takes the action with the highest Q-value.

## 6.3  Neural Networks and Reinforcement Learning

Neural networks and reinforcement learning have been combined in a number of ways. One approach of combining these models is to use a NN as an approximator of the value function used to predict future reward [162, 432]. Another approach uses RL to adjust weights. Both these approaches are discussed in this section.

As already indicated, the LVQ-II (refer to Section 5.1) implements a form of RL. Weights of the winning output unit are positively updated only if that output unit provided the correct response for the corresponding input pattern. If not, weights are penalized through adjustment away from that input pattern. Other approaches to use RL for NN training include RPROP (refer to Section 6.3.1), and gradient descent on the expected reward (refer to Section 6.3.2). Connectionist Q-learning is used to approximate the value function (refer to Section 6.3.3).

### 6.3.1  RPROP

Resilient propagation (RPROP) [727, 728] performs a direct adaptation of the weight step using local gradient information. Weight adjustments are implemented in the form of a reward or punishment, as follows: If the partial derivative, $\frac{\partial E}{\partial v_{ji}}$ (or $\frac{\partial E}{\partial w_{kj}}$), of weight $v_{ji}$ (or $w_{kj}$) changes its sign, the weight update value, $\Delta_{ji}$ ($\Delta_{kj}$), is decreased by the factor, $\eta^-$. The reason for this penalty is because the last weight update was too large, causing the algorithm to jump over a local minimum. On the other hand, if the derivative retains its sign, the update value is increased by factor $\eta^+$ to accelerate convergence.

For each weight, $v_{ji}$ (and $w_{kj}$), the change in weight is determined as

$$\Delta v_{ji}(t) = \begin{cases} -\Delta_{ji}(t) & \text{if } \frac{\partial E}{\partial v_{ji}}(t) > 0 \\ +\Delta_{ji}(t) & \text{if } \frac{\partial E}{\partial v_{ji}}(t) < 0 \\ 0 & \text{otherwise} \end{cases} \tag{6.11}$$

where

$$\Delta_{ji}(t) = \begin{cases} \eta^+ \Delta_{ji}(t-1) & \text{if } \frac{\partial E}{\partial v_{ji}}(t-1)\frac{\partial E}{\partial v_{ji}}(t) > 0 \\ \eta^- \Delta_{ji}(t-1) & \text{if } \frac{\partial E}{\partial v_{ji}}(t-1)\frac{\partial E}{\partial v_{ji}}(t) < 0 \\ \Delta_{ji}(t) & \text{otherwise} \end{cases} \tag{6.12}$$

Using the above,

$$v_{ji}(t+1) = v_{ji}(t) + \Delta v_{ji}(t) \tag{6.13}$$

RPROP is summarized in Algorithm 6.1. The value of $\Delta_0$ indicates the first weight step, and is chosen as a small value, e.g. $\Delta_0 = 0.1$ [728]. It is shown in [728] that the performance of RPROP is insensitive to the value of $\Delta_0$. Parameters $\Delta_{max}$ and $\Delta_{min}$ respectively specify upper and lower limits on update step sizes. It is suggested in [728] that $\eta^- = 0.5$ and $\eta^+ = 1.2$.

---

**Algorithm 6.1** RPROP Neural Network Training Algorithm

---

Initialize NN weights to small random values;
Set $\Delta_{ji} = \Delta_{kj} = \Delta_0$, $\forall i = 1, \ldots, I+1$, $\forall j = 1, \ldots, J+1$, $\forall k = 1, \ldots, K$;
Let $t = 0$;
**while** *stopping condition(s) not true* **do**
    **for** *each $w_{kj}$, $j = 1, \ldots, J+1$, $k = 1, \ldots, K$* **do**
        **if** $\frac{\partial E}{\partial w_{kj}}(t-1)\frac{\partial E}{\partial w_{kj}}(t) > 0$ **then**
            $\Delta_{kj}(t) = \min\{\Delta_{kj}(t-1)\eta^+, \Delta_{max}\}$;
            $\Delta w_{kj}(t) = -\text{sign}\left(\frac{\partial E}{\partial w_{kj}}(t)\right)\Delta_{kj}(t)$;
            $w_{kj}(t+1) = w_{kj}(t) + \Delta w_{kj}(t)$;
        **else if** $\frac{\partial E}{\partial w_{kj}}(t-1)\frac{\partial E}{\partial w_{kj}}(t) < 0$ **then**
            $\Delta_{kj}(t) = \max\{\Delta_{kj}(t-1)\eta^-, \Delta_{min}\}$;
            $w_{kj}(t+1) = w_{kj}(t) - \Delta w_{kj}(t-1)$;
            $\frac{\partial E}{\partial w_{kj}} = 0$;
        **else if** $\frac{\partial E}{\partial w_{kj}}(t-1)\frac{\partial E}{\partial w_{kj}}(t) = 0$ **then**
            $\Delta w_{kj}(t) = -\text{sign}\left(\frac{\partial E}{\partial w_{kj}}(t)\right)\Delta_{kj}(t)$;
            $w_{kj}(t+1) = w_{kj}(t) + \Delta w_{kj}(t)$;
    **end**
    Repeat the above for each $v_{ji}$ weight, $j = 1, \ldots, J$, $i = 1, \ldots, I+1$;
**end**

---

## 6.3.2   Gradient Descent Reinforcement Learning

For problems where only the immediate reward is maximized (i.e. there is no value function, only a reward function), Williams [911] proposed weight update rules that perform a gradient descent on the expected reward. These rules are then integrated with back-propagation. Weights are updated as follows:

$$\Delta w_{kj} = \eta_{kj}(r_p - \theta_k)e_{kj} \tag{6.14}$$

where $\eta_{kj}$ is a non-negative learning rate, $r_p$ is the reinforcement associated with pattern $\mathbf{z}_p$, $\theta_k$ is the reinforcement threshold value, and $e_{kj}$ is the eligibility of weight $w_{kj}$, given as

$$e_{kj} = \frac{\partial}{\partial w_{kj}}[\ln(g_j)] \tag{6.15}$$

where

$$g_j = P(o_{k,p} = t_{k,p}|\mathbf{w}_k, \mathbf{z}_p) \tag{6.16}$$

is the probability density function used to randomly generate actions, based on whether the target was correctly predicted or not. Thus, this NN reinforcement learning rule computes a GD in probability space.

Similar update equations are used for the $v_{ji}$ weights.

### 6.3.3 Connectionist Q-Learning

Neural networks have been used to learn the Q-function in Q-learning [527, 891, 745]. The NN is used to approximate the mapping between states and actions, and even to generalize between states. The input to the NN is the current state of the environment, and the output represents the action to execute. If there are $n_a$ actions, then either one NN with $n_a$ output units can be used [825], or $n_a$ NNs, one for each of the actions, can be used [527, 891, 745].

Assuming that one NN is used per action, Lin [527] used the Q-learning in equation (6.10) to update weights as follows:

$$\Delta \mathbf{w}(t) = \eta[r(t) + \gamma \max_{a \in \mathcal{A}} Q(t-1) - Q(t)] \nabla_w Q(t) \tag{6.17}$$

where $Q(t)$ is used as shorthand notation for $Q(s(t), a(t))$ and $\nabla_w Q(t)$ is a vector of the output gradients, $\frac{\partial Q}{\partial w}(t)$, which are calculated by means of back-propagation. Similar equations are used for the $\mathbf{v}_j$ weights.

Watkins [891] proposed a combination of Q-learning with TD($\lambda$)-learning, in which case,

$$\Delta \mathbf{w}(t) = \eta[r(t) + \gamma \max_{a \in \mathcal{A}} Q(t-1) - Q(t)] \left( \sum_{t'=0}^{t} (\lambda \gamma)^{t-t'} \nabla_w Q(t') \right) \tag{6.18}$$

where the relevance of the current error on earlier Q-value predictions is determined by $\lambda$. The update algorithm is given in Algorithm 6.2.

Rummery and Niranjan [745] proposed an alternative hybrid, where

$$\Delta \mathbf{w}(t) = \eta(r(t) + \gamma Q(t+1) - Q(t)) \left( \sum_{t'=0}^{t} (\lambda \gamma)^{t-t'} \nabla_w Q(t') \right) \tag{6.23}$$

which replaces the greedy $\max_{a \in \mathcal{A}} Q(t+1)$ with $Q(t+1)$.

Peng and Williams [674] proposed the $Q(\lambda)$ method, which combines Q-learning and TD($\lambda$)-learning as follows: A two step approach is followed, where weights are first updated using equation (6.17), followed by

$$\Delta \mathbf{w}(t) = \eta[r(t) + \gamma \max_{a \in \mathcal{A}} Q(t+1) - \max_{q \in \mathcal{A}} Q(t)] \sum_{t'=0}^{t} (\lambda \gamma)^{t-t'} \nabla_w Q(t') \tag{6.24}$$

---

**Algorithm 6.2** Connectionist Q-Learning Update Algorithm

---

Reset all eligibilities, $\mathbf{e}(t) = \mathbf{0}$;
$t = 0$;
**while** *stopping condition(s) not true* **do**
    Select action $a(t)$ as the one with maximum predicted Q-value;
    **if** $t > 0$ **then**

$$\mathbf{w}(t) = \mathbf{w}(t-1) + \eta(r(t-1) + \gamma Q(t) - Q(t-1))\mathbf{e}(t-1) \qquad (6.19)$$

    **end**
    Calculate $\nabla_w Q(t)$ with respect to action $a(t)$;
    Update eligibilities,
$$\mathbf{e}(t) = \nabla_w Q(t) + \gamma\lambda\mathbf{e}(t-1) \qquad (6.20)$$

    Perform action $a(t)$, and receive reward, $r(t)$;
**end**

---

---

**Algorithm 6.3** $Q(\lambda)$ Connectionist Update Algorithm

---

Reset all eligibilities, $\mathbf{e}(t) = \mathbf{0}$;
$t = 0$;
**while** *stopping condition(s) not true* **do**
    Select action $a(t)$ as the one with maximum predicted Q-value;
    **if** $t > 0$ **then**
$$\begin{aligned} \mathbf{w}(t) \;=\;\; & \mathbf{w}(t-1) + \eta([r(t-1) + \gamma \max_{a \in \mathcal{A}} Q(t) - Q(t-1)]\nabla_w Q(t-1) \\ & + [r(t-1) + \gamma \max_{a \in \mathcal{A}} Q(t) - \max_{a \in \mathcal{A}} Q(t-1)]\mathbf{e}(t-1)) \end{aligned} \qquad (6.21)$$

    **end**
    Update eligibilities,
$$\mathbf{e}(t) = \lambda\gamma[\mathbf{e}(t-1) + \lambda_w Q(t-1)] \qquad (6.22)$$

    Calculate $\nabla_w Q(t)$ with respect to action $a(t)$;
    Perform action $a(t)$, and receive reward, $r(t)$;
**end**

---

This gives an overall update of

$$\begin{aligned} \Delta\mathbf{w}(t) \;=\;\; & \eta\Bigg( [r(t) + \gamma \max_{a \in \mathcal{A}} Q(t+1) - Q(t)]\nabla_w Q(t) \\ & + [r(t) + \gamma \max_{a \in \mathcal{A}} Q(t+1) - \max_{a \in \mathcal{A}} Q(t)]\mathbf{e}(t) \Bigg) \end{aligned} \qquad (6.25)$$

where the eligibility is calculated using

$$\mathbf{e}(t) = \sum_{t'=0}^{t} (\lambda\gamma)^{t-t'} \nabla_w Q(t-t') = \nabla_w Q(t) + \lambda\gamma\mathbf{e}(t) \qquad (6.26)$$

Equation (6.26) keeps track of the weighted sum of previous error gradients.

The $Q(\lambda)$ update algorithm is given in Algorithm 6.3.

## 6.4   Assignments

1. Discuss how reinforcement learning can be used to guide a robot out of a room filled with obstacles.

2. Discuss the influence of the reinforcement threshold in equation (6.14) on performance.

3. Contrast reinforcement learning with coevolution (refer to Chapter 15).

4. For the RPROP algorithm, what will be the consequence if

    (a) $\Delta_{max}$ is too small?

    (b) $\eta^+$ is very large?

    (c) $\eta^-$ is very small?

5. Provide a motivation for replacing $\max_{a \in \mathcal{A}} Q(t+1)$ with $Q(t)$ in equation (6.23).

# Chapter 7

# Performance Issues (Supervised Learning)

Performance is possibly the driving force of all organisms. If no attention is given to improve performance, the quality of life will not improve. Similarly, performance is the most important aspect that has to be considered when an artificial neural network is being designed. The performance of an artificial NN is not just measured as the accuracy achieved by the network, but aspects such as computational complexity and convergence characteristics are just as important. These measures and other measures that quantify performance are discussed in Section 7.1, with specific reference to supervised networks.

The design of NNs for optimal performance requires careful consideration of several factors that influence network performance. In the early stages of NN research and applications, the design of NNs was basically done by following the intuitive feelings of an expert user, or by following rules of thumb. The vast number of theoretical analyses of NNs made it possible to better understand the working of NNs – to unravel the "black box". These insights helped to design NNs with improved performance. Factors that influence the performance of NNs are discussed in Section 7.3.

Although the focus of this chapter is on supervised learning, several ideas can be extrapolated to unsupervised learning NNs.

## 7.1 Performance Measures

This section presents NN performance measures under three headings: *accuracy, complexity* and *convergence*.

### 7.1.1 Accuracy

Generalization is a very important aspect of neural network learning. Since it is a measure of how well the network interpolates to points not used during training, the ultimate objective of NN learning is to produce a learner with low generalization error.

*Computational Intelligence: An Introduction*, Second Edition A.P. Engelbrecht
©2007 John Wiley & Sons, Ltd

That is, to minimize the true risk function

$$\mathcal{E}_G(\Omega; \mathbf{W}) = \int (f_{NN}(\mathbf{z}, \mathbf{W}) - \mathbf{t})^2 d\Omega(\mathbf{z}, \mathbf{t}) \tag{7.1}$$

where, from Section 3.2.1, $\Omega(\mathbf{z}, \mathbf{t})$ is the stationary density according to which patterns are sampled, $\mathbf{W}$ describes the network weights, and $\mathbf{z}$ and $\mathbf{t}$ are respectively the input and target vectors. The function $f_{NN}$ is an approximation of the true underlying function. Since $\Omega$ is generally not known, $f_{NN}$ is found through minimization of the empirical error function

$$\mathcal{E}_T(D_T; \mathbf{W}) = \frac{1}{P_T} \sum_{p=1}^{P_T} (f_{NN}(\mathbf{z}_p, W) - \mathbf{t}_p)^2 \tag{7.2}$$

over a finite data set $D_T \sim \Omega$. When $P_T \rightarrow \infty$, then $\mathcal{E}_T \rightarrow \mathcal{E}_G$. The aim of NN learning is therefore to learn the examples presented in the training set well, while still providing good generalization to examples not included in the training set. It is, however, possible that a NN exhibits a very low training error, but bad generalization due to overfitting (memorization) of the training patterns.

The most common measure of accuracy is the mean squared error (MSE), in which case the training error, $\mathcal{E}_T$, is expressed as

$$\mathcal{E}_T = \frac{\sum_{p=1}^{P_T} \sum_{k=1}^{K} (t_{k,p} - o_{k,p})^2}{P_T K} \tag{7.3}$$

where $P_T$ is the total number of training patterns in the training set $D_T$, and $K$ is the number of output units. The generalization error, $\mathcal{E}_G$, is approximated in the same way, but with the first summation over the $P_G$ patterns in the generalization, or test set, $D_G$. Instead of the MSE, the sum squared error (SSE),

$$SSE = \sum_{p=1}^{P} \sum_{k=1}^{K} (t_{k,p} - o_{k,p})^2 \tag{7.4}$$

can also be used, where $P$ is the total number of patterns in the data set considered. However, the SSE is not a good measure when the performance on different data set sizes are compared.

An additional error measure is required for classification problems, since the MSE alone is not a good descriptor of accuracy. In the case of classification problems, the percentage correctly classified (or incorrectly classified) patterns is used as a measure of accuracy. The reason why the MSE is not a good measure, is that the network may have a good accuracy in terms of the number of correct classifications, while having a relatively large MSE. If just the MSE is used to indicate when training should stop, it can result in the network being trained too long in order to reach the low MSE; hence, wasting time and increasing the chances of overfitting the training data (with reference to the number of correct classifications). But when is a pattern classified as correct? When the output class of the NN is the same as the target class – which is not a problem to determine when the ramp or step function is used as the activation function in the output layer. In the case of continuous activation functions, a pattern $\mathbf{z}_p$ is usually considered as being correctly classified if for each output unit $o_k$, $((o_{k,p} \geq 0.5 + \theta$ and $t_{k,p} = 1)$ or $(o_{k,p} \leq 0.5 - \theta$ and $t_{k,p} = 0))$, where $\theta \in [0, 0.5]$ – of course, assuming that the target classes are binary encoded.

An additional measure of accuracy is to calculate the correlation between the output and target values for all patterns. This measure, referred to as the correlation coefficient, is calculated as

$$
\begin{aligned}
r &= \frac{\sum_{i=1}^{n}(x_i - \overline{x})\sum_{i=1}^{n}(y_i - \overline{y})}{\sigma_x \sigma_y} \\
&= \frac{\sum_{i=1}^{n} x_i y_i - \frac{1}{n}\sum_{i=1}^{n} x_i \sum_{i=1}^{n} y_i}{\sqrt{\sum_{i=1}^{n} x_i^2 - \frac{1}{n}(\sum_{i=1}^{n} x_i)^2}\sqrt{\sum_{i=1}^{n} y_i^2 - \frac{1}{n}(\sum_{i=1}^{n} y_i)^2}}
\end{aligned} \tag{7.5}
$$

where $x_i$ and $y_i$ are observations, $\overline{x}$ and $\overline{y}$ are respectively the averages over all observations $x_i$ and $y_i$, and $\sigma_x$ and $\sigma_y$ are the standard deviations of the $x_i$ and $y_i$ observations respectively, and can be used to quantify the linear relationship between variables $x$ and $y$. As measure of learning accuracy, where $x = o_{k,p}$ and $y = t_{k,p}$, the correlation coefficient quantifies the linear relationship between the approximated (learned) function and the true function. A correlation value close to 1 indicates a good approximation to the true function. Therefore, the correlation coefficient

$$
r = \frac{\sum_{p=1}^{P} o_{k,p} t_{k,p} - \frac{1}{P}\sum_{p=1}^{P} o_{k,p} \sum_{p=1}^{P} t_{k,p}}{\sqrt{\sum_{p=1}^{P} o_{k,p}^2 - \frac{1}{P}(\sum_{p=1}^{P} o_{k,p})^2}\sqrt{\sum_{p=1}^{P} t_{k,p}^2 - \frac{1}{P}(\sum_{p=1}^{P} t_{k,p})^2}} \tag{7.6}
$$

is calculated as a measure of how well the NN approximates the true function.

Another very important aspect of NN accuracy is *overfitting*. Overfitting of a training set means that the NN memorizes the training patterns, and consequently loses the ability to generalize. That is, NNs that overfit cannot predict correct output for data patterns not seen during training. Overfitting occurs when the NN architecture is too large, i.e. the NN has too many weights (in statistical terms: too many *free parameters*) – a direct consequence of having too many hidden units and irrelevant input units. If the NN is trained for too long, the excess free parameters start to memorize all the training patterns, and even noise contained in the training set. Remedies for overfitting include optimizing the network architecture and using enough training patterns (discussed in Section 7.3).

Estimations of generalization error during training can be used to detect the point of overfitting. The simplest approach to find the point of overfitting was developed through studies of training and generalization profiles. Figure 7.1 presents a general illustration of training and generalization errors as a function of training epochs. From the start of training, both the training and generalization errors decrease - usually exponentially. In the case of oversized NNs, there is a point at which the training error continues to decrease, while the generalization error starts to increase. This is the point of overfitting. Training should stop as soon an increase in generalization error is observed.

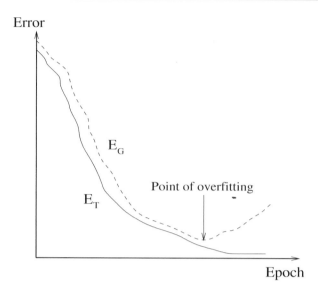

**Figure 7.1** Illustration of Overfitting

In order to detect the point of overfitting, the original data set is divided into three disjoint sets, i.e. the training set $D_T$, the generalization set $D_G$ and the validation set $D_V$. The validation set is then used to estimate the generalization error. Since both the training error and the validation error usually fluctuate, determining the point of overfitting is not straightforward. A moving average of the validation error has to be used. Overfitting is then detected when

$$\mathcal{E}_V > \overline{\mathcal{E}}_V + \sigma_{\mathcal{E}_V} \tag{7.7}$$

where $\mathcal{E}_V$ is the MSE on the validation set, $\overline{\mathcal{E}}_V$ is the average MSE on the validation set since training started, and $\sigma_{\mathcal{E}_V}$ is the standard deviation in validation error.

Röbel suggested the *generalization factor* as an alternative indication of overfitting [732]. Röbel defines the generalization factor $\rho = \frac{\mathcal{E}_V}{\mathcal{E}_T}$, where $\mathcal{E}_V$ and $\mathcal{E}_T$ are the MSE on the validation set $D_V$ and current training subset $D_T$ respectively. The generalization factor indicates the error made in training on $D_T$ only, instead of training on the entire input space. Overfitting is detected when $\rho(t) > \varphi_\rho(t)$, where $\varphi_\rho(t) = \min\{\varphi_\rho(t-1), \overline{\rho} + \sigma_\rho, 1.0\}$; $t$ is the current epoch, $\overline{\rho}$ is the average generalization factor over a fixed number of preceding epochs, and $\sigma_\rho$ is the standard deviation. This test ensures that $\rho \leq 1.0$. Keep in mind that $\rho$ does not give an indication of the accuracy of learning, but only the ratio between the training and validation error. For function approximation problems (as is the case with Röbel's work) where the MSE is used as a measure of accuracy, a generalization factor $\rho < 1$ means that the validation error is smaller than the training error – which is desirable. As $\rho$ becomes large (greater than 1), the difference between the training error and validation error increases, which indicates an increase in validation error with a decrease in training error – an indication of overfitting. For classification problems where the percentage of correctly classified patterns is used as a measure of accuracy, $\rho$ should be larger than 1.

It is important to note that the training error or the generalization error alone is not sufficient to quantify the accuracy of a NN. Both these errors should be considered.

## Additional Reading Material on Accuracy

The trade-off between training error and generalization has prompted much research in the generalization performance of NNs. Average generalization performance has been studied theoretically to better understand the behavior of NNs trained on a finite data set. Research shows a dependence of generalization error on the training set, the network architecture and weight values. Schwartz *et al.* [767] show the importance of training set size for good generalization in the context of ensemble networks. Other research uses the VC-dimension (Vapnik-Chervonenkis dimension) [8, 9, 152, 643] to derive boundaries on the generalization error as a function of network and training set size. Best known are the limits derived by Baum and Haussler [54] and Haussler *et al.* [353]. While these limits are derived for, and therefore limited to, discrete input values, Hole derives generalization limits for real valued inputs [375].

Limits on generalization have also been developed by studying the relationship between training error and generalization error. Based on Akaike's final prediction error and information criterion [15], Moody derived the generalized prediction error which gives a limit on the generalization error as a function of the training error, training set size, the number of effective parameters, and the effective noise variance [603, 604]. Murata *et al.* [616, 617, 618] derived a similar network information criterion. Using a different approach, i.e. Vapnik's Bernoulli theorem, Depenau and Møller [202] derived a bound as a function of training error, the VC-dimension and training set size.

These research results give, sometimes overly pessimistic, limits that help to clarify the behavior of generalization and its relationship with architecture, training set size and training error. Another important issue in the study of generalization is that of overfitting. Overfitting means that the NN learns too much detail, effectively memorizing training patterns. This normally happens when the network complexity does not match the size of the training set, i.e. the number of adjustable weights (free parameters) is larger than the number of independent patterns. If this is the case, the weights learn individual patterns and even capture noise. This overfitting phenomenon is the consequence of training on a finite data set, minimizing the empirical error function given in equation (7.2), which differs from the true risk function given in equation (7.1).

Amari *et al.* developed a statistical theory of overtraining in the asymptotic case of large training set sizes [22, 21]. They analytically determine the ratio in which patterns should be divided into training and test sets to obtain optimal generalization performance and to avoid overfitting. Overfitting effects under large, medium and small training set sizes have been investigated analytically by Amari *et al.* [21] and Müller *et al.* [612].

## 7.1.2  Complexity

The computational complexity of a NN is directly influenced by:

1. **The network architecture**: The larger the architecture, the more feedforward calculations are needed to predict outputs after training, and the more learning calculations are needed per pattern presentation.

2. **The training set size**: The larger the training set size, the more patterns are presented for training. Therefore, the total number of learning calculations per epoch is increased.

3. **Complexity of the optimization method**: As will be discussed in Section 7.3, sophisticated optimization algorithms have been developed to improve the accuracy and convergence characteristics of NNs. The sophistication comes, however, at the cost of increased computational complexity to determine the weight updates.

Training time is usually quantified in terms of the number of epochs to reach specific training or generalization errors. When different learning algorithms are compared, the number of epochs is usually not an accurate estimate of training time or computational complexity. Instead, the total number of pattern presentations, or weight updates are used. A more accurate estimate of computational complexity is to count the total number of calculations made during training.

## 7.1.3  Convergence

The convergence characteristics of a NN can be described by the ability of the network to converge to specified error levels (usually considering the generalization error). The ability of a network to converge to a specific error is expressed as the number of times, out of a fixed number of simulations, that the network succeeded in reaching that error. While this is an empirical approach, rigorous theoretical analysis has been done for some network types.

# 7.2  Analysis of Performance

Any study of the performance of NNs (or any other stochastic algorithm for that matter) and any conclusions based on just one simulation are incomplete and inconclusive. Conclusions on the performance of NNs must be based on the results obtained from several simulations. For each simulation the NN starts with new random initial weights and uses different training, validation and generalization sets, independent of previous sets. Performance results are then expressed as averages over all the simulations, together with variances, or confidence intervals.

Let $\varrho$ denote the performance measure under consideration. Results are then reported as $\bar{\varrho} \pm \sigma_{\varrho}$. The average $\bar{\varrho}$ is an indication of the average performance over all simulations, while $\sigma_\varrho$ gives an indication of the variance in performance. The $\sigma_\varrho$ parameter is very important in decision making. For example, if two algorithms A and B are compared where the MSE for A is $0.001\pm0.0001$, and that of $B$ is $0.0009\pm0.0006$, then algorithm A will be preferred even though B has a smaller MSE. Algorithm A has a smaller variance, having MSE values in the range $[0.0009, 0.0011]$, while $B$ has MSE values in a larger range of $[0.0003, 0.0015]$.

While the above approach to present results is sufficient, results are usually reported with associated confidence intervals. If a confidence level of $\alpha = 0.01$ is used, for example, then 99% of the observations will be within the calculated confidence interval. Before explaining how to compute the confidence intervals, it is important to note that statistical literature suggests that at least 30 independent simulations are needed. This allows the normality assumption as stated by the central limit theorem: the probability distribution governing the variable $\bar\varrho$ approaches a Normal distribution as the number of observations (simulations) tends to infinity. Using this result, the confidence interval associated with confidence level $\alpha$ can be estimated as

$$\bar\varrho \pm t_{\alpha,n-1}\sigma_\varrho \tag{7.8}$$

where $t_{\alpha,n-1}$ is a constant obtained from the $t$-distribution with $n-1$ degrees of freedom ($n$ is the number of simulations) and

$$\sigma_\varrho = \sqrt{\frac{\sum_{i=1}^{n}(\varrho_i - \bar\varrho)^2}{n(n-1)}} \tag{7.9}$$

It should be noted at this point that the t-test assumes that samples are normally distributed. It is, however, not always the case that 30 samples will guarantee a normal distribution. If not normally distributed, nonparametric tests need to be used.

## 7.3 Performance Factors

This section discusses various aspects that have an influence on the performance of supervised NNs. These aspects include data manipulation, learning parameters, architecture selection, and optimization methods.

### 7.3.1 Data Preparation

One of the most important steps in using a NN to solve real-world problems is to collect and transform data into a form acceptable to the NN. The first step is to decide on what the inputs and the outputs are. Obviously irrelevant inputs should be excluded. Section 7.3.5 discusses ways in which the NN can decide itself which inputs are irrelevant. The second step is to process the data in order to remove outliers, handle missing data, transform non-numeric data to numeric data and to scale the data into the active range of the activation functions used. Each of these aspects are discussed in the sections below.

## Missing Values

It is common that real-world data sets have missing values for input parameters. NNs need a value for each of the input parameters. Therefore, something has to be done with missing values. The following options exist:

- Remove the entire pattern if it has a missing value. While pattern removal solves the missing value problem, other problems are introduced: (1) the available information for training is reduced which can be a problem if data is already limited, and (2) important information may be lost.

- Replace each missing value with the average value for that input parameter in the case of continuous values, or with the most frequently occurring value in the case of nominal or discrete values. This replacing of missing values introduces no bias.

- For each input parameter that has a missing value, add an additional input unit to indicate patterns for which parameters are missing. It can then be determined after training whether the missing values had a significant influence on the performance of the network.

While missing values present a problem to supervised neural networks, SOMs do not suffer under these problems. Missing values do not need to be replaced. The BMN for a pattern with missing values is, for example, calculated by ignoring the missing value and the corresponding weight value of the codebook vector in the calculation of the Euclidean distance between the pattern and codebook vector.

## Coding of Input Values

All input values to a NN must be numeric. Nominal values therefore need to be transformed to numerical values. A nominal input parameter that has $n$ different values is coded as $n$ different binary input parameters, where the input parameter that corresponds to a nominal value has the value 1, and the rest of these parameters have the value 0. An alternative is to use just one input parameter and to map each nominal value into an equivalent numerical value. This is, however, not a good idea, since the NN will interpret the input parameter as having continuous values, thereby losing the discrete characteristic of the original data.

## Outliers

Outliers have severe effects on accuracy, especially when gradient descent is used with the SSE as objective function. An outlier is a data pattern that deviates substantially from the data distribution. Because of the large deviation from the norm, outliers result in large errors, and consequently large weight updates. Figure 7.3 shows that larger differences between target and output values cause an exponential increase in the error if the SSE is used as objective function. The fitted function is then pulled toward the outliers in an attempt to reduce the training error. As result, the generalization deteriorates. Figure 7.2 illustrates this effect.

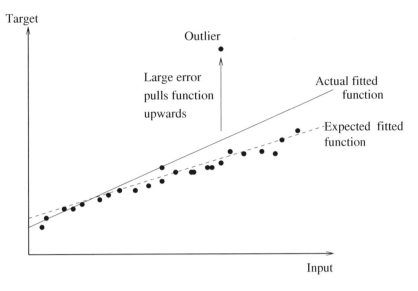

**Figure 7.2** Effect of Outliers

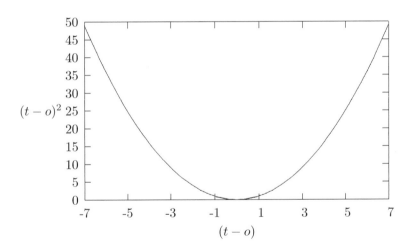

**Figure 7.3** Sum Squared Error Objective Function

The outlier problem can be addressed in the following ways:

- Remove outliers before training starts, using statistical techniques. While such actions will eliminate the outlier problem, it is believed that important information about the data might also be removed at the same time.

- Use a robust objective function that is not influenced by outliers. An example objective function is the Huber function as illustrated in Figure 7.4 [396]. Patterns for which the error is larger than $|\epsilon|$ have a constant value, and have a zero influence when weights are updated (the derivative of a constant is zero).

- Slade and Gedeon [796] and Gedeon *et al.* [311] proposed bimodal distribution removal, where the aim is to remove outliers from training sets during training. Frequency distributions of pattern errors are analyzed during training to identify and remove outliers. If the original training set contains no outliers, the method simply reduces to standard learning.

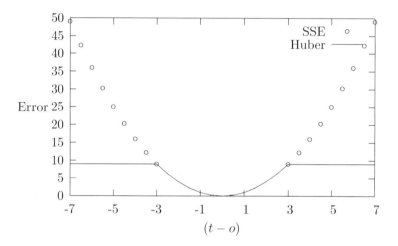

**Figure 7.4** Huber Objective Function

### Scaling and Normalization

Data needs to be scaled to the active range and domain of the activation functions used. While it is not necessary to scale input values, performance can be improved if inputs are scaled to the active domain of the activation functions. For example, consider the sigmoid activation function. Simple mathematical calculations show that the active domain of the sigmoid function is $[-\sqrt{3}, \sqrt{3}]$, corresponding to the parts of the function for which changes in input values have relatively large changes in output. Values near the asymptotic ends of the sigmoid function have a very small influence on weight updates. Changes in these values result in very small changes in output. Furthermore, the derivatives near the asymptotes are approximately zero, causing weight updates to be approximately zero; therefore, achieving no learning in these areas.

When bounded activation functions are used, the target values have to be scaled to the range of the activation function, for example $(0, 1)$ for the sigmoid function and $(-1, 1)$ for the hyperbolic tangent. If $t_{u,max}$ and $t_{u,min}$ are the maximum and minimum values of the unscaled target $t_u$, then,

$$t_s = \frac{t_u - t_{u,min}}{t_{u,max} - t_{u,min}}(t_{s,max} - t_{s,min}) + t_{s,min} \qquad (7.10)$$

where $t_{s,max}$ and $t_{s,min}$ are the new maximum and minimum values of the scaled values, linearly maps the range $[t_{u,min}, t_{u,max}]$ to the range $[t_{s,min}, t_{s,max}]$.

In the case of classification problems, target values are usually elements of the set $\{0.1, 0.9\}$ for the sigmoid function. The value 0.1 is used instead of 0, and 0.9 instead of 1. Since the output of the sigmoid function can only approach 0 and 1, a NN can never converge to the best set of weights if the target values are 0 or 1. In this case the goal of the NN is always out of reach, and the network continues to push weight values toward extreme values until training is stopped.

Scaling of target values into a smaller range does have the disadvantage of increased training time. Engelbrecht *et al.* [244] showed that if target values are linearly scaled using

$$t_s = c_1 t_u + c_2 \qquad (7.11)$$

where $t_s$ and $t_u$ are respectively the scaled and original unscaled target values, the NN must be trained longer until

$$MSE_s = (c_1)^2 MSE_r \qquad (7.12)$$

to reach a desired accuracy, $MSE_r$, on the original unscaled data set.

The hyperbolic tangent will therefore result in faster training times than the sigmoid function, assuming the same initial conditions and training data.

The scaling process above is usually referred to as amplitude scaling, or min-max scaling. Min-max scaling preserves the relationships among the original data. Two other frequently used scaling methods are mean centering and variance scaling. To explain these two scaling methods, assume that $Z \in \mathbb{R}^{I \times P}$ is a matrix containing all input vectors such that input vectors are arranged as columns in $Z$, and $T \in \mathcal{R}^{K \times P}$ is the matrix of associated target vectors, arranged in column format. For the mean centering process, compute

$$\overline{Z}_i \;\; = \;\; \sum_{p=1}^{P} Z_{i,p}/P \qquad (7.13)$$

$$\overline{T}_k \;\; = \;\; \sum_{p=1}^{P} T_{k,p}/P \qquad (7.14)$$

for all $i = 1, \cdots, I$ and $k = 1, \cdots, K$; $\overline{Z}_i$ is the average value for input $z_i$ over all the patterns, and $\overline{T}_k$ is the average target value for the $k$-th output unit over all patterns. Then,

$$Z_{i,p}^{M} \;\; = \;\; Z_{i,p} - \overline{Z}_i \qquad (7.15)$$
$$T_{k,p}^{M} \;\; = \;\; T_{k,p} - \overline{T}_k \qquad (7.16)$$

for all $i = 1, \cdots, I$, $k = 1, \cdots, K$ and $p = 1, \cdots, P$; $Z_{i,p}^{M}$ is the scaled value of the input to unit $z_i$ for pattern $p$, and $T_{k,p}^{M}$ is the corresponding scaled target value.

Variance scaling, on the other hand, computes for each row in each matrix the standard deviations ($I$ deviations for matrix $Z$ and $K$ deviations for matrix $T$) over all $P$ elements in the row. Let $\sigma_{z_i}$ denote the standard deviation of row $i$ of matrix $Z$, and

$\sigma_{t_k}$ is the standard deviation of row $k$ of matrix $T$. Then,

$$Z_{i,p}^V = \frac{Z_{i,p}}{\sigma_{z_i}} \tag{7.17}$$

$$T_{k,p}^V = \frac{T_{k,p}}{\sigma_{t_k}} \tag{7.18}$$

for all $i = 1, \cdots, I$, $k = 1, \cdots, K$ and $p = 1, \cdots, P$.

Mean centering and variance scaling can both be used on the same data set. Mean centering is, however, more appropriate when the data contains no biases, while variance scaling is appropriate when training data are measured with different units.

Both mean centering and variance scaling can be used in situations where the minimum and maximum values are unknown. Z-score normalization is another data transformation scheme that can be used in situations where the range of values is unknown. It is essentially a combination of mean centering and variance scaling, and is very useful when there are outliers in the data. For z-score normalization,

$$Z_{i,p}^{MV} = \frac{Z_{i,p} - \overline{Z}_i}{\sigma_{z_i}} \tag{7.19}$$

$$T_{k,p}^{MV} = \frac{T_{k,p} - \overline{T}_k}{\sigma_{t_k}} \tag{7.20}$$

For some NN types, for example the LVQ, input data is preferred to be normalized to vectors of unit length. The values $z_{i,p}$ of each input parameter $z_i$ are then normalized using

$$z_{i,p}' = \frac{z_{i,p}}{\sqrt{\sum_{i=1}^{I} z_{i,p}^2}} \tag{7.21}$$

The normalization above loses information on the absolute magnitude of the input parameters, since it requires the length of all input vectors (patterns) to be the same. Input patterns with parameter values of different magnitudes are normalized to the same vector, e.g. vectors $(-1, 1, 2, 3)$ and $(-3, 3, 6, 9)$. Z-axis normalization is an alternative approach that preserves the absolute magnitude information of input patterns. Before the normalization step, input values are scaled to the range $[-1, 1]$. Input values are then normalized using

$$z_{i,p}' = \frac{z_{i,p}}{\sqrt{I}} \tag{7.22}$$

and adding an additional input unit $z_0$ to the NN, referred to as the synthetic parameter, with value

$$z_0 = \sqrt{1 - \frac{L^2}{I}} \tag{7.23}$$

where $L$ is the Euclidean length of input vector $\mathbf{z}_p$.

## Noise Injection

For problems with a limited number of training patterns, controlled injection of noise helps to generate new training patterns. Provided that noise is sampled from a normal distribution with a small variance and zero mean, it can be assumed that the resulting changes in the network output will have insignificant consequences [379]. Also, the addition of noise results in a convolutional smoothing of the target function, resulting in reduced training time and increased accuracy [713]. Engelbrecht used noise injection around decision boundaries to generate new training patterns for improved performance [237].

## Training Set Manipulation

Several researchers have developed techniques to control the order in which patterns are presented for learning. These techniques resulted in the improvement of training time and accuracy. A short summary of such training set manipulation techniques is given below.

Ohnishi *et. al.* [634] suggested a method called selective presentation where the original training set is divided into two training sets. One set contains typical patterns, and the other set contains confusing patterns. With "typical pattern" the authors mean a pattern far from decision boundaries, while "confusing pattern" refers to a pattern close to a boundary. The two training sets are created once before training. Generation of these training sets assumes prior knowledge about the problem, i.e. where decision boundaries are located in input space. In many practical applications such prior knowledge is not available, thus limiting the applicability of this approach. The selective presentation strategy alternately presents the learner with typical and then confusing patterns.

Kohara developed selective presentation learning specifically for forecasting applications [471]. Before training starts, the algorithm generates two training sets. The one set contains all patterns representing large next-day changes, while patterns representing small next-day changes are contained in the second set. Large-change patterns are then simply presented more often than small-change patterns (similar to selective presentation).

Cloete and Ludik [137, 537] have done extensive research on *training strategies*. Firstly, they proposed Increased Complexity Training where a NN first learns easy problems, and then the complexity of the problem to be learned is gradually increased. The original training set is split into subsets of increasing complexity before training commences. A drawback of this method is that the complexity measure of training data is problem dependent, thus making the strategy unsuitable for some tasks. Secondly, Cloete and Ludik developed *incremental training strategies*, i.e. incremental subset training [139] and incremental increased complexity training [538]. In incremental subset training, training starts on a random initial subset. During training, random subsets from the original training set are added to the actual training subset. Incremental increased complexity training is a variation of increased complexity training, where the complexity ranked order is maintained, but training is not done on each complete complexity subset. Instead, each complexity subset is further divided into smaller random subsets. Training starts on an initial subset of a complexity subset, and is incrementally increased during training. Finally, delta training strategies were proposed [138]. With delta subset training examples are ordered according to inter-example distance, e.g. Hamming or Euclidean distance. Different strategies of example presentations were investigated: smallest difference examples first, largest difference examples first, and alternating difference.

When vast quantities of data are available, training on all these data can be pro-hibitively slow, and may require reduction of the training set. The problem is which of the data should be selected for training. An easy strategy is to simply sample a smaller data set at each epoch using a uniform random number generator. Alterna-tively, a fast clustering algorithm can be used to group similar patterns together, and to sample a number of patterns from each cluster.

## 7.3.2    Weight Initialization

Gradient-based optimization methods, for example gradient descent, is very sensitive to the initial weight vectors. If the initial position is close to a local minimum, con-vergence will be fast. However, if the initial weight vector is on a flat area in the error surface, convergence is slow. Furthermore, large initial weight values have been shown to prematurely saturate units due to extreme output values with associated zero derivatives [400]. In the case of optimization algorithms such as PSO and GAs, initialization should be uniformly over the entire search space to ensure that all parts of the search space are covered.

A sensible weight initialization strategy is to choose small random weights centered around 0. This will cause net input signals to be close to zero. Activation functions then output midrange values regardless of the values of input units. Hence, there is no bias toward any solution. Wessels and Barnard [898] showed that random weights in the range $[\frac{-1}{\sqrt{fanin}}, \frac{1}{\sqrt{fanin}}]$ is a good choice, where $fanin$ is the number of connections leading to a unit.

Why are weights not initialized to zero in the case of gradient-based optimization? This strategy will work only if the NN has just one hidden unit. For more than one hidden unit, all the units produce the same output, and thus make the same contribution to the approximation error. All the weights are therefore adjusted with the same value. Weights will remain the same irrespective of training time – hence, no learning takes place. Initial weight values of zero for PSO will also fail, since no velocity changes are made; therefore no weight changes. GAs, on the other hand, will work with initial zero weights if mutation is implemented.

### 7.3.3 Learning Rate and Momentum

The convergence speed of NNs is directly proportional to the learning rate $\eta$. Considering stochastic GD, the momentum term added to the weight updates also has the objective of improving convergence time.

**Learning Rate**

The learning rate controls the size of each step toward the minimum of the objective function. If the learning rate is too small, the weight adjustments are correspondingly small. More learning iterations are then required to reach a local minimum. However, the search path will closely approximate the gradient path. Figure 7.5(a) illustrates the effect of small $\eta$. On the other hand, large $\eta$ will have large weight updates. Convergence will initially be fast, but the algorithm will eventually oscillate without reaching the minimum. It is also possible that too large a learning rate will cause "jumping" over a good local minimum proceeding toward a bad local minimum. Figure 7.5(b) illustrates the oscillating behavior, while Figure 7.5(c) illustrates how large learning rates may cause the network to overshoot a good minimum and get trapped in a bad local minimum. Small learning rates also have the disadvantage of being trapped in a bad local minimum as illustrated in Figure 7.5(d). The search path goes down the first local minimum, with no mechanism to move out of it toward the next, better minimum. Of course, the search trajectory depends on the initial starting position. If the second initial point is used, the NN will converge to the better local minimum.

But how should the value of the learning rate be selected? One approach is to find the optimal value of the learning rate through cross-validation, which is a lengthy process. An alternative is to select a small value (e.g. 0.1) and to increase the value if convergence is too slow, or to decrease it if the error does not decrease fast enough. Plaut *et al.* [680] proposed that the learning rate should be inversely proportional to the fanin of a neuron. This approach has been theoretically justified through an analysis of the eigenvalue distribution of the Hessian matrix of the objective function [167].

Several heuristics have been developed to dynamically adjust the learning rate during training. One of the simplest approaches is to assume that each weight has a different learning rate $\eta_{kj}$. The following rule is then applied to each weight before that weight is updated: if the direction in which the error decreases at this weight change is the same as the direction in which it has been decreasing recently, then $\eta_{kj}$ is increased; if not, $\eta_{kj}$ is decreased [410]. The direction in which the error decreases is determined by the sign of the partial derivative of the objective function with respect to the weight. Usually, the average change over a number of pattern presentations is considered and not just the previous adjustment.

An alternative is to use an annealing schedule to gradually reduce a large learning rate to a smaller value (refer to equation 4.40). This allows for large initial steps, and ensures small steps in the region of the minimum.

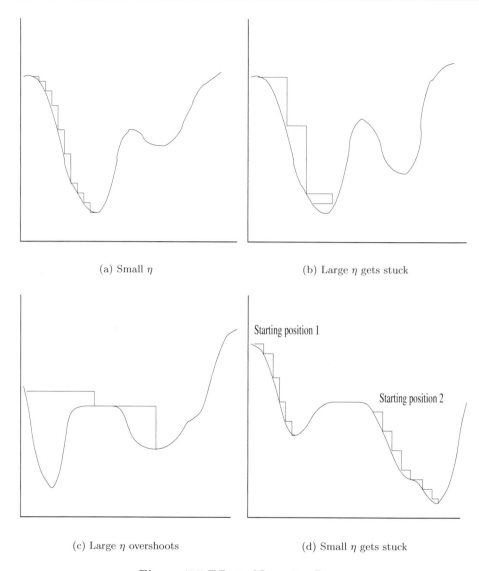

(a) Small $\eta$                        (b) Large $\eta$ gets stuck

(c) Large $\eta$ overshoots          (d) Small $\eta$ gets stuck

**Figure 7.5** Effect of Learning Rate

Of course more complex adaptive learning rate techniques have been developed, with elaborate theoretical analysis. The interested reader is referred to [170, 552, 755, 880].

### Momentum

Stochastic learning, where weights are adjusted after each pattern presentation, has the disadvantage of fluctuating changes in the sign of the error derivatives. The network spends a lot of time going back and forth, unlearning what the previous steps have learned. Batch learning is a solution to this problem, since weight changes are accumulated and applied only after all patterns in the training set have been presented. Another solution is to keep with stochastic learning, and to add a momentum term.

The idea of the momentum term is to average the weight changes, thereby ensuring that the search path is in the average downhill direction. The momentum term is then simply the previous weight change weighted by a scalar value $\alpha$. If $\alpha = 0$, then the weight changes are not influenced by past weight changes. The larger the value of $\alpha$, the longer the change in the steepest descent direction has to be persevered in order to affect the direction in which weights are adjusted. A static value of 0.9 is usually used.

The optimal value of $\alpha$ can also be determined through cross-validation. Strategies have also been developed that use adaptive momentum rates, where each weight has a different momentum rate. Fahlman developed the schedule

$$\alpha_{kj}(t) = \frac{\frac{\partial \mathcal{E}}{\partial w_{kj}(t)}}{\frac{\partial \mathcal{E}}{\partial w_{kj}(t-1)} - \frac{\partial \mathcal{E}}{\partial w_{kj}(t)}} \tag{7.24}$$

This variation to the standard back-propagation algorithm is referred to as *quickprop* [253]. Becker and Le Cun [57] calculated the momentum rate as a function of the second-order error derivatives:

$$\alpha = \left(\frac{\partial^2 \mathcal{E}}{\partial w_{kj}^2}\right)^{-1} \tag{7.25}$$

For more information on other approaches to adapt the momentum rate refer to [644, 942].

### 7.3.4 Optimization Method

The optimization method used to determine weight adjustments has a large influence on the performance of NNs. While GD is a very popular optimization method, GD is plagued by slow convergence and susceptibility to local minima (as introduced and discussed in Section 3.2.2). Improvements of GD have been made to address these problems, for example, the addition of the momentum term. Also, second-order derivatives of the objective function have been used to compute weight updates. In doing so, more information about the structure of the error surface is used to direct weight changes. The reader is referred to [51, 57, 533]. Other approaches to improve NN training are to use global optimization algorithms instead of local optimization algorithms, for example simulated annealing [736], genetic algorithms [247, 412, 494], particle swarm optimization algorithms [157, 229, 247, 862, 864], and LeapFrog optimization [247, 799, 800].

### 7.3.5 Architecture Selection

Referring to one of Ockham's statements, if several networks fit the training set equally well, then the simplest network (i.e. the network that has the smallest number of weights) will on average give the best generalization performance [844]. This hypothesis has been investigated and confirmed by Sietsma and Dow [789]. A network with too many free parameters may actually memorize training patterns and may also accurately fit the noise embedded in the training data, leading to bad generalization. Overfitting can thus be prevented by reducing the size of the network through elimination of individual weights or units. The objective is therefore to balance the complexity of the network with goodness-of-fit of the true function. This process is referred to as *architecture selection*. Several approaches have been developed to select the optimal architecture, i.e. regularization, network construction (growing) and pruning. These approaches will be overviewed in more detail below.

Learning is not just perceived as finding the optimal weight values, but also finding the optimal architecture. However, it is not always obvious what is the best architecture. Finding the ultimate best architecture requires a search of all possible architectures. For large networks an exhaustive search is prohibitive, since the search space consists of $2^w$ architectures, where $w$ is the total number of weights [602]. Instead, heuristics are used to reduce the search space. A simple method is to train a few networks of different architecture and to choose the one that results in the lowest generalization error as estimated from the generalized prediction error [603, 604] or the network information criterion [616, 617, 618]. This approach is still expensive and requires many architectures to be investigated to reduce the possibility that the optimal model is not found. The NN architecture can alternatively be optimized by trial and error. An architecture is selected, and its performance is evaluated. If the performance is unacceptable, a different architecture is selected. This process continues until an architecture is found that produces an acceptable generalization error.

Other approaches to architecture selection are divided into three categories:

- **Regularization**: Neural network regularization involves the addition of a penalty term to the objective function to be minimized. In this case the objective function changes to

$$\mathcal{E} = \mathcal{E}_T + \lambda \mathcal{E}_C \tag{7.26}$$

  where $\mathcal{E}_T$ is the usual measure of data misfit, and $\mathcal{E}_C$ is a penalty term, penalizing network complexity (network size). The constant $\lambda$ controls the influence of the penalty term. With the changed objective function, the NN now tries to find a locally optimal trade-off between data-misfit and network complexity. Neural network regularization has been studied rigorously by Girosi *et al.* [318], and Williams [910].

  Several penalty terms have been developed to reduce network size automatically during training. Weight decay, where $\mathcal{E}_C = \frac{1}{2}\sum w_i^2$, is intended to drive small weights to zero [79, 346, 435, 491]. It is a simple method to implement, but suffers from penalizing large weights at the same rate as small weights. To solve this problem, Hanson and Pratt [346] propose the hyperbolic and exponential penalty functions which penalize small weights more than large weights. Nowlan and Hinton [633] developed a more complicated soft weight sharing, where the distribution of weight values is modeled as a mixture of multiple Gaussian distributions. A narrow Gaussian is responsible for small weights, while a broad Gaussian is responsible for large weights. Using this scheme, there is less pressure on large weights to be reduced.

  Weigend *et al.* [895] propose weight elimination where the penalty function

$\mathcal{E}_C = \sum \frac{w_i^2/w_0^2}{1+w_i^2/w_0^2}$, effectively counts the number of weights. Minimization of this objective function will then minimize the number of weights. The constant $w_0$ is very important to the success of this approach. If $w_0$ is too small, the network ends up with a few large weights, while a large value results in many small weights. The optimal value for $w_0$ can be determined through cross-validation, which is not cost-effective.

Chauvin [116, 117] introduces a penalty term that measures the "energy spent" by the hidden units, where the energy is expressed as a function of the squared activation of the hidden units. The aim is then to minimize the energy spent by hidden units, and in so doing, to eliminate unnecessary units.

Kamimura and Nakanishi [435] show that, in an information theoretical context, weight decay actually minimizes entropy. Entropy can also be minimized directly by including an entropy penalty term in the objective function [434]. Minimization of entropy means that the information about input patterns is minimized, thus improving generalization. For this approach entropy is defined with respect to hidden unit activity. Schittenkopf *et al.* [763] also propose an entropy penalty term and show how it reduces complexity and avoids overfitting.

Yasui [938] develops penalty terms to make minimal and joint use of hidden units by multiple outputs. Two penalty terms are added to the objective function to control the evolution of hidden-to-output weights. One penalty causes weights leading into an output unit to prevent another from growing, while the other causes weights leaving a hidden unit to support another to grow.

While regularization models are generally easy to implement, the value of the constant $\lambda$ in equation (7.26) may present problems. If $\lambda$ is too small, the penalty term will have no effect. If $\lambda$ is too large, all weights might be driven to zero. Regularization therefore requires a delicate balance between the normal error term and the penalty term. Another disadvantage of penalty terms is that they tend to create additional local minima [346], increasing the possibility of converging to a bad local minimum. Penalty terms also increase training time due to the added calculations at each weight update. In a bid to reduce this complexity, Finnoff *et al.* [260] show that the performance of penalty terms is greatly enhanced if they are introduced only after overfitting is observed.

- **Network construction (growing)**: Network construction algorithms start training with a small network and incrementally add hidden units during training when the network is trapped in a local minimum [291, 368, 397, 495]. A small network forms an approximate model of a subset of the training set. Each new hidden unit is trained to reduce the current network error – yielding a better approximation. Crucial to the success of construction algorithms is effective criteria to trigger when to add a new unit, when to stop the growing process, where and how to connect the new unit to the existing architecture, and how to avoid restarting training. If these issues are treated on an *ad hoc* basis, overfitting may occur and training time may be increased.

- **Network pruning**: Neural network pruning algorithms start with an oversized network and remove unnecessary network parameters, either during training or after convergence to a local minimum. Network parameters that are considered for removal are individual weights, hidden units and input units. The decision to prune a network parameter is based on some measure of parameter relevance or significance. A relevance is computed for each parameter and a pruning heuristic is used to decide when a parameter is considered as being irrelevant or not. A large initial architecture allows the network to converge reasonably quickly, with less sensitivity to local minima and the initial network size. Larger networks have more functional flexibility, and are guaranteed to learn the input-output mapping with the desired degree of accuracy. Due to the larger functional flexibility, pruning weights and units from a larger network may give rise to a better fit of the underlying function, hence better generalization [604].

A more elaborate discussion of pruning techniques is given next, with the main objective of presenting a flavor of the techniques available to prune NN architectures. For more detailed discussions, the reader is referred to the given references. The first results in the quest to find a solution to the architecture optimization problem were the derivation of theoretical limits on the number of hidden units to solve a particular problem [53, 158, 436, 751, 759]. However, these results are based on unrealistic assumptions about the network and the problem to be solved. Also, they usually apply to classification problems only. While these limits do improve our understanding of the relationship between architecture and training set characteristics, they do not predict the correct number of hidden units for a general class of problems.

Recent research concentrated on the development of more efficient pruning techniques to solve the architecture selection problem. Several different approaches to pruning have been developed. This chapter groups these approaches in the following general classes: intuitive methods, evolutionary methods, information matrix methods, hypothesis testing methods and sensitivity analysis methods.

- **Intuitive pruning techniques**: Simple intuitive methods based on weight values and unit activation values have been proposed by Hagiwara [342]. The *goodness factor* $G_i^l$ of unit $i$ in layer $l$, $G_i^l = \sum_p \sum_j (w_{ji}^l o_i^l)^2$, where the first sum is over all patterns, and $o_i^l$ is the output of the unit, assumes that an important unit is one that excites frequently and has large weights to other units. The *consuming energy*, $E_i^l = \sum_p \sum_j w_{ji}^l o_j^{l+1} o_j^l$, additionally assumes that unit $i$ excites the units in the next layer. Both methods suffer from the flaw that when a unit's output is more frequently 0 than 1, that unit might be considered as being unimportant, while this is not necessarily the case. Magnitude-based pruning assumes that small weights are irrelevant [342, 526]. However, small weights may be of importance, especially compared to very large weights that cause saturation in hidden and output units. Also, large weights (in terms of their absolute value) may cancel each other out.

- **Evolutionary pruning techniques**: The use of genetic algorithms (GA) to prune NNs provides a biologically plausible approach to pruning [494, 712, 901, 904]. Using GA terminology, the population consists of several pruned versions of the original network, each needed to be trained. Differently pruned networks are created by the application of mutation, reproduction and crossover operators. These pruned networks "compete" for survival, being awarded for using fewer parameters and for improving generalization. GA NN pruning is thus a time-consuming process.

- **Information matrix pruning techniques**: Several researchers have used approximations to the Fisher information matrix to determine the optimal number of hidden units and weights. Based on the assumption that outputs are linearly activated, and that least squares estimators satisfy asymptotic normality, Cottrell *et al.* [160] compute the relevance of a weight as a function of the information matrix, approximated by

$$\mathcal{I} = \frac{1}{P} \sum_{p=1}^{P} \frac{\partial f_{NN}}{\partial w} \left( \frac{\partial f_{NN}}{\partial w} \right)^T \tag{7.27}$$

Weights with a low relevance are removed.

Hayashi [355], Tamura *et al.* [837], Xue *et al.* [929] and Fletcher *et al.* [261] use singular value decomposition (SVD) to analyze the hidden unit activation covariance matrix to determine the optimal number of hidden units. Based on the assumption that outputs are linearly activated, the *rank* of the covariance matrix is the optimal number of hidden units (also see [292]). SVD of this information matrix results in an eigenvalue and eigenvector decomposition where low eigenvalues correspond to irrelevant hidden units. The rank is the number of non-zero eigenvalues. Fletcher *et al.* [261] use the SVD of the conditional Fisher information matrix, as given in equation (7.27), together with likelihood-ratio tests to determine irrelevant hidden units. In this case the conditional Fisher information matrix is restricted to weights between the hidden and output layer only, whereas previous techniques are based on all network weights. Each iteration of the pruning algorithm identifies exactly which hidden units to prune.

Principal Component Analysis (PCA) pruning techniques have been developed that use the SVD of the Fisher information matrix to find the principal components (relevant parameters) [434, 515, 763, 834]. These principal components are linear transformations of the original parameters, computed from the eigenvectors obtained from a SVD of the information matrix. The result of PCA is the orthogonal vectors on which variance in the data is maximally projected. Non-principal components/parameters (parameters that do not account for data variance) are pruned. Pruning using PCA is thus achieved through projection of the original $w$-dimensional space onto a $w'$-dimensional linear subspace ($w' < w$) spanned by the eigenvectors of the data's correlation or covariance matrix corresponding to the largest eigenvalues.

- **Hypothesis testing techniques**: Formal statistical hypothesis tests can be used to test the statistical significance of a subset of weights, or a subset of hidden units. Steppe *et al.* [809] and Fletcher *et al.* [261] use the *likelihood-ratio test statistic* to test the null hypothesis that a subset of weights is zero. Weights associated with a hidden unit are tested to see if they are statistically different from zero. If these weights are not statistically different from zero, the corresponding hidden unit is pruned.

Belue and Bauer [58] propose a method that injects a noisy input parameter into the NN model, and then use statistical tests to decide if the significances of the original NN parameters are higher than that of the injected noisy parameter. Parameters with lower significances than the noisy parameter are pruned.

Similarly, Prechelt [694] and Finnoff *et al.* [260] test the assumption that a weight becomes zero during the training process. This approach is based on the observation that the distribution of weight values is roughly normal. Weights located in the left tail of this distribution are removed.

- **Sensitivity analysis pruning techniques**: Two main approaches to sensitivity analysis exist, namely with regard to the objective function and with regard to the NN output function. Both sensitivity analysis with regard to the objective function and sensitivity analysis with regard to the NN output function resulted in the development of a number of pruning techniques. Possibly the most popular of these are optimal brain damage (OBD) [166] and its variants, optimal brain surgeon (OBS) [351, 352] and optimal cell damage (OCD) [129]. A parameter saliency measure is computed for each parameter, indicating the influence small perturbations to the parameter have on the approximation error. Parameters with a low saliency are removed. These methods are time-consuming due to the calculation of the Hessian matrix. Buntine and Weigend [95] and Bishop [71] derived methods to simplify the calculation of the Hessian matrix in a bid to reduce the complexity of these pruning techniques. In OBD, OBS and OCD, sensitivity analysis is performed with regard to the training error. Pedersen *et al.* [669] and Burrascano [98] develop pruning techniques based on sensitivity analysis with regard to the generalization error. Other objective function sensitivity analysis pruning techniques have been developed by Mozer and Smolensky [611] and Moody and Utans [602].

  NN output sensitivity analysis pruning techniques have been developed that are less complex than objective function sensitivity analysis, and that do not rely on simplifying assumptions. Zurada *et al.* [962] introduced output sensitivity analysis pruning of input units, further investigated by Engelbrecht *et al.* [245]. Engelbrecht and Cloete [238, 240, 246] extended this approach to also prune irrelevant hidden units.

  A similar approach to NN output sensitivity analysis was followed by Dorizzi *et al.* [218] and Czernichow [168] to prune parameters of a RBFNN.

The aim of all architecture selection algorithms is to find the smallest architecture that accurately fits the underlying function. In addition to improving generalization performance and avoiding overfitting (as discussed earlier), smaller networks have the following advantages. Once an optimized architecture has been found, the cost of forward calculations is significantly reduced, since the cost of computation grows almost linearly with the number of weights. From the generalization limits overviewed in section 7.3.7, the number of training patterns required to achieve a certain generalization performance is a function of the network architecture. Smaller networks therefore require less training patterns. Also, the knowledge embedded in smaller networks is more easily described by a set of simpler rules. Viktor *et al.* [879] show that the number of rules extracted from smaller networks is less for pruned networks than that extracted from larger networks. They also show that rules extracted from smaller networks contain only relevant clauses, and that the combinatorics of the rule extraction algorithm is significantly reduced. Furthermore, for smaller networks the function of each hidden unit is more easily visualized. The complexity of decision boundary detection algorithms is also reduced.

With reference to the bias/variance decomposition of the MSE function [313], smaller network architectures reduce the variance component of the MSE. NNs are generally plagued by high variance due to the limited training set sizes. This variance is reduced by introducing bias through minimization of the network architecture. Smaller networks are biased because the hypothesis space is reduced; thus limiting the available functions that can fit the data. The effects of architecture selection on the bias/variance trade-off have been studied by Gedeon *et al.* [311].

## 7.3.6 Adaptive Activation Functions

The performance of NNs can be improved by allowing activation functions to change dynamically according to the characteristics of the training data. One of the first techniques to use adaptive activations functions was developed by Zurada [961], where the slope of the sigmoid activation function is learned together with the weights. A slope parameter $\lambda$ is kept for each hidden and output unit. The lambda-learning algorithm of Zurada was extended by Engelbrecht *et al.* [244] where the sigmoid function is given as

$$f(net, \lambda, \gamma) = \frac{\gamma}{1 + e^{-\lambda net}} \qquad (7.28)$$

where $\lambda$ is the slope of the function and $\gamma$ the maximum range. Engelbrecht *et al.* developed learning equations to also learn the maximum ranges of the sigmoid functions, thereby performing automatic scaling. By using gamma-learning, it is not necessary to scale target values to the range $(0, 1)$. The effect of changing the slope and range of the sigmoid function is illustrated in Figure 7.6.

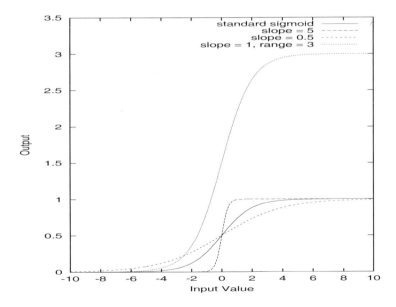

**Figure 7.6** Adaptive Sigmoid

Algorithm 7.1 illustrates the differences between standard GD learning (referred to as delta learning) and the lambda and gamma learning variations. (Note that although the momentum terms are omitted below, a momentum term is usually used for the weight, lambda and gamma updates.)

## 7.3.7 Active Learning

Ockham's razor states that unnecessarily complex models should not be preferred to simpler ones – a very intuitive principle [544, 844]. A neural network (NN) model is described by the network weights. Model selection in NNs consists of finding a set of weights that best performs the learning task. In this sense, the data, and not just the architecture should be viewed as part of the NN model, since the data is instrumental in finding the "best" weights. Model selection is then viewed as the process of designing an optimal NN architecture as well as the implementation of techniques to make optimal use of the available training data. Following from the principle of Ockham's razor is a preference then for both simple NN architectures and optimized training data. Usually, model selection techniques address only the question of which architecture best fits the task.

Standard error back-propagating NNs are passive learners. These networks passively receive information about the problem domain, randomly sampled to form a fixed size training set. Random sampling is believed to reproduce the density of the true distribution. However, more gain can be achieved if the learner is allowed to use current attained knowledge about the problem to guide the acquisition of training examples. As passive learner, a NN has no such control over what examples are presented for learning. The NN has to rely on the teacher (considering supervised learning) to present informative examples.

The generalization abilities and convergence time of NNs are greatly influenced by the training set size and distribution: Literature has shown that to generalize well, the training set must contain enough information to learn the task. Here lies one of the problems in model selection: the selection of concise training sets. Without prior knowledge about the learning task, it is very difficult to obtain a representative training set. Theoretical analysis provides a way to compute worst-case bounds on the number of training examples needed to ensure a specified level of generalization. A widely used theorem concerns the Vapnik-Chervonenkis (VC) dimension [8, 9, 54, 152, 375, 643]. This theorem states that the generalization error, $\mathcal{E}_G$, of a learner with VC-dimension, $d_{VC}$, trained on $P_T$ random examples will, with high confidence, be no worse than a limit of order $d_{VC}/P_T$. For NN learners, the total number of weights in a one hidden layer network is used as an estimate of the VC-dimension. This means that the appropriate number of examples to ensure an $\mathcal{E}_G$ generalization is approximately the number of weights divided by $\mathcal{E}_G$.

The VC-dimension provides overly pessimistic bounds on the number of training examples, often leading to an overestimation of the required training set size [152, 337, 643, 732, 948]. Experimental results have shown that acceptable generalization performances can be obtained with training set sizes much less than that

---

**Algorithm 7.1** Lambda-Gamma Training Rule

---

Choose the values of the learning rates $\eta_1, \eta_2$ and $\eta_3$ according to the learning rule:

| | |
|---|---|
| Delta learning rule | $\eta_1 > 0, \eta_2 = 0, \eta_3 = 0$ |
| Lambda learning rule | $\eta_1 > 0, \eta_2 > 0, \eta_3 = 0$ |
| Gamma learning rule | $\eta_1 > 0, \eta_2 = 0, \eta_3 > 0$ |
| Lambda-gamma learning rule | $\eta_1 > 0, \eta_2 > 0, \eta_3 > 0$ |

Initialize weights to small random values;
Initialize the number of epochs $t = 0$;
Initialize the steepness and range coefficients
$\lambda_{y_j} = \gamma_{y_j} = 1 \ \forall \ j = 1, \ldots, J \quad$ and $\quad \lambda_{o_k} = \gamma_{o_k} = 1 \ \forall \ k = 1, \ldots, K$;
**while** *stopping condition(s) not true* **do**
$\quad$ Let $\mathcal{E}_T = 0$;
$\quad$ **for** *each pattern $p = 1, \ldots, P_T$* **do**
$\quad\quad$ $\mathbf{z} = \mathbf{z}_p$ and $\mathbf{t} = \mathbf{t}_p$;
$\quad\quad$ **for** *each $j = 1, \ldots, J$* **do**
$\quad\quad\quad$ $y_j = f(\gamma_{y_j}, \lambda_{y_j}, \mathbf{v}_j^T \mathbf{z})$;
$\quad\quad$ **end**
$\quad\quad$ **for** *each $k = 1, \ldots, K$* **do**
$\quad\quad\quad$ $o_k = f(\gamma_{o_k}, \lambda_{o_k}, \mathbf{w}_k^T \mathbf{y})$;
$\quad\quad\quad$ $\mathcal{E}_T{+}{=} \frac{1}{2}(t_k - o_k)^2$;
$\quad\quad\quad$ Compute the error signal, $\delta_{o_k}$:

$$\delta_{o_k} = -\frac{\lambda_{o_k}}{\gamma_{o_k}}(t_k - o_k)o_k(\gamma_{o_k} - o_k) \tag{7.29}$$

$\quad\quad\quad$ Adjust output unit weights and gains, $\forall \ j = 1, \ldots, J+1$:

$$w_{kj} = w_{kj} + \eta_1 \delta_{o_k} y_j, \quad \lambda_{o_k} = \lambda_{o_k} + \eta_2 \delta_{o_k} \frac{net_{o_k}}{\lambda_{o_k}} \tag{7.30}$$

$$\gamma_{o_k} = \gamma_{o_k} + \eta_3(t_k - o_k)\frac{1}{\gamma_{o_k}}o_k \tag{7.31}$$

$\quad\quad$ **end**
$\quad\quad$ **for** *each $j = 1, \ldots, J$* **do**
$\quad\quad\quad$ Compute the error signal, $\delta_{y_j}$:

$$\delta_{y_j} = \frac{\lambda_{y_j}}{\gamma_{y_j}}y_j(\gamma_{y_j} - y_j)\sum_{k=1}^{K}\delta_{o_k}w_{kj} \tag{7.32}$$

$\quad\quad\quad$ Adjust hidden unit weights and gains, $\forall \ i = 1, \ldots, I+1$:

$$v_{ji} = v_{ji} + \eta_1 \delta_{y_j} z_i, \quad \lambda_{y_j} = \lambda_{y_j} + \eta_2 \frac{1}{\lambda_{y_j}}\delta_{y_j} net_{y_j} \tag{7.33}$$

$$\gamma_{y_j} = \gamma_{y_j} + \eta_3 \frac{1}{\gamma_{y_j}}f(\gamma_{y_j}, \lambda_{y_j}, net_{y_j})\sum_{k=1}^{K}\delta_{o_k}w_{kj} \tag{7.34}$$

$\quad\quad$ **end**
$\quad\quad$ $t = t + 1$;
$\quad$ **end**
**end**

specified by the VC-dimension [152, 732]. Cohn and Tesauro [152] show that for experiments conducted, the generalization error decreases exponentially with the number of examples, rather than the $1/P_T$ result of the VC bound. Experimental results by Lange and Männer [502] show that more training examples do not necessarily improve generalization. In their paper, Lange and Männer introduce the notion of a critical training set size. Through experimentation they found that examples beyond this critical size do not improve generalization, illustrating that excess patterns have no real gain. This critical training set size is problem dependent.

While enough information is crucial to effective learning, too large training set sizes may be of disadvantage to generalization performance and training time [503, 948]. Redundant training examples may be from uninteresting parts of input space, and do not serve to refine learned weights – it only introduces unnecessary computations, thus increasing training time. Furthermore, redundant examples might not be equally distributed, thereby biasing the learner.

The ideal, then, is to implement structures to make optimal use of available training data. That is, to select only informative examples for training, or to present examples in a way to maximize the decrease in training and generalization error. To this extent, active learning algorithms have been developed.

Cohn *et al.* [151] define *active learning* (also referred to in the literature as example selection, sequential learning, query-based learning) *as any form of learning in which the learning algorithm has some control over what part of the input space it receives information from.* An active learning strategy allows the learner to dynamically select training examples, during training, from a candidate training set as received from the teacher (supervisor). The learner capitalizes on current attained knowledge to select examples from the candidate training set that are most likely to solve the problem, or that will lead to a maximum decrease in error. Rather than passively accepting training examples from the teacher, the network is allowed to use its current knowledge about the problem to have some deterministic control over which training examples to accept, and to guide the search for informative patterns. By adding this functionality to a NN, the network changes from a passive learner to an active learner.

Figure 7.7 illustrates the difference between active learning and passive learning.

With careful dynamic selection of training examples, shorter training times and better generalization may be obtained. Provided that the added complexity of the example selection method does not exceed the reduction in training computations (due to a reduction in the number of training patterns), training time will be reduced [399, 822, 948]. Generalization can potentially be improved, provided that selected examples contain enough information to learn the task. Cohn [153] and Cohn *et al.* [151] show through average case analysis that the expected generalization performance of active learning is significantly better than passive learning. Seung *et al.* [777], Sung and Niyogi [822] and Zhang [948] report similar improvements. Results presented by Seung *et al.* indicate that generalization error decreases more rapidly for active learning than for passive learning [777].

Two main approaches to active learning can be identified, i.e. *incremental learning*

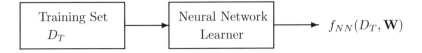

Passive Learning

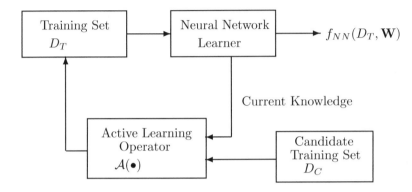

Active Learning

**Figure 7.7** Passive vs Active Learning

and *selective learning*. Incremental learning starts training on an initial subset of a candidate training set. During training, at specified selection intervals (e.g. after a specified number of epochs, or when the error on the current training subset no longer decreases), further subsets are selected from the candidate examples using some criteria or heuristics, and added to the training set. The training set consists of the union of all previously selected subsets, while examples in selected subsets are removed from the candidate set. Thus, as training progresses, the size of the candidate set decreases while the size of the actual training set grows. Note that this chapter uses the term *incremental learning* to denote data selection, and should not be confused with the NN architecture selection growing approach. The term *NN growing* is used in this chapter to denote the process of finding an optimal architecture starting with too few hidden units and adding units during training.

In contrast to incremental learning, selective learning selects a new training subset from the original candidate set at each selection interval. Selected patterns are not removed from the candidate set. At each selection interval, all candidate patterns have a chance to be selected. The subset is selected and used for training until some convergence criteria on the subset is met (e.g. a specified error limit on the subset is reached, the error decrease per iteration is too small, the maximum number of epochs allowed on the subset is exceeded). A new training subset is then selected for the next training period. This process repeats until the NN is trained to satisfaction.

The main difference between these two approaches to active learning is that no examples are discarded by incremental learning. In the limit, all examples in the candidate set will be used for training. With selective learning, training starts on all candidate examples, and uninformative examples are discarded as training progresses.

**Selective Learning**

Not much research has been done in selective learning. Hunt and Deller [399] developed Selective Updating, where training starts on an initial candidate training set. Patterns that exhibit a high influence on weights, i.e. patterns that cause the largest changes in weight values, are selected from the candidate set and added to the training set. Patterns that have a high influence on weights are selected at each epoch by calculating the effect that patterns have on weight estimates. These calculations are based on matrix perturbation theory, where an input pattern is viewed as a perturbation of previous patterns. If the perturbation is expected to cause large changes to weights, the corresponding pattern is included in the training set. The learning algorithm does use current knowledge to select the next training subset, and training subsets may differ from epoch to epoch. Selective Updating has the drawback of assuming uncorrelated input units, which is often not the case for practical applications.

Another approach to selective learning is simply to discard those patterns that have been classified correctly [50]. The effect of such an approach is that the training set will include those patterns that lie close to decision boundaries. If the candidate set contains outlier patterns, these patterns will, however, also be selected. This error selection approach therefore requires a robust estimator (objective function) to be used in the case of outliers.

Engelbrecht *et al.* [241, 242, 239] developed a selective learning approach for classification problems where sensitivity analysis is used to locate patterns close to decision boundaries. Only those patterns that are close to a decision boundary are selected for training. The algorithm resulted in substantial reductions in the number of learning calculations due to reductions in the training set size, while either maintaining performance as obtained from learning from all the training data, or improving performance.

**Incremental learning**

Research on incremental learning is more abundant than for selective learning. Most current incremental learning techniques have their roots in information theory, adapting Fedorov's optimal experiment design for NN learning [153, 295, 544, 681, 822]. The different information theoretic incremental learning algorithms are very similar, and differ only in whether they consider only bias, only variance, or both bias and variance terms in their selection criteria.

Cohn [153] developed neural network optimal experiment design (OED), where the objective is to select, at each iteration, a new pattern from a candidate set which minimizes the expectation of the MSE. This is achieved by minimizing output variance as estimated from the Fisher information matrix [153, 154]. The model assumes an unbiased estimator and considers only the minimization of variance. OED is computationally very expensive because it requires the calculation of the inverse of the information matrix.

MacKay [544] proposed similar information-based objective functions for active learning, where the aim is to maximize the expected information gain by maximizing the change in Shannon entropy when new patterns are added to the actual training set, or by maximizing cross-entropy gain. Similar to OED, the maximization of information gain is achieved by selecting patterns that minimize the expected MSE. Information-based objective functions also ignore bias, by minimizing only variance. The required inversion of the Hessian matrix makes this approach computationally expensive.

Plutowski and White [681] proposed selecting patterns that minimize the integrated squared bias (ISB). At each iteration, a new pattern is selected from a candidate set that maximizes the change, $\Delta ISB$, in the ISB. In effect, the patterns with error gradient most highly correlated with the error gradient of the entire set of patterns is selected. A noise-free environment is assumed and variance is ignored. Drawbacks of this method are the need to calculate the inverse of a Hessian matrix, and the assumption that the target function is known.

Sung and Niyogi [822] proposed an information theoretic approach to active learning that considers both bias and variance. The learning goal is to minimize the expected misfit between the target function and the approximated function. The patterns that minimize the expected squared difference between the target and approximated function are selected to be included in the actual training set. In effect, the net amount of information gained with each new pattern is then maximized. No assumption is made about the target function. This technique is computationally expensive, since it requires computations over two expectations, i.e. the *a-posteriori* distribution over function space, and the *a-posteriori* distribution over the space of targets one would expect given a candidate sample location.

One drawback of the incremental learning algorithms summarized above is that they rely on the inversion of an information matrix. Fukumizu showed that, in relation to pattern selection to minimize the expected MSE, the Fisher information matrix may be singular [295]. If the information matrix is singular, the inverse of that matrix may not exist. Fukumizu continues to show that the information matrix is singular if and only if the corresponding NN contains redundant units. Thus, the information matrix can be made non-singular by removing redundant hidden units. Fukumizu developed an algorithm that incorporates an architecture reduction algorithm with a pattern selection algorithm. This algorithm is complex due to the inversion of the information matrix at each selection interval, but ensures a non-singular information matrix.

Approximations to the information theoretical incremental learning algorithms can be used. Zhang [948] shows that information gain is maximized when a pattern is selected whose addition leads to the greatest decrease in MSE. Zhang developed selective incremental learning where training starts on an initial subset which is increased during training by adding additional subsets, where each subset contains those patterns with largest errors. Selective incremental learning has a very low computational overhead, but is negatively influenced by outlier patterns since these patterns have large errors.

Dynamic pattern selection, developed by Röbel [732], is very similar to Zhang's selective incremental learning. Röbel defines a generalization factor on the current training subset, expressed as $\mathcal{E}_G/\mathcal{E}_T$ where $\mathcal{E}_G$ and $\mathcal{E}_T$ are the MSE of the test set and the training set respectively. As soon as the generalization factor exceeds a certain threshold, patterns with the highest errors are selected from the candidate set and added to the actual training set. Testing against the generalization factor prevents overfitting of the training subset. A low overhead is involved.

Very different from the methods previously described are incremental learning algorithms for classification problems, where decision boundaries are utilized to guide the search for optimal training subsets. Cohn *et al.* [151] developed selective sampling, where patterns are sampled only within a *region of uncertainty*. Cohn *et al.* proposed an SG-network (most specific/most general network) as an approach to compute the region of uncertainty. Two separate networks are trained: one to learn a "most specific" concept $s$ consistent with the given training data, and the other to learn a "most general" concept, $g$. The region of uncertainty is then all patterns $p$ such that $s(p) \neq g(p)$. In other words, the region of uncertainty encapsulates all those patterns for which $s$ and $g$ present a different classification. A new training pattern is selected from this region of uncertainty and added to the training set. After training on the new training set, the region of uncertainty is recalculated, and another pattern is sampled according to some distribution defined over the uncertainty region – a very expensive approach. To reduce complexity, the algorithm is changed to select patterns in batches, rather than individually. An initial pattern subset is drawn, the network is trained on this subset, and a new region of uncertainty is calculated. Then, a new distribution is defined over the region of uncertainty that is zero outside this region. A next subset is drawn according to the new distribution and added to the training set. The process repeats until convergence is reached.

Query-based learning, developed by Hwang *et al.* [402] differs from selective sampling in that query-based learning generates new training data in the region of uncertainty. The aim is to increase the steepness of the boundary between two distinct classes by narrowing the regions of ambiguity. This is accomplished by inverting the NN output function to compute decision boundaries. New data in the vicinity of boundaries are then generated and added to the training set.

Seung *et al.* [777] proposed query by committee. The optimal training set is built by selecting one pattern at a time from a candidate set based on the principle of maximal disagreement among a committee of learners. Patterns classified correctly by half of the committee, but incorrectly by the other half, are included in the actual training set. Query by committee is time-consuming due to the simultaneous training of several networks, but will be most effective for ensemble networks.

Engelbrecht *et al.* [243] developed an incremental learning algorithm where sensitivity analysis is used to locate the most informative patterns. The most informative patterns are viewed as those patterns in the midrange of the sigmoid activation function. Since these patterns have the largest derivatives of the output with respect to inputs, the algorithm incrementally selects from a candidate set of patterns those patterns that have the largest derivatives. Substantial reductions in computational complexity are achieved using this algorithm, with improved accuracy.

The incremental learning algorithms reviewed in this section all make use of the NN learner's current knowledge about the learning task to select those patterns that are most informative. These algorithms start with an initial training set, which is increased during training by adding a single informative pattern, or a subset of informative patterns.

In general, active learning is summarized as in Algorithm 7.2.

---

**Algorithm 7.2** Generic Active Learning Algorithm

---

Initialize the NN architecture;
Construct an initial training subset $D_{S_0}$ from the candidate set $D_C$;
Initialize the current training set $D_T \leftarrow D_{S_0}$;
**while** *stopping condition(s) not true* **do**
    **while** *stopping condition(s) on training subset $D_T$ not true* **do**
        Train the NN on training subset $D_T$ to produce the function $f_{NN}(D_T, \mathbf{W})$;
    **end**
    Apply the active learning operator to generate a new subset $D_{S_s}$ at subset
    selection interval $\tau_s$, using either

$$D_{S_s} \leftarrow \mathcal{A}^-(D_C, f_{NN}(D_T, \mathbf{W})), \quad D_T \leftarrow D_{S_s} \tag{7.35}$$

    for selective learning, or

$$D_{S_s} \leftarrow \mathcal{A}^+(D_C, D_T, f_{NN}(D_T, \mathbf{W})) \tag{7.36}$$
$$D_T \leftarrow D_T \cup D_{S_s}, \quad D_C \leftarrow D_C - D_{S_s} \tag{7.37}$$

    for incremental learning
**end**

---

In Algorithm 7.2, $\mathcal{A}$ denotes the active learning operator, which is defined as follows for each of the active learning classes:

1) $\mathcal{A}^-(D_C, f_{NN}(D_T, \mathbf{W})) = D_S$, where $D_S \subseteq D_C$. The operator $\mathcal{A}^-$ receives as input the candidate set $D_C$, performs some calculations on each pattern $\mathbf{z}_p \in D_C$, and produces the subset $D_S$ with the characteristics $D_S \subseteq D_C$, that is $|D_S| \leq |D_C|$. The aim of this operator is therefore to produce a subset $D_S$ from $D_C$ that is smaller than, or equal to, $D_C$. Then, let $D_T \leftarrow D_S$, where $D_T$ is the actual training set.

2) $\mathcal{A}^+(D_C, D_T, f_{NN}(D_T, \mathbf{W})) = D_S$, where $D_C, D_T$ and $D_S$ are sets such that $D_T \subseteq D_C$, $D_S \subseteq D_C$. The operator $\mathcal{A}^+$ performs calculations on each pattern $\mathbf{z}_p \in D_C$ to determine if that element should be added to the current training set. Selected patterns are added to subset $D_S$. Thus, $D_S = \{\mathbf{z}_p | \mathbf{z}_p \in D_C$, and $\mathbf{z}_p$ satisfies the selection criteria$\}$. Then, $D_T \leftarrow D_T \cup D_S$ (the new

subset is added to the current training subset), and $D_C \leftarrow D_C - D_S$.

Active learning operator $\mathcal{A}^-$ corresponds with selective learning where the training set is "pruned", while $\mathcal{A}^+$ corresponds with incremental learning where the actual training subset "grows". Inclusion of the NN function $f_{NN}$ as a parameter of each operator indicates the dependence on the NN's current knowledge.

# 7.4   Assignments

1. Discuss measures that quantify the performance of unsupervised neural networks.

2. Discuss factors that influence the performance of unsupervised neural networks. Explain how the performance can be improved.

3. Why is the SSE not a good measure to compare the performance of NNs on different data set sizes?

4. Why is the MSE not a good measure of performance for classification problems?

5. One approach to incremental learning is to select from the candidate training set the most informative pattern as the one with the largest error. Justify and criticize this approach. Assume that a new pattern is selected at each epoch.

6. Explain the role of the steepness coefficient in $\frac{1}{1+e^{-\lambda net}}$ in the performance of supervised NNs.

7. Explain how architecture selection can be used to avoid overfitting.

8. Explain how active learning can be used to avoid overfitting.

9. Consider the sigmoid activation function. Discuss how scaling of the training data affects the performance of NNs.

10. Explain how the Huber function makes a NN more robust to outliers.

# Part III

# EVOLUTIONARY COMPUTATION

The world we live in is constantly changing. In order to survive in a dynamically changing environment, individuals must have the ability to adapt. Evolution is this process of adaption with the aim of improving the survival capabilities through processes such as natural selection, survival of the fittest, reproduction, mutation, competition and symbiosis.

This part covers evolutionary computing (EC) – a field of CI that models the processes of natural evolution. Several evolutionary algorithms (EA) have been developed. This text covers genetic algorithms in Chapter 9, genetic programming in Chapter 10, evolutionary programming in Chapter 11, evolutionary strategies in Chapter 12, differential evolution in Chapter 13, cultural algorithms in Chapter 14, and coevolution in Chapter 15. An introduction to basic EC concepts is given in Chapter 8.

# Chapter 8

# Introduction to Evolutionary Computation

Evolution is an optimization process where the aim is to improve the ability of an organism (or system) to survive in dynamically changing and competitive environments. Evolution is a concept that has been hotly debated over centuries, and still causes active debates.[1]  When talking about evolution, it is important to first identify the area in which evolution can be defined, for example, cosmic, chemical, stellar and planetary, organic or man-made systems of evolution. For these different areas, evolution may be interpreted differently. For the purpose of this part of the book, the focus is on biological evolution. Even for this specific area, attempts to define the term *biological evolution* still cause numerous debates, with the Lamarckian and Darwinian views being the most popular and accepted. While Darwin (1809–1882) is generally considered as the founder of both the theory of evolution and the principle of common descent, Lamarck (1744–1829) was possibly the first to theorize about biological evolution.

Jean-Baptiste Lamarck's theory of evolution was that of *heredity*, i.e. the inheritance of acquired traits. The main idea is that individuals adapt during their lifetimes, and transmit their traits to their offspring. The offspring then continue to adapt. According to Lamarckism, the method of adaptation rests on the concept of use and disuse: over time, individuals lose characteristics they do not require, and develop those which are useful by "exercising" them.

It was Charles Darwin's theory of *natural selection* that became the foundation of biological evolution (Alfred Wallace developed a similar theory at the same time, but independently of Darwin). The Darwinian theory of evolution [173] can be summarized as: In a world with limited resources and stable populations, each individual competes with others for survival. Those individuals with the "best" characteristics (traits) are more likely to survive and to reproduce, and those characteristics will be passed on to their offspring. These desirable characteristics are inherited by the following generations, and (over time) become dominant among the population.

A second part of Darwin's theory states that, during production of a child organism,

---

[1]Refer to http:www.johmann.net/book/ciy7-1.html
http://www.talkorigins.org/faqs/evolution-definition.html
http://www.evolutionfairytale.com/articles_debates/evolution-definition.html
http://www.creationdesign.org/ (accessed 05/08/2004).

*Computational Intelligence: An Introduction*, Second Edition A.P. Engelbrecht
©2007 John Wiley & Sons, Ltd

random events cause random changes to the child organism's characteristics. If these new characteristics are a benefit to the organism, then the chances of survival for that organism are increased.

Evolutionary computation (EC) refers to computer-based problem solving systems that use computational models of evolutionary processes, such as natural selection, survival of the fittest and reproduction, as the fundamental components of such computational systems.

This chapter gives an overview of the evolution processes modeled in EC. Section 8.1 presents a generic evolutionary algorithm (EA) and reviews the main components of EAs. Section 8.2 discusses ways in which the computational individuals are represented, and Section 8.3 discusses aspects about the initial population. The importance of fitness functions, and different types of fitness functions are discussed in Section 8.4. Selection and reproduction operators are respectively discussed in Sections 8.5 and 8.6. Algorithm stopping conditions are considered in Section 8.7. A short discussion on the differences between EC and classical optimization is given in Section 8.8.

# 8.1   Generic Evolutionary Algorithm

Evolution via natural selection of a randomly chosen population of individuals can be thought of as a search through the space of possible chromosome values. In that sense, an evolutionary algorithm (EA) is a stochastic search for an optimal solution to a given problem. The evolutionary search process is influenced by the following main components of an EA:

- an **encoding** of solutions to the problem as a chromosome;
- a **function** to evaluate the **fitness**, or survival strength of individuals;
- **initialization** of the initial population;
- **selection** operators; and
- **reproduction** operators.

Algorithm 8.1 shows how these components are combined to form a generic EA.

---

**Algorithm 8.1** Generic Evolutionary Algorithm

---

Let $t = 0$ be the generation counter;
Create and initialize an $n_x$-dimensional population, $\mathcal{C}(0)$, to consist of $n_s$ individuals;
**while** *stopping condition(s) not true* **do**
    Evaluate the fitness, $f(\mathbf{x}_i(t))$, of each individual, $\mathbf{x}_i(t)$;
    Perform reproduction to create offspring;
    Select the new population, $\mathcal{C}(t + 1)$;
    Advance to the new generation, i.e. $t = t + 1$;
**end**

---

The steps of an EA are applied iteratively until some stopping condition is satisfied (refer to Section 8.7). Each iteration of an EA is referred to as a generation.

The different ways in which the EA components are implemented result in diffferent EC paradigms:

- **Genetic algorithms** (GAs), which model genetic evolution.
- **Genetic programming** (GP), which is based on genetic algorithms, but individuals are programs (represented as trees).
- **Evolutionary programming** (EP), which is derived from the simulation of adaptive behavior in evolution (i.e. *phenotypic* evolution).
- **Evolution strategies** (ESs), which are geared toward modeling the strategic parameters that control variation in evolution, i.e. the evolution of evolution.
- **Differential evolution** (DE), which is similar to genetic algorithms, differing in the reproduction mechanism used.
- **Cultural evolution** (CE), which models the evolution of culture of a population and how the culture influences the genetic and phenotypic evolution of individuals.
- **Co-evolution** (CoE), where initially "dumb" individuals evolve through cooperation, or in competition with one another, acquiring the necessary characteristics to survive.

These paradigms are discussed in detail in the chapters that follow in this part of the book.

With reference to Algorithm 8.1, both parts of Darwin's theory are encapsulated within this algorithm:

- Natural selection occurs within the reproduction operation where the "best" parents have a better chance of being selected to produce offspring, and to be selected for the new population.
- Random changes are effected through the mutation operator.

## 8.2 Representation – The Chromosome

In nature, organisms have certain characteristics that influence their ability to survive and to reproduce. These characteristics are represented by long strings of information contained in the chromosomes of the organism. Chromosomes are structures of compact intertwined molecules of DNA, found in the nucleus of organic cells. Each chromosome contains a large number of genes, where a gene is the unit of heredity. Genes determine many aspects of anatomy and physiology through control of protein production. Each individual has a unique sequence of genes. An alternative form of a gene is referred to as an *allele*.

In the context of EC, each individual represents a candidate solution to an optimization problem. The characteristics of an individual is represented by a *chromosome*, also referred to as a *genome*. These characteristics refer to the variables of the optimization problem, for which an optimal assignment is sought. Each variable that needs to be optimized is referred to as a *gene*, the smallest unit of information. An assignment of a value from the allowed domain of the corresponding variable is referred to as an *allele*. Characteristics of an individual can be divided into two classes of evolutionary information: genotypes and phenotypes. A *genotype* describes the genetic composition of an individual, as inherited from its parents; it represents which allele the individual possesses. A *phenotype* is the expressed behavioral traits of an individual in a specific environment; it defines what an individual looks like. Complex relationships exist between the genotype and phenotype [570]:

- *pleiotropy*, where random modification of genes causes unexpected variations in the phenotypic traits, and

- *polygeny*, where several genes interact to produce a specific phenotypic trait.

An important step in the design of an EA is to find an appropriate representation of candidate solutions (i.e. chromosomes). The efficiency and complexity of the search algorithm greatly depends on the representation scheme. Different EAs from the different paradigms use different representation schemes. Most EAs represent solutions as vectors of a specific data type. An exception is genetic programming (GP) where individuals are represented in a tree format.

The classical representation scheme for GAs is binary vectors of fixed length. In the case of an $n_x$-dimensional search space, each individual consists of $n_x$ variables with each variable encoded as a bit string. If variables have binary values, the length of each chromosome is $n_x$ bits. In the case of nominal-valued variables, each nominal value can be encoded as an $n_d$-dimensional bit vector where $2^{n_d}$ is the total number of discrete nominal values for that variable. To solve optimization problems with continuous-valued variables, the continuous search space problem can be mapped into a discrete programming problem. For this purpose mapping functions are needed to convert the space $\{0,1\}^{n_b}$ to the space $\mathbb{R}^{n_x}$. For such mapping, each continuous-valued variable is mapped to an $n_d$-dimensional bit vector, i.e.

$$\phi : \mathbb{R} \to (0,1)^{n_d} \tag{8.1}$$

The domain of the continuous space needs to be restricted to a finite range, $[\mathbf{x}_{min}, \mathbf{x}_{max}]$. A standard binary encoding scheme can be used to transform the individual $\mathbf{x} = (x_1, \ldots, x_j, \ldots, x_{n_x})$, with $x_j \in \mathbb{R}$ to the binary-valued individual, $\mathbf{b} = (\mathbf{b}_1, \ldots, \mathbf{b}_j, \ldots, \mathbf{b}_{n_x})$, where $\mathbf{b}_j = (b_{(j-1)n_d+1}, \ldots, b_{jn_d})$, with $b_l \in \{0,1\}$ and the total number of bits, $n_b = n_x n_d$. Decoding each $\mathbf{b}_j$ back to a floating-point representation can be done using the function, $\Phi_j : \{0,1\}^{n_d} \to [x_{min,j}, x_{max,j}]$, where [39]

$$\Phi_j(\mathbf{b}) = x_{min,j} + \frac{x_{max,j} - x_{min,j}}{2^{n_d} - 1} \left( \sum_{l=1}^{n_d-1} b_{j(n_d-l)} 2^l \right) \tag{8.2}$$

Holland [376] and De Jong [191] provided the first applications of genetic algorithms to solve continuous-valued problems using such a mapping scheme. It should be noted that if a bitstring representation is used, a grid search is done in a discrete search space. The EA may therefore fail to obtain a precise optimum. In fact, for a conversion form a floating-point value to a bitstring of $n_d$ bits, the maximum attainable accuracy is

$$\frac{x_{max,j} - x_{min,j}}{2^{n_d} - 1} \tag{8.3}$$

for each vector component, $j = 1, \ldots, n_x$.

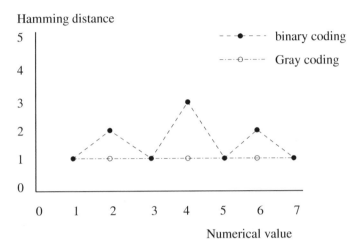

**Figure 8.1** Hamming Distance for Binary and Gray Coding

While binary coding is frequently used, it has the disadvantage of introducing Hamming cliffs as illustrated in Figure 8.1. A Hamming cliff is formed when two numerically adjacent values have bit representations that are far apart. For example, consider the decimal numbers 7 and 8. The corresponding binary representations are (using a 4-bit representation) $7 = 0111$ and $8 = 1000$, with a Hamming distance of 4 (the Hamming distance is the number of corresponding bits that differ). This presents a problem when a small change in variables should result in a small change in fitness. If, for example, 7 represents the optimal solution, and the current best solution has a fitness of 8, many bits need to be changed to cause a small change in fitness value.

An alternative bit representation is to use Gray coding, where the Hamming distance between the representation of successive numerical values is one (as illustrated in Figure 8.1). Table 8.1 compares binary and Gray coding for a 3-bit representation.

Binary numbers can easily be converted to Gray coding using the conversion

$$
\begin{aligned}
g_1 &= b_1 \\
g_l &= b_{l-1}\bar{b}_l + \bar{b}_{l-1}b_l
\end{aligned} \tag{8.4}
$$

where $b_l$ is bit $l$ of the binary number $b_1 b_2 \cdots b_{n_b}$, with $b_1$ the most significant bit; $\bar{b}_l$ denotes *not* $b_l$, $+$ means logical OR, and multiplication implies logical AND.

A Gray code representation, $\mathbf{b}_j$ can be converted to a floating-point representation

**Table 8.1** Binary and Gray Coding

|   | Binary | Gray |
|---|--------|------|
| 0 | 000    | 000  |
| 1 | 001    | 001  |
| 2 | 010    | 011  |
| 3 | 011    | 010  |
| 4 | 100    | 110  |
| 5 | 101    | 111  |
| 6 | 110    | 101  |
| 7 | 111    | 100  |

using

$$\Phi_j(\mathbf{b}) = x_{min,j} + \frac{x_{max,j} - x_{min,j}}{2^{n_d} - 1} \left( \sum_{l=1}^{n_d-1} \left( \sum_{q=1}^{n_d-l} b_{(j-1)n_d+q} \right) \bmod 2 \right) 2^l \qquad (8.5)$$

Real-valued representations have been used for a number of EAs, including GAs. Although EP (refer to Chapter 11) was originally developed for finite-state machine representations, it is now mostly applied to real-valued representations where each vector component is a floating-point number, i.e. $x_j \in \mathbb{R}, j = 1, \ldots, n_x$. ESs and DE, on the other hand, have been developed for floating-point representation (refer to Chapters 12 and 13). Real-valued representations have also been used for GAs [115, 178, 251, 411, 918]. Michalewicz [583] indicated that the original floating-point representation outperforms an equivalent binary representation, leading to more accurate, faster obtained solutions.

Other representation schemes that have been used include integer representations [778], permutations [778, 829, 905, 906], finite-state representations [265, 275], tree representations (refer to Chapter 10), and mixed-integer representations [44].

## 8.3    Initial Population

Evolutionary algorithms are stochastic, population-based search algorithms. Each EA therefore maintains a population of candidate solutions. The first step in applying an EA to solve an optimization problem is to generate an initial population. The standard way of generating an initial population is to assign a random value from the allowed domain to each of the genes of each chromosome. The goal of random selection is to ensure that the initial population is a uniform representation of the entire search space. If regions of the search space are not covered by the initial population, chances are that those parts will be neglected by the search process.

The size of the initial population has consequences in terms of computational complexity and exploration abilities. Large numbers of individuals increase diversity, thereby improving the exploration abilities of the population. However, the more the individuals, the higher the computational complexity per generation. While the execution time per generation increases, it may be the case that fewer generations are needed to locate an acceptable solution. A small population, on the other hand will represent a small part of the search space. While the time complexity per generation is low, the EA may need more generations to converge than for a large population.

In the case of a small population, the EA can be forced to explore more of the search space by increasing the rate of mutation.

## 8.4 Fitness Function

In the Darwinian model of evolution, individuals with the best characteristics have the best chance to survive and to reproduce. In order to determine the ability of an individual of an EA to survive, a mathematical function is used to quantify how good the solution represented by a chromosome is. The *fitness function*, $f$, maps a chromosome representation into a scalar value:

$$f : \Gamma^{n_x} \rightarrow \mathbb{R} \tag{8.6}$$

where $\Gamma$ represents the data type of the elements of an $n_x$-dimensional chromosome.

The fitness function represents the objective function, $\Psi$, which describes the optimization problem. It is not necessarily the case that the chromosome representation corresponds to the representation expected by the objective function. In such cases, a more detailed description of the fitness function is

$$f : \mathcal{S}_C \xrightarrow{\Phi} \mathcal{S}_X \xrightarrow{\Psi} \mathbb{R} \xrightarrow{\Upsilon} \mathbb{R}_+ \tag{8.7}$$

where $\mathcal{S}_C$ represents the search space of the objective function, and $\Phi$, $\Psi$ and $\Upsilon$ respectively represent the chromosome decoding function, the objective function, and the scaling function. The (optional) scaling function is used in proportional selection to ensure positive fitness values (refer to Section 8.5). As an example,

$$f : \{0,1\}^{n_b} \xrightarrow{\Phi} \mathbb{R}^{n_x} \xrightarrow{\Psi} \mathbb{R} \xrightarrow{\Upsilon} \mathbb{R}_+ \tag{8.8}$$

where an $n_b$-bitstring representation is converted to a floating-point representation using either equation (8.2) or (8.5).

For the purposes of the remainder of this part on EC, it is assumed that $\mathcal{S}_C = \mathcal{S}_X$ for which $f = \Psi$.

Usually, the fitness function provides an absolute measure of fitness. That is, the solution represented by a chromosome is directly evaluated using the objective function. For some applications, for example game learning (refer to Chapter 11) it is not possible to find an absolute fitness function. Instead, a relative fitness measure is used to quantify the performance of an individual in relation to that of other individuals in the population or a competing population. Relative fitness measures are used in coevolutionary algorithms (refer to Chapter 15).

It is important to realize at this point that different types of optimization problems exist (refer to Section A.3), which have an influence on the formulation of the fitness function:

- **Unconstrained** optimization problems as defined in Definition A.4, where, assuming that $S_C = S_X$, the fitness function is simply the objective function.

- **Constrained** optimization problems as defined in Definition A.5. To solve constrained problems, some EAs change the fitness function to contain two objectives: one is the original objective function, and the other is a constraint penalty function (refer to Section A.6).

- **Multi-objective** optimization problems (MOP) as defined in Definition A.10. MOPs can be solved by using a weighted aggregation approach (refer to Section A.8), where the fitness function is a weighted sum of all the sub-objectives (refer to equation (A.44)), or by using a Pareto-based optimization algorithm.

- **Dynamic** and **noisy** problems, where function values of solutions change over time. Dynamic fitness functions are time-dependent whereas noisy functions usually have an added Gaussian noise component. Dynamic problems are defined in Definition A.16. Equation (A.58) gives a noisy function with an additive Gaussian noise component.

As a final comment on the fitness function, it is important to emphasize its role in an EA. The evolutionary operators, e.g. selection, crossover, mutation and elitism, usually make use of the fitness evaluation of chromosomes. For example, selection operators are inclined towards the most-fit individuals when selecting parents for crossover, while mutation leans towards the least-fit individuals.

## 8.5 Selection

Selection is one of the main operators in EAs, and relates directly to the Darwinian concept of survival of the fittest. The main objective of selection operators is to emphasize better solutions. This is achieved in two of the main steps of an EA:

- **Selection of the new population**: A new population of candidate solutions is selected at the end of each generation to serve as the population of the next generation. The new population can be selected from only the offspring, or from both the parents and the offspring. The selection operator should ensure that good individuals do survive to next generations.

- **Reproduction**: Offspring are created through the application of crossover and/or mutation operators. In terms of crossover, "superior" individuals should have more opportunities to reproduce to ensure that offspring contain genetic material of the best individuals. In the case of mutation, selection mechanisms should focus on "weak" individuals. The hope is that mutation of weak individuals will result in introducing better traits to weak individuals, thereby increasing their chances of survival.

Many selection operators have been developed. A summary of the most frequently used operators is given in this section. Preceding this summary is a discussion of selective pressure in Section 8.5.1.

### 8.5.1 Selective Pressure

Selection operators are characterized by their *selective pressure*, also referred to as the *takeover time*, which relates to the time it requires to produce a uniform population. It is defined as the speed at which the best solution will occupy the entire population by repeated application of the selection operator alone [38, 320]. An operator with a high selective pressure decreases diversity in the population more rapidly than operators with a low selective pressure, which may lead to premature convergence to suboptimal solutions. A high selective pressure limits the exploration abilities of the population.

### 8.5.2 Random Selection

Random selection is the simplest selection operator, where each individual has the same probability of $\frac{1}{n_s}$ (where $n_s$ is the population size) to be selected. No fitness information is used, which means that the best and the worst individuals have exactly the same probability of surviving to the next generation. Random selection has the lowest selective pressure among the selection operators discussed in this section.

### 8.5.3 Proportional Selection

Proportional selection, proposed by Holland [376], biases selection towards the most-fit individuals. A probability distribution proportional to the fitness is created, and individuals are selected by sampling the distribution,

$$\varphi_s(\mathbf{x}_i(t)) = \frac{f_{\Upsilon}(\mathbf{x}_i(t))}{\sum_{l=1}^{n_s} f_{\Upsilon}(\mathbf{x}_l(t))} \tag{8.9}$$

where $n_s$ is the total number of individuals in the population, and $\varphi_s(\mathbf{x}_i)$ is the probability that $\mathbf{x}_i$ will be selected. $f_{\Upsilon}(\mathbf{x}_i)$ is the scaled fitness of $\mathbf{x}_i$, to produce a positive floating-point value. For minimization problems, possible choices of scaling function, $\Upsilon$, are

- $f_{\Upsilon}(\mathbf{x}_i(t)) = \Upsilon(\mathbf{x}_i(t)) = f_{max} - f_{\Psi}(\mathbf{x}_i(t))$ where $f_{\Psi}(\mathbf{x}_i(t)) = \Psi(\mathbf{x}_i(t))$ is the raw fitness value of $\mathbf{x}_i(t)$. However, knowledge of $f_{max}$ (the maximum possible fitness) is usually not available. An alternative is to use $f_{max}(t)$, which is the maximum fitness observed up to time step $t$.

- $f_\Upsilon(\mathbf{x}_i(t)) = \Upsilon(\mathbf{x}_i(t)) = \frac{1}{1+f_\Psi(\mathbf{x}_i(t))-f_{min}(t)}$, where $f_{min}(t)$ is the minimum observed fitness up to time step $t$. Here, $f_\Upsilon(\mathbf{x}_i(t)) \in (0,1]$.

In the case of a maximization problem, the fitness values can be scaled to the range $(0,1]$ using

$$f_\Upsilon(\mathbf{x}_i(t)) = \Upsilon(\mathbf{x}_i(t)) = \frac{1}{1+f_{max}(t)-f(\mathbf{x}_i(t))} \qquad (8.10)$$

Two popular sampling methods used in proportional selection is roulette wheel sampling and stochastic universal sampling.

Assuming maximization, and normalized fitness values, roulette wheel selection is summarized in Algorithm 8.2. Roulette wheel selection is an example proportional selection operator where fitness values are normalized (e.g. by dividing each fitness by the maximum fitness value). The probability distribution can then be seen as a roulette wheel, where the size of each slice is proportional to the normalized selection probability of an individual. Selection can be likened to the spinning of a roulette wheel and recording which slice ends up at the top; the corresponding individual is then selected.

---

**Algorithm 8.2** Roulette Wheel Selection

---

Let $i = 1$, where $i$ denotes the chromosome index;
Calculate $\varphi_s(\mathbf{x}_i)$ using equation (8.9);
$sum = \varphi_s(\mathbf{x}_i)$;
Choose $r \sim U(0,1)$;
**while** $sum < r$ **do**
$\quad i = i + 1$, i.e. advance to the next chromosome;
$\quad sum = sum + \varphi_s(\mathbf{x}_i)$;
**end**
Return $\mathbf{x}_i$ as the selected individual;

---

When roulette wheel selection is used to create offspring to replace the entire population, $n_s$ independent calls are made to Algorithm 8.2. It was found that this results in a high variance in the number of offspring created by each individual. It may happen that the best individual is not selected to produce offspring during a given generation. To prevent this problem, Baker [46] proposed stochastic universal sampling (refer to Algorithm 8.3), used to determine for each individual the number of offspring, $\lambda_i$, to be produced by the individual with only one call to the algorithm.

Because selection is directly proportional to fitness, it is possible that strong individuals may dominate in producing offspring, thereby limiting the diversity of the new population. In other words, proportional selection has a high selective pressure.

---

**Algorithm 8.3** Stochastic Universal Sampling

---

**for** $i = 1, \ldots, n_s$ **do**
    $\lambda_i(t) = 0$;
**end**
$r \sim U(0, \frac{1}{\lambda})$, where $\lambda$ is the total number of offspring;
$sum = 0.0$;
**for** $i = 1, \ldots, n_s$ **do**
    $sum = sum + \varphi_s(\mathbf{x}_i(t))$;
    **while** $r < sum$ **do**
        $\lambda_i + +$;
        $r = r + \frac{1}{\lambda}$;
    **end**
**end**
return $\lambda = (\lambda_1, \ldots, \lambda_{n_s})$;

---

### 8.5.4 Tournament Selection

Tournament selection selects a group of $n_{ts}$ individuals randomly from the population, where $n_{ts} < n_s$ ($n_s$ is the total number of individuals in the population). The performance of the selected $n_{ts}$ individuals is compared and the best individual from this group is selected and returned by the operator. For crossover with two parents, tournament selection is done twice, once for the selection of each parent.

Provided that the tournament size, $n_{ts}$, is not too large, tournament selection prevents the best individual from dominating, thus having a lower selection pressure. On the other hand, if $n_{ts}$ is too small, the chances that bad individuals are selected increase.

Even though tournament selection uses fitness information to select the best individual of a tournament, random selection of the individuals that make up the tournament reduces selective pressure compared to proportional selection. However, note that the selective pressure is directly related to $n_{ts}$. If $n_{ts} = n_s$, the best individual will always be selected, resulting in a very high selective pressure. On the other hand, if $n_{ts} = 1$, random selection is obtained.

### 8.5.5 Rank-Based Selection

Rank-based selection uses the rank ordering of fitness values to determine the probability of selection, and not the absolute fitness values. Selection is therefore independent of actual fitness values, with the advantage that the best individual will not dominate in the selection process.

Non-deterministic linear sampling selects an individual, $\mathbf{x}_i$, such that $i \sim U(0, U(0, n_s - 1))$, where the individuals are sorted in decreasing order of fitness value. It is also assumed that the rank of the best individual is 0, and that of the worst individual is $n_s - 1$.

Linear ranking assumes that the best individual creates $\hat{\lambda}$ offspring, and the worst individual $\tilde{\lambda}$, where $1 \leq \hat{\lambda} \leq 2$ and $\tilde{\lambda} = 2 - \hat{\lambda}$. The selection probability of each individual is calculated as

$$\varphi_s(\mathbf{x}_i(t)) = \frac{\tilde{\lambda} + (f_r(\mathbf{x}_i(t))/(n_s - 1))(\hat{\lambda} - \tilde{\lambda})}{n_s} \tag{8.11}$$

where $f_r(\mathbf{x}_i(t))$ is the rank of $\mathbf{x}_i(t)$.

Nonlinear ranking techniques calculate the selection probabilities, for example, as follows:

$$\varphi_s(\mathbf{x}_i(t)) = \frac{1 - e^{-f_r(\mathbf{x}_i(t))}}{\beta} \tag{8.12}$$

or

$$\varphi_s(\mathbf{x}_i) = \nu(1 - \nu)^{n_p - 1 - f_r(\mathbf{x}_i)} \tag{8.13}$$

where $f_r(\mathbf{x}_i)$ is the rank of $\mathbf{x}_i$ (i.e. the individual's position in the ordered sequence of individuals), $\beta$ is a normalization constant, and $\nu$ indicates the probability of selecting the next individual.

Rank-based selection operators may use any sampling method to select individuals, e.g. roulette wheel selection (Algorithm 8.2) or stochastic universal sampling (Algorithm 8.3).

## 8.5.6   Boltzmann Selection

Boltzmann selection is based on the thermodynamical principles of simulated annealing (refer to Section A.5.2). It has been used in different ways, one of which computes selection probabilities as follows:

$$\varphi(\mathbf{x}_i(t)) = \frac{1}{1 + e^{f(\mathbf{x}_i(t))/T(t)}} \tag{8.14}$$

where $T(t)$ is the temperature parameter. A temperature schedule is used to reduce $T(t)$ from its initial large value to a small value.

The initial large value ensures that all individuals have an equal probability of being selected. As $T(t)$ becomes smaller, selection focuses more on the good individuals. The sampling methods discussed in Section 8.5.3 can be used to select individuals.

Alternatively, Boltzmann selection can be used to select between two individuals, for example, to decide if a parent, $\mathbf{x}_i(t)$, should be replaced by its offspring, $\mathbf{x}_i'(t)$. If

$$U(0, 1) > \frac{1}{1 + e^{(f(\mathbf{x}_i(t)) - f(\mathbf{x}_i'(t)))/T(t)}} \tag{8.15}$$

then $\mathbf{x}_i'(t)$ is selected; otherwise, $\mathbf{x}_i(t)$ is selected.

### 8.5.7 $(\mu \dagger \lambda)$-Selection

The $(\mu, \lambda)$- and $(\mu + \lambda)$-selection methods are deterministic rank-based selection methods used in evolutionary strategies (refer to Chapter 12). For both methods $\mu$ indicates the number of parents (which is the size of the population), and $\lambda$ is the number of offspring produced from each parent. After production of the $\lambda$ offspring, $(\mu, \lambda)$-selection selects the best $\mu$ offspring for the next population. This process of selection is very similar to beam search (refer to Section A.5.2). $(\mu + \lambda)$-selection, on the other hand, selects the best $\mu$ individuals from both the parents and the offspring.

### 8.5.8 Elitism

Elitism refers to the process of ensuring that the best individuals of the current population survive to the next generation. The best individuals are copied to the new population without being mutated. The more individuals that survive to the next generation, the less the diversity of the new population.

### 8.5.9 Hall of Fame

The hall of fame is a selection scheme similar to the list of best players of an arcade game. For each generation, the best individual is selected to be inserted into the hall of fame. The hall of fame will therefore contain an archive of the best individuals found from the first generation. The hall of fame can be used as a parent pool for the crossover operator, or, at the last generation, the best individual is selected as the best one in the hall of fame.

## 8.6 Reproduction Operators

Reproduction is the process of producing offspring from selected parents by applying crossover and/or mutation operators. Crossover is the process of creating one or more new individuals through the combination of genetic material randomly selected from two or more parents. If selection focuses on the most-fit individuals, the selection pressure may cause premature convergence due to reduced diversity of the new populations.

Mutation is the process of randomly changing the values of genes in a chromosome. The main objective of mutation is to introduce new genetic material into the population, thereby increasing genetic diversity. Mutation should be applied with care not to distort the good genetic material in highly fit individuals. For this reason, mutation is usually applied at a low probability. Alternatively, the mutation probability can be made proportional to the fitness of individuals: the less fit the individual, the more it is mutated. To promote exploration in the first generations, the mutation probability can be initialized to a large value, which is then reduced over time to allow for exploitation during the final generations.

Reproduction can be applied with replacement, in which case newly generated individuals replace parent individuals only if the fitness of the new offspring is better than that of the corresponding parents.

Since crossover and mutation operators are representation and EC paradigm dependent, the different implementations of these operators are covered in chapters that follow.

## 8.7  Stopping Conditions

The evolutionary operators are iteratively applied in an EA until a stopping condition is satisfied. The simplest stopping condition is to limit the number of generations that the EA is allowed to execute, or alternatively, a limit is placed on the number of fitness function evaluations. This limit should not be too small, otherwise the EA will not have sufficient time to explore the search space.

In addition to a limit on execution time, a convergence criterion is usually used to detect if the population has converged. Convergence is loosely defined as the event when the population becomes stagnant. In other words, when there is no genotypic or phenotypic change in the population. The following convergence criteria can be used:

- **Terminate when no improvement is observed over a number of consecutive generations**. This can be detected by monitoring the fitness of the best individual. If there is no significant improvement over a given time window, the EA can be stopped. Alternatively, if the solution is not satisfactory, mechanisms can be applied to increase diversity in order to force further exploration. For example, the mutation probability and mutational step sizes can be increased.

- **Terminate when there is no change in the population**. If, over a number of consecutive generations, the average change in genotypic information is too small, the EA can be stopped.

- **Terminate when an acceptable solution has been found**. If $\mathbf{x}^*(t)$ represents the optimum of the objective function, then if the best individual, $\mathbf{x}_i$, is such that $f(\mathbf{x}_i) \leq |f(\mathbf{x}) - \epsilon|$, an acceptable solution is found; $\epsilon$ is the error threshold. If $\epsilon$ is too large, solutions may be bad. Too small values of $\epsilon$ may cause the EA never to terminate if a time limit is not imposed.

- **Terminate when the objective function slope is approximately zero**, as defined in equation (16.16) of Chapter 16.

## 8.8  Evolutionary Computation versus Classical Optimization

While classical optimization algorithms have been shown to be very successful (and more efficient than EAs) in linear, quadratic, strongly convex, unimodal and other specialized problems, EAs have been shown to be more efficient for discontinuous, non-differentiable, multimodal and noisy problems.

EC and classical optimization (CO) differ mainly in the search process and information about the search space used to guide the search process:

- **The search process:** CO uses deterministic rules to move from one point in the search space to the next point. EC, on the other hand, uses probabilistic transition rules. Also, EC applies a parallel search of the search space, while CO uses a sequential search. An EA search starts from a diverse set of initial points, which allows parallel search of a large area of the search space. CO starts from one point, successively adjusting this point to move toward the optimum.

- **Search surface information:** CO uses derivative information, usually first-order or second-order, of the search space to guide the path to the optimum. EC, on the other hand, uses no derivative information. The fitness values of individuals are used to guide the search.

## 8.9  Assignments

1. Discuss the importance of the fitness function in EC.

2. Discuss the difference between genetic and phenotypic evolution.

3. In the case of a small population size, how can we ensure that a large part of the search space is covered?

4. How can premature convergence be prevented?

5. In what situations will a high mutation rate be of advantage?

6. Is the following statement valid? *"A genetic algorithm is assumed to have converged to a local or global solution when the ratio $\overline{f}/f_{max}$ is close to 1, where $f_{max}$ and $\overline{f}$ are the maximum and average fitness of the evolving population respectively."*

7. How can an EA be used to train a NN? In answering this question, focus on
   - (a) the representation scheme, and
   - (b) fitness function.

8. Show how an EA can be used to solve systems of equations, by illustrating how
   - (a) solutions are represented, and
   - (b) the fitness is calculated.

   What problem can be identified in using an EA to solve systems of equations?

9. How can the effect of a high selective pressure be countered?

10. Under which condition will stochastic universal sampling behave like tournament selection?

11. Identify disadvantages of fitness-based selection operators.

12. For the nonlinear ranking methods given in equations (8.12) and (8.13), indicate if these assume a minimization or maximization problem.

13. Critisize the following stopping condition: Stop execution of the EA when there is no significant change in the average fitness of the population over a number of consecutive generations.

# Chapter 9

# Genetic Algorithms

Genetic algorithms (GA) are possibly the first algorithmic models developed to simulate genetic systems. First proposed by Fraser [288, 289], and later by Bremermann [86] and Reed *et al.* [711], it was the extensive work done by Holland [376] that popularized GAs. It is then also due to his work that Holland is generally considered the father of GAs.

GAs model genetic evolution, where the characteristics of individuals are expressed using genotypes. The main driving operators of a GA is selection (to model survival of the fittest) and recombination through application of a crossover operator (to model reproduction). This section discusses in detail GAs and their evolution operators, organized as follows: Section 9.1 reviews the canonical GA as proposed by Holland. Crossover operators for binary and floating-point representations are discussed in Section 9.2. Mutation operators are covered in Section 9.3. GA control parameters are discussed in Section 9.4. Different GA implementations are reviewed in Section 9.5, while advanced topics are considered in Section 9.6. A summary of GA applications is given in Section 9.7.

## 9.1 Canonical Genetic Algorithm

The canonical GA (CGA) as proposed by Holland [376] follows the general algorithm as given in Algorithm 8.1, with the following implementation specifics:

- A bitstring representation was used.
- Proportional selection was used to select parents for recombination.
- One-point crossover (refer to Section 9.2) was used as the primary method to produce offspring.
- Uniform mutation (refer to Section 9.3) was proposed as a background operator of little importance.

It is valuable to note that mutation was not considered as an important operator in the original GA implementations. It was only in later implementations that the explorative power of mutation was used to improve the search capabilities of GAs.

*Computational Intelligence: An Introduction*, Second Edition A.P. Engelbrecht
©2007 John Wiley & Sons, Ltd

Since the CGA, several variations of the GA have been developed that differ in representation scheme, selection operator, crossover operator, and mutation operator. Some implementations introduce other concepts from nature such as mass extinction, culling, population islands, amongst others. While it is impossible to provide a complete review of these alternatives, this chapter provides a good flavor of these approaches to illustrate the richness of GAs.

## 9.2 Crossover

Crossover operators can be divided into three main categories based on the arity (i.e. the number of parents used) of the operator. This results in three main classes of crossover operators:

- **asexual**, where an offspring is generated from one parent.
- **sexual**, where two parents are used to produce one or two offspring.
- **multi-recombination**, where more than two parents are used to produce one or more offspring.

Crossover operators are further categorized based on the representation scheme used. For example, binary-specific operators have been developed for binary string representations (refer to Section 9.2.1), and operators specific to floating-point representations (refer to Section 9.2.2).

Parents are selected using any of the selection schemes discussed in Section 8.5. It is, however, not a given that selected parents will mate. Recombination is applied probabilistically. Each pair (or group) of parents have a probability, $p_c$, of producing offspring. Usually, a high crossover probability (also referred to as the crossover rate) is used.

In selection of parents, the following issues need to be considered:

- Due to probabilistic selection of parents, it may happen that the same individual is selected as both parents, in which case the generated offspring will be a copy of the parent. The parent selection process should therefore incorporate a test to prevent such unnecessary operations.
- It is also possible that the same individual takes part in more than one application of the crossover operator. This becomes a problem when fitness-proportional selection schemes are used.

In addition to parent selection and the recombination process, the crossover operator considers a replacement policy. If one offspring is generated, the offspring may replace the worst parent. Such replacement can be based on the restriction that the offspring must be more fit than the worst parent, or it may be forced. Alternatively, Boltzmann selection (refer to Section 8.5.6) can be used to decide if the offspring should replace the worst parent. Crossover operators have also been implemented where the offspring replaces the worst individual of the population. In the case of two offspring, similar replacement strategies can be used.

### 9.2.1 Binary Representations

Most of the crossover operators for binary representations are sexual, being applied to two selected parents. If $\mathbf{x}_1(t)$ and $\mathbf{x}_2(t)$ denote the two selected parents, then the recombination process is summarized in Algorithm 9.1. In this algorithm, $\mathbf{m}(t)$ is a mask that specifies which bits of the parents should be swapped to generate the offspring, $\tilde{\mathbf{x}}_1(t)$ and $\tilde{\mathbf{x}}_2(t)$. Several crossover operators have been developed to compute the mask:

- **One-point crossover:** Holland [376] suggested that segments of genes be swapped between the parents to create their offspring, and not single genes. A one-point crossover operator was developed that randomly selects a crossover point, and the bitstrings after that point are swapped between the two parents. One-point crossover is illustrated in Figure 9.1(a). The mask is computed using Algorithm 9.2.

- **Two-point crossover:** In this case two bit positions are randomly selected, and the bitstrings between these points are swapped as illustrated in Figure 9.1(b). The mask is calculated using Algorithm 9.3. This operator can be generalized to an $n$-point crossover [85, 191, 250, 711].

- **Uniform crossover:** The $n_x$-dimensional mask is created randomly [10, 828] as summarized in Algorithm 9.4. Here, $p_x$ is the bit-swapping probability. If $p_x = 0.5$, then each bit has an equal chance to be swapped. Uniform crossover is illustrated in Figure 9.1(c).

---

**Algorithm 9.1** Generic Algorithm for Bitstring Crossover

---

Let $\tilde{\mathbf{x}}_1(t) = \mathbf{x}_1(t)$ and $\tilde{\mathbf{x}}_2(t) = \mathbf{x}_2(t)$;
**if** $U(0,1) \leq p_c$ **then**
    Compute the binary mask, $\mathbf{m}(t)$;
    **for** $j = 1, \ldots, n_x$ **do**
        **if** $m_j = 1$ **then**
            //swap the bits
            $\tilde{\mathbf{x}}_{1j}(t) = \mathbf{x}_{2j}(t)$ ;
            $\tilde{\mathbf{x}}_{2j}(t) = \mathbf{x}_{1j}(t)$;
        **end**
    **end**
**end**

---

---

**Algorithm 9.2** One-Point Crossover Mask Calculation

---

Select the crossover point, $\xi \sim U(1, n_x - 1)$;
Initialize the mask: $m_j(t) = 0$, for all $j = 1, \ldots, n_x$;
**for** $j = \xi + 1$ $to$ $n_x$ **do**
    $m_j(t) = 1$;
**end**

---

---

**Algorithm 9.3** Two-Point Crossover Mask Calculation

---

Select the two crossover points, $\xi_1, \xi_2 \sim U(1, n_x)$;
Initialize the mask: $m_j(t) = 0$, for all $j = 1, \ldots, n_x$;
**for** $j = \xi_1 + 1$ *to* $\xi_2$ **do**
   $m_j(t) = 1$;
**end**

---

**Algorithm 9.4** Uniform Crossover Mask Calculation

---

Initialize the mask: $m_j(t) = 0$, for all $j = 1, \ldots, n_x$;
**for** $j = 1$ *to* $n_x$ **do**
   **if** $U(0, 1) \leq p_x$ **then**
      $m_j(t) = 1$;
   **end**
**end**

---

Bremermann *et al.* [85] proposed the first multi-parent crossover operators for binary representations. Given $n_\mu$ parent vectors, $\mathbf{x}_1(t), \ldots, \mathbf{x}_{n_\mu}(t)$, majority mating generates one offspring using

$$\tilde{\mathbf{x}}_{ij}(t) = \begin{cases} 0 & \text{if } n'_\mu \geq n_\mu/2, \ l = 1, \ldots, n_\mu \\ 1 & \text{otherwise} \end{cases} \tag{9.1}$$

where $n'_\mu$ is the number of parents with $x_{lj}(t) = 0$.

A multiparent version of $n$-point crossover was also proposed by Bremermann *et al.* [85], where $n_\mu - 1$ identical crossover points are selected in the $n_\mu$ parents. One offspring is generated by selecting one segment from each parent.

Jones [427] developed a crossover hillclimbing operator that can be applied to any representation. Crossover hillclimbing starts with two parents, and continues to produce offspring from this pair of parents until either a maximum number of crossover attempts has been exceeded, or a pair of offspring is found where one of the offspring has a better fitness than the best parent. Crossover hillclimbing then continues reproduction using these two offspring as the new parent pair. If a better parent pair cannot be found within the specified time limit, the worst parent is replaced by a randomly selected parent.

## 9.2.2 Floating-Point Representation

The crossover operators discussed above (excluding majority mating) can also be applied to floating-point representations as discrete recombination strategies. In contrast

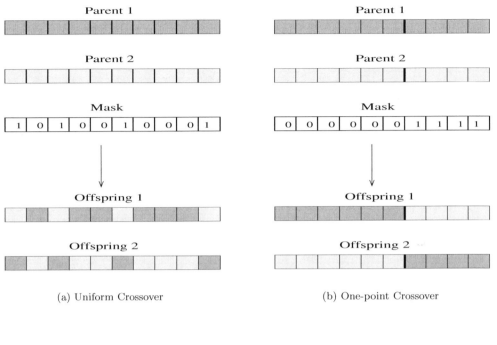

(a) Uniform Crossover

(b) One-point Crossover

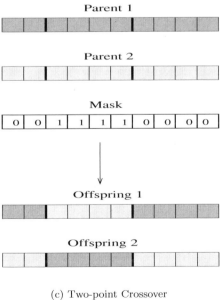

(c) Two-point Crossover

**Figure 9.1** Crossover Operators for Binary Representations

to these discrete operators where information is swapped between parents, intermediate recombination operators, developed specifically for floating-point representations, blend components across the selected parents.

One of the first floating-point crossover operators is the linear operator proposed by Wright [918]. From the parents, $\mathbf{x}_1(t)$ and $\mathbf{x}_2(t)$, three candidate offspring are generated as $(\mathbf{x}_1(t) + \mathbf{x}_2(t)), (1.5\mathbf{x}_1(t) - 0.5\mathbf{x}_2(t))$ and $(-0.5\mathbf{x}_1(t) + 1.5\mathbf{x}_2(t))$. The two best solutions are selected as the offspring. Wright [918] also proposed a directional heuristic crossover operator where one offspring is created from two parents using

$$\tilde{x}_{ij}(t) = U(0,1)(\mathbf{x}_{2j}(t) - \mathbf{x}_{1j}(t)) + \mathbf{x}_{2j}(t) \tag{9.2}$$

subject to the constraint that parent $\mathbf{x}_2(t)$ cannot be worse than parent $\mathbf{x}_1(t)$.

Michalewicz [586] coined the arithmetic crossover, which is a multiparent recombination strategy that takes a weighted average over two or more parents. One offspring is generated using

$$\tilde{x}_{ij}(t) = \sum_{l=1}^{n_\mu} \gamma_l x_{lj}(t) \tag{9.3}$$

with $\sum_{l=1}^{n_\mu} \gamma_l = 1$. A specialization of the arithmetic crossover operator is obtained for $n_\mu = 2$, in which case

$$\tilde{x}_{ij}(t) = (1-\gamma)x_{1j}(t) + \gamma x_{2j}(t) \tag{9.4}$$

with $\gamma \in [0,1]$. If $\gamma = 0.5$, the effect is that each component of the offspring is simply the average of the corresponding components of the parents.

Eshelman and Schaffer [251] developed a variation of the weighted average given in equation (9.4), referred to as the blend crossover (BLX-$\alpha$), where

$$\tilde{x}_{ij}(t) = (1-\gamma_j)x_{1j}(t) + \gamma_j x_{2j}(t) \tag{9.5}$$

with $\gamma_j = (1+2\alpha)U(0,1) - \alpha$. The BLX-$\alpha$ operator randomly picks, for each component, a random value in the range

$$[x_{1j}(t) - \alpha(x_{2j}(t) - x_{1j}(t)), x_{2j}(t) + \alpha(x_{2j}(t) - x_{1j}(t))] \tag{9.6}$$

BLX-$\alpha$ assumes that $x_{1j}(t) < x_{2j}(t)$. Eshelman and Schaffer found that $\alpha = 0.5$ works well.

The BLX-$\alpha$ has the property that the location of the offspring depends on the distance that the parents are from one another. If this distance is large, then the distance between the offpsring and its parents will be large. The BLX-$\alpha$ allows a bit more exploration than the weighted average of equation (9.3), due to the stochastic component in producing the offspring.

Michalewicz et al. [590] developed the two-parent geometrical crossover to produce a single offspring as follows:

$$\tilde{x}_{ij}(t) = (x_{1j}x_{2j})^{0.5} \tag{9.7}$$

The geometrical crossover can be generalized to multi-parent recombination as follows:

$$\tilde{x}_{ij}(t) = (x_{1j}^{\alpha_1} x_{2j}^{\alpha_2} \ldots x_{n_\mu j}^{\alpha_{n_\mu}}) \tag{9.8}$$

where $n_\mu$ is the number of parents, and $\sum_{l=1}^{n_\mu} \alpha_l = 1$.

Deb and Agrawal [196] developed the simulated binary crossover (SBX) to simulate the behavior of the one-point crossover operator for binary representations. Two parents, $\mathbf{x}_1(t)$ and $\mathbf{x}_2(t)$ are used to produce two offspring, where for $j = 1, \ldots, n_x$

$$\tilde{x}_{1j}(t) = 0.5[(1+\gamma_j)x_{1j}(t) + (1-\gamma_j)x_{2j}(t)] \tag{9.9}$$
$$\tilde{x}_{2j}(t) = 0.5[(1-\gamma_j)x_{1j}(t) + (1+\gamma_j)x_{2j}(t)] \tag{9.10}$$

where

$$\gamma_j = \begin{cases} (2r_j)^{\frac{1}{\eta+1}} & \text{if } r_j \leq 0.5 \\ \left(\frac{1}{2(1-r_j)}\right)^{\frac{1}{\eta+1}} & \text{otherwise} \end{cases} \tag{9.11}$$

where $r_j \sim U(0,1)$, and $\eta > 0$ is the distribution index. Deb and Agrawal suggested that $\eta = 1$.

The SBX operator generates offspring symmetrically about the parents, which prevents bias towards any of the parents. For large values of $\eta$ there is a higher probability that offspring will be created near the parents. For small $\eta$ values, offspring will be more distant from the parents.

While the above focused on sexual crossover operators (some of which can also be extended to multiparent operators), the remainder of this section considers a number of multiparent crossover operators. The main objective of these multiparent opera-tors is to intensify the explorative capabilities compared to two-parent operators. By aggregating information from multiple parents, more disruption is achieved with the resemblance between offspring and parents on average smaller compared to two-parent operators.

Ono and Kobayashi [642] developed the unimodal distributed (UNDX) operator where two or more offspring are generated using three parents. The offspring are created from an ellipsoidal probability distribution, with one of the axes formed along the line that connects two of the parents. The extent of the orthogonal direction is determined from the perpendicular distance of the third parent from the axis. The UNDX operator can be generalized to work with any number of parents, with $3 \leq n_\mu \leq n_s$. For the generalization, $n_\mu - 1$ parents are randomly selected and their center of mass (mean), $\overline{\mathbf{x}}(t)$, is calculated, where

$$\overline{x}_j(t) = \sum_{l=1}^{n_\mu - 1} x_{lj}(t) \tag{9.12}$$

From the mean, $n_\mu - 1$ direction vectors, $\mathbf{d}_l(t) = \mathbf{x}_l(t) - \overline{\mathbf{x}}(t)$ are computed, for $l = 1, \ldots, n_\mu - 1$. Using the direction vectors, the direction cosines are computed as $\mathbf{e}_l(t) = \mathbf{d}_l(t)/|\mathbf{d}_l(t)|$, where $|\mathbf{d}_l(t)|$ is the length of vector $\mathbf{d}_l(t)$. A random parent, with index $n_\mu$ is selected. Let $\mathbf{x}_{n_\mu}(t) - \overline{\mathbf{x}}(t)$ be the vector orthogonal to all $\mathbf{e}_l(t)$, and

$\delta = |\mathbf{x}_{n_\mu}(t) - \overline{\mathbf{x}}(t)|$. Let $\mathbf{e}_l(t), l = n_\mu, \dots, n_s$ be the orthonormal basis of the subspace orthogonal to the subspace spanned by the direction cosines, $\mathbf{e}_l(t), l = 1, \dots, n_\mu - 1$. Offspring are then generated using

$$\tilde{\mathbf{x}}_i(t) = \overline{\mathbf{x}}(t) + \sum_{l=1}^{n_\mu - 1} N(0, \sigma_1^2)|\mathbf{d}_l|\mathbf{e}_l + \sum_{l=n_\mu}^{n_s} N(0, \sigma_2^2)\delta\mathbf{e}_l(t) \tag{9.13}$$

where $\sigma_1 = \dfrac{1}{\sqrt{n_\mu - 2}}$ and $\sigma_2 = \dfrac{0.35}{\sqrt{n_s - n_\mu - 2}}$.

Using equation (9.13) any number of offspring can be created, sampled around the center of mass of the selected parents. A higher probability is assigned to create offspring near the center rather than near the parents. The effect of the UNDX operator is illustrated in Figure 9.2(a) for $n_\mu = 4$.

Tsutsui and Goldberg [857] and Renders and Bersini [714] proposed the simplex crossover (SPX) operator as another center of mass approach to recombination. Renders and Bersini selects $n_\mu > 2$ parents, and determines the best and worst parent, say $\mathbf{x}_1(t)$ and $\mathbf{x}_2(t)$ respectively. The center of mass, $\overline{\mathbf{x}}(t)$ is computed over the selected parents, but with $\mathbf{x}_2(t)$ excluded. One offspring is generated using

$$\tilde{\mathbf{x}}(t) = \overline{\mathbf{x}}(t) + (\mathbf{x}_1(t) - \mathbf{x}_2(t)) \tag{9.14}$$

Tsutsui and Goldberg followed a similar approach, selecting $n_\mu = n_x + 1$ parents independent from one another for an $n_x$-dimensional search space. These $n_\mu$ parents form a simplex. The simplex is expanded in each of the $n_\mu$ directions, and offspring sampled from the expanded simplex as illustrated in Figure 9.2(b). For $n_x = 2, n_\mu = 3$, and

$$\overline{\mathbf{x}}(t) = \sum_{l=1}^{n_\mu} \mathbf{x}_l(t) \tag{9.15}$$

the expanded simplex is defined by the points

$$(1 + \gamma)(\mathbf{x}_l(t) - \overline{\mathbf{x}}(t)) \tag{9.16}$$

for $l = 1, \dots, n_\mu = 3$ and $\gamma \geq 0$. Offspring are obtained by uniform sampling of the expanded simplex.

Deb et al. [198] proposed a variation of the UNDX operator, which they refer to as parent-centric crossover (PCX). Instead of generating offspring around the center of mass of the selected parents, offspring are generated around selected parents. PCX selects $n_\mu$ parents and computes their center of mass, $\overline{\mathbf{x}}(t)$. For each offspring to be generated one parent is selected uniformly from the $n_\mu$ parents. A direction vector is calculated for each offspring as

$$\mathbf{d}_i(t) = \mathbf{x}_i(t) - \overline{\mathbf{x}}(t)$$

where $\mathbf{x}_i(t)$ is the randomly selected parent. From the other $n_\mu - 1$ parents perpendicular distances, $\delta_l$, for $i \neq l = 1, \dots, n_\mu$, are calculated to the line $\mathbf{d}_i(t)$. The average

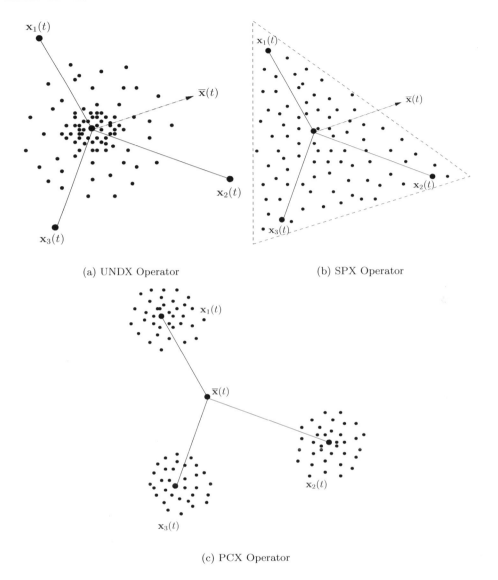

(a) UNDX Operator                    (b) SPX Operator

(c) PCX Operator

Figure 9.2 Illustration of Multi-parent Center of Mass Crossover Operators (dots represent potential offpsring)

over these distances is calculated, i.e.

$$\bar{\delta} = \frac{\sum_{l=1,l\neq i}^{n_\mu} \delta_l}{n_\mu - 1} \tag{9.17}$$

Offspring is generated using

$$\tilde{\mathbf{x}}_i(t) = \mathbf{x}_i(t) + N(0,\sigma_1^2)|\mathbf{d}_i(t)| + \sum_{l=1,i\neq l}^{n_\mu} N(0,\sigma_2^2)\bar{\delta}\mathbf{e}_l(t) \tag{9.18}$$

where $\mathbf{x}_i(t)$ is the randomly selected parent of offspring $\tilde{\mathbf{x}}_i(t)$, and $\mathbf{e}_l(t)$ are the $n_\mu - 1$

orthonormal bases that span the subspace perpendicular to $\mathbf{d}_i(t)$.

The effect of the PCX operator is illustrated in Figure 9.2(c).

Eiben *et al.* [231, 232, 233] developed a number of gene scanning techniques as multi-parent generalizations of $n$-point crossover. For each offspring to be created, the gene scanning operator is applied as summarized in Algorithm 9.5. The algorithm contains two main procedures:

- A scanning strategy, which assigns to each selected parent a probability that the offspring will inherit the next component from that parent. The component under consideration is indicated by a marker.

- A marker update strategy, which updates the markers of parents to point to the next component of each parent.

Marker initialization and updates depend on the representation method. For binary representations the marker of each parent is set to its first gene. The marker update strategy simply advances the marker to the next gene.

Eiben *et al.* proposed three scanning strategies:

- **Uniform scanning** creates only one offspring. The probability, $p_s(\mathbf{x}_l(t))$, of inheriting the gene from parent $\mathbf{x}_l(t), l = 1, \ldots, n_\mu$, as indicated by the marker of that parent is computed as

$$p_s(\mathbf{x}_l(t+1)) = \frac{1}{n_\mu} \tag{9.19}$$

  Each parent has an equal probability of contributing to the creation of the offspring.

- **Occurrence-based scanning** bases inheritance on the premise that the allele that occur most in the parents for a particular gene is the best possible allele to inherit by the offspring (similar to the majority mating operator). Occurrence-based scanning assumes that fitness-proportional selection is used to select the $n_\mu$ parents that take part in recombination.

- **Fitness-based scanning**, where the allele to be inherited is selected proportional to the fitness of the parents. Considering maximization, the probability to inherit from parent $\mathbf{x}_l(t)$ is

$$p_s(\mathbf{x}_l(t)) = \frac{f(\mathbf{x}_l(t))}{\sum_{i=1}^{n_\mu} f(\mathbf{x}_i(t))} \tag{9.20}$$

  Roulette-wheel selection is used to select the parent to inherit from.

For each of these scanning strategies, the offspring inherits $p_s(\mathbf{x}_l(t+1))n_x$ genes from parent $\mathbf{x}_l(t)$.

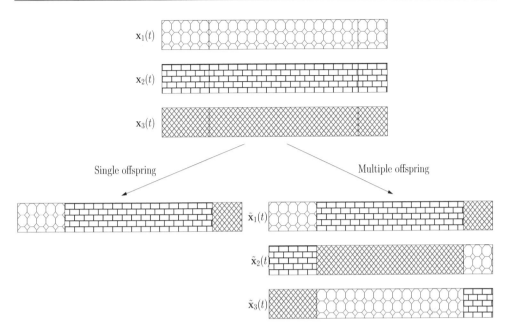

$\mathbf{x}_1(t)$

$\mathbf{x}_2(t)$

$\mathbf{x}_3(t)$

Single offspring

Multiple offspring

$\tilde{\mathbf{x}}_1(t)$

$\tilde{\mathbf{x}}_2(t)$

$\tilde{\mathbf{x}}_3(t)$

**Figure 9.3** Diagonal Crossover

---

**Algorithm 9.5** Gene Scanning Crossover Operator

---

Initialize parent markers;
**for** $j = 1, \ldots, n_x$ **do**
    Select the parent, $\mathbf{x}_l(t)$, to inherit from;
    $\tilde{x}_j(t) = x_{lj}(t)$;
    Update parent markers;
**end**

---

The diagonal crossover operator developed by Eiben *et al.* [232] is a generalization of $n$-point crossover for more than two parents: $n \geq 1$ crossover points are selected and applied to all of the $n_\mu = n + 1$ parents. One or $n + 1$ offspring can be generated by selecting segments from the parents along the diagonals as illustrated in Figure 9.3, for $n = 2, n_\mu = 3$.

## 9.3   Mutation

The aim of mutation is to introduce new genetic material into an existing individual; that is, to add diversity to the genetic characteristics of the population. Mutation is used in support of crossover to ensure that the full range of allele is accessible for each gene. Mutation is applied at a certain probability, $p_m$, to each gene of the offspring, $\tilde{\mathbf{x}}_i(t)$, to produce the mutated offspring, $\mathbf{x}'_i(t)$. The mutation probability, also referred to as the mutation rate, is usually a small value, $p_m \in [0, 1]$, to ensure that good solutions are not distorted too much.

Given that each gene is mutated at probability $p_m$, the probability that an individual will be mutated is given by

$$Prob(\tilde{\mathbf{x}}_i(t) \text{ is mutated}) = 1 - (1 - p_m)^{n_x} \tag{9.21}$$

where the individual contains $n_x$ genes.

Assuming binary representations, if $H(\tilde{\mathbf{x}}_i(t), \mathbf{x}'_i(t))$ is the Hamming distance between offspring, $\tilde{\mathbf{x}}_i(t)$, and its mutated version, $\mathbf{x}'_i(t)$, then the probability that the mutated version resembles the original offspring is given by

$$Prob(\mathbf{x}'_i(t)) \approx \tilde{\mathbf{x}}_i(t)) = p_m^{H(\tilde{\mathbf{x}}_i(t), \mathbf{x}'_i(t))}(1 - p_m)^{n_x - H(\tilde{\mathbf{x}}_i(t), \mathbf{x}'_i(t))} \tag{9.22}$$

This section describes mutation operators for binary and floating-point representations in Sections 9.3.1 and 9.3.2 respectively. A macromutation operator is described in Section 9.3.3.

### 9.3.1 Binary Representations

For binary representations, the following mutation operators have been developed:

- **Uniform (random) mutation** [376], where bit positions are chosen randomly and the corresponding bit values negated as illustrated in Figure 9.4(a). Uniform mutation is summarized in Algorithm 9.6.

- **Inorder mutation**, where two mutation points are randomly selected and only the bits between these mutation points undergo random mutation. Inorder mutation is illustrated in Figure 9.4(b) and summarized in Algorithm 9.7.

- **Gaussian mutation:** For binary representations of floating-point decision variables, Hinterding [366] proposed that the bitstring that represents a decision variable be converted back to a floating-point value and mutated with Gaussian noise. For each chromosome random numbers are drawn from a Poisson distribution to determine the genes to be mutated. The bitstrings representing these genes are then converted. To each of the floating-point values is added the stepsize $N(0, \sigma_j)$, where $\sigma_j$ is 0.1 of the range of that decision variable. The mutated floating-point value is then converted back to a bitstring. Hinterding showed that Gaussian mutation on the floating-point representation of decision variables provided superior results to bit flipping.

For large dimensional bitstrings, mutation may significantly add to the computational cost of the GA. In a bid to reduce computational complexity, Birru [69] divided the bitstring of each individual into a number of bins. The mutation probability is applied to the bins, and if a bin is to be mutated, one of its bits are randomly selected and flipped.

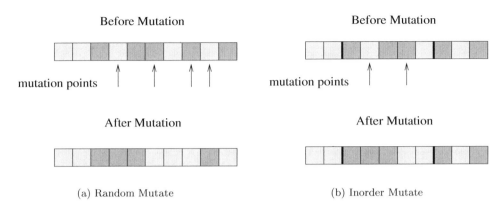

Figure 9.4 Mutation Operators for Binary Representations

---

**Algorithm 9.6** Uniform/Random Mutation

---

**for** $j = 1, \ldots, n_x$ **do**
    **if** $U(0, 1) \leq p_m$ **then**
        $x'_{ij}(t) = \neg \tilde{x}_{ij}(t)$, where $\neg$ denotes the boolean NOT operator;
    **end**
**end**

---

---

**Algorithm 9.7** Inorder Mutation

---

Select mutation points, $\xi_1, \xi_2 \sim U(1, \ldots, n_x)$;
**for** $j = \xi_1, \ldots, \xi_2$ **do**
    **if** $U(0, 1) \leq p_m$ **then**
        $x'_{ij}(t) = \neg \tilde{x}_{ij}(t)$;
    **end**
**end**

---

## 9.3.2 Floating-Point Representations

As indicated by Hinterding [366] and Michalewicz [586], better performance is obtained by using a floating-point representation when decision variables are floating-point values and by applying appropriate operators to these representations, than to convert to a binary representation. This resulted in the development of mutation operators for floating-point representations. One of the first proposals was a uniform mutation, where [586]

$$x'_{ij}(t) = \begin{cases} \tilde{x}_{ij}(t) + \Delta(t, x_{max,j} - \tilde{x}_{ij}(t)) & \text{if a random digit is 0} \\ \tilde{x}_{ij}(t) + \Delta(t, \tilde{x}_{ij}(t) - x_{min,j}(t)) & \text{if a random digit is 1} \end{cases} \qquad (9.23)$$

where $\Delta(t, x)$ returns random values from the range $[0, x]$.

Any of the mutation operators discussed in Sections 11.2.1 (for EP) and 12.4.3 (for ES) can be applied to GAs.

### 9.3.3   Macromutation Operator – Headless Chicken

Jones [427] proposed a macromutation operator, referred to as the headless chicken operator. This operator creates an offspring by recombining a parent individual with a randomly generated individual using any of the previously discussed crossover operators. Although crossover is used to combine an individual with a randomly generated individual, the process cannot be referred to as a crossover operator, as the concept of inheritence does not exist. The operator is rather considered as mutation due to the introduction of new randomly generated genetic material.

## 9.4   Control Parameters

In addition to the population size, the performance of a GA is influenced by the mutation rate, $p_m$, and the crossover rate, $p_c$. In early GA studies very low values for $p_m$ and relatively high values for $p_c$ are propagated. Usually, the values for $p_m$ and $p_c$ are kept static. It is, however, widely accepted that these parameters have a significant influence on performance, and that optimal settings for $p_m$ and $p_c$ can significantly improve performance. To obtain such optimal settings through empirical parameter tuning is a time consuming process. A solution to the problem of finding best values for these control parameters is to use dynamically changing parameters.

Although dynamic, and self-adjusting parameters have been used for EP and ES (refer to Sections 11.3 and 12.3) as early as the 1960s, Fogarty [264] provided one of the first studies of dynamically changing mutation rates for GAs. In this study, Fogarty concluded that performance can be significantly improved using dynamic mutation rates. Fogarty used the following schedules where the mutation rate exponentially decreases with generation number:

$$p_m(t) = \frac{1}{240} + \frac{0.11375}{2^t} \tag{9.24}$$

As an alternative, Fogarty also proposed for binary representations a mutation rate per bit, $j = 1, \ldots, n_b$, where $n_b$ indicates the least significant bit:

$$p_m(j) = \frac{0.3528}{2^{j-1}} \tag{9.25}$$

The two schedules above were combined to give

$$p_m(j, t) = \frac{28}{1905 \times 2^{j-1}} = \frac{0.4026}{2^{t+j-1}} \tag{9.26}$$

A large initial mutation rate favors exploration in the initial steps of the search, and with a decrease in mutation rate as the generation number increases, exploitation is facilitated. Different schedules can be used to reduce the mutation rate. The schedule above results in an exponential decrease. An alternative may be to use a linear schedule, which will result in a slower decrease in $p_m$, allowing more exploration. However, a slower decrease may be too disruptive for already found good solutions. A good strategy is to base the probability of being mutated on the fitness of the individual: the more fit the individual is, the lower the probability that its genes will be mutated; the more unfit the individual, the higher the probability of mutation.

Annealing schedules similar to those used for the learning rate of NNs (refer to equation (4.40)), and to adjust control parameters for PSO and ACO can be applied to $p_m$ (also refer to Section A.5.2).

For floating-point representations, performance is also influenced by the mutational step sizes. An ideal strategy is to start with large mutational step sizes to allow larger, stochastic jumps within the search space. The step sizes are then reduced over time, so that very small changes result near the end of the search process. Step sizes can also be proportional to the fitness of an individual, with unfit individuals having larger step sizes than fit individuals. As an alternative to deterministic schedules to adapt step sizes, self-adaptation strategies as for EP and ES can be used (refer to Sections 11.3.3 and 12.3.3).

The crossover rate, $p_c$, also bears significant influence on performance. With its optimal value being problem dependent, the same adaptive strategies as for $p_m$ can be used to dynamically adjust $p_c$.

In addition to $p_m$ (and mutational step sizes in the case of floating-point representations) and $p_c$, the choice of the best evolutionary operators to use is also problem dependent. While a best combination of crossover, mutation, and selection operators together with best values for the control parameters can be obtained via empirical studies, a number of adaptive methods can be found as reviewed in [41]. These methods adaptively switch between different operators based on search progress. Ultimately, finding the best set of operators and control parameter values is a multi-objective optimization problem by itself.

## 9.5 Genetic Algorithm Variants

Based on the general GA, different implementations of a GA can be obtained by using different combinations of selection, crossover, and mutation operators. Although different operator combinations result in different behaviors, the same algorithmic flow as given in Algorithm 8.1 is followed. This section discusses a few GA implementations that deviate from the flow given in Algorithm 8.1. Section 9.5.1 discusses generation gap methods. The messy GA is described in Section 9.5.2. A short discussion on interactive evolution is given in Section 9.5.3. Island (or parallel) GAs are discussed in Section 9.5.4.

## 9.5.1   Generation Gap Methods

The GAs as discussed thus far differ from biological models of evolution in that population sizes are fixed. This allows the selection process to be described by two steps:

- Parent selection, and

- a replacement strategy  that decides if offspring will replace parents, and which parents to replace.

Two main classes of GAs are identified based on the replacement strategy used, namely generational genetic algorithms (GGA) and steady state genetic algorithms (SSGA), also referred to as incremental GAs. For GGAs  the replacement strategy replaces all parents with their offspring after all offpsring have been created and mutated. This results in no overlap between the current population and the new population (assuming that elitism is not used). For SSGAs,  a decision is made immediately after an offspring is created and mutated as to whether the parent or the offspring survives to the next generation. Thus, there exists an overlap between the current and new populations.

The amount of overlap between the current and new populations is referred to as the generation gap  [191]. GGAs have a zero generation gap, while SSGAs generally have large generation gaps.

A number of replacement strategies have been developed for SSGAs:

- **Replace worst** [192], where the offspring replaces the worst individual of the current population.

- **Replace random** [192, 829], where the offspring replaces a randomly selected individual of the current population.

- **Kill tournament** [798], where a group of individuals is randomly selected, and the worst individual of this group is replaced with the offspring. Alternatively, a tournament size of two is used, and the worst individual is replaced with a probability, $0.5 \leq p_r \leq 1$.

- **Replace oldest**, where a first-in-first-out strategy is followed by replacing the oldest individual of the current population. This strategy has a high probability of replacing one of the best individuals.

- **Conservative selection** [798] combines a first-in-first-out replacement strategy with a modified deterministic binary tournament selection. A tournament size of two individuals is used of which one is always the oldest individual of the current population. The worst of the two is replaced by the offspring. This approach ensures that the oldest individual will not be lost if it is the fittest.

- **Elitist** strategies of the above replacement strategies have also been developed, where the best individual is excluded from selection.

- **Parent-offspring** competition, where a selection strategy is used to decide if an offspring replaces one of its own parents.

Theoretical and empirical studies of steady state GAs can be found in [734, 797, 798, 872].

## 9.5.2  Messy Genetic Algorithms

Standard GAs use populations where all individuals are of the same fixed size. For an $n_x$-dimensional search space, a standard GA finds a solution through application of the evolutionary operators to the complete $n_x$-dimensional individuals. It may happen that good individuals are found, but some of the genes of a good individual are non-optimal. It may be difficult to find optimal allele for such genes through application of crossover and mutation on the entire individual. It may even happen that crossover looses optimized genes, or groups of optimized genes.

Goldberg *et al.* [321, 323, 324] developed the messy GA (mGA), which finds solutions by evolving optimal building blocks and combining building blocks. Here a building block refers to a group of genes. In a messy GA individuals are of variable length, and specified by a list of position-value pairs. The position specifies the gene index, and the value specifies the allele for that gene. These pairs are referred to as messy genes. As an example, if $n_x = 4$, then the individual, $((1,0)(3,1),(4,0)(1,1))$, represents the individual $0 * 10$.

The messy representation may result in individuals that are over-specified or under-specified. The example above illustrates both cases. The individual is over-specified because gene 1 occurs twice. It is under-specified because gene 2 does not occur, and has no value assigned. Fitness evaluation of messy individuals requires strategies to cope with such individuals. For over-specified individuals, a first-come-first-served approach is followed where the first specified value is assigned to the repeating gene. For under-specified individuals, a missing gene's allele is obtained from a competitive template. The competitive template is a locally optimal solution. As an example, if 1101 is the template, the fitness of $0 * 10$ is evaluated as the fitness of 0101.

The objective of mGAs is to evolve optimal building blocks, and to incrementally combine optimized building blocks to form an optimal solution. An mGA is implemented using two loops as shown in Algorithm 9.8. The inner loop consists of three steps:

- **Initialization** to create a population of building blocks of a specified length, $n_m$.

- **Primordial**, which aims to generate small, promising building blocks.

- **Juxtapositional**, to combine building blocks.

The outer loop specifies the size of the building blocks to be considered, starting with the smallest size of one, and incrementally increasing the size until a maximum size is reached, or an acceptable solution is found. The outer loop also sets the best solution obtained from the juxtaposition step as the competitive template for the next iteration.

---

**Algorithm 9.8** Messy Genetic Algorithm

---

Initialize the competitive template;
**for** $n_m = 1$ *to* $n_{m,max}$ **do**
    Initialize the population to contain building blocks of size $n_m$;
    Apply the primordial step;
    Apply the juxtaposition step;
    Set the competitive template to the best solution from the juxtaposition step;
**end**

---

The initialization step creates all possible combinations of building blocks of length $n_m$. For $n_x$-dimensional solutions, this results in a population size of

$$n_s = 2^{n_m} \begin{pmatrix} n_x \\ n_m \end{pmatrix} \tag{9.27}$$

where

$$\begin{pmatrix} n_x \\ n_m \end{pmatrix} = \frac{n_x!}{n_m!(n_x - n_m)!} \tag{9.28}$$

This leads to one of the major disadvantages of mGAs, in that computational complexity explodes with increase in $n_m$ (i.e. building block size). The fast mGA addresses this problem by starting with larger building block sizes and adding a gene deletion operator to the primordial step to prune building blocks [322].

The primordial step is executed for a specified number of generations, applying only selection to find the best building blocks. At regular intervals the population is halved, with the worst individuals (building blocks) discarded. No crossover or mutation is used. While any selection operator can be used, fitness proportional selection is usually used. Because individuals in an mGA may contain different sets of genes (as specified by the building blocks), thresholding selection has been proposed to apply selection to "similar" individuals. Thresholding selection applies tournament selection between two individuals that have in common a number of genes greater than a specified threshold. The effect achieved via the primordial step is that poor building blocks are eliminated, while good building blocks survive to the juxtaposition step.

The juxtaposition step applies cut and splice operators. The cut operator is applied to selected individuals at a probability proportional to the length of the individual (i.e. the size of the building block). The objective of the cut operator is to reduce the size of building blocks by splitting the individual at a randomly selected gene. The splicing operator combines two individuals to form a larger building block. Since the probability of cutting is proportional to the length of the individual, and the mGA starts with small building blocks, splicing occurs more in the beginning. As $n_m$ increases, cutting occurs more. Cutting and splicing then resembles crossover.

### 9.5.3   Interactive Evolution

In standard GAs (and all EAs for that matter), the human user plays a passive role. Selection is based on an explicitly defined analytical function, used to quantify the quality of a candidate solution. It is, however, the case that such a function cannot be defined for certain application areas, for example, evolving art, music, animations, etc. For such application areas subjective judgment is needed, based on human intuition, aesthetical values or taste. This requires interaction of a human evaluator as the "fitness function".

Interactive evolution (IE) [48, 179, 792] involves a human user online into the selection and variation processes. The search process is now directed through interactive selection of solutions by the human user instead of an absolute fitness function. Dawkins [179] was the first to consider IE to evolve biomorphs, which are tree-like representations of two-dimensional graphical forms. Todd and Latham [849] used IE to evolve computer sculptures. Sims [792] provides further advances in the application of IE to evolve complex simulated structures, textures, and motions.

Algorithm 9.9 provides a summary of the standard IE algorithm. The main component of the IE algorithm is the interactive selection step. This step requires that the phenotype of individuals be generated from the genotype, and visualized. Based on the visual representations of candidate solutions, the user selects those individuals that will take part in reproduction, and that will survive to the next generation. Some kind of fitness function can be defined (if possible) to order candidate solutions and to perform a pre-selection to reduce the number of solutions to be evaluated by the human user.

In addition to act as the selection mechanism, the user can also interactively specify the reproduction operators and population parameters.

Instead of the human user performing selection, interaction may be of the form where the user assigns a fitness score to individuals. Automatic selection is then applied, using these user assigned quality measures.

---

**Algorithm 9.9** Interactive Evolution Algorithm

---

Set the generation counter, $t = 0$;
Initialize the control parameters;
Create and initialize the population, $\mathcal{C}(0)$, of $n_s$ individuals;
**while** *stopping condition(s) not true* **do**
    Determine reproduction operators, either automatically or via interaction;
    Select parents via interaction;
    Perform crossover to produce offspring;
    Mutate offspring;
    Select new population via interaction;
**end**

---

Although the section on IE is provided as part of the chapter on GAs, IE can be applied to any of the EAs.

## 9.5.4   Island Genetic Algorithms

GAs lend themselves to parallel implementation. Three main categories of parallel GA have been identified [100]:

- Single-population master-slave GAs, where the evaluation of fitness is distributed over several processors.

- Single-population fine-grained GAs, where each individual is assigned to one processor, and each processor is assigned only one individual. A small neighborhood is defined for each individual, and selection and reproduction are restricted to neighborhoods. Whitley [903] refers to these as cellular GAs.

- Multi-population, or island GAs, where multiple populations are used, each on a separate processor. Information is exchanged among populations via a migration policy. Although developed for parallel implementation, island GAs can be implemented on a single processor system.

The remainder of this section focuses on island GAs. In an island GA, a number of subpopulations are evolved in parallel, in a cooperative framework [335, 903, 100]. In this GA model, a number of islands occurs, where each island represents one population. Selection, crossover and mutation occur in each subpopulation independently from the other subpopulations. In addition, individuals are allowed to migrate between islands (or subpopulations), as illustrated in Figure 9.5.

An integral part of an island GA is the migration policy which governs the exchange of information between islands. A migration policy specifies [100, 102, 103, 104]:

- A **communications topology**, which determines the migration paths between islands. For example, a ring topology (such as illustrated in Figure 16.4(b)) allows exchange of information between neighboring islands. The communication topology determines how fast (or slow) good solutions disseminate to other subpopulations. For a sparsely connected structure (such as the ring topology), islands are more isolated from one another, and the spread of information about good solutions is slower. Sparse topologies also facilitate the appearance of multiple solutions. Densely connected structures have a faster spread of information, which may lead to premature convergence.

- A **migration rate**, which determines the frequency of migration. Tied with the migration rate is the question of when migration should occur. If migration occurs too early, the number of good building blocks in the migrants may be too small to have any influence at their destinations. Usually, migration occurs when each population has converged. After exchange of individuals, all populations are restarted.

- A **selection mechanism** to decide which individuals will migrate.

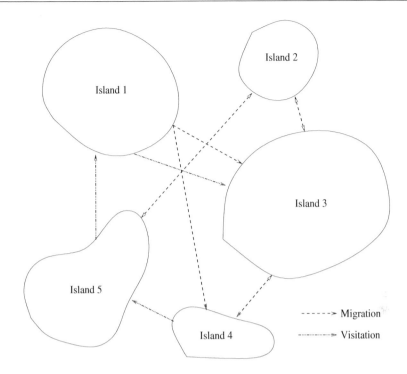

**Figure 9.5** An Island GA Model

- A **replacement strategy** to decide which individual of the destination island will be replaced by the migrant.

Based on the selection and replacement strategies, island GAs can be grouped into two classes of algorithms, namely static island GAs and dynamic island GAs. For static island GAs, deterministic selection and replacement strategies are followed, for example [101],

- a good migrant replaces a bad individual,
- a good migrant replaces a randomly selected individual,
- a randomly selected migrant replaces a bad individual, or
- a randomly selected migrant replaces a randomly selected individual.

To select the good migrant, any of the fitness-proportional selection operators given in Section 8.5 can be used. For example, an elitist strategy will have the best individual of a population move to another population. Gordon [329] uses tournament selection, considering only two randomly selected individuals. The best of the two will migrate, while the worst one will be replaced by the winning individual from the neighboring population.

Dynamic models do not use a topology to determine migration paths. Instead, migration decisions are made probabilistically. Migration occurs at a specified probability.

If migration from an island does occur, the destination island is also decided probabilistically. Tournament selection may be used, based on the average fitness of the subpopulations. Additionally, an acceptance strategy can be used to decide if an immigrant should be accepted. For example, an immigrant is probabilistically accepted if its fitness is better than the average fitness of the island (using, e.g. Boltzmann selection).

Another interesting aspect to consider for island GAs is how subpopulations should be initialized. Of course a pure random approach can be used, which will cause different populations to share the same parts of the search space. A better approach would be to initialize subpopulations to cover different parts of the search space, thereby covering a larger search space and facilitating a kind of niching by individuals islands. Also, in multicriteria optimization, each subpopulation can be allocated the task to optimize one criterion. A meta-level step is then required to combine the solutions from each island (refer to Section 9.6.3).

A different kind of "island" GA is the cooperative coevolutionary GA (CCGA) of Potter [686, 687]. In this case, instead of distributing entire individuals over several subpopulations, each subpopulation is given one or a few genes (one decision variable) to optimize. The subpopulations are mutually exclusive, each having the task of evolving a single (or limited set of) gene(s). A subpopulation therefore optimizes one parameter (or a limited number of parameters) of the optimization problem. Thus, no single subpopulation has the necessary information to solve the problem itself. Rather, information of all the subpopulations must be combined to construct a solution.

Within the CCGA, a solution is constructed by adding together the best individual from each subpopulation. The main problem is how to determine the best individual of a subpopulation, since individuals do not represent complete solutions. A simple solution to this problem is to keep all other components (genes) within a complete chromosome fixed and to change just the gene that corresponds to the current subpopulation for which the best individual is sought. For each individual in the subpopulation, the value of the corresponding gene in the complete chromosome is replaced with that of the individual. Values of the other genes of the complete chromosome are usually kept fixed at the previously determined best values.

The constructed complete chromosome is then a candidate solution to the optimization problem.

It has been shown that such a cooperative approach substantially improves the accuracy of solutions, and the convergence speed compared to non-cooperative, noncoevolutionary GAs.

## 9.6 Advanced Topics

This section shows how GAs can be used to find multiple solutions (Section 9.6.1), to solve multi-objective optimization problems (Section 9.6.3), to cope with constraints (Section 9.6.2), and to track dynamically changing optima (Section 9.6.4). For each of these problem types, only a few of the GA approaches are discussed.


## 9.6.1  Niching Genetic Algorithms

Section A.7 defines niching and different classes of niching methods. This section provides a short summary of GA implementations with the ability to locate multiple solutions to optimization problems.


## Fitness Sharing

Fitness sharing is one of the earliest GA niching techniques, originally introduced as a population diversity maintenance technique [325]. It is a parallel, explicit niching approach. The algorithm regards each niche as a finite resource, and shares this resource among all individuals in the niche. Individuals are encouraged to populate a particular area of the search space by adapting their fitness based on the number of other individuals that populate the same area. The fitness $f(\mathbf{x}_i(t))$ of individual $\mathbf{x}_i$ is adapted to its shared fitness:

$$f_s(\mathbf{x}_i(t)) = \frac{f(\mathbf{x}_i(t))}{\sum_j sh(d_{ab})} \qquad (9.29)$$

where $\sum_j sh(d_{ab})$ is an estimate of how crowded a niche is. A common sharing function is the triangular sharing function,

$$sh(d) = \begin{cases} 1 - (d/\sigma_{share})^\alpha & \text{if } d < \sigma_{share} \\ 0 & \text{otherwise.} \end{cases} \qquad (9.30)$$

The symbol $d_{ab}$ represents the distance between individuals $\mathbf{x}_a$ and $\mathbf{x}_b$. The distance measure may be genotypic or phenotypic, depending on the optimization problem. If the sharing function finds that $d_{ab}$ is less than $\sigma_{share}$, it returns a value in the range $[0, 1]$, which increases as $d_{ab}$ decreases. The more similar $\mathbf{x}_a$ and $\mathbf{x}_b$ are, the lower their individual fitnesses will become. Individuals within $\sigma_{share}$ of one another will reduce each other's fitness. Sharing assumes that the number of niches can be estimated, i.e. it must be known prior to the application of the algorithm how many niches there are. It is also assumed that niches occur at least a minimum distance, $2\sigma_{share}$, from each other.


## Dynamic Niche Sharing

Miller and Shaw [593] introduced dynamic niche sharing as an optimized version of fitness sharing. The same assumptions are made as with fitness sharing. Dynamic niche sharing attempts to classify individuals in a population as belonging to one of the emerging niches, or to a non-niche category. Fitness calculation for individuals belonging to the non-niche category is the same as in the standard fitness sharing technique above. The fitness of individuals found to belong to one of the developing niches is diluted by dividing it by the size of the developing niche. Dynamically finding niches is a simple process of iterating through the population of individuals and constructing a set of non-overlapping areas in the search space. Dynamic sharing is computationally less expensive than 'normal' sharing. Miller and Shaw [593] presented results showing that dynamic sharing has improved performance when compared to fitness sharing.

## Sequential Niching

Sequential niching (SN), introduced by Beasley *et al.* [55], identifies multiple solutions by adapting an optimization problem's objective function's fitness landscape through the application of a *derating* function at a position where a potential solution was found. A derating function is designed to lower the fitness appeal of previously located solutions. By repeatedly running the algorithm, all optima are removed from the fitness landscape. Sample derating functions, for a previous maximum $\mathbf{x}^*$, include:

$$G_1(\mathbf{x}, \mathbf{x}^*) = \begin{cases} \left(\frac{\|\mathbf{x}-\mathbf{x}^*\|}{R}\right)^\alpha & \text{if } \|\mathbf{x} - \mathbf{x}^*\| < R \\ 1 & \text{otherwise} \end{cases} \tag{9.31}$$

and

$$G_2(\mathbf{x}, \mathbf{x}^*) = \begin{cases} e^{\log m \frac{R-\|\mathbf{x}-\mathbf{x}^*\|}{R}} & \text{if } \|\mathbf{x} - \mathbf{x}^*\| < R \\ 1 & \text{otherwise} \end{cases} \tag{9.32}$$

where $R$ is the radius of the derating function's effect. In $G_1$, $\alpha$ determines whether the derating function is concave ($\alpha > 1$) or convex ($\alpha < 1$). For $\alpha = 1$, $G_1$ is a linear function. For $G_2$, $m$ determines 'concavity'. Noting that $\lim_{x \to 0} \log(x) = -\infty$, $m$ must always be larger than 0. Smaller values for $m$ result in a more concave derating function. The fitness function $f(\mathbf{x})$ is then redefined to be

$$M_{n+1}(\mathbf{x}) \equiv M_n(\mathbf{x}) \times G(\mathbf{x}, \hat{\mathbf{x}}_n) \tag{9.33}$$

where $M_0(\mathbf{x}) \equiv f(\mathbf{x})$ and $\hat{\mathbf{x}}_n$ is the best individual found during run $n$ of the algorithm. $G$ can be any derating function, such as $G_1$ and $G_2$.

## Crowding

Crowding (or the crowding factor model), as introduced by De Jong [191], was originally devised as a diversity preservation technique. Crowding is inspired by a naturally occurring phenomenon in ecologies, namely competition amongst similar individuals for limited resources. Similar individuals compete to occupy the same ecological niche, while dissimilar individuals do not compete, as they do not occupy the same ecological niche. When a niche has reached its carrying capacity (i.e. being occupied by the maximum number of individuals that can exist within it) older individuals are replaced by newer (younger) individuals. The carrying capacity of the niche does not change, so the population size will remain constant.

For a genetic algorithm, crowding is performed as follows: It is assumed that a population of GA individuals evolve over several generational steps. At each step, the crowding algorithm selects only a portion of the current generation to reproduce. The selection strategy is fitness proportionate, i.e. more fit individuals are more likely to be chosen. After the selected individuals have reproduced, individuals in the current population are replaced by their offspring. For each offspring, a random sample is taken from the current generation, and the most similar individual is replaced by the offspring individual.

Deterministic crowding (DC) is based on De Jong's crowding technique, but with the following improvements as suggested by Mahfoud [553]:

- Phenotypic similarity measures are used instead of genotypic measures. Phenotypic metrics embody domain specific knowledge that is most useful in multimodal optimization, as several different spatial positions can contain equally optimal solutions.

- It was shown that there exists a high probability that the most similar individuals to an offspring are its parents. Therefore, DC compares an offspring only to its parents and not to a random sample of the population.

- Random selection is used to select individuals for reproduction. Offspring replace parents only if the offspring perform better than the parents.

Probabilistic crowding, , introduced by Mengshoel *et al.* [578], is based on Mahfoud's deterministic crowding, but employs a probabilistic replacement strategy. Where the original crowding and DC techniques replaced an individual $\mathbf{x}_a$ with $\mathbf{x}_b$ if $\mathbf{x}_b$ was more fit than $\mathbf{x}_a$, probabilistic crowding uses the following rule: If individuals $\mathbf{x}_a$ and $\mathbf{x}_b$ are competing against each other, the probability of $\mathbf{x}_a$ winning is given by

$$P(\mathbf{x}_a(t) \text{ wins}) = \frac{f(\mathbf{x}_a(t))}{f(\mathbf{x}_a(t)) + f(\mathbf{x}_b(t))} \tag{9.34}$$

where $f(\mathbf{x}_a(t))$ is the fitness of individual $\mathbf{x}_a(t)$. The core of the algorithm is therefore to use a probabilistic tournament replacement strategy. Experimental results have shown it to be both fast and effective.

**Coevolutionary Shared Niching**

Goldberg and Wang [326] introduced coevolutionary shared niching (CSN). CSN locates niches by co-evolving two different populations of individuals in the same search space, in parallel. Let the two parallel populations be designated by $\mathcal{C}_1$ and $\mathcal{C}_2$, respectively. Population $\mathcal{C}_1$ can be thought of as a normal population of candidate solutions, and it evolves as a normal population of individuals. Individuals in population $\mathcal{C}_2$ are scattered throughout the search space. Each individual in population $\mathcal{C}_1$ associates with itself a member of $\mathcal{C}_2$ that lies the closest to it using a genotypic metric. The fitness calculation of the $i^{th}$ individual in population $\mathcal{C}_1$, $\mathcal{C}_1.\mathbf{x}_i$, is then adapted to $f'(\mathcal{C}_1.\mathbf{x}_i) = \frac{f(\mathcal{C}_1.\mathbf{x}_i)}{\mathcal{C}_2.n_2}$, where $f(\cdot)$ is the fitness function; $\mathcal{C}_2.n_2$ designates the cardinality of the set of individuals associated with individual $\mathcal{C}_2.\mathbf{x}_i$ and $\mathcal{C}_2.n_2$ is the index of the closest individual in population $\mathcal{C}_2$ to individual $\mathcal{C}_1.\mathbf{x}_i$ in population $\mathcal{C}_1$. The fitness of individuals in population $\mathcal{C}_2$ is simply the average fitness of all the individuals associated to it in population $\mathcal{C}_1$, multiplied by $\mathcal{C}_2.\mathbf{x}_i$. Goldberg and Wang also developed the *imprint* CSN technique, that allows for the transfer of good performing individuals from the $\mathcal{C}_1$ to the $\mathcal{C}_2$ population.

CSN overcomes the limitation imposed by fixed inter-niche distances assumed in the original fitness sharing algorithm [325] and its derivate, dynamic fitness sharing [593]. The concept of a niche radius is replaced by the association made between individuals from the different populations.

### Dynamic Niche Clustering

Dynamic niche clustering (DNC) is a fitness sharing based, cluster driven niching technique [305, 306]. It is distinguished from all other niching techniques by the fact that it supports 'fuzzy' clusters, i.e. clusters may overlap. This property allows the algorithm to distinguish between different peaks in a multi-modal function that may lie extremely close together. In most other niching techniques, a more general inter-niche radius (such as the $\sigma_{share}$ parameter in fitness sharing) would prohibit this.

The algorithm constructs a *nicheset*, which is a list of niches in a population. The nicheset persists over multiple generations. Initially, each individual in a population is regarded to be in its own niche. Similar niches are identified using Euclidean distance and merged. The population of individuals is then evolved over a pre-determined number of generational steps. Before selection takes place, the following process occurs:

- The midpoint of each niche in the nicheset is updated, using the formula

$$\overline{\mathbf{x}}_u = \overline{\mathbf{x}}_u + \frac{\sum_{i=1}^{n_u} (\mathbf{x}_i - \overline{\mathbf{x}}_u) \cdot f(\mathbf{x}_i)}{\sum_{i=1}^{n_u} f(\mathbf{x}_i)} \tag{9.35}$$

  where $\overline{\mathbf{x}}_u$ is the midpoint of niche $u$, initially set to be equal to the position of the individual from which it was constructed, as described above. $n_u$ is the niche count, or the number of individuals in the niche, $f(\mathbf{x}_i)$ is the fitness of individual $\mathbf{x}_i$ in niche $u$.

- A list of inter-niche distances is calculated and sorted. Niches are then merged.

- Similar niches are merged. Each niche is associated with a minimum and maximum niche radius. If the midpoints of two niches lie within the minimum radii of each other, they are merged.

- If any niche has a population size greater than 10% of the total population, random checks are done on the niche population to ensure that all individuals are focusing on the same optima. If this is not the case, such a niche may be split into sub-niches, which will be optimized individually in further generational steps.

Using the above technique, Gan and Warwick [307] also suggested a *niche linkage* extension to model niches of arbitrary shape.

## 9.6.2   Constraint Handling

Section A.6 summarizes different classes of methods to solve constrained optimization problems, as defined in Definition A.5. Standard genetic algorithms cannot be applied as is to solve constrained optimization problems. Most GA approaches to solve constrained problems require a change in the fitness function, or in the behavior of the algorithm itself.

Penalty methods are possibly one of the first approaches to address constraints [726]. As shown in Definition A.7, unfeasible solutions are penalized by adding a penalty function. A popular approach to implement penalties is given in equations (A.25) and (A.26) [584]. This approach basically converts the constrained problem to a penalized unconstrained problem.

Homaifar *et al.* [380] proposed a multi-level penalty function, where the magnitude of a penalty is proportional to the severity of the constraint violation. The multi-level function assumes that a set of intervals (or penalty levels) are defined for each constraint. An appropriate penalty value, $\lambda_{mq}$, is assigned to each level, $q = 1, \ldots, n_q$ for each constraint, $m$. The penalty function then changes to

$$p(\mathbf{x}_i, t) = \sum_{m=1}^{n_g+n_h} \lambda_{mq}(t) p_m(\mathbf{x}_i) \tag{9.36}$$

As an example, the following penalties can be used:

$$\lambda_{mq}(t) = \begin{cases} 10\sqrt{t} & \text{if } p_m(\mathbf{x}_i) < 0.001 \\ 20\sqrt{t} & \text{if } p_m(\mathbf{x}_i) \le 0.1 \\ 100\sqrt{t} & \text{if } p_m(\mathbf{x}_i) \le 1.0 \\ 300\sqrt{t} & \text{otherwise} \end{cases} \tag{9.37}$$

The multi-level function approach has the weakness that the number of parameters that has to be maintained increases significantly with increase in the number of levels, $n_q$, and the number of constraints, $n_g + n_h$.

Joines and Houck [425] proposed dynamic penalties, where

$$p(\mathbf{x}_i, t) = (\gamma \times t)^\alpha \sum_{m=1}^{n_g+n_h} p_m^\beta(\mathbf{x}_i) \tag{9.38}$$

where $\gamma$, $\alpha$ and $\beta$ are constants. The longer the search continues, the higher the penalty for constraint violations. This allows for better exploration.

Other penalty methods can be found in [587, 588, 691].

Often referred to as the "death penalty" method, unfeasible solutions can be rejected. However, Michalewicz [585] shows that the method performs badly when the feasible region is small compared to the entire search space.

The interested reader is referred to [584] for a more complete survey of constraint handling methods.

## 9.6.3   Multi-Objective Optimization

Extensive research has been done to solve multi-objective optimization problems
(MOP) as defined in Definition A.10 [149, 195]. This section summarizes only a few
of these GA approaches to multi-objective optimization (MOO).

GA approaches for solving MOPs can be grouped into three main categories [421]:

- **Weighted aggregation** approaches where the objective is defined as a weighted
  sum of sub-objectives.

- **Population-based non-Pareto** approaches, which do not make use of the
  dominance relation as defined in Section A.8.

- **Pareto-based** approaches, which apply the dominance relation to find an approximation of the Pareto front.

Examples from the first and last classes are considered below.

## Weighted Aggregation

One of the simplest approaches to deal with MOPs is to define an aggregate objective
function as a weighted sum of sub-objectives:

$$f(\mathbf{x}) = \sum_{k=1}^{n_k} \omega_k f_k(\mathbf{x}) \tag{9.39}$$

where $n_k \geq 2$ is the total number of sub-objectives, and $\omega_k \in [0,1], k = 1, \ldots, n_k$ with
$\sum_{k=1}^{n_k} \omega_k = 1$. While the aggregation approach above is very simple to implement and
computationally efficient, it suffers from the following problems:

- It is difficult to get the best values for the weights, $\omega_k$, since these are problem
  dependent.

- These methods have to be re-applied to find more than one solution, since only
  one solution can be obtained with a single run of an aggregation algorithm.
  However, even for repeated applications, there is no guarantee that different
  solutions will be found.

- The conventional weighted aggregation as given above cannot solve MOPs with
  a concave Pareto front [174].

To address these problems, Jin *et al.* [421, 422], proposed aggregation methods with
dynamically changing weights (for $n_k = 2$) and an approach to maintain an archive
of nondominated solutions. The following approaches have been used to dynamically
adapt weights:

- **Random distribution of weights**, where for each individual,

$$\omega_{1,i}(t) = U(0, n_s)/n_s \tag{9.40}$$
$$\omega_{2,i}(t) = 1 - \omega_{1,i}(t) \tag{9.41}$$

- **Bang-bang weighted aggregation**, where

$$\omega_1(t) = \text{sign}(\sin(2\pi t/\tau)) \qquad (9.42)$$
$$\omega_2(t) = 1 - \omega_1(t) \qquad (9.43)$$

where $\tau$ is the weights' change frequency. Weights change abruptly from 0 to 1 each $\tau$ generation.

- **Dynamic weighted aggregation**, where

$$\omega_1(t) = |\sin(2\pi t/\tau)| \qquad (9.44)$$
$$\omega_2(t) = 1 - \omega_1(t) \qquad (9.45)$$

With this approach, weights change more gradually.

Jin *et al.* [421, 422] used Algorithm 9.10 to produce an archive of nondominated solutions. This algorithm is called after the reproduction (crossover and mutation) step.

---

**Algorithm 9.10** Algorithm to Maintain an Archive of Nondominated Solutions

---

**for** *each offspring,* $\mathbf{x}'_i(t)$ **do**
  **if** $\mathbf{x}'_i(t)$ *dominates an individual in the current population,* $\mathcal{C}(t)$, *and* $\mathbf{x}'_i(t)$ *is not dominated by any solutions in the archive and* $\mathbf{x}'_i(t)$ *is not similar to any solutions in the archive* **then**
    **if** *archive is not full* **then**
      Add $\mathbf{x}'_i(t)$ to the archive;

    **else if** $\mathbf{x}'_i(t)$ *dominates any solution* $\mathbf{x}_a$ *in the archive* **then**
      Replace $\mathbf{x}_a$ with $\mathbf{x}'_i(t)$;

    **else if** *any* $\mathbf{x}_{a_1}$ *in the archive dominates another* $\mathbf{x}_{a_2}$ *in the archive* **then**
      Replace $\mathbf{x}_{a_2}$ with $\mathbf{x}'_i(t)$;

    **else**
      Discard $\mathbf{x}'_i(t)$;
    **end**
  **end**
  **else**
    Discard $\mathbf{x}'_i(t)$;
  **end**
  **for** *each solution* $\mathbf{x}_{a_1}$ *in the archive* **do**
    **if** $\mathbf{x}_{a_1}$ *dominates* $\mathbf{x}_{a_2}$ *in the archive* **then**
      Remove $\mathbf{x}_{a_2}$ from the archive;
    **end**
  **end**
**end**

---

## Vector Evaluated Genetic Algorithm

The vector evaluated GA (VEGA) [760, 761] is one of the first algorithms to solve MOPs using multiple populations. One subpopulation is associated with each objective. Selection is applied to each subpopulation to construct a mating pool. The result of this selection process is that the best individuals with respect to each objective are included in the mating pool. Crossover then continues by selecting parents from the mating pool.

## Niched Pareto Genetic Algorithm

Horn *et al.* [382] developed the niched Pareto GA (NPGA), where an adapted tournament selection operator is used to find nondominated solutions. The Pareto domination tournament selection operator randomly selects two candidate individuals, and a comparison set of randomly selected individuals. Each candidate is compared against each individual in the comparison set. If one candidate is dominated by an individual in the comparison set, and the other candidate is not dominated, then the latter is selected. If neither or both are dominated equivalence class sharing is used to select one individual: The individual with the lowest niche count is selected, where the niche count is the number of individuals within a niche radius, $\sigma_{share}$, from the candidate. This strategy will prefer a solution on a less populated part of the Pareto front.

## Nondominated Sorting Genetic Algorithm

Srinivas and Deb [807] developed the nondominated sorting GA (NSGA), where only the selection operator is changed. Individuals are Pareto-ranked into different Pareto fronts as described in Section 12.6.2. Fitness proportionate selection is used based on the shared fitness assigned to each solution. The NSGA is summarized in Algorithm 9.11.

Deb *et al.* [197] pointed out that the NSGA has a very high computational complexity of $O(n_k n_s^3)$. Another issue with the NSGA is the reliance on a sharing parameter, $\sigma_{share}$. To address these problems, a fast nondominated sorting strategy was proposed and a crowding comparison operator defined. The fast nondominated sorting algorithm calculates for each solution, $\mathbf{x}_a$, the number of solutions, $n_a$, which dominates $\mathbf{x}_a$, and the set, $\mathcal{X}_a$, of solutions dominated by $\mathbf{x}_a$. All those solutions with $n_a = 0$ are added to a list, referred to as the current front. For each solution in the current front, each element, $\mathbf{x}_b$, of the set $\mathcal{X}_b$ has its counter, $n_b$, decremented. When $n_b = 0$, the corresponding solution is added to a temporary list. When all the elements of the current front have been processed, its elements form the first front, and the temporary list becomes the new current list. The process is repeated to form the other fronts.

To eliminate the need for a sharing parameter, solutions are sorted for each subobjective. For each subobjective, the average distance of the two points on either side of $\mathbf{x}_a$ is calculated. The sum of the average distances over all subobjectives gives the

---

**Algorithm 9.11** Nondominated Sorting Genetic Algorithm

---

Set the generation counter, $t = 0$;
Initialize all control parameters;
Create and initialize the population, $\mathcal{C}(0)$, of $n_s$ individuals;
**while** *stopping condition(s) not true* **do**
    Set the front counter, $p = 1$;
    **while** *there are individuals not assigned to a front* **do**
        Identify nondominated individuals;
        Assign fitness to each of these individuals;
        Apply fitness sharing to individuals in the front;
        Remove individuals;
        $p = p + 1$;
    **end**
    Apply reproduction using rank-based, shared fitness values;
    Select new population;
    $t = t + 1$;
**end**

---

crowding distance, $d_a$. If $R_a$ indicates the nondomination rank of $\mathbf{x}_a$, then a crowding comparison operator is defined as: $a \leq_* b$ if $(R_a < R_b)$ or $((R_a = R_b)$ and $(d_a > d_b))$. For two solutions with differing nondomination ranks, the one with the lower rank is preferred. If both solutions have the same rank, the one located in the less populated region of the Pareto front is preferred.

## 9.6.4 Dynamic Environments

A very simple approach to track solutions in a dynamic environment as defined in Definition A.16 is to restart the GA when a change is detected. A restart approach can be quite inefficient if changes in the landscape are small. For small changes, the question arises if a changing optimum can be tracked by simply continuing with the search. This will be possible only if the population has some degree of diversity to enable further exploration. An ability to maintain diversity is therefore an important ingredient in tracking changing optima. This is even more so for large changes in the landscape, which may cause new optima to appear and existing ones to disappear.

A number of approaches have been developed to maintain population diversity. The hyper-mutation strategy of Cobb [141] drastically increases the rate of mutation for a number of generations when a change is detected. An increased rate of mutation as well as an increased mutational step size allow for further exploration. The variable local search strategy [873] gradually increases the mutational step sizes and rate of mutation after a change is detected. Mutational step sizes are increased if no improvement is obtained over a number of generations for the smaller step sizes.

Grefenstette [336] proposed the random immigrants strategy where, for each generation, part of the population is replaced by randomly generated individuals.

Mori *et al.* [607] proposed that a memory of every generation's best individual be stored (in a kind of hall-of-fame). Individuals stored in the memory serve as candidate parents to the reproduction operators. The memory has a fixed size, which requires some kind of replacement strategy. Branke [82] proposed that the best individual of the current generation replaces the individual that is most similar.

For a more detailed treatment of dynamic environments, the reader is referred to [83].

## 9.7 Applications

Genetic algorithms have been applied to solve many real-world optimization problems. The reader is referred to http://www.doc.ic.ac.uk/~nd/surprise_96/journal/vol4/tcw2/report.html for a good summary of and references to applications of GAs. The rest of this section describes how a GA can be applied to routing optimization in telecommunications networks [778]. Given a network of $n_x$ switches, an origin switch and a destination switch, the objective is to find the best route to connect a call between the origin and destination switches. The design of the GA is done in the following steps:

1. **Chromosome representation**: A chromosome consists of a maximum of $n_x$ switches. Chromosomes can be of variable length, since telecommunication routes can differ in length. Each gene represents one switch. Integer values representing switch numbers are used as gene values - no binary encoding is used. The first gene represents the origin switch and the last gene represents the destination switch. Example chromosomes are

$$(1 \quad 3 \quad 6 \quad 10)$$

$$(1 \quad 5 \quad 2 \quad 5 \quad 10) = (1 \quad 5 \quad 2 \quad 10)$$

   Duplicate switches are ignored. The first chromosome represents a route from switch 1 to switch 3 to switch 6 to switch 10.

2. **Initialization of population**: Individuals are generated randomly, with the restriction that the first gene represents the origin switch and the last gene represents the destination switch. For each gene, the value of that gene is selected as a uniform random value in the range $[1, n_x]$.

3. **Fitness function:** The multi-criteria objective function

$$f(\mathbf{x}_i) = \omega_1 f_{Switch}(\mathbf{x}_i) + \omega_2 f_{Block}(\mathbf{x}_i) + \omega_3 f_{Util}(\mathbf{x}_i) + \omega_4 f_{Cost}(\mathbf{x}_i) \qquad (9.46)$$

   is used where

$$f_{Switch}(\mathbf{x}_i) = \frac{|\mathbf{x}_i|}{n_x} \qquad (9.47)$$

   represents the minimization of route length, where $\mathbf{x}_i$ denotes the route and $|\mathbf{x}_i|$ is the total number of switches in the route,

$$f_{Block}(\mathbf{x}_i) = 1 - \prod_{ab \in \mathbf{x}_i}^{|\mathbf{x}_i|} (1 - B_{ab} + \alpha_{ab}) \qquad (9.48)$$

with

$$\alpha_{ab} = \begin{cases} 1 & \text{if } ab \text{ does not exist} \\ 0 & \text{if } ab \text{ does exist} \end{cases} \tag{9.49}$$

represent the objective to select routes with minimum congestion, where $B_{ab}$ denotes the blocking probability on the link between switches $a$ and $b$,

$$f_{Util}(\mathbf{x}_i) = \min_{ab \in \mathbf{x}_i} \{1 - U_{ab}\} + \alpha_{ab} \tag{9.50}$$

maximizes utilization, where $U_{ab}$ quantifies the level of utilization of the link between $a$ and $b$, and

$$f_{Cost}(\mathbf{x}_i) = \sum_{ab \in \mathbf{x}_i}^{|\mathbf{x}_i|} C_{ab} + \alpha_{ab} \tag{9.51}$$

ensures that minimum cost routes are selected, where $C_{ab}$ represents the financial cost of carrying a call on the link between $a$ and $b$. The constants $\omega_1$ to $\omega_4$ control the influence of each criterion.

4. Use any **selection** operator.

5. Use any **crossover** operator.

6. **Mutation**: Mutation consists of replacing selected genes with a uniformly random selected switch in the range $[1, n_x]$.

This example is an illustration of a GA that uses a numeric representation, and variable length chromosomes with constraints placed on the structure of the initial individuals.


# 9.8   Assignments

1. Discuss the importance of the crossover rate, by considering the effect of different values in the range $[0,1]$.

2. Compare the following replacement strategies for crossover operators that produce only one offspring:

   (a) The offspring always replaces the worst parent.
   (b) The offspring replaces the worst parent only when its fitness is better than the worst parent.
   (c) The offspring always replaces the worst individual in the population.
   (d) Boltzmann selection is used to decide if the offspring should replace the worst parent.

3. Show how the heuristic crossover operator incorporates search direction.

4. Propose a multiparent version of the geometrical crossover operator.

5. Propose a marker initialization and update strategy for gene scanning applied to order-based representations

6. Propose a random mutation operator for discrete-valued decision variables.

7. Show how a GA can be used to train a FFNN.

8. In the context of GAs, when is a high mutation rate an advantage?

9. Is the following strategy sensible? Explain your answer. *"Start evolution with a large mutation rate, and decrease the mutation rate with an increase in generation number."*

10. Discuss how a GA can be used to cluster data.

11. For floating-point representations, devise a deterministic schedule to dynamically adjust mutational step sizes. Discuss the merits of your proposal.

12. Suggest ways in which the competitive template can be initialized for messy GAs.

13. Discuss the consequences of migrating the best individuals before islands have converged.

14. Discuss the influence that the size of the comparison set has on the performance of the niched Pareto GA.

# Chapter 10

# Genetic Programming

Genetic programming (GP) is viewed by many researchers as a specialization of genetic algorithms. Similar to GAs, GP concentrates on the evolution of genotypes. The main difference between the two paradigms is in the representation scheme used. Where GAs use string (or vector) representations, GP uses a tree representation. Originally, GP was developed by Koza [478, 479] to evolve computer programs. For each generation, each evolved program (individual) is executed to measure its performance within the problem domain. The result obtained from the evolved computer program is then used to quantify the fitness of that program.

This chapter provides a very compact overview of basic GP implementations to solve specific problems. More detail about GP can be found in the books by Koza [482, 483]. The chapter is organized as follows: The tree-based representation scheme is discussed in Section 10.1. Section 10.2 discusses initialization of the GP population, and the fitness function is covered in Section 10.3. Crossover and mutation operators are described in Sections 10.4 and 10.5. A building-block approach to GP is reviewed in Section 10.6. A summary of GP applications is given in Section 10.7.

## 10.1  Tree-Based Representation

GP was developed to evolve executable computer programs [478, 479]. Each individual, or chromosome, represents one computer program, represented using a tree structure. Tree-based representations have a number of implications that the reader should be aware of:

- **Adaptive individuals**: Contrary to GAs where the size of individuals are usually fixed, a GP population will usually have individuals of different size, shape and complexity. Here size refers to the tree depth, and shape refers to the branching factor of nodes in the tree. The size and shape of a specific individual are also not fixed, but may change due to application of the reproduction operators.

- **Domain-specific grammar**: A grammar needs to be defined that accurately reflects the problem to be solved. It should be possible to represent any possible solution using the defined grammar.

*Computational Intelligence: An Introduction*, Second Edition A.P. Engelbrecht
©2007 John Wiley & Sons, Ltd

**Table 10.1** XOR Truth Table

| $x_1$ | $x_2$ | **Target Output** |
|---|---|---|
| 0 | 0 | 0 |
| 0 | 1 | 1 |
| 1 | 0 | 1 |
| 1 | 1 | 0 |

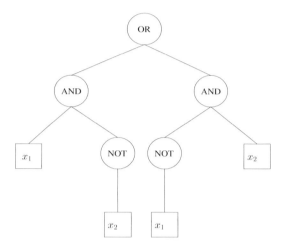

**Figure 10.1** Tree-Representation of XOR

As mentioned above, a grammar forms an important part of chromosome representation. As part of the grammar, a terminal set, function set, and semantic rules need to be defined. The terminal set specifies all the variables and constants, while the function set contains all the functions that can be applied to the elements of the terminal set. These functions may include mathematical, arithmetic and/or Boolean functions. Decision structures such as *if-then-else* and loops can also be included in the function set. Using tree terminology, elements of the terminal set form the leaf nodes of the evolved tree, and elements of the function set form the non-leaf nodes. For a specific problem, the search space consists of the set of all possible trees that can be constructed using the defined grammar.

Two examples are given next to illustrate GP representations. One of the first applications of GP was to evolve Boolean expressions. Consider the expression,

$$(x_1 \text{ AND NOT } x_2) \text{ OR } (\text{NOT } x_1 \text{ AND } x_2)$$

and given a data set of interpretations and their associated target outputs (as given in Table 10.1), the task is to evolve this expression. The solution is represented in Figure 10.1. For this problem, the function set is defined as $\{AND, OR, NOT\}$, and the terminal set is $\{x_1, x_2\}$ where $x_1, x_2 \in \{0, 1\}$.

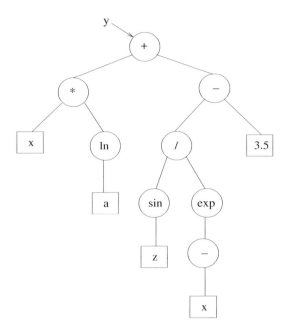

**Figure 10.2** Tree-Representation for Mathematical Expressions

The next example considers the problem of evolving a mathematical expression. Consider the task of evolving the program,

```
y:=x*ln(a)+sin(z)/exp(-x)-3.4;
```

The terminal set is specified as $\{a, x, z, 3.4\}$ with $a, x, z \in \mathbb{R}$. The minimal function set is given as $\{-, +, *, /, \sin, \exp, \ln\}$. The global optimum is illustrated in Figure 10.2.

In addition to the terminal and function sets, rules can be specified to ensure the construction of semantically correct trees. For example, the logarithmic function, ln, can take only positive values. Similarly, the second parameter of the division operator can not be zero.

## 10.2   Initial Population

The initial population  is generated randomly within the restrictions of a maximum depth and semantics as expressed by the given grammar. For each individual, a root is randomly selected from the set of function elements. The branching factor (the number of children) of the root, and each non-terminal node, are determined by the arity of the selected function. For each non-root node, the initialization algorithm randomly selects an element either from the terminal set or the function set. As soon as an element from the terminal set is selected, the corresponding node becomes a leaf node and is no longer considered for expansion.

Instead of initializing individuals as large trees, individuals can be initialized to be as simple as possible. During the evolutionary process these individuals will grow if increased complexity is necessary (refer to Section 10.6). This facilitates creation of simple solutions.

## 10.3  Fitness Function

The fitness function used for GP is problem-dependent. Because individuals usually represent a program, calculation of fitness requires the program to be evaluated against a number of test cases. Its performance on the test cases is then used to quantify the individual's fitness. For example, refer to the problems considered in Section 10.1. For the Boolean expression, fitness is calculated as the number of correctly predicted target outputs. For the mathematical expression a data set of sample input patterns and associated target output is needed. Each pattern contains a value for each of the variables $(a, x$ and $z)$ and the corresponding value of $y$. For each pattern the output of the expression represented by the individual is determined by executing the program. The output is compared with the target output to compute the error for that pattern. The MSE over the errors for all the patterns gives the fitness of the individual.

As will be shown in Section 10.6, GP can also be used to evolve decision trees. For this application each individual represents a decision tree. The fitness of individuals is calculated as the classification accuracy of the corresponding decision tree. If the objective is to evolve a game strategy in terms of a computer program [479, 481], the fitness of an individual can be the number of times that the individual won the game out of a total number of games played.

In addition to being used as a measure of the performance of individuals, the fitness function can also be used to penalize individuals with undesirable structural properties. For example, instead of having a predetermined depth limit, the depth of a tree can be penalized by adding an appropriate penalty term to the fitness function. Similarly, bushy trees (which result when nodes have a large branching factor) can be penalized by adding a penalty term to the fitness function. The fitness function can also be used to penalize semantically incorrect individuals.

## 10.4  Crossover Operators

Any of the previously discussed selection operators (refer to Section 8.5) can be used to select two parents to produce offspring. Two approaches can be used to generate offspring, each one differing in the number of offspring generated:

- **Generating one offspring**: A random node is selected within each of the parents. Crossover then proceeds by replacing the corresponding subtree in the one parent by that of the other parent. Figure 10.3(a) illustrates this operator.

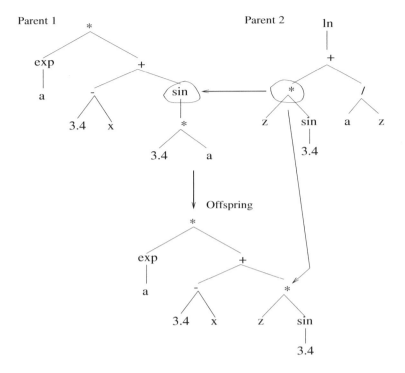

(a) Creation of one offspring

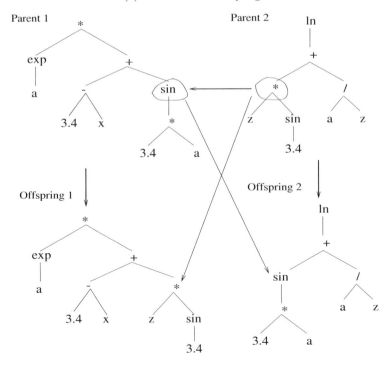

(b) Creation of two offspring

**Figure 10.3** Genetic Programming Crossover

- **Generating two offspring**: Again, a random node is selected in each of the two parents. In this case the corresponding subtrees are swapped to create the two offspring as illustrated in Figure 10.3(b).

## 10.5  Mutation Operators

Mutation operators are usually developed to suit the specific application. However, many of the mutation operators developed for GP are applicable to general GP representations. With reference to Figure 10.4(a), the following mutation operators can be applied:

- **Function node mutation**: A non-terminal node, or function node, is randomly selected and replaced with a node of the same arity, randomly selected from the function set. Figure 10.4(b) illustrates that function node '+' is replaced with function node '−'.

- **Terminal node mutation**: A leaf node, or terminal node, is randomly selected and replaced with a new terminal node, also randomly selected from the terminal set. Figure 10.4(c) illustrates that terminal node $a$ has been replaced with terminal node $z$.

- **Swap mutation**: A function node is randomly selected and the arguments of that node are swapped as illustrated in Figure 10.4(d).

- **Grow mutation**: With grow mutation a node is randomly selected and replaced by a randomly generated subtree. The new subtree is restricted by a predetermined depth. Figure 10.4(e) illustrates that the node 3.4 is replaced with a subtree.

- **Gaussian mutation**: A terminal node that represents a constant is randomly selected and mutated by adding a Gaussian random value to that constant. Figure 10.4(f) illustrates Gaussian mutation.

- **Trunc mutation**: A function node is randomly selected and replaced by a random terminal node. This mutation operator performs a pruning of the tree. Figure 10.4(g) illustrates that the + function node is replaced by the terminal node $a$.

Individuals to be mutated are selected according to a mutation probability $p_m$. In addition to a mutation probability, nodes within the selected tree are mutated according to a probability $p_n$. The larger the probability $p_n$, the more the genetic build-up of that individual is changed. On the other hand, the larger the mutation probability $p_m$, the more individuals will be mutated.

All of the mutation operators can be implemented, or just a subset thereof. If more than one mutation operator is implemented, then either one operator is selected randomly, or more than one operator is selected and applied in sequence.

In addition to the mutation operators above, Koza [479] proposed the following asexual operators:

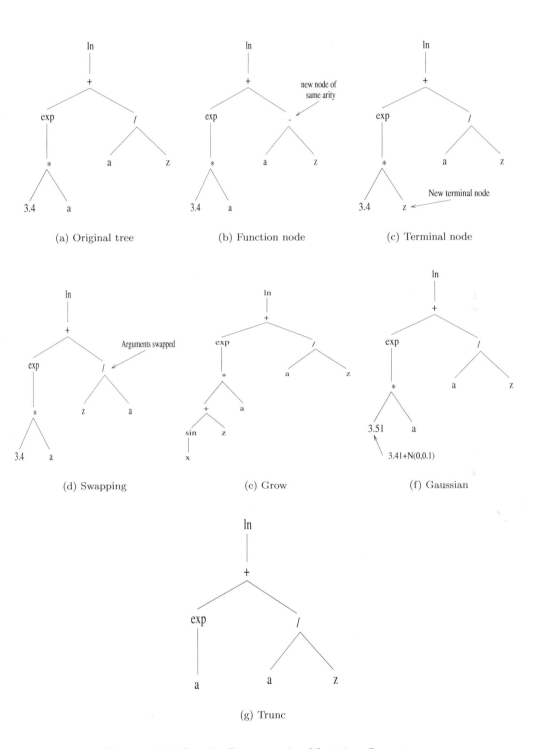

**Figure 10.4** Genetic Programming Mutation Operators

- **Permutation operator**: This operator is similar to the swap mutation. If a function has $n$ parameters, the permutation operator generates a random permutation from the possible $n!$ permutations of parameters. The arguments of the function are then permutated according to this randomly generated permutation.

- **Editing operator**: This operator is used to restructure individuals according to predefined rules. For example, a subtree that represents the Boolean expression, $x$ AND $x$ is replaced with the single node, $x$. Editing rules can also be used to enforce semantic rules.

- **Building block operator**: The objective of the building block operator is to automatically identify potentially useful building blocks. A new function node is defined for an identified building block and is used to replace the subtree represented by the building block. The advantage of this operator is that good building blocks will not be altered by reproduction operators.

## 10.6   Building Block Genetic Programming

The GP process discussed thus far generates an initial population of individuals where each individual represents a tree consisting of several nodes and levels. An alternative approach has been developed in [248, 742] – specifically for evolving decision trees – referred to as a building-block approach to GP (BGP). In this approach, initial individuals consist of only a root and the immediate children of that node. Evolution starts on these "small" initial trees. When the simplicity of the population's individuals can no longer account for the complexity of the problem to be solved, and no improvement in the fitness of any of the individuals within the population is observed, individuals are expanded. Expansion occurs by adding a randomly generated building block (i.e. a new node) to individuals. In other words, grow mutation is applied. This expansion occurs at a specified expansion probability, $p_e$, and therefore not all of the individuals are expanded. Described more formally, the building-block approach starts with models with a few degrees of freedom – most likely too few to solve the problem to the desired degree of accuracy. During the evolution process, more degrees of freedom are added when no further improvements are observed. In between the triggering of expansion, crossover and mutation occur as for normal GP.

This approach to GP helps to reduce the computational complexity of the evolution process, and helps to produce smaller individuals.

## 10.7   Applications

GP was developed to evolve computer programs [478, 479]. Programs have been evolved for a wide range of problem types as illustrated in [479]. These problem types

**Table 10.2** Genetic Programming Applications

| Application | References |
|---|---|
| Decision trees | [248, 479, 685, 742] |
| Game-playing | [479, 481] |
| Bioinformatics | [484, 485] |
| Data mining | [648, 741, 917] |
| Robotics | [206, 486, 347] |

include Boolean expressions, planning, symbolic function identification, empirical discovery, solving systems of equations, concept formation, automatic programming, pattern recognition, game-playing strategies, and neural network design. Table 10.2 provides a summary of other applications of GP.

A very complete list of GP publications and applications can be found at http://www.cs.bham.ac.uk/~ubl/biblio/gp-html/

## 10.8 Assignments

1. Explain how a GP can be used to evolve a program to control a robot, where the objective of the robot is to move out of a room (through the door) filled with obstacles.

2. First explain what a decision tree is, and then show how GP can be used to evolve decision trees.

3. Is it possible to use GP for adaptive story telling?

4. Given a pre-condition and a post-condition of a function, is it possible to evolve the function using GP?

5. Explain why BGP is computationally less expensive than GP.

6. Show how a GP can be used to evolve polynomial expressions.

7. Discuss how GP can be used to evolve the evaluation function used to evaluate the desirability of leaf nodes in a game tree.

# Chapter 11

# Evolutionary Programming

Evolutionary programming (EP) originated from the research of L.J. Fogel in 1962 [275] on using simulated evolution to develop artificial intelligence. While EP shares the objective of imitating natural evolutionary processes with GAs and GP, it differs substantially in that EP emphasizes the development of behavioral models and not genetic models: EP is derived from the simulation of adaptive behavior in evolution. That is, EP considers phenotypic evolution. EP iteratively applies two evolutionary operators, namely variation through application of mutation operators, and selection. Recombination operators are not used within EP.

This chapter provides an overview of EP, organized as follows: The basic EP is described in Section 11.1. Different mutation and selection operators are discussed in Section 11.2. Self-adaptation and strategy parameters are discussed in Section 11.3. Variations of EP that combine aspects from other optimization paradigms are reviewed in Section 11.4. A compact treatment of a few advanced topics is given in Section 11.5, including constraint handling, multi-objective optimization, niching, and dynamic environments.

## 11.1  Basic Evolutionary Programming

Evolutionary programming (EP) was conceived by Laurence Fogel in the early 1960s [275, 276] as an alternative approach to artificial intelligence (AI), which, at that time, concentrated on models of human intelligence. From his observation that intelligence can be viewed as "that property which allows a system to adapt its behavior to meet desired goals in a range of environments" [267], a model has been developed that imitates evolution of behavioral traits. The evolutionary process, first developed to evolve finite state machines (FSM), consists of finding a set of optimal behaviors from a space of observable behaviors. Therefore, in contrast to other EAs, the fitness function measures the "behavioral error" of an individual with respect to the environment of that individual.

As an approach to evolve FSMs, individuals in an EP population use a representation of ordered sequences, which differs significantly from the bitstring representation proposed by Holland for GAs (refer to Chapter 9). EP is, however, not limited to an ordered sequence representation. David Fogel *et al.* [271, 272] extended EP for

*Computational Intelligence: An Introduction*, Second Edition A.P. Engelbrecht
©2007 John Wiley & Sons, Ltd

real-valued vector representations, with application to the optimization of continuous functions.

As summarized in Algorithm 11.1, EP utilizes four main components of EAs:

- **Initialization:** As with other EC paradigms, a population of individuals is initialized to uniformly cover the domain of the optimization problem.

- **Mutation:** The mutation operator's main objective is to introduce variation in the population, i.e. to produce new candidate solutions. Each parent produces one or more offspring through application of the mutation operator. A number of EP mutation operators have been developed, as discussed in Section 11.2.1.

- **Evaluation:** A fitness function is used to quantify the "behavioral error" of individuals. While the fitness function provides an absolute fitness measure to indicate how well the individual solves the problem being optimized, survival in EP is usually based on a relative fitness measure (refer to Chapter 15). A score is computed to quantify how well an individual compares with a randomly selected group of competing individuals. Individuals that survive to the next generation are selected based on this relative fitness. The search process in EP is therefore driven by a relative fitness measure, and not an absolute fitness measure as is the case with most EAs.

- **Selection:** The main purpose of the selection operator is to select those individuals that survive to the next generation. Selection is a competitive process where parents and their offspring compete to survive, based on their performance against a group of competitors. Different selection strategies are discussed in Section 11.2.2.

Mutation and selection operators are applied iteratively until a stopping condition is satisfied. Any of the stopping conditions given in Section 8.7 can be used.

---

**Algorithm 11.1** Basic Evolutionary Programming Algorithm

---

Set the generation counter, $t = 0$;
Initialize the strategy parameters;
Create and initialize the population, $\mathcal{C}(0)$, of $n_s$ individuals;
**for** *each individual,* $\mathbf{x}_i(t) \in \mathcal{C}(t)$ **do**
    Evaluate the fitness, $f(\mathbf{x}_i(t))$;
**end**
**while** *stopping condition(s) not true* **do**
    **for** *each individual,* $\mathbf{x}_i(t) \in \mathcal{C}(t)$ **do**
        Create an offspring, $\mathbf{x}'_i(t)$, by applying the mutation operator;
        Evaluate the fitness, $f(\mathbf{x}'_i(t))$;
        Add $\mathbf{x}'_i(t)$ to the set of offspring, $\mathcal{C}'(t)$;
    **end**
    Select the new population, $\mathcal{C}(t+1)$, from $\mathcal{C}(t) \cup \mathcal{C}'(t)$, by applying a selection operator;
    $t = t + 1$;
**end**

---

In comparison with GAs and GP, there are a few differences between these EAs and EP, some of which have already been discussed:

- EP emphasizes phenotypic evolution, instead of genotypic evolution. The focus is on behaviors.

- Due to the above, EP does not make use of any recombination operator. There is no exchange of genetic material.

- EP uses a relative fitness function to quantify performance with respect to a randomly chosen group of individuals.

- Selection is based on competition. Those individuals that perform best against a group of competitors have a higher probability of being included in the next generation.

- Parents and offspring compete for survival.

- The behavior of individuals is influenced by strategy parameters, which determine the amount of variation between parents and offspring. Section 11.3 discusses strategy parameters in more detail.

## 11.2 Evolutionary Programming Operators

The search process of an EP algorithm is driven by two main evolutionary operators, namely mutation and selection. Different implementations of these operators are discussed in Sections 11.2.1 and 11.2.2 respectively.

### 11.2.1 Mutation Operators

As mutation is the only means of introducing variation in an EP population, it is very important that the design of a mutation operator considers the exploration–exploitation trade-off. The variation process should facilitate exploration in the early stages of the search to ensure that as much of the search space is covered as possible. After an initial exploration phase, individuals should be allowed to exploit obtained information about the search space to fine tune solutions. A number of mutation operators have been developed, which addressses this trade-off to varying degrees.

For this discussion, assume that the task is to minimize a continuous, unconstrained function, $f : \mathbb{R}^{n_x} \to \mathbb{R}$. If $\mathbf{x}_i(t)$ denotes a candidate solution (as represented by the $i$-th individual) at generation $t$, then each $x_{ij}(t) \in \mathbb{R}, j = 1, \ldots, n_x$.

In general, mutation is defined as

$$x'_{ij}(t) = x_{ij}(t) + \Delta x_{ij}(t) \tag{11.1}$$

where $\mathbf{x}'_i(t)$ is the offspring created from parent $\mathbf{x}_i(t)$ by adding a step size $\Delta \mathbf{x}_i(t)$ to the parent. The step size is noise sampled from some probability distribution, where the deviation of the noise is determined by a strategy parameter, $\sigma_{ij}$. Generally, the step size is calculated as

$$\Delta x_{ij}(t) = \Phi(\sigma_{ij}(t))\eta_{ij}(t) \tag{11.2}$$

where $\Phi : \mathbb{R} \to \mathbb{R}$ is a function that scales the contribution of the noise, $\eta_{ij}(t)$.

Based on the characteristics of the scaling function, $\Phi$, EP algorithms can be grouped into three main categories of algorithms:

- **non-adaptive** EP, in which case $\Phi(\sigma) = \sigma$. In other words, the deviations in step sizes remain static.

- **dynamic** EP, where the deviations in step sizes change over time using some deterministic function, $\Phi$, usually a function of the fitness of individuals.

- **self-adaptive** EP, in which case deviations in step sizes change dynamically. The best values for $\sigma_{ij}$ are learned in parallel with the decision variables, $x_{ij}$.

Since the deviations, $\sigma_{ij}$, have an influence on the behavior of individuals in the case of dynamic and self-adaptive EP, these deviations are referred to as strategy parameters. Each individual has its own strategy parameters, in which case an individual is represented as the tuple,

$$\chi_i(t) = (\mathbf{x}_i(t), \sigma_i(t)) \tag{11.3}$$

While deviations are the most popular choice for strategy parameters, Fogel [265, 266], extended EP to use correlation coefficients between components of the individual as strategy parameters, similar to their use in evolution strategies (refer to Chapter 12). Strategy parameters are discussed in more detail in Section 11.3.

As is the case with all EAs, EP follows a stochastic search process. Stochasticity is introduced by computing step sizes as a function of noise, $\eta_{ij}$, sampled from some probability distribution. The following distributions have been used for EP:

- **Uniform:** Noise is sampled from a uniform distribution [580]

$$\eta_{ij}(t) \sim U(x_{min,j}, x_{max,j}) \tag{11.4}$$

where $\mathbf{x}_{min}$ and $\mathbf{x}_{max}$ provide lower and upper bounds for the values of $\eta_{ij}$. It is important to note that

$$E[\eta_{ij}] = 0 \tag{11.5}$$

to prevent any bias induced by the noise. Here $E[\bullet]$ denotes the expectation operator.

Wong and Yuryevich [916] proposed a uniform mutation operator where

$$\Delta x_{ij}(t) = U(0,1)(\hat{y}_j(t) - x_{ij}(t)) \tag{11.6}$$

with $\hat{\mathbf{y}}(t)$ the best individual from the current population, $\mathcal{C}(t)$. This mutation operator directs all individuals to make random movements towards the best individual (very similar to the social component used in particle swarm optimization; refer to Chapter 16). Note that the best individual does not change.

- **Gaussian:** For the Gaussian mutation operators, noise is sampled from a zero-mean, normal distribution [266, 265]:

$$\eta_{ij}(t) \sim N(0, \sigma_{ij}(t)) \tag{11.7}$$

For completeness sake, and comparison with other distributions, the Gaussian density function is given as (assuming a zero mean)

$$f_G(x) = \frac{1}{\sigma\sqrt{2\pi}} e^{-x^2/(2\sigma^2)} \tag{11.8}$$

where $\sigma$ is the deviation of the distribution.

- **Cauchy:** For the Cauchy mutation operators [934, 932, 936],

$$\eta_{ij}(t) \sim C(0, \nu) \tag{11.9}$$

where $\nu$ is the scale parameter.

The Cauchy density function centered at the origin is defined by

$$f_C(x) = \frac{1}{\pi} \frac{\nu}{\nu + x^2} \tag{11.10}$$

for $\nu > 0$. The corresponding distribution function is

$$F_C(x) = \frac{1}{2} + \frac{1}{\pi} \arctan(\frac{x}{\nu}) \tag{11.11}$$

The Cauchy distribution has wider tails than the Gaussian distribution, and therefore produces more, larger mutations than the Gaussian distribution.

- **Lévy:** For the Lévy distribution [505],

$$\eta_{ij}(t) \sim L(\nu) \tag{11.12}$$

The Lévy probability function, centered around the origin, is given as

$$F_{L,\nu,\gamma}(x) = \frac{1}{\pi} \int_0^\infty e^{-\gamma q^\nu} \cos(qx) dq \tag{11.13}$$

where $\gamma > 0$ is the scaling factor, and $0 < \nu < 2$ controls the shape of the distribution. If $\nu = 1$, the Cauchy distribution is obtained, and if $\nu = 2$, the Gaussian distribution is obtained.

For $|x| \gg 1$, the Lévy density function can be approximated by

$$f_L(x) \propto x^{-(\nu+1)} \tag{11.14}$$

An algorithm for generating Lévy random numbers is given in [505].

- **Exponential:** In this case [621],

$$\eta_{ij}(t) \sim E(0, \xi) \tag{11.15}$$

The density function of the double exponential probability distribution is given as

$$f_{E,\xi}(x) = \frac{\xi}{2} e^{-\xi|x|} \qquad (11.16)$$

where $\xi > 0$ controls the variance (which is equal to $\frac{2}{\xi^2}$). Random numbers can be calculated as follows:

$$x = \begin{cases} \frac{1}{\xi} \ln(2y) & \text{if } y \leq 0.5 \\ -\frac{1}{\xi} \ln(2(1-y)) & \text{if } y > 0.5 \end{cases} \qquad (11.17)$$

where $y \sim U(0,1)$. It can be noted that $E(0,\xi) = \frac{1}{\xi} E(0,1)$.

- **Chaos:** A chaotic distribution is used to sample noise [417]:

$$\eta_{ij}(t) \sim R(0,1) \qquad (11.18)$$

where $R(0,1)$ represents a chaotic sequence within the space $(-1,1)$. The chaotic sequence can be generated using

$$x_{t+1} = \sin(2/x_t)x_t, \quad t = 0,1\ldots \qquad (11.19)$$

- **Combined distributions:** Chellapilla [118] proposed the mean mutation operator (MMO), which uses a linear combination of Gaussian and Cauchy distributions. In this case,

$$\eta_{ij}(t) = \eta_{N,ij}(t) + \eta_{C,ij}(t) \qquad (11.20)$$

where

$$\eta_{N,ij} \sim N(0,1) \qquad (11.21)$$
$$\eta_{C,ij} \sim C(0,1) \qquad (11.22)$$

The resulting distribution generates more very small and large mutations compared to the Gaussian distribution. It generates more very small and small mutations compared to the Cauchy distribution. Generally, this convoluted distribution produces larger mutations than the Gaussian distribution, and smaller mutations than the Cauchy distribution.

Chellapilla also proposed an adaptive MMO, where

$$\Delta x_{ij}(t) = \gamma_{ij}(t)(C_{ij}(0,1) + \nu_{ij}(t)N_{ij}(0,1)) \qquad (11.23)$$

where $\gamma_{ij}(t) = \sigma_{2,ij}(t)$ is an overall scaling parameter, and $\nu_{ij} = \sigma_{1,ij}/\sigma_{2,ij}$ determines the shape of the probability distribution function; $\sigma_{1,ij}$ and $\sigma_{2,ij}$ are deviation strategy parameters. For low values of $\nu_{ij}$, the Cauchy distribution is approximated, while large values of $\nu_{ij}$ resemble the Gaussian distribution.

The question now is how these distributions address the exploration–exploitation trade-off. Recall that a balance of small and large mutations is needed. The Cauchy distribution, due to its wider tail, creates more, and larger mutations than the Gaussian distribution. The Cauchy distribution therefore facilitates better exploration than the Gaussian distribution. The Lévy distribution have tails in-between that of the Gaussian and Cauchy distributions, and therefore also provides better exploration than for the Gaussian distribution. While the Cauchy distribution does result in larger mutations, care should be taken in applying Cauchy mutations. As pointed out by Yao *et al.* [936], the smaller peak of the Cauchy distribution implies less time for exploitation. The Cauchy mutation operators therefore are weaker than the Gaussian operators in fine-tuning solutions. Yao *et al.* [932, 936] also show that the large mutations caused by Cauchy operators are beneficial only when candidate solutions are far from the optimum. It is due to these advantages and disadvantages that the Lévy distribution and convolutions such as those given in equations (11.20) and (11.23) offer good alternatives for balancing exploration and exploitation.

Another factor that plays an important role in balancing exploration and exploitation is the way in which strategy parameters are calculated and managed, since step sizes are directly influenced by these parameters. The next section discusses strategy parameters in more detail.

## 11.2.2 Selection Operators

Selection operators are applied in EP to select those individuals that will survive to the next generation. In the original EP, and most variations of it, the new population is selected from all the parents and their offspring. That is, parents and offspring compete to survive. Differing from other EAs, competition is based on a relative fitness measure and not an absolute fitness measure. An absolute fitness measure refers to the actual fitness function that quantifies how optimal a candidate solution is. On the other hand, the relative fitness measure expresses how well an individual performs compared to a group of randomly selected competitors (selected from the parents and offspring).

As suggested by Fogel [275], this is possibly the first hint towards coevolutionary optimization. For more detail on coevolution and relative fitness measures, refer to Chapter 15. This section only points out those methods that have been applied to EP.

For the purposes of this section, notation is changed to correspond with that of EP literature. In this light, $\mu$ is used to indicate the number of parent individuals (i.e. population size, $n_s$), and $\lambda$ is used to indicate the number of offspring.

The first step in the selection process is to calculate a score, or relative fitness, for each parent, $\mathbf{x}_i(t)$, and offspring, $\mathbf{x}_i'(t)$. Define $\mathcal{P}(t) = \mathcal{C}(t) \cup \mathcal{C}'(t)$ to be the competition pool, and let $\mathbf{u}_i(t) \in \mathcal{P}(t), i = 1, \ldots, \mu + \lambda$ denote an individual in the competition pool. Then, for each $\mathbf{u}_i(t) \in \mathcal{P}(t)$ a group of $n_\mathcal{P}$ competitors is randomly selected from the remainder of individuals (i.e. from $\mathcal{P}(t) \backslash \{\mathbf{u}_i(t)\}$). A score is calculated for each $\mathbf{u}_i(t)$ as follows

$$s_i(t) = \sum_{l=1}^{n_\mathcal{P}} s_{il}(t) \tag{11.24}$$

where (assuming minimization)

$$s_{il}(t) = \begin{cases} 1 & \text{if } f(\mathbf{u}_i(t)) < f(\mathbf{u}_l(t)) \\ 0 & \text{otherwise} \end{cases} \tag{11.25}$$

Wong and Yuryevich [916], and Ma and Lai [542] proposed an alternative scoring strategy where

$$s_{il}(t) = \begin{cases} 1 & \text{if } r_1 < \frac{f(\mathbf{u}_l(t))}{f(\mathbf{u}_l(t)) + f(\mathbf{u}_i(t))} \\ 0 & \text{otherwise} \end{cases} \tag{11.26}$$

where the $n_{\mathcal{P}}$ opponents are selected as $l = \lfloor 2\mu r_2 + 1 \rfloor$, with $r_1, r_2 \sim U(0,1)$.

In this case, if $f(\mathbf{u}_i(t)) << f(\mathbf{u}_l(t))$, in which case the fitness of $\mathbf{u}_i$ is significantly better than that of $\mathbf{u}_l$, then $\mathbf{u}_i$ will have a high probability of being assigned a winning score of 1. This approach is less strict than the requirement that $f(\mathbf{u}_i(t)) < f(\mathbf{u}_l(t))$, somewhat reducing the effects of selection pressure.

Based on the score assigned to each individual, $\mathbf{u}_i(t)$, any of a number of selection methods can be used (as summarized in Section 8.5):

- **Elitism:** the best $\mu$ individuals from $\mathcal{P}(t)$ are selected to form the new population, $\mathcal{C}(t+1)$.

- **Tournament** selection: The best $\mu$ individuals are stochastically selected using tournament selection.

- **Proportional** selection: Each individual is assigned a probability of being selected:

$$p_s(\mathbf{u}_i(t)) = \frac{s_i(t)}{\sum_{l=1}^{2\mu} s_l(t)} \tag{11.27}$$

  Roulette-wheel selection can then be used to select the $\mu$ individuals for the next generation.

- **Nonlinear ranking** selection [933]: Individuals are sorted in ascending order of score and then ranked. Then,

$$p_s\left(\mathbf{u}_{(2\mu-i)}(t)\right) = \frac{i}{\sum_{l=1}^{2\mu} l} \tag{11.28}$$

  Instead of using a stochastic selection, ranking can be used to find the $\mu$ elite individuals to form the new population.

Different methods have also been proposed to decide which of the parent or its offspring will survive to the next generation. Wei *et al.* [894] proposed that each parent produces more than one offspring, where the number of offspring produced is determined by the fitness of the individual. The more fit the parent is, the more offspring are generated. The best offspring generated from a parent is selected (based on absolute fitness measure), and competes with the parent for survival. Competition between the parent and offspring is based on simulated annealing [894]. The offspring, $\mathbf{x}_i'(t)$, survives to the next generation if $f(\mathbf{x}_i'(t)) < f(\mathbf{x}_i(t))$ or if

$$e^{(-(f(\mathbf{x}_i'(t)) - f(\mathbf{x}_i(t)))/\tau(t))} > U(0,1) \tag{11.29}$$

where $\tau$ is the temperature coefficient, with $\tau(t) = \gamma\tau(t-1), 0 < \gamma < 1$; otherwise the parent survives.

The above metropolis selection has the advantage that the offspring has a chance of surviving even if it has a worse fitness than the parent, which reduces selection pressure and improves exploration.

## 11.3 Strategy Parameters

As hinted in equation (11.2), step sizes are dependent on strategy parameters, which form an integral part of EP mutation operators. Although Section 11.2.1 indicated that a strategy parameter is associated with each component of an individual, it is totally possible to use one strategy parameter per individual. However, the latter approach limits the degrees of freedom in addressing the exploration – exploitation trade-off. For the purposes of this section, it is assumed that each component has its own strategy parameter, and that individuals are represented as given in equation (11.3).

### 11.3.1 Static Strategy Parameters

The simplest approach to handling strategy parameters is to fix the values of deviations. In this case, the strategy parameter function is linear, i.e.

$$\Phi(\sigma_{ij}(t)) = \sigma_{ij}(t) = \sigma_{ij} \tag{11.30}$$

where $\sigma_{ij}$ is a small value. Offspring are then calculated as (assuming a Gaussian distribution)

$$x'_{ij}(t) = x_{ij}(t) + N_{ij}(0, \sigma_{ij}) \tag{11.31}$$

with $\Delta x_{ij}(t) = N_{ij}(0, \sigma_{ij})$. The notation $N_{ij}(\bullet, \bullet)$ indicates that a new random value is sampled for each component of each individual.

A disadvantage of this approach is that a too small value for $\sigma_{ij}$ limits exploration and slows down convergence. On the other hand, a too large value for $\sigma_{ij}$ limits exploitation and the ability to fine-tune a solution.

### 11.3.2 Dynamic Strategies

One of the first approaches to change the values of strategy parameters over time, was to set them to the fitness of the individual [271, 265]:

$$\sigma_{ij}(t) = \sigma_i(t) = \gamma f(\mathbf{x}_i(t)) \tag{11.32}$$

in which case offspring is generated using

$$\begin{aligned} x'_{ij}(t) &= x_{ij}(t) + N(0, \sigma_i(t)) \\ &= x_{ij}(t) + \sigma_i(t)N(0, 1) \end{aligned} \tag{11.33}$$

In the above, $\gamma \in (0, 1]$.

If knowledge of the global optimum exists, the error of an individual can be used instead of absolute fitness. However, such information is usually not available. Alternatively, the phenotypic distance from the best individual can be used as follows:

$$\sigma_{ij}(t) = \sigma_i(t) = |f(\hat{\mathbf{y}}) - f(\mathbf{x}_i)| \tag{11.34}$$

where $\hat{\mathbf{y}}$ is the most fit individual. Distance in decision space can also be used [827]:

$$\sigma_{ij}(t) = \sigma_i(t) = \mathcal{E}(\hat{\mathbf{y}}, \mathbf{x}_i) \tag{11.35}$$

where $\mathcal{E}(\bullet, \bullet)$ gives the Euclidean distance between the two vectors.

The advantage of this approach is that the weaker an individual is, the more that individual will be mutated. The offspring then moves far from its weak parent. On the other hand, the stronger an individual is, the less the offspring will be removed from its parent, allowing the current good solution to be refined. This approach does have some disadvantages:

- If fitness values are very large, step sizes may be too large, causing individuals to overshoot a good minimum.

- The problem is even worse if the function value of the optimum is a large, non-zero value. If the fitness values of good individuals are large, large step sizes result, causing individuals to move away from good solutions. In such cases, if knowledge of the optimum is available, using an error measure will be more appropriate.

A number of proposals have been made to control step sizes as a function of fitness. A non-extensive list of these methods is given below (unless otherwise stated, these methods assume a minimization problem):

- Fogel [266] proposed an additive approach, where

$$x'_{ij}(t) = x_{ij}(t) + \sqrt{\beta_{ij}(t)f(\mathbf{x}_i) + \gamma_{ij}} + N_{ij}(0, 1) \tag{11.36}$$

  where $\beta_{ij}$ and $\gamma_{ij}$ are respectively the proportionality constant and offset parameter.

- For the function $f(x_1, x_2) = x_1^2 + x_2^2$, Bäck and Schwefel [45] proposed that

$$\sigma_{ij}(t) = \sigma_i(t) = \frac{1.224\sqrt{f(\mathbf{x}_i(t))}}{n_x} \tag{11.37}$$

  where $n_x$ is the dimension of the problem (in this case, $n_x = 2$).

- For training recurrent neural networks, Angeline et al. [28] proposed that

$$x'_{ij}(t) = x_{ij}(t) + \beta\sigma_{ij}(t)N_{ij}(0, 1) \tag{11.38}$$

  where $\beta$ is the proportionality constant, and

$$\sigma_{ij}(t) = U(0, 1)\left[1 - \frac{f(\mathbf{x}_i(t))}{f_{max}(t)}\right] \tag{11.39}$$

with $f_{max}(t)$ the maximum fitness of the current population. Take note that the objective here is to maximize $f$, and that $f(\mathbf{x}_i(t))$ returns a positive value. If $f(\mathbf{x}_i(t))$ is a small value, then $\sigma_{ij}(t)$ will be large (bounded above by 1), which results in large mutations. Deviations are scaled by a uniform number in the range $[0, 1]$ to ensure a mix of small and large step sizes.

- Ma and Lai [542] proposed that deviations be proportional to normalized fitness values:

$$x'_{ij}(t) = x_{ij}(t) + \beta_{ij}\sigma_i(t)N_{ij}(0, 1) \tag{11.40}$$

where $\beta_{ij}$ is the proportionality constant, and deviations are calculated as

$$\sigma_i(t) = \frac{f(\mathbf{x}_i(t))}{\sum_{l=1}^{n_s} f(\mathbf{x}_l(t))} \tag{11.41}$$

with $n_s$ the size of the population. This approach assumes $f$ is minimized.

- Yuryevich and Wong [943] proposed that

$$\sigma_{ij}(t) = (x_{max,j} - x_{min,j})\left(\frac{f_{max}(t) - f(\mathbf{x}_i(t))}{f_{max}(t)} + \gamma\right) \tag{11.42}$$

to combine both boundary information and fitness information. In the above $\mathbf{x}_{min}$ and $\mathbf{x}_{max}$ specify the bounds in decision space, and $\gamma > 0$ is an offset parameter to ensure non-zero deviations. Usually, $\gamma$ is a small value.

This approach assumes that $f$ is maximized. The inclusion of boundary constraints forces large mutations for components with a large domain, and small mutations if the domain is small.

- Swain and Morris [827] set deviations proportional to the distance from the best individual, i.e.

$$\sigma_{ij}(t) = \beta_{ij}|\hat{y}_j(t) - x_{ij}(t)| + \gamma \tag{11.43}$$

where $\gamma > 0$, and the proportionality constant is calculated as

$$\beta_{ij} = \beta\frac{\sqrt{\mathcal{E}(\mathbf{x}_{min}, \mathbf{x}_{max})}}{\pi} \tag{11.44}$$

with $\beta \in [0, 2]$, and $\mathcal{E}(\mathbf{x}_{min}, \mathbf{x}_{max})$ gives the width of the search space as the Euclidean distance between the vectors $\mathbf{x}_{min}$ and $\mathbf{x}_{max}$. The parameter, $\gamma$, defines a search neighborhood. Larger values of $\gamma$ promote exploration, while smaller values promote exploitation. A good idea is to adapt $\gamma$ over time, starting with large values that are decreased over time.

Offspring are generated using

$$x'_{ij}(t) = x_{ij}(t) - \text{dir}(x_{ij})\sigma_{ij}(t)N_{ij}(0, 1) \tag{11.45}$$

where the direction of the update is

$$\text{dir}(x_{ij}) = \text{sign}(\hat{y}_j - x_{ij}) \tag{11.46}$$

- Gao [308] suggested that

$$\sigma_{ij}(t) = \left[ \frac{1}{\sqrt{\beta_j f(\mathbf{x}_i(t)) + \gamma_j}} \right] \left[ \frac{\gamma}{f_{max}(t) - f_{min}(t)} \right] \tag{11.47}$$

where it is proposed that $\gamma = 2.5$; $f_{max}$ and $f_{min}$ refer to the largest and smallest fitness values of the current population.

### 11.3.3   Self-Adaptation

The emphasis of EP is on developing behavioral models. EP is derived from simulations of adaptive behavior. Previous sections have already indicated the strong influence that strategy parameters have on the behavior of individuals, as quantified via the fitness function. Two of the major problems concerning strategy parameters are the amount of mutational noise that should be added, and the severity (i.e. step sizes) of such noise. To address these problems, and to produce truly self-organizing behavior, strategy parameters can be "evolved" (or "learned") in parallel with decision variables. An EP that utilizes such mechanisms is referred to as a self-adaptive EP.

Self-adaptation is not unique to EP. According to Fogel *et al.* [277], the idea of self-adaptation stretches back as far as 1967 with proposals by Rechenberg. However, Schwefel [769] provided the first detailed account of self-adaptation in the context of evolution strategies (ES) (also refer to Chapter 12). With reference to EP, Fogel *et al.* [271] provided the first suggestions for self-adaptive EP. Since then, a number of self-adaptation methods have been proposed. These methods can be divided into three broad categories [40]:

- **Additive** methods: The first self-adaptive EP as proposed by Fogel *et al.* [265] is an additive method where

$$\sigma_{ij}(t + 1) = \sigma_{ij}(t) + \eta \sigma_{ij}(t) N_{ij}(0, 1) \tag{11.48}$$

  with $\eta$ referred to as the learning rate. In the first application of this approach, $\eta = 1/6$. If $\sigma_{ij}(t) \leq 0$, then $\sigma_{ij}(t) = \gamma$, where $\gamma$ is a small positive constant (typically, $\gamma = 0.001$) to ensure positive, non-zero deviations.

  As an alternative, Fogel [266] proposed

$$\sigma_{ij}(t + 1) = \sigma_{ij}(t) + \sqrt{f_\sigma(\sigma_{ij}(t))} N_{ij}(0, 1) \tag{11.49}$$

  where

$$f_\sigma(a) = \begin{cases} a & \text{if } a > 0 \\ \gamma & \text{if } a \leq 0 \end{cases} \tag{11.50}$$

  ensures that the square root is applied to a positive, non-zero value.

- **Multiplicative** methods: Jiang and Wang [418] proposed a multiplicative adjustment, where

$$\sigma_{ij}(t + 1) = \sigma(0)(\lambda_1 e^{-\lambda_2 \frac{t}{n_t}} + \lambda_3) \tag{11.51}$$

  where $\lambda_1, \lambda_2$ and $\lambda_3$ are control parameters, and $n_t$ is the maximum number of iterations.

- **Lognormal** methods: Borrowed from the ES literature [277],

$$\sigma_{ij}(t+1) = \sigma_{ij}(t)e^{(\tau N_i(0,1) + \tau' N_{ij}(0,1))} \tag{11.52}$$

with

$$\tau' = \frac{1}{\sqrt{2\sqrt{n_x}}} \tag{11.53}$$

$$\tau = \frac{1}{\sqrt{2n_x}} \tag{11.54}$$

Offspring are produced using

$$x'_{ij}(t) = x_{ij}(t) + \sigma_{ij}(t)N_{ij}(0,1) \tag{11.55}$$

Self-adaptive EP showed the undesirable behavior of stagnation due to the tendency that strategy parameters converge too fast. The consequence is that deviations become small too fast, thereby limiting exploration. The search stagnates for some time until strategy parameters grow sufficiently large due to random variation.

One solution to this problem is to impose a lower bound on the values of $\sigma_{ij}$. However, this triggers another problem of deciding when $\sigma_{ij}$ values are to be considered as too small. Liang *et al.* [524] provided a solution by considering dynamic lower bounds:

$$\sigma_{min}(t+1) = \sigma_{min}(t)\left(\frac{n_m(t)}{\xi}\right) \tag{11.56}$$

where $\sigma_{min}(t)$ is the lower bound at time step (generation) $t$, $\xi \in [0.25, 0.45]$ is the reference rate, and $n_m(t)$ is the number of successful consecutive mutations (i.e. the number of mutations that results in improved fitness values). This approach is based on the 1/5 success rule of Rechenberg [709] (refer to Chapter 12).

Matsumura *et al.* [565] developed the robust EP (REP) where the representation of each individual is expanded to allow for $n_\sigma$ strategy parameter vectors to be associated with each individual, as follows

$$(\mathbf{x}_i(t), \sigma_{i0}, \ldots, \sigma_{ik}, \ldots \sigma_{in_\sigma}) \tag{11.57}$$

where $\sigma_{i0}$ is referred to as the active strategy parameter vector, obtained through application of three mutation operators on the other strategy parameter vectors. Component values of the strategy parameter vectors are mutated as follows:

- **Duplication**:

$$\sigma'_{i0j}(t) = \sigma_{i0j}(t) \tag{11.58}$$

$$\sigma'_{ilj}(t) = \sigma_{i(l-1)j}(t) \tag{11.59}$$

for $l \in \{1, 2, \ldots, n_\sigma\}$. Then $\sigma_{ikj}(t)$ is self-adapted by application of the lognormal method of equation (11.52) on the $\sigma'_{ikj}(t)$ for $k = 0, 1, \ldots, n_\sigma$.

- **Deletion**:

$$\sigma'_{i(l-1)j}(t) \;=\; \sigma_{ilj}(t) \tag{11.60}$$

$$\sigma_{in_\sigma j}(t) \;=\; \min\{\sigma_{max}(t), \sum_{k=0}^{n_\sigma-1} \sigma_{ikj}(t)\} \tag{11.61}$$

for $l \in \{1, 2, \dots, n_\sigma\}$. Then $\sigma_{ikj}(t)$ is self-adapted by application of the lognormal method of equation (11.52) on the $\sigma'_{ikj}(t)$ for $k = 0, 1, \dots, n_\sigma$.

- **Invert**:

$$\sigma'_{i0j}(t) \;=\; \sigma_{ilj}(t) \tag{11.62}$$

$$\sigma'_{ilj}(t) \;=\; \sigma_{i0j}(t) \tag{11.63}$$

for $l \in \{1, 2, \dots, n_\sigma\}$. The lognormal self-adaptation method of equation (11.52) is applied to $\sigma'_{i0j}(t)$ and $\sigma'_{ilj}(t)$ to produce $\sigma_{i0j}(t)$ and $\sigma_{ilj}(t)$ respectively.

After application of the mutation operators, offspring is created using

$$x'_{ij}(t) = x_{ij}(t) + \sigma_{i0j}(t)C(0,1) \tag{11.64}$$

In a similar way, Fogel and Fogel [269] proposed multiple-vector self-adaptation. In their strategy, at each iteration and before offspring is generated, the active strategy parameter vector has a probability of $p_\sigma$ of changing to one of the other $n_\sigma - 1$ vectors. The problem is then to determine the best values for $n_\sigma$ and $p_\sigma$, which are problem dependent.

At this point it should be noted that offspring is first generated, and then strategy parameters are updated. This differs from ES where strategy parameters are updated first, and then offspring is generated. The order should not have a significant influence, as use of new values for strategy parameters is delayed for just one generation of EP.

## 11.4   Evolutionary Programming Implementations

This section gives a short overview of a number of EP algorithm implementations. Note that this is not an exhaustive review of different EP implementations.

### 11.4.1   Classical Evolutionary Programming

Yao *et al.* [934, 936] coined the term classical EP (CEP) to refer to EP with Gaussian mutation. More specifically, CEP uses the lognormal self-adaptation given in equation (11.52), and produces offspring using equation (11.55). Elitism selection is used to construct the new population from the current parent population and generated offspring.

## 11.4.2  Fast Evolutionary Programming

Yao *et al.* [934, 936] and Wei *et al.* [894] adapted the CEP to produce the fast EP (FEP) by changing the distribution from which mutational noise is sampled to the Cauchy mutation as given in equation (11.9) with $\nu = 1$. Offspring is generated using

$$x'_{ij}(t) = x_{ij}(t) + \sigma_{ij}(t)C_{ij}(0,1) \tag{11.65}$$

where the lognormal self-adaptation (refer to equation (11.52)) is used. Elitism is used to select the new population.

The wider tails of the Cauchy distribution provide larger step sizes, and therefore result in faster convergence. An analysis of FEP showed that step sizes may be too large for proper exploitation [932, 936], while Gaussian mutations showed a better ability to fine-tune solutions. This prompted a proposal for the improved FEP (IFEP). For each parent, IFEP generates two offspring, one using Gaussian mutation and one using Cauchy mutation. The best offspring is chosen as the surviving offspring, which will compete with the parent for survival. An alternative approach would be to start the search using Cauchy mutations, and to switch to Gaussian mutation at a later point. However, such a strategy introduces the problem of when the optimal switching point is reached. Diversity measures provide a solution here, where the switch can occur when diversity is below a given threshold to indicate that exploitation should be favored.

The mean mutation operators of Chellapilla [118] (refer to equations (11.20) and (11.23)) provide a neat solution by using a convolution of Gaussian and Cauchy distributions.

## 11.4.3  Exponential Evolutionary Programming

Narihisa *et al.* [621] proposed that the double exponential probability distribution as defined in equation (11.16) be used to sample mutational noise. Offspring are generated using

$$x'_{ij}(t) = x_{ij}(t) + \sigma_{ij}(t)\frac{1}{\xi}E_{ij}(0,1) \tag{11.66}$$

where $\sigma_{ij}$ is self-adapted, and the variance of the distribution is controlled by $\xi$. The smaller the value of $\xi$, the greater the variance. Larger values of $\xi$ result in smaller step sizes. To ensure initial exploration and later exploitation, $\xi$ can be initialized to a small value that increases with time.

## 11.4.4  Accelerated Evolutionary Programming

In an attempt to improve the convergence speed of EP, Kim *et al.* [462] proposed the accelerated EP (AEP), which uses two variation operators:

- A directional operator to determine the direction of the search based fitness scores, and

- the Gaussian mutation operator given in equation (11.7).

Individuals are represented as

$$\chi_i(t) = (\mathbf{x}_i(t), \rho_i(t), a_i(t)) \tag{11.67}$$

where $\rho_{ij} \in \{-1, 1\}, j = 1, \ldots, n_x$ gives the search direction for each component of the $i$-th individual, and $a_i$ represents the age of the individual. Age is used to force wider exploration if offspring are worse than their parents.

Offspring generation consists of two steps. The first step updates age parameters for each individual, and determines search directions (assuming minimization):

$$a_i(t) = \begin{cases} 1 & \text{if } f(\mathbf{x}_i(t)) < f(\mathbf{x}_i(t-1)) \\ a_i(t-1)+1 & \text{otherwise} \end{cases} \tag{11.68}$$

and

$$\rho_{ij}(t) = \begin{cases} \text{sign}(x_{ij}(t) - x_{ij}(t-1)) & \text{if } f(\mathbf{x}_i(t)) < f(\mathbf{x}_i(t-1)) \\ \rho_{ij}(t-1) & \text{otherwise} \end{cases} \tag{11.69}$$

If the fitness of an individual improved, the search will continue in the direction of the improvement. If the fitness does not improve, the age is incremented, which will result in larger step sizes as follows: If $a_i(t) = 1$, then

$$\sigma_i(t) = \gamma_1 f(\mathbf{x}_i(t)) \tag{11.70}$$
$$x'_{ij}(t) = x_{ij}(t) + \rho_{ij}(t)|N(0, \sigma_i(t))| \tag{11.71}$$

Otherwise, if $a_i(t) > 1$,

$$\sigma_i(t) = \gamma_2 f(\mathbf{x}_i(t)) a_i(t) \tag{11.72}$$
$$x'_{ij}(t) = x_{ij}(t) + N(0, \sigma_i(t)) \tag{11.73}$$

where $\gamma_1$ and $\gamma_2$ are positive constants.

Selection occurs by having an offspring compete directly with its parent using absolute fitness.

Wen *et al.* [896] used a similar approach, but using the dynamic strategy parameter approach given in equation (11.41).

## 11.4.5 Momentum Evolutionary Programming

Choi and Oh [126] proposed an EP algorithm based on backpropagation learning of feedforward neural networks (refer to Section 3.2.2). The best individual, $\hat{\mathbf{y}}(t)$, of the current population, $\mathcal{C}(t)$, calculated as

$$\hat{\mathbf{y}}(t) = \mathbf{x}_i(t) : f(\mathbf{x}_i(t)) = \min_{i=1,\ldots,\mu} \{f(\mathbf{x}_i(t))\} \tag{11.74}$$

is taken as the target. The temporal error between this target, $\hat{\mathbf{y}}(t)$, and the individual, $\mathbf{x}_i(t)$, is then used by the mutation operator to improve exploration. For each parent, $\mathbf{x}_i(t)$, an offspring is generated as follows (assuming minimization):

$$x'_{ij}(t) = x_{ij}(t) + \eta\Delta x_{ij}(t) + \alpha\tilde{x}_{ij}(t) \tag{11.75}$$

where

$$
\begin{aligned}
\Delta x_{ij}(t) &= (\hat{y}_j(t) - x_{ij}(t))|N_{ij}(0,1)| & (11.76)\\
\tilde{x}_{ij}(t) &= \eta\rho_i(t)\Delta x_{ij}(t-1) + \alpha\tilde{x}_{ij}(t-1) & (11.77)
\end{aligned}
$$

with $\eta > 0$ the learning rate, $\alpha > 0$ the momentum rate, and

$$\rho_i(t) = \begin{cases} 1 & \text{if } f(\mathbf{x}'_i(t-1)) < f(\mathbf{x}_i(t-1)) \\ 0 & \text{otherwise} \end{cases} \tag{11.78}$$

## 11.4.6 Evolutionary Programming with Local Search

A very simple approach to improve the exploitation ability of EP, is to add a hill-climbing facility to generated offspring. While a better fitness can be obtained, hill-climbing is applied to each offspring [235]. Alternatively, gradient descent has been used to regenerate offspring [920, 779]. For each offspring, $\mathbf{x}'_i(t)$, recalculate the offspring using

$$x'_{ij}(t) = x'_{ij}(t) - \eta_i(t)\frac{\partial f}{\partial x_{ij}(t)} \tag{11.79}$$

where the learning rate is calculated as

$$\eta_i(t) = \frac{\sum_{j=1}^{n_x}\frac{\partial f}{\partial x_{ij}(t)}}{\sum_{h=1}^{n_x}\sum_{j=1}^{n_x}\frac{\partial^2 f}{\partial x_{ih}(t)\partial x_{ij}(t)}\frac{\partial f}{\partial x_{ih}(t)}\frac{\partial f}{\partial x_{ij}(t)}} \tag{11.80}$$

As an alternative to gradient descent, Birru *et al.* [70] used conjugate gradient search (refer to Section 3.2.3), where line searches are performed for each component of the offspring. The initial search direction is the downhill gradient, with subsequent search directions chosen along subsequent gradient components that are orthogonal to all previous search directions.

Birru *et al.* [70] also proposed a derivitive-free local search method to refine offspring. The stochastic search developed by Solis and Wets [802] is applied to each offspring at a specified probability. Based on this probability, if the local search is performed, a limited number of steps is done as summarized in Algorithm 11.2.

## 11.4.7 Evolutionary Programming with Extinction

Fogel *et al.* [274] incorporated concepts of mass extinction into EP. The outcome of an extinction event is that a significant portion of populations is killed, after which

---

**Algorithm 11.2** Solis and Wets Random Search Algorithm for Function Minimization

---

Initialize the candidate solution, $\mathbf{x}(0)$, with $x_j(0) \sim U(x_{min,j}, x_{max,j})$, $j = 1, \ldots, n_x$;
Let $t = 0$;
Let $\rho(0) = 1$;
**while** *stopping condition(s) not true* **do**
    $t = t + 1$;
    Generate a new candidate solution as $\mathbf{x}'(t) = \mathbf{x}(t) + \rho(t)\mathbf{N}(0, \sigma)$;
    **if** $f(\mathbf{x}'(t)) < f(\mathbf{x}(t))$ **then**
        $\mathbf{x}(t) = \mathbf{x}'(t)$;
    **end**
    **else**
        $\rho(t) = -\rho(t - 1)$;
        $\mathbf{x}'(t) = \mathbf{x}(t) + \rho(t)N(0, \sigma)$;
        **if** $f(\mathbf{x}'(t)) < f(\mathbf{x}(t))$ **then**
            $\mathbf{x}(t) = \mathbf{x}'(t)$;
        **end**
        **else**
            **for** $j = 1, \ldots, n_x$ **do**
                $x_j(t) \sim U(x_{min,j}, x_{max,j})$;
            **end**
        **end**
    **end**
**end**

---

reproduction produces totally new populations with different survival behaviors than populations that existed before extinction. Central to the EP with extinction is the concept of environmental stress, which is a random variable sampled from a uniform distribution, i.e.

$$\delta(t) \sim U(0, 0.96) \tag{11.81}$$

If the normalized fitness of an individual is less than the environmental stress, then that individual is killed. The fitness of each individual is normalized as follows:

$$\tilde{f}(\mathbf{x}_i(t)) = \alpha + (1 - \alpha)\left[\frac{f(\mathbf{x}_i(t)) - f_{max}(t)}{f_{min}(t) - f_{max}(t)}\right] \tag{11.82}$$

where $f_{min}(t)$ and $f_{max}(t)$ are respectively the lowest and largest fitness values of the current generation, and $\alpha \in [0, 1]$ provides a lower limit on the percentage killed.

The EP with extinction is summarized in Algorithm 11.3.

## 11.4.8 Hybrid with Particle Swarm Optimization

A number of suggestions have been made to combine EP with particle swarm optimization (PSO) (refer to Chapter 16). Wei *et al.* [893], and Sinha and Purkayastha

---

**Algorithm 11.3** Extinction Evolutionary Programming for Function Minimization

---

Set the generation counter, $t = 0$;
Initialize the strategy parameters;
Create and initialize the population, $\mathcal{C}(0)$;
**while** *stopping condition(s) not true* **do**
    $t = t + 1$;
    Let $\mathcal{C}(t) = \mathcal{C}(t-1)$;
    $\delta(t) \sim U(0, 0.96)$;
    $n_\delta = 0$;
    **for** $i = 1, \ldots, \mu$ **do**
        **if** $\tilde{f}(\mathbf{x}_i(t)) < \delta(t)$ **then**
            $\mathcal{C}(t) = \mathcal{C}(t) \backslash \{\mathbf{x}_i(t)\}$;
            $n_\delta = n_\delta + 1$;
        **end**
    **end**
    **if** $n_\delta > 0$ **then**
        Let $\tilde{n}_s = n_s - n_\delta$ be the number of survivors;
        **for** *each of the top 10% survivors* **do**
            Generate $\frac{n_\delta}{0.1\tilde{n}_s}$ offspring;
        **end**
        Calculate the fitness of all offspring;
        Select $n_\delta$ of the offspring using tournament selection;
        Add selected offspring to $\mathcal{C}(t)$;
    **end**
    **else**
        Mutate the top 10% individuals of $\mathcal{C}(t)$;
        **for** *each offspring, $x_i'(t)$, generated* **do**
            **if** $f(\mathbf{x}_i'(t)) < f(\mathbf{x}_i(t))$ **then**
                $\mathbf{x}_i(t) = \mathbf{x}_i'(t)$;
            **end**
        **end**
    **end**
**end**

---

[794] applies the PSO position update (refer to Chapter 16),

$$\mathbf{x}_i(t+1) = \mathbf{x}_i(t) + \mathbf{v}_i(t) \tag{11.83}$$

and then mutate the new position using an EP mutation operator. Wei *et al.* [893] uses the mutation operator and self-adaptation of CEP, while Sinha and Purkayastha [794] uses a variation of the dynamic strategy parameter approach of equation (11.42), where

$$\sigma_i(t) = \gamma \left[ \frac{f(\mathbf{x}_i(t))}{f_{min}(t)} \right] (x_{max,j} - x_{min,j}) \tag{11.84}$$

with Gaussian mutational noise. That is,

$$x_{ij}(t+1) = x_{ij}(t) + \nu_{ij}(t) + \sigma_i N_{ij}(0, 1) \tag{11.85}$$

# 11.5   Advanced Topics

This section provides a very compact review of some approaches to apply EP to problems more difficult than unconstrained problems.

## 11.5.1   Constraint Handling Approaches

Any of a number of methods from the EC literature can be used to evolve feasible solutions that satisfy all constraints (with reference to problems as defined in Definition A.5). With reference to Section A.6, the following approaches have been used in EP literature:

- Penalty methods (refer to Section A.6.2), where a penalty is added to the objective function to penalize an individual for constraint violation [445, 795, 463].

- The constrained problem is converted to an unconstrained dual Lagrangian problem, where Lagrangian multipliers are optimized in parallel with decision variables [463]. Kim and Myung [463] developed a two-phase EP for constrained problems. Phase one uses a penalty function. The best individual from phase one is then used to generate a new population for phase two, which optimizes the dual Lagrangian problem.

- Mutation operators are adapted to ensure that only feasible offspring are generated [943]. El-Sharkh and El-Keib [235] applied hill-climbing to offspring to reduce the number of constraints violated. If the hill-climbing search fails in producing a feasible solution, mutation is applied again. Ma and Lai [542] used a simple, but inefficient approach by setting components that violate constraints to boundary values.

## 11.5.2   Multi-Objective Optimization and Niching

Multi-objective optimization (MOO) techniques that can be found in the general EA literature can be applied to EP to solve multi-objective problems as defined in Definition A.10. Simple approaches are to use weight aggregation methods as summarized in Section A.8.2. Pareto-based methods have been used in [953].

To implement a niching EP algorithm, Li *et al.* [519] utilized crowding and fitness sharing as used in GAs. Damavandi and Safavi-Nacini [169] used a clustering algorithm applied to individuals to facilitate niche formation.

## 11.5.3   Dynamic Environments

Not much has been done to analyze the performance of EP for dynamically changing, or noisy landscapes. Ma and Lai [542] used Gaussian mutations with dynamic strategy parameters as defined in equation (11.41) with success, while Matsumura *et al.* [566]

analyzed the performance of CEP, FEP and robust EP on noisy-environments. Bäck [40] concluded that EP with additive strategy parameters fails for dynamic landscapes, while lognormal self-adaptation succeeded.

## 11.6 Applications

The first application of EP was to evolve finite-state machines. Section 11.6.1 shows how this can be done, while Section 11.6.2 illustrates how EP can be used to optimize a continuous function. Section 11.6.3 shows how an EP can be used to train a NN. A summary of real-world applications of EP is given in Section 11.6.4.

### 11.6.1 Finite-State Machines

EP was originally developed to evolve finite-state machines (FSM). The aim of this application type is to evolve a program to predict the next symbol (of a finite alphabet) based on a sequence of previously observed symbols.

A finite-state machine is essentially a computer program that represents a sequence of actions that must be executed, where each action depends on the current state of the machine and an input. Formally, a FSM is defined as

$$FSM = (\mathcal{S}, \mathcal{I}, \mathcal{O}, \rho, \phi) \qquad (11.86)$$

where $\mathcal{S}$ is a finite set of machine states, $\mathcal{I}$ is a finite set of input symbols, $\mathcal{O}$ is a finite set of output symbols (the alphabet of the FSM), $\rho : \mathcal{S} \times \mathcal{I} \rightarrow \mathcal{S}$ is the next state function, and $\phi : \mathcal{S} \times \mathcal{I} \rightarrow \mathcal{O}$ is the next output function. An example of a 3-state FSM is given in Figure 11.1 (taken from [278]). The response of the FSM to a given string of symbols is given in Table 11.1, presuming an initial state $C$.

**Table 11.1** Response of Finite-State Machine

| Present state | C | B | C | A | A | B |
|---|---|---|---|---|---|---|
| Input symbol | 0 | 1 | 1 | 1 | 0 | 0 |
| Next state | B | C | A | A | B | C |
| Output symbol | $\beta$ | $\alpha$ | $\gamma$ | $\beta$ | $\beta$ | $\gamma$ |

**Representation**

Each state can be represented by a 6-bit string. The first bit represents the activation of the corresponding state (0 indicates not active, and 1 indicates active). The second bit represents the input symbol, the next two bits represent the next state, and the last two bits represent the output symbol. Each individual therefore consists of 18 bits. The initial population is randomly generated, with the restriction that the output symbol and next state bits represent only valid values.

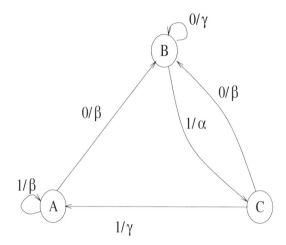

**Figure 11.1** Finite-State Machine [278]

**Fitness Evaluation**

The fitness of each individual is measured as the individual's ability to correctly predict the next output symbol. A sequence of symbols is used for this purpose. The first symbol from the sequence is presented to each individual, and the predicted symbol compared to the next symbol in the sequence. The second symbol is then presented as input, and the process iterates over the entire sequence. The individual with the most correct predictions is considered the most fit individual.

**Mutation**

The following mutation operations can be applied: The initial state can be changed, a state can be deleted, a state can be added, a state transition can be changed, or an output symbol for a given state and input symbol can be changed.

These operators are applied probabilistically, in one of the following ways:

- Select a uniform random number between 1 and 5. The corresponding mutation operator is then applied with probability $p_m$.
- Generate a Poisson number, $\xi$ with mean $\lambda$. Select $\xi$ mutation operators uniformly from the set of operators, and apply them in sequence.

## 11.6.2   Function Optimization

The next example application of EP is in function optimization. Consider, for example, finding the minimum of the function $\sin(2\pi x)e^{-x}$ in the range $[0, 2]$.

## Representation

The function has one parameter. Each individual is therefore represented by a vector consisting of one floating-point element (not binary encoded). The initial population is generated randomly, with each individual's parameter $x_{ij}$ selected such that $x_{ij} \sim U(0,2)$.

## Fitness Evaluation

In the case of minimization, the fittest individual is the one with the smallest value for the function being optimized; that is, the individual with the smallest value for the function $\sin(2\pi x)e^{-x}$. For maximization, it is the largest value.

# Mutation

Any of the mutation operators discussed in Section 11.2.1 can be used to produce offspring.

## 11.6.3 Training Neural Networks

One of the first applications as an approach to optimize unconstrained functions was to train supervised feedforward neural networks [272]. Since then, EP has been applied to many neural network problems [28, 571, 933].

# Representation

Each individual represents one neural network (NN), where a component represents a single weight or bias.

# Fitness Evaluation

The mean squared error (MSE), or sum squared error (SSE), can be used to quantify the performance of a NN. In the case of a classification task, the percentage of incorrectly classified patterns can be used. Fitness evaluation involves conversion of the vector representation used for individuals to a layered NN structure in order to perform feedforward passes for patterns in the given data set.

**Table 11.2** Real-World Applications of Evolutionary Programming

| Application Class | References |
|---|---|
| Bayesian networks | [519] |
| Controller design | [889] |
| Robotics | [350, 445, 465] |
| Games | [267, 273] |
| Image processing | [546] |
| Power systems | [110, 111, 417, 543, 779, 916, 943] |
| Scheduling and routing | [235, 270, 795] |
| Model selection | [542, 620] |
| Design | [169, 381, 819] |

## Mutation

Any of the mutation operators discussed in Section 11.2.1 can be used to adjust weight and bias values. Special mutation operators are available if the NN architecture is optimized simultaneously with weight values [28, 933]. Optimizing NN architecture is a discrete-valued optimization problem. Architecture mutation operators include node deletion and addition, as well as removing or adding a connection between two nodes.

### 11.6.4   Real-World Applications

Table 11.2 summarizes some applications of EP. This table should not be considered as an exhaustive list.

## 11.7   Assignments

1. Show if EP can be used to evolve the regular expression of a sequence of characters.

2. Use unconstrained functions from Section A.5.3 to show which probability distribution results in step sizes that maximize exploration.

3. Develop an EP to train an LVQ-I network.

4. The representation scheme used in Section 11.6.1 to evolve FSMs can be reduced to use less bits. Suggest a way in which this can be accomplished.

5. How can premature convergence be prevented in EP?

6. With reference to Chapter 12 discuss the similarities and differences between EP and ES.

7. With reference to Chapter 15 discuss the influence of different fitness sampling methods on EP performance.

8. With reference to Chapter 15 discuss the influence of different fitness sharing methods on EP performance.

9. Evaluate the performance of different mutation operators on the following deceptive function [169]:

$$f(x_1, x_2) = \left[1 - \left|\frac{\sin(\pi(x_1 - 2))\sin(\pi(x_2 - 2))}{\pi^2(x_1 - 2)(x_2 - 2)}\right|^5\right]$$
$$\times [2 + (x_1 - 7)^2 + 2(x_2 - 7)^2] \tag{11.87}$$

for $x_1, x_2, \sim U(0, 12)$. The global minimum is $f^*(x_1, x_2) = 0$ at $(x_1, x_2) = (2, 2)$, and the local minimum of $f^*(x_1, x_2) = 2$ at $(x_1, x_2) = (7, 7)$.

10. Explain why mutational noise should have a mean of 0.

11. Propose a method to ensure that the uniform mutation operator in equation (11.6) does not prematurely stagnate.

12. Compare the characteristics of the Lévy distribution with that of the Gaussian and Cauchy distributions in relation to the exploration–exploitation trade-off.

13. Propose ways in which strategy parameters can be initialized.

14. Propose an approach to self-adapt the $\alpha$ parameter of the Lévy distribution.

15. Discuss the merits of the following approach to calculate dynamic strategy parameters:

$$\sigma_{ij}(t) = \sigma_i(t) = |f(\hat{\mathbf{y}}(t)) - f(\mathbf{x}_i(t))| \tag{11.88}$$

where $\hat{\mathbf{y}}(t)$ is the best individual of the current generation.

# Chapter 12

# Evolution Strategies

Rechenberg reasoned that, since biological processes have been optimized by evolution, and evolution is a biological process itself, then it must be the case that evolution optimizes itself [710]. Evolution strategies (ES), piloted by Rechenberg in the 1960s [708, 709] and further explored by Schwefel [768], are then based on the concept of *the evolution of evolution*. While ESs consider both genotypic and phenotypic evolution, the emphasis is toward the phenotypic behavior of individuals. Each individual is represented by its genetic building blocks and a set of strategy parameters that models the behavior of that individual in its environment. Evolution then consists of evolving both the genetic characteristics and the strategy parameters, where the evolution of the genetic characteristics is controlled by the strategy parameters. An additional difference between ESs and other EC paradigms is that changes due to mutation are only accepted in the case of success. In other words, mutated individuals are only accepted if the mutation resulted in improving the fitness of the individual. Also interesting in ESs is that offspring can also be produced from more than two parents.

The rest of this chapter is organized as follows: An overview of the first ES is given in Section 12.1. A generic framework for ES algorithms is given in Section 12.2, and the main components of ES are discussed. Section 12.3 discusses strategy parameters – one of the most distinguishing aspects of ES. Evolutionary operators for ES are described in Section 12.4. A few ES variants are described in Section 12.5. Advanced topics are addressed in Section 12.6, including constraint handling, multi-objective optimization, niching, and dynamic environments.

## 12.1   $(1 + 1)$-ES

The first ES was developed for experimental optimization, applied to hydrodynamical problems [708]. This ES, referred to as the $(1 + 1)$-ES, does not make use of a population. A single individual is used from which one offspring is produced through application of a mutation operator. The $(1 + 1)$-ES is one of the first evolutionary algorithms that represents an individual as a tuple to consist of the decision vector, $\mathbf{x}$, to be optimized and a vector of strategy parameters, $\sigma$. The strategy parameter vector represents the mutational step size for each dimension, which is adapted dynamically according to performance.

*Computational Intelligence: An Introduction*, Second Edition A.P. Engelbrecht
©2007 John Wiley & Sons, Ltd

The individual is represented as the tuple,

$$\chi(t) = (\mathbf{x}(t), \sigma(t)) \tag{12.1}$$

According to the biological observation that offspring are similar to their parents, and that smaller deviations from the parent occur more often than larger ones, the offspring,

$$\chi'(t) = (\mathbf{x}'(t), \sigma'(t)) \tag{12.2}$$

is created (very similar to the CEP in Chapter 11) by adding Gaussian noise as follows:

$$
\begin{aligned}
x_j'(t) &= x_j(t) + N_j(0, \sigma_j(t)) \\
&= x_j(t) + \sigma_j(t) N_j(0, 1)
\end{aligned}
\tag{12.3}
$$

Strategy parameters are adapted based on the 1/5 success rule proposed by Rechenberg: Increase deviations, $\sigma_j$, if the relative frequency of successful mutations over a certain period is larger than 1/5; otherwise, deviations are decreased. Schwefel [769, 770] proposed that, after $t > 10n_x$, if $t \bmod n_x = 0$, the number of successful mutations, $n_m$, that have occurred during steps $t - 10n_x$ to $t - 1$ is calculated. The deviations are then updated using,

$$
\sigma_j'(t) = \begin{cases}
\alpha \sigma_j(t) & \text{if } n_m < 2n_x \\
\sigma_j(t)/\alpha & \text{if } n_m > 2n_x \\
\sigma_j(t) & \text{if } n_m = 2n_x
\end{cases}
\tag{12.4}
$$

where $\alpha = 0.85$. A successful mutation produces an offspring with a fitness that is better than the fitness of the parent.

Note that the original $(1+1)$-ES as proposed by Rechenburg did not adapt deviations. Variations have also been proposed where $\sigma_j(t) = \sigma$, $j = 1, \ldots, n_x$.

The selection operator selects the best between the parent and the offspring. That is, assuming minimization,

$$
\mathbf{x}(t+1) = \begin{cases}
\mathbf{x}'(t) & \text{if } f(\mathbf{x}'(t)) < f(\mathbf{x}(t)) \\
\mathbf{x}(t) & \text{otherwise}
\end{cases}
\tag{12.5}
$$

and

$$
\sigma(t+1) = \begin{cases}
\sigma'(t) & \text{if } f(\mathbf{x}'(t)) < f(\mathbf{x}(t)) \\
\sigma(t) & \text{otherwise}
\end{cases}
\tag{12.6}
$$

Rechenberg [709] suggested that the $(1 + 1)$-ES can be extended to a multimembered ES, denoted as the $(\mu + 1)$-ES. This strategy uses a population of $\mu > 1$ parents. Two parents are randomly selected and recombined by discrete, multipoint crossover to produce one offspring. If $\mathbf{x}_1(t)$ and $\mathbf{x}_2(t)$ denote the two parents, then

$$
x_j'(t) = \begin{cases}
x_{1j}(t) & \text{if } r_j \leq 0.5 \\
x_{2j}(t) & \text{otherwise}
\end{cases}
\tag{12.7}
$$

and
$$\sigma_j(t) = \left\{ \begin{array}{ll} \sigma_{1j}(t) & \text{if } r_j \leq 0.5 \\ \sigma_{2j}(t) & \text{otherwise} \end{array} \right. \tag{12.8}$$

where $r_j \sim U(0,1), j = 1, \ldots, n_x$.

The offspring is mutated as for $(1+1)$-ES. An elitist approach is followed to select the new population: the best $\mu$ individuals out of the $\mu + 1$ (parents and offspring) survive to the next generation.

Due to problems with self-adaptation of step sizes, $(\mu + 1)$-ES (also referred to as the steady-state ES) have not been regularly used.

## 12.2   Generic Evolution Strategy Algorithm

A generic framework for the implementation of an ES is given in Algorithm 12.1. Parameters $\mu$ and $\lambda$ respectively indicate the number of parents and the number of offspring.

---

**Algorithm 12.1** Evolution Strategy Algorithm

---

Set the generation counter, $t = 0$;
Initialize the strategy parameters;
Create and initialize the population, $\mathcal{C}(0)$, of $\mu$ individuals;
**for** *each individual,* $\chi_i(t) \in \mathcal{C}(t)$ **do**
    Evaluate the fitness, $f(\mathbf{x}_i(t))$;
**end**
**while** *stopping condition(s) not true* **do**
    **for** $i = 1, \ldots, \lambda$ **do**
        Choose $\rho \geq 2$ parents at random;
        Create offspring through application of crossover operator on parent
        genotypes and strategy parameters;
        Mutate offspring strategy parameters and genotype;
        Evaluate the fitness of the offspring;
    **end**
    Select the new population, $\mathcal{C}(t + 1)$;
    $t = t + 1$;
**end**

---

As summarized in Algorithm 12.1, an ES uses the following main components:

- **Initialization:** For each individual, its genotype is initialized to fall within the problem boundary constraints. The strategy parameters are also initialized.

- **Recombination:** Offspring are produced through application of a crossover operator on two or more parents. ES crossover operators are discussed in Section 12.4.2.

- **Mutation:** Offspring are mutated, where mutational step sizes are determined from self-adaptive strategy parameters. Mutation operators for ES are discussed in Section 12.4.3.

- **Evaluation:** An absolute fitness function is used to determine the quality of the solution represented by the genotype of the individual.

- **Selection:** Selection operators are used for two purposes in an ES. Firstly, to select parents for recombination, and secondly, to determine which individuals survive to the next generation. Selection methods for ES are discussed in Section 12.4.1.

Any of the stopping conditions discussed in Section 8.7 can be used to terminate execution of an ES.

## 12.3 Strategy Parameters and Self-Adaptation

As with EP, strategy parameters are associated with each individual. These strategy parameters are self-adapted in order to determine the best search direction and maximum step size per dimension. In essence, the strategy parameters define the mutation distribution from which mutational step sizes are sampled. The main goal of a self-adaptation strategy is to refine the mutation distribution such that maximal search progress is maintained. This section discusses strategy parameters and self-adaptation in relation to ES. Since much of what has been discussed about self-adaptation in EP (refer to Chapter 11) is also applicable to ES, this section emphasizes aspects related to ES.

Section 12.3.1 discusses different types of strategy parameters, while variations in which these parameters can be used are described in Section 12.3.2. Different self-adaptation strategies proposed in ES literature is overviewed in Section 12.3.3.

### 12.3.1 Strategy Parameter Types

First implementations of ES used one type of strategy parameter, i.e. the deviation of the Gaussian distributed noise used by the mutation operator [708, 709, 769]. In this case, individuals are represented as

$$\chi_i(t) = (\mathbf{x}_i(t), \sigma_i(t)) \tag{12.9}$$

where $\mathbf{x}_i \in \mathbb{R}^{n_x}$ represents the genotype (i.e. the vector of decision variables), and $\sigma_i$ represents the deviation strategy parameter vector. Usually, $\sigma_i \in \mathbb{R}_+^{n_x}$. However, ES have been tested using one deviation for all components of the genotype, i.e. $\sigma_{ij} = \sigma_i, j = 1, \ldots, n_x$, in which case $\sigma_i \in \mathbb{R}_+$ [42, 39].

Using more strategy parameters provide more degrees of freedom to individuals to fine tune their mutation distribution in all dimensions (refer to Section 12.3.2 for visual illustrations of this point).

If deviations are used as the only strategy parameters, best search directions are determined along the axes of the coordinate system in which the search space resides. It is not always the case that the best search direction (i.e. the gradient) is aligned with the axes. In such cases, the search trajectory have been shown to fluctuate along the gradient, decreasing the rate of progress toward the optimum [43]. More information about the search is needed to speed up convergence for such cases. More information about the fitness function, which defines the search space, can be obtained by the Hessian matrix of the fitness function. If the Hessian is used as strategy parameter, mutations are done as follows:

$$\mathbf{x}_i'(t) = \mathbf{x}_i(t) + N(\mathbf{0}, \mathbf{H}^{-1}) \tag{12.10}$$

where $\mathbf{H}$ is the Hessian matrix.

It is, however, not feasible to use the Hessian matrix. Fitness (objective) functions are not always guaranteed to have a second-order derivative. Even if a second-order derivative does exist, it is computationally expensive to calculate the Hessian.

Schwefel [769] proposed that the covariance matrix, $\mathbf{C}^{-1}$, described by the deviation strategy parameters of the individual, be used as additional information to determine optimal step sizes and directions. In this case,

$$\mathbf{x}_i'(t) = \mathbf{x}_i(t) + N(\mathbf{0}, \mathbf{C}) \tag{12.11}$$

where $N(\mathbf{0}, \mathbf{C})$ refers to a normally distributed random vector $\mathbf{r}$ with expectation zero and probability density [43],

$$f_G(\mathbf{r}) = \frac{\det \mathbf{C}}{(2\pi)^{n}_x} e^{-\frac{1}{2}\mathbf{r}^T \mathbf{C}\mathbf{r}} \tag{12.12}$$

The diagonal elements of $\mathbf{C}^{-1}$ are the variances, $\sigma_j^2$, while the off-diagonal elements are the covariances of the mutational step sizes.

Covariances are given by rotation angles which describe the rotations that need to be done to transform an uncorrelated mutation vector to a correlated vector. If $\omega_i(t)$ denotes the vector of rotational angles for individual $i$, then individuals are represented as the triplet,

$$\chi_i(t) = (\mathbf{x}_i(t), \sigma_i(t), \omega_i(t)) \tag{12.13}$$

where $\mathbf{x}_i(t) \in \mathbb{R}^{n_x}, \sigma_i(t) \in \mathbb{R}_+^{n_x}, \omega_i(t) \in \mathbb{R}^{n_x(n_x-1)/2}$, and $\omega_{ik}(t) \in (0, 2\pi], k = 1, \ldots, n_x(n_x - 1)/2$.

The rotational angles are used to represent the covariances among the $n_x$ genetic variables in the genetic vector $\mathbf{x}_i$. Because the covariance matrix is symmetric, a vector can be used to represent the rotational angles instead of a matrix. The rotational angles are used to calculate an orthogonal rotation matrix, $T(\omega_i)$, as

$$T(\omega_i) = \prod_{l=1}^{n_x-1} \prod_{j=i+1}^{n_x} R_{lj}(\omega_i) \tag{12.14}$$

which is the product of $n_x(n_x - 1)/2$ rotation matrices. Each rotation matrix $R_{lj}(\omega_i)$ is a unit matrix with $r_{ll} = \cos(\omega_{ik})$ and $r_{lj} = -r_{jl} = -\sin(\omega_{ik})$, with $k = 1 \Leftrightarrow (l = 1, j = 2), k = 2 \Leftrightarrow (l = 1, j = 3), \cdots$. The rotational matrix is used by the mutation operator as described in Section 12.4.3.

## 12.3.2   Strategy Parameter Variants

As discussed in Section 12.3.1, the two types of strategy parameters that have been used are the standard deviation of mutational step sizes, and rotational angles that represent covariances of mutational step sizes. These strategy parameters have resulted in a number of self-adaptation variants [39, 364]. For the discussion below, let $n_\sigma$ denote the number of deivation parameters used, and $n_\omega$ the number of rotational angles. The following cases have been used:

- $n_\sigma = 1, n_\omega = 0$, i.e. only one deviation parameter is used ($\sigma_j = \sigma \in \mathbb{R}_+, j = 1, \ldots, n_x$) for all components of the genotype, and no rotational angles. The mutation distribution has a circular shape as illustrated in Figure 12.1(a). The middle of the circle indicates the position of the parent, $\mathbf{x}_i$, while the boundary indicates the deviation in step sizes. Keep in mind that this distribution indicates the probability of the position of the offspring, $\mathbf{x}'_i$, with the highest probability at the center.

  The strategy parameter is adjusted as follows:

  $$\sigma'_i(t) = \sigma_i(t)e^{\tau N(0,1)} \tag{12.15}$$

  where $\tau = \frac{1}{\sqrt{n_x}}$.

  While adjustment of the single parameter is computationally fast, the approach is not flexible when the coordinates have different gradients.

- $n_\sigma = n_x, n_\omega = 0$, in which each component has its own deviation parameter. The mutation distribution has an elliptic shape as illustrated in Figure 12.1(b), where $\sigma_1 < \sigma_2$. In this case the increased number of parameters causes a linear increase in computational complexity, but the added degrees of freedom provide for better flexibility. Different gradients along the coordinate axes can now be taken into consideration.

  Strategy parameters are updated as follows:

  $$\sigma'_{ij}(t) = \sigma_{ij}(t)e^{\tau' N(0,1)+\tau N_j(0,1)} \tag{12.16}$$

  where $\tau' = \frac{1}{\sqrt{2n_x}}$ and $\tau = \frac{1}{\sqrt{2\sqrt{n_x}}}$.

- $n_\sigma = n_x, n_\omega = n_x(n_x - 1)/2$, where in addition to the deviations, rotational angles are used. The elliptical mutation distribution is rotated with respect to the coordinate axes as illustrated in Figure 12.1(c). Such rotations allow better approximation of the contours of the search space.

  Deviation parameters are updated using equation (12.16), while rotational angles are updated using,

  $$\omega'_{ik}(t) = \omega_{ik}(t) + \gamma N_j(0, 1) \bmod 2\pi \tag{12.17}$$

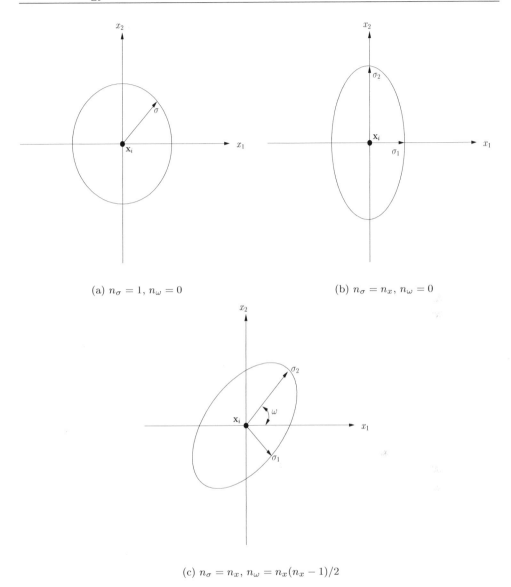

(a) $n_\sigma = 1$, $n_\omega = 0$

(b) $n_\sigma = n_x$, $n_\omega = 0$

(c) $n_\sigma = n_x$, $n_\omega = n_x(n_x - 1)/2$

**Figure 12.1** Illustration of Mutation Distributions for ES

where $\gamma \approx 0.0873$ [39].

Adding the rotational angles improves flexibility, but at the cost of a quadratic increase in computational complexity.

- $1 < n_\sigma < n_x$: This approach allows for different degrees of freedom. For all $j > n_\sigma$, deviation $\sigma_{n_\sigma}$ is used.

## 12.3.3  Self-Adaptation Strategies

The most frequently used approach to self-adapt strategy parameters is the lognormal self-adaptation mechanism used above in Section 12.3.2. Additive methods as discussed in Section 11.3.3 for EP can also be used.

Lee *et al.* [507] and Müller *et al.* [614] proposed that reinforcement learning be used to adapt strategy parameters, as follows:

$$\sigma'_{ij}(t) = \sigma_{ij}(t)e^{\Theta_i(t)|\tau' N(0,1)+\tau N_j(0,1)|} \tag{12.18}$$

where $\Theta_i(t)$ is the sum of temporal rewards over the last $n_\Theta$ generations for individual $i$, i.e.

$$\Theta_i(t) = \frac{1}{n_\Theta} \sum_{t'=0}^{n_\Theta} \theta_i(t - t') \tag{12.19}$$

Different methods can be used to calculate the reward for each individual at each time step. Lee *et al.* [507] proposed that

$$\theta_{ij}(t) = \begin{cases} 0.5 & \text{if } \Delta f(\mathbf{x}_i(t)) > 0 \\ 0 & \text{if } \Delta f(\mathbf{x}_i(t)) = 0 \\ -1 & \text{if } \Delta f(\mathbf{x}_i(t)) < 0 \end{cases} \tag{12.20}$$

where deterioration in fitness is heavily penalized. In equation (12.20),

$$\Delta f(\mathbf{x}_i(t)) = f(\mathbf{x}_i(t)) - f(\mathbf{x}_i(t - 1)) \tag{12.21}$$

Müller *et al.* [614] suggested a reward of $+1, 0$ or $-1$ depending on performance. Alternatively, they suggested that

- $\theta_{ij}(t) = f(\mathbf{x}_i(t)) - f(\mathbf{x}_i(t - \Delta t))$, with $0 < \Delta t < t$. This approach bases rewards on changes in phenotypic behavior, as quantified by the fitness function. The more an individual improves its current fitness, the greater the reward. On the other hand the worse the fitness of the individual becomes, the greater the penalty for that individual.
- $\theta_{ij}(t) = \text{sign}(f(\mathbf{x}_i(t)) - f(\mathbf{x}_i(t - \Delta t)))$. This scheme results in $+1, 0, -1$ rewards.
- $\theta_{ij}(t) = ||\mathbf{x}_i(t) - \mathbf{x}_i(t - \Delta t)||\text{sign}(f(\mathbf{x}_i(t)) - f(\mathbf{x}_i(t - \Delta t)))$. Here the reward (or punishment) is proportional to the step size in decision (genotypic) space.

Ostermeier and Hansen [645] considered a self-adaptation scheme where $n_\sigma = 1$ and where a covariance matrix is used. In this scheme, the deviation of an offspring is calculated as a function of the deviations of those parents from which the offspring has been derived. For each offspring, $\mathbf{x}'_l(t), l = 1, \ldots, \lambda$,

$$\sigma'_l(t) = \left( \sqrt[\rho]{\prod_{i \in \Omega_l(t)} \sigma_i(t)} \right) e^\xi \tag{12.22}$$

where $\Omega_l(t)$ is the index set of the $\rho$ parents of offspring $\mathbf{x}'_l(t)$, and the distribution of $\xi$ is such that $\text{prob}(\xi = 0.4) = \text{prob}(\xi = -0.4) = 0.5$. Section 12.4.3 shows how this self-adaptation scheme is used in a coordinate system independent mutation operator.

Kursawe [492] used a self-adaptation scheme where $1 \leq n_\sigma \leq n_x$, and each individual uses a different number of deviation parameters, $n_{\sigma_i}(t)$. At each generation, $t$, the number of deviation parameters can be increased or decreased at a probability of 0.05. If the number of deviation parameters increases, i.e. $n_{\sigma_i}(t) = n_{\sigma_i}(t-1)$, then the new deviation parameter is initialized as

$$\sigma_{in_{\sigma,i}(t)}(t) = \frac{1}{n_{\sigma,i}(t-1)} \sum_{k=1}^{n_{\sigma,i}(t-1)} \sigma_{ik}(t) \tag{12.23}$$

## 12.4 Evolution Strategy Operators

Evolution strategies use the three main operators of EC, namely selection, crossover, and mutation. These operators are discussed in Sections 12.4.1, 12.4.2, and 12.4.3 respectively.

### 12.4.1 Selection Operators

Selection is used for two tasks in an ES: (1) to select parents that will take part in the recombination process and (2) to select the new population. For selecting the $\rho$ parents for the crossover operator, any of the selection methods reviewed in Section 8.5 can be used. Usually, parents are randomly selected.

For each generation, $\lambda$ offspring are generated from $\mu$ parents and mutated. After crossover and mutation, the individuals for the next generation are selected. Two main strategies have been developed:

- $(\mu + \lambda) - ES$: In this case (also referred to as the plus strategies) the ES generates $\lambda$ offspring from $\mu$ parents, with $1 \leq \mu \leq \lambda < \infty$. The next generation consists of the $\mu$ best individuals selected from $\mu$ parents and $\lambda$ offspring. The $(\mu+\lambda)-ES$ strategy implements elitism to ensure that the fittest parents survive to the next generation.

- $(\mu, \lambda) - ES$: In this case (also referred to as the comma strategies), the next generation consists of the $\mu$ best individuals selected from the $\lambda$ offspring. Elitism is not used, and therefore this approach exhibits a lower selective pressure than the plus strategies. Diversity is therefore larger than for the plus strategies, which results in better exploration. The $(\mu, \lambda) - ES$ requires that $1 \leq \mu < \lambda < \infty$.

Using the above notation, ES are collectively referred to as $(\mu \overset{+}{,} \lambda)$-ES. The $(\mu + \lambda)$ notation has been extended to $(\mu, \kappa, \lambda)$, where $\kappa$ denotes the maximum lifespan of an individual. If an individual exceeds its lifespan, it is not selected for the next population. Note that $(\mu, \lambda)$-ES is equivalent to $(\mu, 1, \lambda)$-ES.

The best selection strategy to use depends on the problem being solved. Highly convoluted search spaces need more exploration, for which the $(\mu, \lambda)$-ES are more applicable.

Because information about the characteristics of the search space is usually not available, it is not possible to say which selection scheme will be more appropriate for an arbitrary function. For this reason, Huang and Chen [392] developed a fuzzy controller to decide on the number of parents that may survive to the next generation. The fuzzy controller receives population diversity measures as input, and attempts to balance exploration against exploitation.

Runarsson and Yao [746] developed a continuous selection method for ES, which is essentially a continuous version of $(\mu, \lambda)$-ES. The basis of this selection method is that the population changes continuously, and not discretely after each generation. There is no selection of a new population at discrete generational intervals. Selection is only used to select parents for recombination, based on a fitness ranking of individuals. As soon as a new offspring is created, it is inserted in the population and the ranking is immediately updated. The consequence is that, at each creation of an offspring, the worst individual among the $\mu$ parents and offspring is eliminated.

## 12.4.2   Crossover Operators

In order to introduce recombination in ES, Rechenberg [709] proposed that the $(1+1)$-ES be extended to a $(\mu + 1)$-ES (refer to Section 12.1). The $(\mu + 1)$-ES is therefore the first ES that utilized a crossover operator. In ES, crossover is applied to both the genotype (vector of decision variables) and the strategy parameters. Crossover is implemented somewhat differently from other EAs.

Crossover operators differ in the number of parents used to produce a single offspring and in the way that the genetic material and strategy parameters of the parents are combined to form the offspring. In general, the notation $(\mu/\rho, \stackrel{+}{,} \lambda)$ is used to indicate that $\rho$ parents are used per application of the crossover operator. Based on the value of $\rho$, the following two approaches can be found:

- **Local crossover** $(\rho = 2)$, where one offspring is generated from two randomly selected parents.

- **Global crossover** $(2 < \rho \le \mu)$, where more than two randomly selected parents are used to produce one offspring. The larger the value of $\rho$, the more diverse the generated offspring is compared to smaller $\rho$ values. Global crossover with large $\rho$ improves the exploration ability of the ES.

In both local and global crossover, recombination is done in one of two ways:

- **Discrete recombination**, where the actual allele of parents are used to construct the offspring. For each component of the genotype or strategy parameter vectors, the corresponding component of a randomly selected parent is used. The notation $(\mu/\rho_D \stackrel{+}{,} \lambda)$ is used to denote discrete recombination.

- **Intermediate recombination**, where allele for the offspring is a weighted average of the allele of the parents (remember that floating-point representations are assumed for the genotype). The notation $(\mu/\rho_I \stackrel{+}{,} \lambda)$ is used to denote intermediate recombination.

Based on the above, five main types of recombination have been identified for ES:

- **No recombination:** If $\chi_i(t)$ is the parent, the offspring is simply $\tilde{\chi}_l(t) = \chi_i(t)$.
- **Local, discrete recombination**, where

$$\tilde{\chi}_{lj}(t) = \begin{cases} \chi_{i_1 j}(t) & \text{if } U_j(0,1) \leq 0.5 \\ \chi_{i_2 j}(t) & \text{otherwise} \end{cases} \tag{12.24}$$

The offspring, $\tilde{\chi}_l(t) = (\tilde{\mathbf{x}}_l(t), \tilde{\sigma}_l(t), \tilde{\omega}_l(t))$ inherits from both parents, $\chi_{i_1}(t) = (\mathbf{x}_{i_1}(t), \sigma_{i_1}(t), \omega_1(t))$ and $\chi_{i_2}(t) = (\mathbf{x}_{i_2}(t), \sigma_{i_2}(t), \omega_{i_2}(t))$.

- **Local, intermediate recombination**, where

$$\tilde{x}_{lj}(t) = r x_{i_1 j}(t) + (1-r) x_{i_2 j}(t), \ \forall j = 1, \ldots, n_x \tag{12.25}$$

and

$$\tilde{\sigma}_{lj}(t) = r \sigma_{i_1 j}(t) + (1-r) \sigma_{i_2 j}(t), \ \forall j = 1, \ldots, n_x \tag{12.26}$$

with $r \sim U(0,1)$. If rotational angles are used, then

$$\omega_{lk}(t) = [r\omega_{i_1 k}(t) + (1-r)\sigma_{i_2 k}(t)] \bmod 2\pi, \ \forall k = 1, \ldots, n_x(n_x - 1) \tag{12.27}$$

- **Global, discrete recombination**, where

$$\tilde{\chi}_{lj}(t) = \begin{cases} \chi_{i_1 j}(t) & \text{if } U_j(0,1) \leq 0.5 \\ \chi_{r_j j}(t) & \text{otherwise} \end{cases} \tag{12.28}$$

with $r_j \sim \Omega_l$; $\Omega_l$ is the set of indices of the $\rho$ parents selected for crossover.

- **Global, intermediate recombination**, which is similar to the local recombination above, except that the index $i_2$ is replaced with $r_j \sim \Omega_l$. Alternatively, the average of the parents can be calculated to form the offspring [62],

$$\tilde{\chi}_l(t) = \left( \frac{1}{\rho} \sum_{i=1}^{\rho} \mathbf{x}_i(t), \frac{1}{\rho} \sum_{i=1}^{\rho} \sigma_i(t), \frac{1}{\rho} \sum_{i=1}^{\rho} \omega_i(t) \right) \tag{12.29}$$

Izumi *et. al.* [409] proposed an arithmetic recombination between the best individual and the average over all the parents:

$$\tilde{x}_l(t) = r\hat{\mathbf{y}}(t) + (1-r)\frac{1}{\rho} \sum_{i \in \Omega_l} \mathbf{x}_i(t) \tag{12.30}$$

where $\hat{\mathbf{y}}(t)$ is the best individual of the current generation. The same can be applied to the strategy parameters. This strategy ensures that offspring are located around the best individual. However, care must be taken as this operator may cause premature stagnation, especially for large $r$.

## 12.4.3    Mutation Operators

Offspring produced by the crossover operator are all mutated with probability one. The mutation operator executes two steps for each offspring:

- The first step self-adapts strategy parameters as discussed in Sections 12.3.2 and 12.3.3.

- The second step mutates the offspring, $\tilde{\chi}_l$, to produce a mutated offspring, $\chi'_l$, as follows

$$\mathbf{x}'_l(t) = \tilde{x}_l(t) + \Delta \mathbf{x}_l(t) \tag{12.31}$$

  The $\lambda$ mutated offspring, $\chi'_l(t) = (\mathbf{x}'_l(t), \tilde{\sigma}_l(t), \tilde{\omega}_l(t))$ take part in the selection process, together with the parents depending on whether a $(\mu + \lambda)$-ES or a $(\mu, \lambda)$-ES is used.

This section considers only mutation of the genotype, as mutation (self-adaptation) of the strategy parameters has been discussed in previous sections.

If only deviations are used as strategy parameters, the genotype, $\tilde{\mathbf{x}}_l(t)$, of each offspring, $\tilde{\chi}_l(t), l = 1, \ldots, \lambda$, is mutated as follows:

- If $n_\sigma = 1$, $\Delta x_{lj}(t) = \sigma_l(t) N_j(0, 1), \forall j = 1, \ldots, n_x$.
- If $n_\sigma = n_x$, $\Delta x_{lj}(t) = \sigma_{lj}(t) N_j(0, 1), \forall j = 1, \ldots, n_x$
- If $1 < n_\sigma < n_x$, $\Delta x_{lj}(t) = \sigma_{lj}(t) N_j(0, 1), \forall j = 1, \ldots, n_\sigma$ and $\Delta x_{lj}(t) = \sigma_{l n_\sigma}(t) N_j(0, 1), \forall j = n_\sigma + 1, \ldots, n_x$

If deviations and rotational angles are used, assuming that $n_\sigma = n_x$, then

$$\Delta \mathbf{x}_l(t) = \mathbf{T}(\tilde{\omega}_l(t)) \mathbf{S}(\tilde{\sigma}_l(t)) \mathbf{N}(0, 1) \tag{12.32}$$

where $\mathbf{T}(\tilde{\omega}_l(t))$ is the orthogonal rotation matrix,

$$\mathbf{T}(\tilde{\omega}_l(t)) = \prod_{a=1}^{n_x - 1} \prod_{b=a+1}^{n_x} \mathbf{R}_{ab}(\tilde{\omega}_l(t)) \tag{12.33}$$

which is a product of $n_x(n_x - 1)/2$ rotation matrices. Each rotation matrix, $\mathbf{R}_{ab}(\tilde{\omega}_l(t))$, is a unit matrix with each element defined as follows: $r = \cos(\tilde{\omega}_{lk})$ and $r_{ab} = -r_{ba} = -\sin(\tilde{\omega}_{lk})$, for $k = 1, \ldots, n_x(n_x - 1)/2$ and $k = 1 \Leftrightarrow (a = 1, b = 2), k = 2 \Leftrightarrow (a = 1, b = 3), \ldots$. $\mathbf{S}(\tilde{\sigma}_l(t)) = \operatorname{diag}(\tilde{\sigma}_{l1}(t), \tilde{\sigma}_{l2}(t), \ldots, \tilde{\sigma}_{l n_x}(t))$ is the diagonal matrix representation of deviations.

Based on similar reasoning as for EP (refer to Section 11.2.1), Yao and Liu [935] replaced the Gaussian distribution with a Cauchy distribution to produce the fast ES. Huband et al. [395] developed a probabilistic mutation as used in GAs and GP, where each component of the genotype is mutated at a given probability. It is proposed that the probability of mutation be $1/n_x$. This approach imposes a smoothing effect on search trajectories.

Hildebrand et al. [364] proposed a directed mutation, where preference can be given to specific coordinate directions. As illustrated in Figure 12.2, the directed mutation

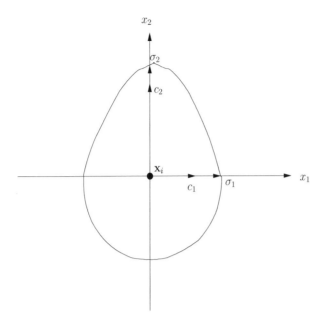

**Figure 12.2** Directed Mutation Operator for ES

results in an asymmetrical mutation probability distribution. Here the step size is larger for the $x_2$ axis than for the $x_1$ axis, and positive directions are preferred. As each component of the genotype is mutated independently, it is sufficient to define a 1-dimensional asymmetrical probability density function. Hildebrand *et al.* proposed the function,

$$f_D(x) = \begin{cases} \frac{2}{\sqrt{\pi}\sigma(1+\sqrt{1+c})} \left(e^{-\frac{x^2}{\sigma}}\right) & \text{if } x < 0 \\ \frac{2}{\sqrt{\pi}\sigma(1+\sqrt{1+c})} \left(e^{-\frac{x^2}{\sigma(1+c)}}\right) & \text{if } x \geq 0 \end{cases} \tag{12.34}$$

where $c > 0$ is the positive directional value.

The directional mutation method uses only deviations as strategy parameters, but associates a directional value, $c_j$, with each deviation, $\sigma_j$. Both $\sigma$ and $\mathbf{c}$ are self-adapted, giving a total of $2n_x$ strategy parameters. This is computationally more efficient than using a $n_x(n_x - 1)/2$-sized rotational vector, and provides more information about preferred search directions and step sizes than deviations alone.

If $D(c, \sigma)$ denotes the asymmetric distribution, then $\Delta x_{ij}(t) = D_j(c_{ij}(t), \sigma_{ij}(t))$.

Ostermeier and Hansen [645] developed a coordinate system invariant mutation operator, with self-adaptation as discussed in Section 12.3.3. Genotypes are mutated using both deviations and correlations, as follows:

$$\mathbf{x}'_l(t) = \frac{1}{\rho} \sum_{i \in \Omega_l(t)} \mathbf{x}_i(t) + \tilde{\sigma}_l N(\mathbf{0}, \mathbf{C}_l(t)) \tag{12.35}$$

where

$$\mathbf{C}_l(t) = \sum_{k=1}^{n_m} \xi_{lk}(t)\xi_{lk}^T(t) \tag{12.36}$$

with $\xi_{lk}(t) \sim N(\mathbf{0}, \frac{1}{\rho}\sum_{i\in\Omega_l(t)} \mathbf{C}_i(t))$ and $n_m$ is the mutation strength. For large values of $n_m$, the mutation strength is small, because a large sample, $\xi_1, \xi_2, \ldots, \xi_{n_m}$, provides a closer approximation to the original distribution than a smaller sample. Ostermeier and Hansen suggested that $n_m = n_x$.

## 12.5  Evolution Strategy Variants

Previous sections have already discussed a number of different self-adaptation and mutation strategies for ES. This section describes a few ES implementations that differ somewhat from the generic ES algorithm summarized in Algorithm 12.1.

### 12.5.1  Polar Evolution Strategies

Bian *et al.* [66], and Sierra and Echeverría [788] independently proposed that the components of genotype be transformed to polar coordinates. Instead of the original genotype, the "polar genotypes" are evolved.

For an $n_x$-dimensional Cartesian coordinate, the corresponding polar coordinate is given as

$$(r, \theta_{n_x-2}, \ldots, \theta_1, \phi) \tag{12.37}$$

where $0 \le \phi < 2\pi$, $0 \le \theta_q \le \pi$ for $q = 1, \ldots, n_x - 2$, and $r > 0$. Each individual is therefore represented as

$$\chi_i(t) = (\mathbf{x}_i^p(t), \sigma_i(t)) \tag{12.38}$$

where , $\mathbf{x}_i^p = (r, \theta_{n_x} - 2, \ldots, \theta_1, \phi)$. Polar coordinates are transformed back to Cartesian coordinates as follows:

$$
\begin{aligned}
x_1 &= r\cos\phi\sin\theta_1\sin\theta_2\ldots\sin\theta_{n_x-2} \\
x_2 &= r\sin\phi\sin\theta_1\sin\theta_2\ldots\sin\theta_{n_x-2} \\
x_3 &= r\cos\theta_1\sin\theta_2\ldots\sin\theta_{n_x-2} \\
&\vdots \\
x_i &= r\cos\theta_{i-2}\sin\theta_{i-1}\ldots\sin\theta_{n_x-2} \\
&\vdots \\
x_n &= r\cos\theta_{n_x-2}
\end{aligned}
\tag{12.39}
$$

The mutation operator uses deviations to adjust the $\phi$ and $\theta_q$ angles:

$$\phi_l' = (\tilde{\phi}_l(t) + \tilde{\sigma}_{\phi,l}(t)N(0,1)) \bmod 2\pi \tag{12.40}$$

$$\theta_{lq}(t) = \pi - (\tilde{\theta}_{lq}(t) + \tilde{\sigma}_{\theta_{lq}}(t)N_q(0,1)) \bmod \pi \tag{12.41}$$

---

**Algorithm 12.2** Polar Evolution Strategy

---

Set the generation counter, $t = 0$;
Initialize the strategy parameters, $\sigma_\phi, \sigma_{\theta,q}, q = 1, \ldots, n_x - 2$;
Create and initialize the population, $\mathcal{C}(0)$, as follows:;
**for** $i = 1, \ldots, \mu$ **do**
    $r = 1$;
    $\phi_i(0) \sim U(0, 2\pi)$;
    $\theta_{iq}(0) \sim U(0, \pi), \forall q = 1, \ldots, n_x - 2$;
    $\mathbf{x}_i^p(0) = (r, \theta_i(0), \phi_i(0))$;
    $\chi_i(0) = (\mathbf{x}_i^p(0), \sigma_i(0))$;
**end**
**for** *each individual,* $\chi_l(0) \in \mathcal{C}(0)$ **do**
    Transform polar coordinate $\mathbf{x}_i^p(0)$ to Cartesian coordinate $\mathbf{nx}_i(0)$;
    Evaluate the fitness, $f(\mathbf{x}_i(0))$;
**end**
**while** *stopping condition(s) not true* **do**
    **for** $l = 1, \ldots, \lambda$, *generate offspring* **do**
        Randomly choose two parents;
        Create offspring, $\tilde{\chi}_l(t)$, using local, discrete recombination;
        Mutate $\tilde{\chi}_l(t)$ to produce $\chi_l'(t)$;
        Transform $\mathbf{x}_l^p(t)$ back to Cartesian $\mathbf{x}_l(t)$;
        Evaluate the fitness, $f(\mathbf{x}_i(t))$;
    **end**
    Select $\mu$ individuals from the $\lambda$ offspring to form $\mathcal{C}(t+1)$;
    $t = t + 1$;
**end**

---

where $\tilde{\phi}_l(t)$ and $\tilde{\theta}_{lq}(t), q = 1, \ldots, n_x - 2$ refer to the components of the off-spring, $\tilde{\chi}_l(t), l = 1, \ldots, \lambda$ produced by the crossover operator, and $\tilde{\sigma}_l(t) = (\tilde{\sigma}_{\phi,l}(t), \tilde{\sigma}_{\theta,l1}(t), \tilde{\sigma}_{\theta,l2}(t), \ldots, \tilde{\sigma}_{\theta,l(n_x-2)}(t))$, is its strategy parameter vector. Note that $r = 1$ is not mutated.

The polar ES as used in [788] is summarized in Algorithm 12.2.

## 12.5.2 Evolution Strategies with Directed Variation

A direction-based mutation operator has been discussed in Section 12.4.3. Zhou and Li [960] proposed a different approach to bias certain directions within the mutation operator, and presented two alternative implementations to utilize directional variation.

The approach is based on intervals defined over the range of each decision variable, and interval fitnesses. For each component of each genotype, the direction of mutation is towards a neighboring interval with highest interval fitness. Assume that the $j$-th component is bounded by the range $[x_{min,j}, x_{max,j}]$. This interval is divided into $n_I$

subintervals of equal length, where the $s$-th interval is computed as

$$I_{js} = \left[ x_{min,j} + (s-1)\left(\frac{x_{max,j} - x_{min,j}}{n_I}\right), x_{min,j} + s\left(\frac{x_{max,j} - x_{min,j}}{n_I}\right)\right] \quad (12.42)$$

The fitness of interval $I_{js}$ is defined by

$$f(I_{js}) = \sum_{i=1}^{\mu} f_I(x_{ij}(t) \in I_{js})\tilde{f}(\mathbf{x}_i(t)) \quad (12.43)$$

where

$$f_I(x_{ij}(t) \in I_{js}) = \begin{cases} 1 & \text{if } x_{ij}(t) \in I_{js} \\ 0 & \text{if } x_{ij}(t) \notin I_{js} \end{cases} \quad (12.44)$$

and $\tilde{f}(\mathbf{x}_i(t))$ is the normalized fitness of $\mathbf{x}_i(t)$,

$$\tilde{f}(\mathbf{x}_i(t)) = \frac{f(\mathbf{x}_i(t)) - f_{min}(t)}{f_{max}(t) - f_{min}(t)} \quad (12.45)$$

The minimum and maximum fitness of the current population is indicated by $f_{min}(t)$ and $f_{max}(t)$ respectively.

Directed variation is applied to each component of each individual as follows. For component $x_{ij}(t)$, the direction of mutation is determined by $f(I_{js})$, $f(I_{j,s-1})$ and $f(I_{j,s+1})$, where $x_{ij}(t) \in I_{js}$. If $f(I_{js}) > f(I_{j,s-1})$ and $f(I_{js}) > f(I_{j,s+1})$, no directed variation will be applied. If $f(I_{j,s-1}) > f(I_{js}) > f(I_{j,s+1})$, then $x_{ij}(t)$ moves toward subinterval $I_{j,s-1}$ with probability $1 - \frac{f(I_{js})}{f(I_{j,s-1})}$. The move is implemented by replacing $x_{ij}(t)$ with a random number uniformly distributed between $x_{ij}(t)$ and the middle-point of the interval $I_{j,s-1}$. A similar approach is followed when $f(I_{j,s-1}) < f(I_{js}) < f(I_{j,s+1})$. If $f(I_{js}) < f(I_{j,s-1})$ and $f(I_{js}) < f(I_{j,s+1})$, then $x_{ij}(t)$ moves toward any of its neighboring intervals with equal probability.

For the above, $f(I_{j0}) = f(I_{j,n_I+1}) = 0$.

Two approaches have been proposed to apply directed variation. For the first, directed variation is applied, after selection, to the $\mu$ members of the new population. For the second strategy, each parent produces one offspring using directed variation. Crossover is then applied as usual to create the remaining $\lambda - \mu$ offspring. The selection operator is applied to the $\mu$ parents, the $\mu$ offspring produced by directed variation, and the $\lambda - \mu$ offspring produced by crossover.

## 12.5.3   Incremental Evolution Strategies

Incremental ES search for an optimal solution by dividing the search process into $n_x$ phases – one phase for each decision variable [597]. Each phase consists of two steps. The first step applies a single variable evolution on the one decision variable, while the second step applies a multi-variable evolution after the first phase. For phase numbers less than $n_x$, a context vector is needed to evaluate the fitness of the partially evolved individual (similar to the CCGA and CPSO discussed in Sections 15.3 and 16.5.4).

### 12.5.4   Surrogate Evolution Strategy

Surrogate methods have been developed specifically for problems where the fitness function is computationally expensive to evaluate. The fitness function is approximated by a set of basis functions, called surrogates. Evaluation of the surrogate model is computationally less expensive than the original function. The reader is referred to [860] for more detail on surrogate models for ES.

## 12.6   Advanced Topics

This section shows how ES can be used to solve constrained problems (Section 12.6.1), multi-objective optimization problems (Section 12.6.2), problems with dynamically changing optima (Section 12.6.3), and to locate multiple optima (Section 12.6.4).

### 12.6.1   Constraint Handling Approaches

While a number of ES variations have been developed to cope with constraints, this section discusses only some of these approaches.

Tahk and Sun [830] converted the constrained problem to an unconstrained problem using the augmented Lagrangian approach given in Section A.6.2. A coevolutionary approach is used to find the saddle point, $(\mathbf{x}^*, \lambda_g^*, \lambda_h^*)$, of the Lagrangian given in equation (A.27). Two populations are used, each with different objectives, both evolved in parallel. Assuming a minimization problem, the one population minimizes the fitness function,

$$f(\mathbf{x}) = \max_{\lambda_g, \lambda_h} L(\mathbf{x}, \lambda_g, \lambda_h) \tag{12.46}$$

where $L(\mathbf{x}, \lambda_g, \lambda_h)$ is defined in equation (A.27). The second population maximizes the fitness function,

$$f(\lambda_g, \lambda_h) = \min_{\mathbf{x}} L(\mathbf{x}, \lambda_g, \lambda_h) \tag{12.47}$$

Both populations use an ES as search algorithm.

Kramer *et al.* [488] developed a biased mutation operator to lead the search to more promising, feasible areas. The mean of the Gaussian distribution, from which mutational step sizes are sampled, is biased to shift the center of the mutation distribution as illustrated in Figure 12.3.

Let $\xi_i(t) = (\xi_{i1}(t), \ldots, \xi_{in_x}(t))$ be the bias coefficient vector, with $\xi_{ij}(0) \sim U(-1, 1)$, for all $j = 1, \ldots, n_x$. The bias vector, $\beta_i(t)$, is then defined as $\beta_{ij}(t) = \sigma_{ij}(t)\xi_{ij}(t)$. Mutational step sizes are calculated as

$$\Delta x_{ij}(t) = \sigma_{ij}(t)N_j(0, 1) + \beta_{ij}(t) = N_j(\xi_{ij}(t), \sigma_{ij}(t)) \tag{12.48}$$

Bias coefficients are self-adapted using

$$\xi'_{ij}(t) = \xi_{ij}(t) + \alpha N(0, 1) \tag{12.49}$$

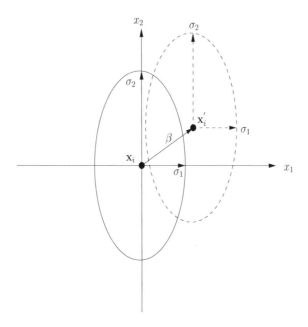

**Figure 12.3** Biased Mutation for Evolution Strategies

with $\alpha = 0.1$ (suggested in [488]).

A very simple approach to handle constraints is to change the selection operator to select the next population as follows: Until $\mu$ individuals have been selected,

- First select the best feasible individuals.
- If all feasible solutions have been selected, select those that violate the fewest constraints.
- As last resort, when individuals are infeasible, and they violate the same number of constraints, select the most fit individuals.

When a selection operator is applied to select one of two individuals, the following rules can be applied:

- If both are feasible, select the one with the best fitness.
- If one is feasible, and the other infeasible, select the feasible solution.
- If both are infeasible, select the one that violates the fewest constraints. If constraint violation is the same, select the most fit individual.

## 12.6.2  Multi-Objective Optimization

One of the first, simple ES for solving multi-objective (MOO) problems was developed by Knowles and Corne [469]. The Pareto archived evolution strategy (PAES) consists of three parts: (1) a candidate solution generator, (2) the candidate solution acceptance function, and (3) the nondominated-solutions archive.

The candidate solution generator is an $(1 + 1)$-ES, where the individual that survives to the next generation is based on dominance. If the parent dominates the offspring, the latter is rejected, and the parent survives. If the offspring dominates the parent, the offspring survives to the next generation. If neither the parent nor the offspring is dominating, the offspring is compared with the nondominated solutions in the archive, as summarized in Algorithm 12.4.

The archive maintains a set of nondominated solutions to the MOO. The size of the archive is restricted. When an offspring dominates the current solutions in the archive, it is included in the archive. When the offspring is dominated by any of the solutions in the archive, the offspring is not included in the archive. When the offspring and the solutions in the archive are nondominating, the offspring is accepted and included in the archive based on the degree of crowding in the corresponding area of objective space.

To keep track of crowding, a grid is defined over objective space, and for each cell of the grid a counter is maintained to keep track of the number of nondominated solutions for that part of the Pareto front. When an offspring is accepted into the archive, and the archive has reached its capacity, the offspring replaces one of the solutions in the highest populated grid cell (provided that the grid cell corresponding to the offspring has a lower frequency count). When the parent and its offspring are nondominating, the one with the lower frequency count is accepted in the archive.

The PAES is summarized in Algorithm 12.3.

Costa and Oliveira [159] developed a different approach to ensure that nondominated solutions survive to next generations, and to produce diverse solutions with respect to objective space. Fitnesses of individuals are based on a Pareto ranking, where individuals are grouped into a number of Pareto fronts. An individual's fitness is determined based on the Pareto front in which the individual resides. At each generation, the Pareto ranking process proceeds as follows. All of the nondominated solutions from the $\lambda$ offspring, or $\mu + \lambda$ parents and offspring, form the first Pareto front. These individuals are then removed, and the nondominating solutions from the remainder of the individuals form the second Pareto front. This process of forming Pareto fronts continues until all $\lambda$ (or $\mu + \lambda$) individuals are assigned to a Pareto front. Individuals of the first Pareto front is assigned a fitness of $1/n_c$, where $n_c$ is the niche count. The niche count is the number of individuals in this front that lies within a distance of $\sigma_{share}$ from the individual (distance is measured with respect to objective space). The threshold, $\sigma_{share}$, is referred to as the niche radius. Individuals of the next Pareto front is assigned a fitness of $(1 + f_{worst})/n_c$, where $f_{worst}$ is the worst fitness from the previous front. This process of fitness assignment continues until all individuals have been assigned a fitness.

The fitness sharing approach described above promotes diversity of nondominated solutions. The process of creating Pareto fronts ensures that dominated individuals are excluded from future generations as follows: Depending on the ES used, $\mu$ individuals

---

**Algorithm 12.3** Pareto Archived Evolution Strategy

---

Set generation counter;
Generate a random parent solution;
Evaluate the parent;
Add the parent to the archive;
**while** *stopping condition(s) not true* **do**
    Create offspring through mutation;
    Evaluate offspring;
    **if** *offspring is not dominated by parent* **then**
        Compare offspring with solutions in the archive;
        Update the archive using Algorithm 12.4;
        **if** *offspring dominated parent* **then**
            Offspring survives to new generation to become the parent;
        **end**
        **else**
            //nondominating case
            Select the individual that maps to the grid cell with lowest frequency count;
        **end**
    **end**
**end**

---

---

**Algorithm 12.4** Archive Update Algorithm used by PAES

---

Input: A candidate solution and an archive of nondominated solutions;
Output: An updated archive;
**if** *candidate is dominated by any member of the archive* **then**
    Reject the candidate;
**end**
**else**
    **if** *candidate dominates any solutions in the archive* **then**
        Remove all dominated members from the archive;
        Add candidate to the archive;
    **end**
    **else**
        **if** *archive is full* **then**
            **if** *candidate will increase diversity in the archive* **then**
                Remove the solution with the highest frequency from the archive;
            **end**
            Add candidate to the archive;
        **end**
    **end**
**end**

---

are selected from $\lambda$ or $\mu + \lambda$ individuals to form the new population. Individuals are sorted according to their fitness values in ascending order. If the number of individuals in the first front is greater than $\mu$, the next population is selected from the first front individuals using tournament selection. Otherwise the best $\mu$ individuals are selected. An archive of fixed size is maintained to contain a set of nondominated solutions, or elite. A specified number of these solutions are randomly selected and included in the next population.

At each generation, each Pareto optimal solution of the current population is tested for inclusion in the archive. The candidate solution is included if

- all solutions in the archive are different from the candidate solution, thereby further promoting diversity,

- none of the solutions in the archive dominates the candidate solution, and

- the distance between the candidate and any of the solutions in the archive is larger than a defined threshold.

Other ES approaches to solve MOPs can be found in [395, 464, 613, 637].

### 12.6.3   Dynamic and Noisy Environments

Beyer [61, 62, 63, 64] provided the first theoretical analyses of ES. In [61] Beyer made one of the first statements claiming that ES are robust against disturbances of the search landscape. Bäck and Hammel [42] provided some of the first empirical results to show that low levels of noise in the objective function do not have an influence on performance. Under higher noise levels, they recommended that the $(\mu, \lambda)$-ES, with $\mu > 1$ be used, due to its ability to maintain diversity. In a more elaborate study, Bäck [40] reconfirmed these statements and showed that the $(\mu, \lambda)$-ES was perfectly capable to track dynamic optima for a number of different dynamic environments.

Markon *et al.* [562] showed that $(\mu + \lambda)$-ES with threshold-selection is capable of optimizing noisy functions. With threshold-selection an offspring is selected only if its performance is better than its parents by a given margin.

Arnold and Beyer [32] provided a study of the performance of ES on noisy functions, where noise is generated from a Gaussian, Cauchy, or $\chi_1^2$ distribution. Beyer [65] and Arnold [31] showed that rescaled mutations improve the performance of ES on noisy functions. Under rescaled mutations, mutational step sizes are multiplied by a scaling factor (which is greater than one) to allow larger step sizes, while preventing that these large step sizes are inherited by offspring. Instead, the smaller step sizes are inherited.

### 12.6.4   Niching

In order to locate multiple solutions, Aichholzer *et al.* [12] developed a multipopulation ES, referred to as the $\tau(\mu/\rho, \kappa, \lambda)$-ES, where $\tau$ is the number of subpopulations and $\kappa$ specifies the number of generations that each individual is allowed to survive. Given $\tau$

clusters, recombination proceeds as follows: One of the clusters is randomly selected, and roulette wheel selection is used to select one of the individuals from this cluster as one of the parents. The second parent is selected from any cluster, but such that individuals from remote clusters are selected with a low probability. Over time, clusters tend to form over local minima.

Shir and Bäck [785] proposed a dynamic niching algorithm for ES that identifies fitness-peaks using a peak identification algorithm. Each fitness-peak will correspond to one of the niches. For each generation, the dynamic niche ES applies the following steps: The mutation operator is applied to each individual. After determination of the fitness of each individual, the dynamic peak identification algorithm is applied to identify $n_K$ niches. Individuals are assigned to their closest niche, after which recombination is applied within niche boundaries. If required, niches are populated with randomly generated individuals.

If $n_K$ niches have to be formed, they are created such that their peaks are a distance of at least $\frac{r}{n_x\sqrt{n_K}}$ from one another, where

$$r = \frac{1}{2}\sqrt{\sum_{j=1}^{n_x}(x_{max,j} - x_{min,j})^2} \qquad (12.50)$$

Peaks are formed from the current individuals of the population, by sequentially considering each as a fitness peak. If an individual is $\frac{r}{n_x\sqrt{n_K}}$ from already defined fitness peaks, then that individual is considered as the next peak. This process continues until $n_K$ peaks have been found or until all individuals have been considered.

Recombination considers a uniform distribution of resources. Each peak is allowed $\tilde{\mu} = \frac{\mu}{n_K}$ parents, and produces $\tilde{\lambda} = \frac{\lambda}{n_K}$ offspring in every generation. For each peak, $\tilde{\lambda}$ offspring are created as follows: One of the parents is selected from the individuals of that niche using tournament selection. The other parent, which should differ from the first parent, is selected as the best individual of that niche. If a niche contains only one individual, the second parent will be the best individual from another niche. Intermediate recombination is used for the strategy parameters, and discrete recombination for the genotypes.

The selection operator selects $\tilde{\mu}$ individuals for the next generation by selecting $n_{\tilde{\lambda}}$ of the best $\tilde{\lambda}$ offspring, along with the best $\tilde{\mu} - n_{\tilde{\lambda}}$ parents from that niche. If there are not enough parents to make up a total of $\tilde{\mu}$ individuals, the remainder is filled up by generating random individuals.

New niches are generated at each generation. Repeated application of the above process results in niches, where each niche represents a different solution.

Shir and Bäck [785] expanded the approach above to a $(1, \lambda)$-ES with corrrelated mutations.

Table 12.1 Applications of Evolution Strategies

| Application Class | References |
|---|---|
| Parameter optimization | [256, 443, 548, 679] |
| Controller design | [59, 456, 646] |
| Induction motor design | [464] |
| Neural network training | [59, 550] |
| Transformer design | [952] |
| Computer security | [52] |
| Power systems | [204] |

## 12.7 Applications of Evolution Strategies

The first application of ES was in experimental optimization as applied to hydrodynamical problems [708]. Since then most new ES implementations were tested on functional optimization problems. ES have, however, been applied to a number of real-world problems, as summarized in Table 12.1 (note that this is not an exhaustive list of applications).

## 12.8 Assignments

1. Discuss the differences and similarities between ES and EP.

2. Can an ES that utilizes strategy parameters be considered a cultural algorithm? Motivate your answer.

3. Determine if the reward scheme as given in equation (12.20) is applicable to minimization or maximization tasks. If it is for minimization (maximization) show how it can be changed for maximization (minimization).

4. Identify problems with the reinforcement learning approach where the reward is proportional to changes in phenotypic space. Do the same for the approach where reward is proportional to step sizes in decision (genetic) space.

5. Implement an $(1 + \lambda)$-ES to train a FFNN on any problem of your choice, and compare its performance with an $(\mu + 1)$-ES, $(\mu + \lambda)$-ES, and $(\mu, \lambda)$-ES.

6. Evaluate the following approaches to initialize deviation strategy parameters:

   (a) For all individuals, the same initial value is used for all components of the genotype, i.e.

   $$\sigma_{ij}(0) = \sigma(0), \ \forall i = 1, \ldots, n_s, \forall j = 1, \ldots, n_x$$

   where $\sigma(0) = \alpha \min_{j=1,\ldots,n_x} \{|x_{max,j} - x_{min,j}|\}$ with $\alpha = 0.9$.

   (b) The same as above, but $\alpha = 0.1$.

(c) For each individual, a different initial value is used, but the same for each component of the genotype, i.e.

$$\sigma_{ij}(0) = \sigma_i(0), \ \forall j = 1, \ldots, n_x$$

where $\sigma_i(0) \sim U(0, \alpha \min_{j=1,\ldots,n_x}\{|x_{max,j} - x_{min,j}|\})$ for $\alpha = 0.9$ and $\alpha = 0.1$.

(d) The same as above, but with $\sigma_i(0) \sim |N(0, \alpha \min_{j=1,\ldots,n_x}\{|x_{max,j} - x_{min,j}|\})|$ for $\alpha = 0.9$ and $\alpha = 0.1$.

(e) Each component of each genotype uses a different initial value, i.e.

$$\sigma_{ij}(0) \sim |N(0, \alpha \min_{j=1,\ldots,n_x} \{|x_{max,j} - x_{min,j}|\})|, \forall i = 1, \ldots, n_s, \forall j = 1, \ldots, n_x$$

for $\alpha \sim U(0, 1)$.

7. Discuss the advantages of using a lifespan within $(\mu, \kappa, \lambda)$-ES compared to $(\mu, \lambda)$-ES.

8. True or false: $(\mu + \lambda)$-ES implements a hill-climbing search. Motivate your answer.

# Chapter 13

# Differential Evolution

Differential evolution (DE) is a stochastic, population-based search strategy developed by Storn and Price [696, 813] in 1995. While DE shares similarities with other evolutionary algorithms (EA), it differs significantly in the sense that distance and direction information from the current population is used to guide the search process. Furthermore, the original DE strategies were developed to be applied to continuous-valued landscapes.

This chapter provides an overview of DE, organized as follows: Section 13.1 discusses the most basic DE strategy and illustrates the method of adaptation. Alternative DE strategies are described in Sections 13.2 and 13.3. Section 13.4 shows how the original DE can be applied to discrete-valued and binary-valued landscapes. A number of advanced topics are covered in Section 13.5, including multi-objective optimization (MOO), constraint handling, and dynamic environments. Some applications of DE are summarized in Section 13.6.

## 13.1 Basic Differential Evolution

For the EAs covered in the previous chapters, variation from one generation to the next is achieved by applying crossover and/or mutation operators. If both these operators are used, crossover is usually applied first, after which the generated offspring are mutated. For these algorithms, mutation step sizes are sampled from some probability distribution function. DE differs from these evolutionary algorithms in that

- mutation is applied first to generate a trial vector, which is then used within the crossover operator to produce one offspring, and

- mutation step sizes are not sampled from a prior known probability distribution function.

In DE, mutation step sizes are influenced by differences between individuals of the current population.

Section 13.1.1 discusses the concept of difference vectors, used to determine mutation step sizes. The mutation, crossover, and selection operators are described in Sections 13.1.2 to 13.1.4. Section 13.1.5 summarizes the DE algorithm, and control

*Computational Intelligence: An Introduction*, Second Edition A.P. Engelbrecht
©2007 John Wiley & Sons, Ltd

parameters are discussed in Section 13.1.6. A geometric illustration of the DE variation approach is given in Section 13.1.7.

## 13.1.1   Difference Vectors

The positions of individuals provide valuable information about the fitness landscape. Provided that a good uniform random initialization method is used to construct the initial population, the initial individuals will provide a good representation of the entire search space, with relatively large distances between individuals. Over time, as the search progresses, the distances between individuals become smaller, with all individuals converging to the same solution. Keep in mind that the magnitude of the initial distances between individuals is influenced by the size of the population. The more individuals in a population, the smaller the magnitude of the distances.

Distances between individuals are a very good indication of the diversity of the current population, and of the order of magnitude of the step sizes that should be taken in order for the population to contract to one point. If there are large distances between individuals, it stands to reason that individuals should make large step sizes in order to explore as much of the search space as possible. On the other hand, if the distances between individuals are small, step sizes should be small to exploit local areas. It is this behaviour that is achieved by DE in calculating mutation step sizes as weighted differences between randomly selected individuals. The first step of mutation is therefore to first calculate one or more difference vectors, and then to use these difference vectors to determine the magnitude and direction of step sizes.

Using vector differentials to achieve variation has a number of advantages. Firstly, information about the fitness landscape, as represented by the current population, is used to direct the search. Secondly, due to the central limit theorem [177], mutation step sizes approaches a Gaussian (Normal) distribution, provided that the population is sufficiently large to allow for a good number of difference vectors [811].[1] The mean of the distribution formed by the difference vectors are always zero, provided that individuals used to calculate difference vectors are selected uniformly from the population [695, 164]. Under the condition that individuals are uniformly selected, this characteristic follows from the fact that difference vectors $(\mathbf{x}_{i_1} - \mathbf{x}_{i_2})$ and $(\mathbf{x}_{i_2} - \mathbf{x}_{i_1})$ occur with equal frequency, where $\mathbf{x}_{i_1}$ and $\mathbf{x}_{i_2}$ are two randomly selected individuals. The zero mean of the resulting step sizes ensures that the population will not suffer from genetic drift. It should also be noted that the deviation of this distribution is determined by the magnitude of the difference vectors. Eventually, differentials will become infinitesimal, resulting in very small mutations.

Section 13.2 shows that more than one differential can be used to determine the mutation step size. If $n_v$ is the number of differentials used, and $n_s$ is the population size, then the total number of differential perturbations is given by [429]

$$\begin{pmatrix} n_s \\ 2n_v \end{pmatrix} 2n_v! \approx \mathcal{O}(n_s^{2n_v}) \tag{13.1}$$

---

[1]The central limit theorem states that the probability distribution governing a random variable approaches the Normal distribution as the number of samples of that random variable tends to infinity.

Equation (13.1) expresses the total number of directions that can be explored per generation. To increase the exploration power of DE, the number of directions can be increased by increasing the population size and/or the number of differentials used.

At this point it is important to emphasize that the original DE was developed for searching through continuous-valued landscapes. The sections that follow will show that exploration of the search space is achieved using vector algebra, applied to the individuals of the current population.

## 13.1.2   Mutation

The DE mutation operator produces a trial vector for each individual of the current population by mutating a target vector with a weighted differential. This trial vector will then be used by the crossover operator to produce offspring. For each parent, $\mathbf{x}_i(t)$, generate the trial vector, $\mathbf{u}_i(t)$, as follows: Select a target vector, $\mathbf{x}_{i_1}(t)$, from the population, such that $i \neq i_1$. Then, randomly select two individuals, $\mathbf{x}_{i_2}$ and $\mathbf{x}_{i_3}$, from the population such that $i \neq i_1 \neq i_2 \neq i_3$ and $i_2, i_3 \sim U(1, n_s)$. Using these individuals, the trial vector is calculated by perturbing the target vector as follows:

$$\mathbf{u}_i(t) = \mathbf{x}_{i_1}(t) + \beta(\mathbf{x}_{i_2}(t) - \mathbf{x}_{i_3}(t)) \tag{13.2}$$

where $\beta \in (0, \infty)$ is the scale factor, controlling the amplication of the differential variation.

Different approaches can be used to select the target vector and to calculate differentials as discussed in Section 13.2.

## 13.1.3   Crossover

The DE crossover operator implements a discrete recombination of the trial vector, $\mathbf{u}_i(t)$, and the parent vector, $\mathbf{x}_i(t)$, to produce offspring, $\mathbf{x}'_i(t)$. Crossover is implemented as follows:

$$x'_{ij}(t) = \begin{cases} u_{ij}(t) & \text{if } j \in \mathcal{J} \\ x_{ij}(t) & \text{otherwise} \end{cases} \tag{13.3}$$

where $x_{ij}(t)$ refers to the $j$-th element of the vector $\mathbf{x}_i(t)$, and $\mathcal{J}$ is the set of element indices that will undergo perturbation (or in other words, the set of crossover points). Different methods can be used to determine the set, $\mathcal{J}$, of which the following two approaches are the most frequently used [811, 813]:

- **Binomial crossover:**  The crossover points are randomly selected from the set of possible crossover points, $\{1, 2, \ldots, n_x\}$, where $n_x$ is the problem dimension. Algorithm 13.1 summarizes this process. In this algorithm, $p_r$ is the probability that the considered crossover point will be included. The larger the value of $p_r$, the more crossover points will be selected compared to a smaller value. This means that more elements of the trial vector will be used to produce the offspring, and less of the parent vector. Because a probabilistic decision is made as to the inclusion of a crossover point, it may happen that no points may be selected, in which case the offspring will simply be the original parent, $\mathbf{x}_i(t)$. This problem becomes more evident for low dimensional search spaces. To enforce that at least one element of the offspring differs from the parent, the set of crossover points, $\mathcal{J}$, is initialized to include a randomly selected point, $j^*$.

- **Exponential crossover:** From a randomly selected index, the exponential crossover operator selects a sequence of adjacent crossover points, treating the list of potential crossover points as a circular array. The pseudocode in Algorithm 13.2 shows that at least one crossover point is selected, and from this index, selects the next until $U(0, 1) \geq p_r$ or $|\mathcal{J}| = n_x$.

---

**Algorithm 13.1** Differential Evolution Binomial Crossover for Selecting Crossover Points

---

$j^* \sim U(1, n_x)$;
$\mathcal{J} \leftarrow \mathcal{J} \cup \{j^*\}$;
**for** *each* $j \in \{1, \ldots, n_x\}$ **do**
    **if** $U(0, 1) < p_r$ *and* $j \neq j^*$ **then**
        $\mathcal{J} \leftarrow \mathcal{J} \cup \{j\}$;
    **end**
**end**

---

**Algorithm 13.2** Differential Evolution Exponential Crossover for Selecting Crossover Points

---

$\mathcal{J} \leftarrow \{\}$;
$j \sim U(0, n_x - 1)$;
**repeat**
    $\mathcal{J} \leftarrow \mathcal{J} \cup \{j + 1\}$;
    $j = (j + 1) \mod n_x$;
**until** $U(0, 1) \geq p_r$ *or* $|\mathcal{J}| = n_x$;

---

### 13.1.4 Selection

Selection is applied to determine which individuals will take part in the mutation operation to produce a trial vector, and to determine which of the parent or the offspring will survive to the next generation. With reference to the mutation operator, a number of selection methods have been used. Random selection is usually used to select the individuals from which difference vectors are calculated. For most DE implementations the target vector is either randomly selected or the best individual is selected (refer to Section 13.2).

To construct the population for the next generation, deterministic selection is used: the offspring replaces the parent if the fitness of the offspring is better than its parent; otherwise the parent survives to the next generation. This ensures that the average fitness of the population does not deteriorate.

### 13.1.5    General Differential Evolution Algorithm

Algorithm 13.3 provides a generic implementation of the basic DE strategies. Initialization of the population is done by selecting random values for the elements of each individual from the bounds defined for the problem being solved. That is, for each individual, $\mathbf{x}_i(t)$, $x_{ij}(t) \sim U(x_{min,j}, x_{max,j})$, where $\mathbf{x}_{min}$ and $\mathbf{x}_{max}$ define the search boundaries.

Any of the stopping conditions given in Section 8.7 can be used to terminate the algorithm.

---

**Algorithm 13.3** General Differential Evolution Algorithm

---

Set the generation counter, $t = 0$;
Initialize the control parameters, $\beta$ and $p_r$;
Create and initialize the population, $\mathcal{C}(0)$, of $n_s$ individuals;
**while** *stopping condition(s) not true* **do**
    **for** *each individual,* $\mathbf{x}_i(t) \in \mathcal{C}(t)$ **do**
        Evaluate the fitness, $f(\mathbf{x}_i(t))$;
        Create the trial vector, $\mathbf{u}_i(t)$ by applying the mutation operator;
        Create an offspring, $\mathbf{x}'_i(t)$, by applying the crossover operator;
        **if** $f(\mathbf{x}'_i(t))$ *is better than* $f(\mathbf{x}_i(t))$ **then**
            Add $\mathbf{x}'_i(t)$ to $\mathcal{C}(t+1)$;
        **end**
        **else**
            Add $\mathbf{x}_i(t)$ to $\mathcal{C}(t+1)$;
        **end**
    **end**
**end**
Return the individual with the best fitness as the solution;

---

### 13.1.6    Control Parameters

In addition to the population size, $n_s$, the performance of DE is influenced by two control parameters, the scale factor, $\beta$, and the probability of recombination, $p_r$. The effects of these parameters are discussed below:

- **Population size:** As indicated in equation (13.1), the size of the population has a direct influence on the exploration ability of DE algorithms. The more individuals there are in the population, the more differential vectors are available, and the more directions can be explored. However, it should be kept in mind that the computational complexity per generation increases with the size of the population. Empirical studies provide the guideline that $n_s \approx 10n_x$. The nature of the mutation process does, however, provide a lower bound on the number of individuals as $n_s > 2n_v + 1$, where $n_v$ is the number of differentials used. For $n_v$ differentials, $2n_v$ different individuals are required, 2 for each differential. The additional individual represents the target vector.

- **Scaling factor:** The scaling factor, $\beta \in (0, \infty)$, controls the amplification of the differential variations, $(\mathbf{x}_{i_2} - \mathbf{x}_{i_3})$. The smaller the value of $\beta$, the smaller the mutation step sizes, and the longer it will be for the algorithm to converge. Larger values for $\beta$ facilitate exploration, but may cause the algorithm to overshoot good optima. The value of $\beta$ should be small enough to allow differentials to explore tight valleys, and large enough to maintain diversity. As the population size increases, the scaling factor should decrease. As explained in Section 13.1.1, the more individuals in the population, the smaller the magnitude of the difference vectors, and the closer individuals will be to one another. Therefore, smaller step sizes can be used to explore local areas. More individuals reduce the need for large mutation step sizes. Empirical results suggest that large values for both $n_s$ and $\beta$ often result in premature convergence [429, 124], and that $\beta = 0.5$ generally provides good performance [813, 164, 19].

- **Recombination probability:** The probability of recombination, $p_r$, has a direct influence on the diversity of DE. This parameter controls the number of elements of the parent, $\mathbf{x}_i(t)$, that will change. The higher the probability of recombination, the more variation is introduced in the new population, thereby increasing diversity and increasing exploration. Increasing $p_r$ often results in faster convergence, while decreasing $p_r$ increases search robustness [429, 164].

Most implementations of DE strategies keep the control parameters constant. Although empirical results have shown that DE convergence is relatively insensitive to different values of these parameters, performance (in terms of accuracy, robustnes, and speed) can be improved by finding the best values for control parameters for each new problem. Finding optimal parameter values can be a time consuming exercise, and for this reason, self-adaptive DE strategies have been developed. These methods are discussed in Section 13.3.3.

### 13.1.7   Geometrical Illustration

Figure 13.1(a) illustrates the mutation operator of the DE as described in Section 13.1.2. The optimum is indicated by $\mathbf{x}^*$, and it is assumed that $\beta = 1.5$. The crossover operator is illustrated in Figure 13.1(b). For this illustration the offspring consists of the first element of the trial vector, $\mathbf{u}_i(t)$, and the second element of the parent, $\mathbf{x}_i(t)$.

## 13.2   DE/$x$/$y$/$z$

A number of variations to the basic DE as discussed in Section 13.1 have been developed. The different DE strategies differ in the way that the target vector is selected, the number of difference vectors used, and the way that crossover points are determined. In order to characterize these variations, a general notation was adopted in the DE literature, namely DE/$x$/$y$/$z$ [811, 813]. Using this notation, $x$ refers to the

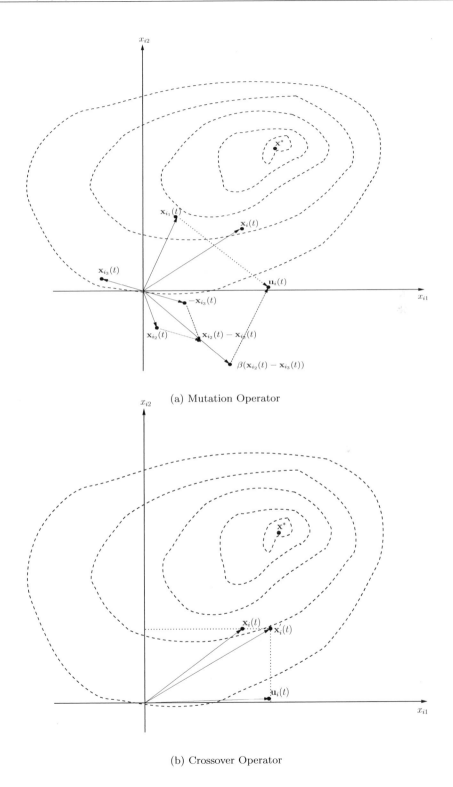

(a) Mutation Operator

(b) Crossover Operator

**Figure 13.1** Differential Evolution Mutation and Crossover Illustrated

method of selecting the target vector, $y$ indicates the number of difference vectors used, and $z$ indicates the crossover method used. The DE strategy discussed in Section 13.1 is referred to as DE/rand/1/bin for binomial crossover, and DE/rand/1/exp for exponential crossover. Other basic DE strategies include [429, 811, 813]:

- **DE/best/1/$z$:** For this strategy, the target vector is selected as the best individual, $\hat{\mathbf{x}}(t)$, from the current population. In this case, the trial vector is calculated as

$$\mathbf{u}_i(t) = \hat{\mathbf{x}}(t) + \beta(\mathbf{x}_{i_2}(t) - \mathbf{x}_{i_3}(t)) \tag{13.4}$$

  Any of the crossover methods can be used.

- **DE/$x$/$n_v$/$z$:** For this strategy, more than one difference vector is used. The trial vector is calculated as

$$\mathbf{u}_i(t) = \mathbf{x}_{i_1}(t) + \beta \sum_{k=1}^{n_v}(\mathbf{x}_{i_2,k}(t) - \mathbf{x}_{i_3,k}(t)) \tag{13.5}$$

  where $\mathbf{x}_{i_2,k}(t) - \mathbf{x}_{i_3,k}(t)$ indicates the $k$-th difference vector, $\mathbf{x}_{i_1}(t)$ can be selected using any suitable method for selecting the target vector, and any of the crossover methods can be used. With reference to equation (13.1), the larger the value of $n_v$, the more directions can be explored per generation.

- **DE/rand-to-best/$n_v$/$z$:** This strategy combines the *rand* and *best* strategies to calculate the trial vector as follows:

$$\mathbf{u}_i(t) = \gamma\hat{\mathbf{x}}(t) + (1 - \gamma)\mathbf{x}_{i_1}(t) + \beta \sum_{k=1}^{n_v}(\mathbf{x}_{i_2,k}(t) - \mathbf{x}_{i_3,k}(t)) \tag{13.6}$$

  where $\mathbf{x}_{i_1}(t)$ is randomly selected, and $\gamma \in [0, 1]$ controls the greediness of the mutation operator. The closer $\gamma$ is to 1, the more greedy the search process becomes. In other words, $\gamma$ close to 1 favors exploitation while a value close to 0 favors exploration. A good strategy will be to use an adaptive $\gamma$, with $\gamma(0) = 0$. The value of $\gamma(t)$ increases with each new generation towards the value 1.

  Note that if $\gamma = 0$, the DE/rand/$y$/$z$ strategies are obtained, while $\gamma = 1$ gives the DE/best/$y$/$z$ strategies.

- **DE/current-to-best/$1+n_v$/$z$:** With this strategy, the parent is mutated using at least two difference vectors. One difference vector is calculated from the best vector and the parent vector, while the rest of the difference vectors are calculated using randomly selected vectors:

$$\mathbf{u}_i(t) = \mathbf{x}_i(t) + \beta(\hat{\mathbf{x}}(t) - \mathbf{x}_i(t)) + \beta \sum_{k=1}^{n_v}(\mathbf{x}_{i_1,k}(t) - \mathbf{x}_{i_2,k}(t)) \tag{13.7}$$

Empirical studies have shown that DE/rand/1/bin maintains good diversity, while DE/current-to-best/2/bin shows good convergence characteristics [698]. Due to this observation, Qin and Suganthan [698] developed a DE algorithm that dynamically switch between these two strategies. Each of these strategies is assigned a probability of being applied. If $p_{s,1}$ is the probability that DE/rand/1/bin will be applied, then $p_{s,2} = 1 - p_{s,1}$ is the probability that DE/current-to-best/2/bin will be applied. Then,

$$p_{s,1} = \frac{n_{s,1}(n_{s,2} + n_{f,2})}{n_{s,2}(n_{s,1} + n_{f,1}) + n_{s,1}(n_{s,2} + n_{f,2})} \tag{13.8}$$

where $n_{s,1}$ and $n_{s,2}$ are respectively the number of offspring that survive to the next generation for DE/rand/1/bin, and $n_{f,1}$ and $n_{f,2}$ represent the number of discarded offspring for each strategy. The more offspring that survive for a specific strategy, the higher the probability for selecting that strategy for the next generation.

## 13.3    Variations to Basic Differential Evolution

The basic DE strategies have been shown to be very efficient and robust [811, 813, 811, 813]. A number of adaptations of the original DE strategies have been developed in order to further improve performance. This section reviews some of these DE variations. Section 13.3.1 describe hybrid DE methods, a population-based DE is described in Section 13.3.2, and self-adaptive DE strategies are discussed in Section 13.3.3.

### 13.3.1    Hybrid Differential Evolution Strategies

DE has been combined with other EAs, particle swarm optimization (PSO), and gradient-based techniques. This section summarizes some of these hybrid methods.

#### Gradient-Based Hybrid Differential Evolution

One of the first DE hybrids was developed by Chiou and Wang [124], referred to as the hybrid DE. As indicated in Algorithm 13.4, the hybrid DE introduces two new operations: an acceleration operator to improve convergence speed – without decreasing diversity – and a migration operator to provide the DE with the improved ability to escape local optima.

The acceleration operator uses gradient descent to adjust the best individual toward obtaining a better position if the mutation and crossover operators failed to improve the fitness of the best individual. Let $\hat{\mathbf{x}}(t)$ denote the best individual of the current population, $\mathcal{C}(t)$, before application of the mutation and crossover operators, and let $\hat{\mathbf{x}}(t + 1)$ be the best individual for the next population after mutation and crossover have been applied to all individuals. Then, assuming a minimization problem, the acceleration operator computes the vector

$$\mathbf{x}(t) = \begin{cases} \hat{\mathbf{x}}(t + 1) & \text{if } f(\hat{\mathbf{x}}(t + 1)) < f(\hat{\mathbf{x}}(t)) \\ \hat{\mathbf{x}}(t + 1) - \eta(t)\nabla f & \text{otherwise} \end{cases} \tag{13.9}$$

where $\eta(t) \in (0, 1]$ is the learning rate, or step size; $\nabla f$ is the gradient of the objective function, $f$. The new vector, $\mathbf{x}(t)$, replaces the worst individual in the new population, $\mathcal{C}(t)$.

---

**Algorithm 13.4** Hybrid Differential Evolution with Acceleration and Migration

---

Set the generation counter, $t = 0$;
Initialize the control parameters, $\beta$ and $p_r$;
Create and initialize the population, $\mathcal{C}(0)$, of $n_s$ individuals;
**while** *stopping condition(s) not true* **do**
    Apply the migration operator if necessary;
    **for** *each individual,* $\mathbf{x}_i(t) \in \mathcal{C}(t)$ **do**
        Evaluate the fitness, $f(\mathbf{x}_i(t))$;
        Create the trial vector, $\mathbf{u}_i(t)$ by applying the mutation operator;
        Create an offspring, $\mathbf{x}_i'(t)$ by applying the crossover operator;
        **if** $f(\mathbf{x}_i'(t))$ *is better than* $f(\mathbf{x}_i(t))$ **then**
            Add $\mathbf{x}_i'(t)$ to $\mathcal{C}(t+1)$;
        **end**
        **else**
            Add $\mathbf{x}_i(t)$ to $\mathcal{C}(t+1)$;
        **end**
    **end**
    Apply the acceleration operator if necessary;
**end**
Return the individual with the best fitness as the solution;

---

The learning rate is initialized to one, i.e. $\eta(0) = 1$. If the gradient descent step failed to create a new vector, $\mathbf{x}(t)$, with better fitness, the learning rate is reduced by a factor. The gradient descent step is then repeated until $\eta(t)\nabla f$ is sufficiently close to zero, or a maximum number of gradient descent steps have been executed.

While use of gradient descent can significantly speed up the search, it has the disadvantage that the DE may get stuck in a local minimum, or prematurely converge. The migration operator addresses this problem by increasing population diversity. This is done by spawning new individuals from the best individual, and replacing the current population with these new individuals. Individuals are spawned as follows:

$$
x_{ij}'(t) = \begin{cases} \hat{x}_j(t) + r_{ij}(x_{min,j} - \hat{x}_j) & \text{if } U(0,1) < \frac{\hat{x}_j - x_{min,j}}{x_{max,j} - x_{min,j}} \\ \hat{x}_j(t) + r_{ij}(x_{max,j} - \hat{x}_j) & \text{otherwise} \end{cases} \tag{13.10}
$$

where $r_{ij} \sim U(0,1)$. Spawned individual $\mathbf{x}_i'(t)$ becomes $\mathbf{x}_i(t+1)$.

The migration operator is applied only when the diversity of the current population becomes too small; that is, when

$$
\left( \sum_{\substack{i=1 \\ \mathbf{x}_i(t) \neq \hat{\mathbf{x}}(t)}}^{n_s} \mathcal{I}_{ij}(t) \right) / (n_x(n_s - 1)) < \epsilon_1 \tag{13.11}
$$

with

$$\mathcal{I}_{ij}(t) = \begin{cases} 1 & \text{if } |(x_{ij}(t) - \hat{x}_j(t))/\hat{x}_j(t)| > \epsilon_2 \\ 0 & \text{otherwise} \end{cases} \qquad (13.12)$$

where $\epsilon_1$ and $\epsilon_2$ are respectively the tolerance for the population diversity and gene diversity with respect to the best individual, $\hat{\mathbf{x}}(t)$. If $\mathcal{I}_{ij}(t) = 0$, then the value of the $j$-th element of individual $i$ is close to the value of the $j$-th element of the best individual.

Magoulas *et al.* [550] combined a stochastic gradient descent (SGD) [549] and DE in a sequential manner to train artificial neural networks (NN). Here, SGD is first used to find a good approximate solution using the process outlined in Algorithm 13.5. A population of DE individuals is then created, with individuals in the neighborhood of the solution returned by the SGD step. As outlined in Algorithm 13.6, the task of DE is to refine the solution obtained from SGD by using then DE to perform a local search.

---

**Algorithm 13.5** Stochastic Gradient Descent for Neural Network Training

---

Initialize the NN weight vector, $\mathbf{w}(0)$;
Initialize the learning rate, $\eta(0)$, and the meta-step size, $\eta_m$;
Set the pattern presentation number, $t = 0$;
**repeat**
    **for** *each training pattern, p* **do**
        Calculate $\mathcal{E}(\mathbf{w}(t))$;
        Calculate $\nabla\mathcal{E}(\mathbf{w}(t))$;
        Update the weights using

$$\mathbf{w}(t + 1) = \mathbf{w}(t) + \eta(t)\nabla\mathcal{E}(\mathbf{w}(t)) \qquad (13.13)$$

        Calculate the new step size using

$$\eta(t + 1) = \eta(t) + \eta_m < \nabla\mathcal{E}(\mathbf{w}(t - 1)), \nabla\mathcal{E}(\mathbf{w}(t)) > \qquad (13.14)$$

        $t = t + 1$;
    **end**
    Return $\mathbf{w}(t + 1)$ as the solution;
**until** *until a termination condition is satisfied*;

---

In Algorithms 13.5 and 13.6, $< \bullet, \bullet >$ denotes the inner product between the two given vectors, $\mathcal{E}$ is the NN training objective function (usually the sum-squared error), $\sigma$ is the standard deviation of mutations to $\mathbf{w}$ used to create DE individuals in the neighborhood of $\mathbf{w}$, and $D_T$ is the training set. The DE algorithm uses the objective function, $\mathcal{E}$, to assess the fitness of individuals.

---

**Algorithm 13.6** Differential Evolution with Stochastic Gradient Descent

---

$\mathbf{w} = SGD(D_T)$;
Set the individual counter, $i = 0$;
Set $\mathcal{C}(0) = \{\}$;
**repeat**
    $i = i + 1$;
    $\mathbf{x}_i(0) = \mathbf{w} + \mathbf{N}(0, \sigma)$;
    $\mathcal{C}(0) \leftarrow \mathcal{C}(0) + \{\mathbf{x}_i(0)\}$;
**until** $i = n_s$;
Apply any DE strategy;
Return the best solution from the final population;

---

## Evolutionary Algorithm-Based Hybrids

Due to the efficiency of DE, Hrstka and Kucerová [384] used the DE reproduction process as a crossover operator in a simple GA.

Chang and Chang [113] used standard mutation operators to increase DE population diversity by adding noise to the created trial vectors. In [113], uniform noise is added to each component of trial vectors, i.e.

$$u_{ij}(t) = u_{ij}(t) + U(u_{min,j}, u_{max,j}) \tag{13.15}$$

where $u_{min,j}$ and $u_{max,j}$ define the boundaries of the added noise. However, the approach above should be considered carefully, as the expected mean of the noise added is

$$\frac{u_{min,j} + u_{max,j}}{2} \tag{13.16}$$

If this mean is not zero, the population may suffer genetic drift. An alternative is to sample the noise from a Gaussian or Cauchy distribution with zero mean and a small deviation (refer to Section 11.2.1).

Sarimveis and Nikolakopoulos [758] use rank-based selection to decide which individuals will take part to calculate difference vectors. At each generation, after the fitness of all individuals have been calculated, individuals are arranged in descending order, $\mathbf{x}_1(t), \mathbf{x}_2(t), \ldots, \mathbf{x}_{n_s}(t)$ where $\mathbf{x}_{i_1}(t)$ precedes $\mathbf{x}_{i_2}(t)$ if $f(\mathbf{x}_{i_1}(t)) > f(\mathbf{x}_{i_2}(t))$. The crossover operator is then applied as summarized in Algorithm 13.7 assuming minimization. After application of crossover on all the individuals, the resulting population is again ranked in descending order. The mutation operator in Algorithm 13.8 is then applied.

With reference to Algorithm 13.8, $p_{m,i}$ refers to the probability of mutation, with each individual assigned a different probability based on its rank. The lower the rank of an individual, the more unfit the individual is, and the higher the probability that the individual will be mutated. Mutation step sizes are initially large, decreasing over time due to the exponential term used in equations (13.17) and (13.18). The direction of the mutation is randomly decided, using the random variable, $r_2$.

---

**Algorithm 13.7** Rank-Based Crossover Operator for Differential Evolution

---

Rank all individuals in decreasing order of fitness;
**for** $i = 1, \ldots, n_s$ **do**
    $r \sim U(0,1)$;
    $\mathbf{x}_i'(t) = \mathbf{x}_i(t) + r(\mathbf{x}_{i+1}(t) - \mathbf{x}_i(t))$;
    **if** $f(\mathbf{x}_i'(t)) < f(\mathbf{x}_{i+1}(t))$ **then**
        $\mathbf{x}_i(t) = \mathbf{x}_i'(t)$;
    **end**
**end**

---

**Algorithm 13.8** Rank-Based Mutation Operator for Differential Evolution

---

Rank all individuals in decreasing order of fitness;
**for** $i = 1, \ldots, n_s$ **do**
    $p_{m,i} = \frac{n_s - i + 1}{n_s}$;
    **for** $j = 1, \ldots, n_x$ **do**
        $r_1 \sim U(0,1)$;
        **if** $(r_1 > p_{m,i})$ **then**
            $r_2 \sim \{0,1\}$;
            $r_3 \sim U(0,1)$;
            **if** $(r_2 = 0)$ **then**

$$x_{ij}'(t) = x_{ij}(t) + (x_{max,j} - x_{ij}(t))r_3 e^{-2t/n_t} \tag{13.17}$$

            **end**
            **if** $(r_2 = 1)$ **then**

$$x_{ij}'(t) = x_{ij}(t) - (x_{ij}(t) - x_{min,j})r_3 e^{-2t/n_t} \tag{13.18}$$

            **end**
        **end**
    **end**
    **if** $f(\mathbf{x}_i'(t)) < f(\mathbf{x}_i(t))$ **then**
        $\mathbf{x}_i(t) = \mathbf{x}_i'(t)$;
    **end**
**end**

---

## Particle Swarm Optimization Hybrids

A few studies have combined DE with particle swarm optimization(PSO) (refer to Chapter 16).

Hendtlass [360] proposed that the DE reproduction process be applied to the particles in a PSO swarm at specified intervals. At the specified intervals, the PSO swarm serves as the population for a DE algorithm, and the DE is executed for a number of generations. After execution of the DE, the evolved population is then further optimized using PSO. Kannan *et al.* [437] apply DE to each particle for a number of iterations, and replaces the particle with the best individual obtained from the DE process.

Zhang and Xie [954], and Talbi and Batouche [836] follow a somewhat different approach. Only the personal best positions are changed using

$$
y'_{ij}(t+1) = \begin{cases} \hat{y}_{ij}(t) + \delta_j & \text{if } j \in \mathcal{J}_i(t) \\ y_{ij}(t) & \text{otherwise} \end{cases} \tag{13.19}
$$

where $\delta$ is the general difference vector defined as

$$
\delta_j = \frac{y_{1j}(t) - y_{2j}(t)}{2} \tag{13.20}
$$

with $\mathbf{y}_1(t)$ and $\mathbf{y}_2(t)$ randomly selected personal best positions; the notations $\mathbf{y}_i(t)$ and $\hat{\mathbf{y}}_i(t)$ are used to indicate a personal best and neighborhood best respectively (refer to Chapter 16). The offspring, $\mathbf{y}'_i(t+1)$, replaces the current personal best, $\mathbf{y}_i(t)$, only if the offspring has a better fitness.

## 13.3.2   Population-Based Differential Evolution

In order to improve the exploration ability of DE, Ali and Törn [19] proposed to use two population sets. The second population, referred to as the auxiliary population, $\mathcal{C}_a(t)$, serves as an archive of those offspring rejected by the DE selection operator. During the initialization process, $n_s$ pairs of vectors are randomly created. The best of the two vectors is inserted as an individual in the population, $\mathcal{C}(0)$, while the other vector, $\mathbf{x}_i^a(0)$, is inserted in the auxiliary population, $\mathcal{C}_a(0)$. At each generation, for each offspring created, if the fitness of the offspring is not better than the parent, instead of discarding the offspring, $\mathbf{x}'_i(t)$, it is considered for inclusion in the auxiliary population. If $f(\mathbf{x}'_i(t))$ is better than $\mathbf{x}_i^a(t)$, then $\mathbf{x}'_i(t)$ replaces $\mathbf{x}_i^a(t)$. The auxiliary set is periodically used to replace the worst individuals in $\mathcal{C}(t)$ with the best individuals from $\mathcal{C}_a(t)$.

## 13.3.3   Self-Adaptive Differential Evolution

Although empirical studies have shown that DE convergence is relatively insensitive to control parameter values, performance can be greatly improved if parameter values are optimized. For the DE strategies discussed thus far, values of control parameters are static, and do not change over time. These strategies require an additional search process to find the best values for control parameters for each different problem – a process that is usually time consuming. It is also the case that different values for a control parameter are optimal for different stages of the optimization process. As an alternative, a number of DE strategies have been developed where values for control parameters adapt dynamically. This section reviews these approaches.

## Dynamic Parameters

One of the first proposals for dynamically changing the values of the DE control parameters was proposed by Chang and Xu [112], where the probability of recombination is linearly decreased from 1 to 0.7, and the scale factor is linearly increased from 0.3 to 0.5:

$$p_r(t) = p_r(t-1) - (p_r(0) - 0.7)/n_t \qquad (13.21)$$
$$\beta(t) = \beta(t-1) - (0.5 - \beta(0))/n_t \qquad (13.22)$$

where $p_r(0) = 1.0$ and $\beta(0) = 0.3$; $n_t$ is the maximum number of iterations.

Abbass *et al.* [3] proposed an approach where a new value is sampled for the scale factor for each application of the mutation operator. The scale factor is sampled from a Gaussian distribution, $\beta \sim N(0,1)$. This approach is also used in [698, 735]. In [698], the mean of the distribution was changed to 0.5 and the deviation to 0.3 (i.e. $\beta \sim N(0.5, 0.3)$), due to the empirical results that suggest that $\beta = 0.5$ provides on average good results. Abbass [2] extends this to the probability of recombination, i.e. $p_r \sim N(0,1)$. Abbass refers incorrectly to the resulting DE strategy as being self-adaptive. For self-adaptive strategies, values of control parameters are evolved over time; this is not the case in [2, 3].

## Self-Adaptive Parameters

Self-adaptive strategies usually make use of information about the search space as obtained from the current population (or a memory of previous populations) to self-adjust values of control parameters.

Ali and Törn [19] use the fitness of individuals in the current population to determine a new value for the scale factor. That is,

$$\beta(t) = \begin{cases} \max\left\{\beta_{min}, 1 - \left|\frac{f_{max}(t)}{f_{min}(t)}\right|\right\} & \text{if } \left|\frac{f_{max}(t)}{f_{min}(t)}\right| < 1 \\ \max\left\{\beta_{min}, 1 - \left|\frac{f_{min}(t)}{f_{max}(t)}\right|\right\} & \text{otherwise} \end{cases} \qquad (13.23)$$

which ensures that $\beta(t) \in [\beta_{min}, 1)$, where $\beta_{min}$ is a lower bound on the scaling factor; $f_{min}(t)$ and $f_{max}(t)$ are respectively the minimum and maximum fitness values for the current population, $\mathcal{C}(t)$. As $f_{min}$ approaches $f_{max}$, the diversity of the population decreases, and the value of $\beta(t)$ approaches $\beta_{min}$ – ensuring smaller step sizes when the population starts to converge. On the other hand, the smaller the ratio $\left|\frac{f_{max}(t)}{f_{min}(t)}\right|$ (for minimization problems) or $\left|\frac{f_{min}(t)}{f_{max}(t)}\right|$ (for maximization problems), the more diverse the population and the larger the step sizes will be – favoring exploration.

Qin and Suganthan [698] propose that the probability of recombination be self-adapted as follows:

$$p_r(t) \sim N(\mu_{p_r}(t), 0.1) \tag{13.24}$$

where $\mu_{p_r}(0) = 0.5$, and $\mu_{p_r}(t)$ is calculated as the average over successful values of $p_r(t)$. A $p_r(t)$ value can be considered as being successful if the fitness of the best individual improved under that value of $p_r(t)$. It is not clear if one probability is used in [698] for the entire population, or if each individual has its own probability, $p_{r,i}(t)$. This approach to self-adaptation can, however, be applied for both scenarios.

For the self-adaptive Pareto DE, Abbass [2] adapts the probability of recombination dynamically as

$$p_{r,i}(t) = p_{r,i_1}(t) + N(0,1)[p_{r,i_2}(t) - p_{r,i_3}(t)] \tag{13.25}$$

where $i_1 \neq i_2 \neq i_3 \neq i \sim U(1, \ldots, n_s)$, while sampling the scale factor from $N(0,1)$. Note that equation (13.25) implies that each individual has its own, learned probability of recombination.

Omran $et\ al.$ [641] propose a self-adaptive DE strategy that makes use of the approach in equation (13.25) to dynamically adapt the scale factor. That is, for each individual,

$$\beta_i(t) = \beta_{i_4}(t) + N(0, 0.5)[\beta_{i_5}(t) - \beta_{i_6}(t)] \tag{13.26}$$

where $i_4 \neq i_5 \neq i_6 \neq i \sim U(1, \ldots, n_s)$. The mutation operator as given in equation (13.2) changes to

$$\mathbf{u}_i(t) = \mathbf{x}_{i_1}(t) + \beta_i(t)[\mathbf{x}_{i_2}(t) + \mathbf{x}_{i_3}(t)] \tag{13.27}$$

The crossover probability can be sampled from a Gaussian distribution as discussed above, or adapted according to equation (13.25).

# 13.4 Differential Evolution for Discrete-Valued Problems

Differential evolution has been developed for optimizing continuous-valued parameters. However, a simple discretization procedure can be used to convert the floating-point solution vectors into discrete-valued vectors. Such a procedure has been used by a number of researchers in order to apply DE to integer and mixed-integer programming [258, 390, 499, 531, 764, 817]. The approach is quite simple: each floating-point value of a solution vector is simply rounded to the nearest integer. For a discrete-valued parameter where an ordering exists among the values of the parameter, Lampinen and Zelinka [499] and Feoktistov and Janaqi [258] take the index number in the ordered sequence as the discretized value.

Pampará *et al.* [653] proposed an approach to apply DE to binary-valued search spaces: The angle modulated DE (AMDE) [653] uses the standard DE to evolve a generating function to produce bitstring solutions. This chapter proposes an alternative, the binary DE (binDE) which treats each floating-point element of solution vectors as a probability of producing either a bit 0 or a bit 1. These approaches are respectively discussed in Sections 13.4.1 and 13.4.2.

## 13.4.1 Angle Modulated Differential Evolution

Pampará *et al.* [653] proposed a DE algorithm to evolve solutions to binary-valued optimization problems, without having to change the operation of the original DE. This is achieved by using a homomorphous mapping [487] to abstract a problem (defined in binary-valued space) into a simpler problem (defined in continuous-valued space), and then to solve the problem in the abstracted space. The solution obtained in the abstracted space is then transformed back into the original space in order to solve the problem. The angle modulated DE (AMDE) makes use of angle modulation (AM), a technique derived from the telecommunications industry [697], to implement such a homomorphous mapping between binary-valued and continuous-valued space.

The objective is to evolve, in the abstracted space, a bitstring generating function, which will be used in the original space to produce bit-vector solutions. The generating function as used in AM is

$$g(x) = \sin(2\pi(x - a) \times b \times \cos(2\pi(x - a) \times c)) + d \qquad (13.28)$$

where $x$ is a single element from a set of evenly separated intervals determined by the required number of bits that need to be generated (i.e. the dimension of the original, binary-valued space).

The coefficients in equation (13.28) determine the shape of the generating function: $a$ represents the horizontal shift of the generating function, $b$ represents the maximum frequency of the sin function, $c$ represents the frequency of the cos function, and $d$ represents the vertical shift of the generating function. Figure 13.2 illustrates the function for $a = 0, b = 1, c = 1$, and $d = 0$, with $x \in [-2, 2]$. The AMDE evolves values for the four coefficients, $a, b, c$, and $d$. Solving a binary-valued problem thus reverts to solving a 4-dimensional problem in a continuous-valued space. After each iteration of the AMDE, the fitness of each individual in the population is determined by substituting the evolved values for the coefficients (as represented by the individual) into equation (13.28). The resulting function is sampled at evenly spaced intervals and a bit value is recorded for each interval. If the output of the function in equation (13.28) is positive, a bit-value of 1 is recorded; otherwise, a bit-value of 0 is recorded. The resulting bit string is then evaluated by the fitness function defined in the original binary-valued space in order to determine the quality of the solution.

The AMDE is summarized in Algorithm 13.9.

Pampará *et al.* [653] show that the AMDE is very efficient and provides accurate solutions to binary-valued problems. Furthermore, the AMDE has the advantage that

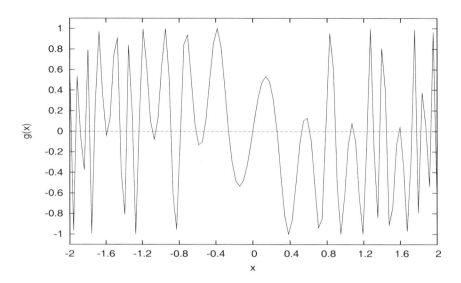

**Figure 13.2** Angle Modulation Illustrated

---

**Algorithm 13.9** Angle Modulated Differential Evolution

---

Generate a population of 4-dimensional individuals;
**repeat**
    Apply any DE strategy for one iteration;
    **for** *each individual* **do**
        Substitute evolved values for coefficients $a, b, c$ and $d$ into equation (13.28);
        Produce $n_x$ bit-values to form a bit-vector solution;
        Calculate the fitness of the bit-vector solution in the original bit-valued space;
    **end**
**until** *a convergence criterion is satisfied*;

---

an $n_x$-dimensional binary-valued problem is transformed into a smaller 4-dimensional continuous-valued problem.

## 13.4.2 Binary Differential Evolution

The binary DE (binDE) borrows concepts from the binary particle swarm optimizer (binPSO), developed by Kennedy and Eberhart [450] (also refer to Section 16.5.7). As with DE, particle swarm optimization (PSO; refer to Chapter 16) uses vector algebra to calculate new search positions, and was therefore developed for continuous-valued problems. In PSO, a velocity vector represents the mutation step sizes as stochastically weighted difference vectors (i.e. the social and cognitive components). The binPSO does not interpret the velocity as a step size vector. Rather, each component of the velocity vector is used to compute the probability that the corresponding component of the solution vector is bit 0 or bit 1.

In a similar way, the binDE uses the floating-point DE individuals to determine a probability for each component. These probabilities are then used to generate a bitstring solution from the floating-point vector. This bitstring is used by the fitness function to determine its quality. The resulting fitness is then associated with the floating-point representation of the individual.

Let $\mathbf{x}_i(t)$ represent a DE individual, with each $x_{ij}(t)$ ($j = 1, \ldots, n_x$, where $n_x$ is the dimension of the binary-valued problem) floating-point number. Then, the corresponding bitstring solution, $\mathbf{y}_i(t)$, is calcualted using

$$y_{ij} = \begin{cases} 0 & \text{if } f(x_{ij}(t)) \geq 0.5 \\ 1 & \text{if } f(x_{ij}(t)) < 0.5 \end{cases} \tag{13.29}$$

where $f$ is the sigmoid function,

$$f(x) = \frac{1}{1 + e^{-x}} \tag{13.30}$$

The fitness of the individual $\mathbf{x}_i(t)$ is then simply the fitness obtained using the binary representation, $\mathbf{y}_i(t)$.

The binDE algorithm is summarized in Algorithm 13.10.

---

**Algorithm 13.10** Binary Differential Evolution Algorithm

---

Initialize a population and set control parameter values;
$t = 0$;
**while** *stopping condition(s) not true* **do**
$\quad$ $t = t + 1$;
$\quad$ Select parent $\mathbf{x}_i(t)$;
$\quad$ Select individuals for reproduction;
$\quad$ Produce one offspring, $\mathbf{x}'(t)$;
$\quad$ $\mathbf{y}_i(t) =$ generated bitstring from $\mathbf{x}_i(t)$;
$\quad$ $\mathbf{y}'_i(t) =$ generated bitstring from $\mathbf{x}'_i(t)$;
$\quad$ **if** $f(\mathbf{y}'_i(t))$ *is better than* $f(\mathbf{x}'_i(t))$ **then**
$\quad\quad$ Replace parent, $\mathbf{x}_i(t)$, with offspring, $\mathbf{x}'_i(t)$;
$\quad$ **end**
$\quad$ **else**
$\quad\quad$ Retain parent, $\mathbf{x}_i(t)$;
$\quad$ **end**
**end**

---

# 13.5   Advanced Topics

The discussions in the previous sections considered application of DE to unconstrained, single-objective optimization problems, where the fitness landscape remains static. This section provides a compact overview of adaptations to the DE such that different types of optimization problems as summarized in Appendix A can be solved using DE.

## 13.5.1　Constraint Handling Approaches

With reference to Section A.6, the following methods have been used to apply DE to solve constrained optimization problems as defined in Definition A.5:

- Penalty methods (refer to Section A.6.2), where the objective function is adapted by adding a function to penalize solutions that violate constraints [113, 394, 499, 810, 884].

- Converting the constrained problem to an unconstrained problem by embedding constraints in an augmented Lagrangian (refer to Section A.6.2) [125, 390, 528, 758]. Lin *et al.* [529] combines both the penalty and the augmented Lagrangian functions to convert a constrained problem to an unconstrained one.

- In order to preserve the feasibility of initial solutions, Chang and Wu [114] used feasible directions to determine step sizes and search directions.

- By changing the selection operator of DE, infeasible solutions can be rejected, and the repair of infeasible solutions facilitated. In order to achieve this, the selection operator accepts an offspring, $\mathbf{x}_i'$, under the following conditions [34, 56, 498]:

  - if $\mathbf{x}_i'$ satisfies all the constraints, and $f(\mathbf{x}_i') \le f(\mathbf{x}_i)$, then $\mathbf{x}_i'$ replaces the parent, $\mathbf{x}_i$ (assuming minimization);
  - if $\mathbf{x}_i'$ is feasible and $\mathbf{x}_i$ is infeasible, then $\mathbf{x}_i'$ replaces $\mathbf{x}_i$;
  - if both $\mathbf{x}_i'$ and $\mathbf{x}_i$ are infeasible, then if the number of constraints violated by $\mathbf{x}_i'$ is less than or equal to the number of constraints violated by $\mathbf{x}_i$, then $\mathbf{x}_i'$ replaces $\mathbf{x}_i$.

  In the case that both the parent and the offspring represent infeasible solutions, there is no selection pressure towards better parts of the fitness landscape; rather, towards solutions with the smallest number of violated constraints.

Boundary constraints are easily enforced by clamping offspring to remain within the given boundaries [34, 164, 498, 499]:

$$x_{ij}'(t) = \begin{cases} x_{min,j} + U(0,1)(x_{max,j} - x_{min,j}) & \text{if } x_{ij}'(t) < x_{min,j} \text{ or } x_{ij}' > x_{max,j} \\ x_{ij}'(t) & \text{otherwise} \end{cases}$$

(13.31)

This restarts the offspring to a random position within the boundaries of the search space.

## 13.5.2　Multi-Objective Optimization

As defined in Definition A.10, multi-objective optimization requires multiple, conflicting objectives to be simultaneously optimized. A number of adaptations have been made to DE in order to solve multiple objectives, most of which make use of the concept of dominance as defined in Definition A.11.

Multi-objective DE approaches include:

- Converting the problem into a minimax problem [390, 925].

- Weight aggregation methods [35].

- Population-based methods, such as the vector evaluated DE (VEDE) [659], based on the vector evaluated GA (VEGA) [761] (also refer to Section 9.6.3). If $K$ objectives have to be optimized, $K$ sub-populations are used, where each sub-population optimizes one of the objectives. These sub-populations are organized in a ring topology (as illustrated in Figure 16.4(b)). At each iteration, before application of the DE reproduction operators, the best individual, $\mathcal{C}_k.\hat{\mathbf{x}}(t)$, of population $\mathcal{C}_k$ migrates to population $\mathcal{C}_{k+1}$ (that of $\mathcal{C}_{k+1}$ migrates to $\mathcal{C}_0$), and is used in population $\mathcal{C}_{k+1}$ to produce the trial vectors for that population.

- Pareto-based methods, which change the DE operators to include the dominance concept.

  **Mutation:** Abbass *et al.* [2, 3] applied mutation only on non-dominated solutions within the current generation. Xue *et al.* [928] computed the differential as the difference between a randomly selected individual, $\mathbf{x}_{i_1}$, and a randomly selected vector, $\mathbf{x}_{i_2}$, that dominates $\mathbf{x}_{i_1}$; that is, $\mathbf{x}_{i_1} \preceq \mathbf{x}_{i_2}$. If $\mathbf{x}_{i_1}$ is not dominated by any other individual of the current generation, the differential is set to zero.

  **Selection:** A simple change to the selection operator is to replace the parent, $\mathbf{x}_i$, with the offspring $\mathbf{x}_i'$, only if $\mathbf{x}_i' \preceq \mathbf{x}_i$ [3, 2, 659]. Alternatively, ideas from non-dominated sorting genetic algorithms [197] can be used, where non-dominated sorting and ranking is applied to parents and offspring [545, 928]. The next population is then selected with preference to those individuals with a higher rank.

### 13.5.3 Dynamic Environments

Not much research has been done in applying DE to dynamically changing landscapes (refer to Section A.9). Chiou and Wang [125] applied the DE with acceleration and migration (refer to Algorithm 13.4) to dynamic environments, due to the improved exploration as provided by the migration phase. Magoulas *et al.* [550] applied the SGDDE (refer to Algorithm 13.6) to slowly changing fitness landscapes.

Mendes and Mohais [577] develop a DE algorithm, referred to as DynDE, to locate and maintain multiple solutions in dynamically changing landscapes. Firstly, it is important to note the following assumptions:

1. It is assumed that the number of peaks, $n_\chi$, to be found are known, and that these peaks are evenly distributed through the search space.

2. Changes in the fitness landscape are small and gradual.

DynDE uses multiple populations, with each population maintaining one of the peaks. To ensure that each peak represents a different solution, an exclusion strategy is followed: At each iteration, the best individuals of each pair of sub-populations are compared. If these global best positions are too close to one another, the sub-population with the worst global best solution is re-initialized. DynDE re-initializes the one sub-population when

$$\mathcal{E}(\mathcal{C}_{k_1}.\hat{\mathbf{x}}(t), \mathcal{C}_{k_2}.\hat{\mathbf{x}}(t)) < \frac{X}{2n_{\mathcal{X}}^{1/n_x}} \tag{13.32}$$

where $\mathcal{E}(\mathcal{C}_{k_1}.\hat{\mathbf{x}}(t), \mathcal{C}_{k_2}.\hat{\mathbf{x}}(t))$ is the Euclidean distance between the best individuals of sub-populations $\mathcal{C}_{k_1}$ and $\mathcal{C}_{k_2}$, $X$ represents the extent of the search space, $n_{\mathcal{X}}$ is the number of peaks, and $n_x$ is the search space dimension. It is this condition that requires assumption 1, which suffers from obvious problems. For example, peaks are not necessarily evenly distributed. It may also be the case that two peaks exist with a distance less than $\frac{X}{2n_{\mathcal{X}}^{1/n_x}}$ from one another. Also, it is rarely the case that the number of peaks is known.

After a change is detected, a strategy is followed to increase diversity. This is done by assigning a different behavior to some of the individuals of the affected sub-population. The following diversity increasing strategies have been proposed [577]:

- Re-initialize the sub-populations: While this strategy does maximize diversity, it also leads to a severe loss of knowledge obtained about the search space.

- Use quantum individuals: Some of the individuals are re-initialized to random points inside a ball centered at the global best individual, $\hat{\mathbf{x}}(t)$, as outlined in Algorithm 13.11. In this algorithm, $R_{max}$ is the maximum radius from $\hat{\mathbf{x}}(t)$.

- Use Brownian individuals: Some positions are initialized to random positions around $\hat{\mathbf{x}}(t)$, where the random step sizes from $\hat{\mathbf{x}}(t)$ are sampled from a Gaussian distribution. That is,

$$\mathbf{x}_i(t) = \hat{\mathbf{x}}(t) + \mathbf{N}(0, \sigma) \tag{13.33}$$

- Introduce some form of entropy: Some individuals are simply added noise, sampled from a Gaussian distribution. That is,

$$\mathbf{x}_i(t) = \mathbf{x}_i(t) + \mathbf{N}(0, \sigma) \tag{13.34}$$

---

**Algorithm 13.11** Initialization of Quantum Individuals

---

**for** *each individual, $\mathbf{x}_i(t)$, to be re-initialized* **do**
    Generate a random vector, $\mathbf{r}_i \sim \mathbf{N}(0, 1)$;
    Compute the distance of $\mathbf{r}_i$ from the origin, i.e.

$$\mathcal{E}(\mathbf{r}_i, \mathbf{0}) = \sqrt{\sum_{j=1}^{n_x} r_{ij}} \tag{13.35}$$

    Find the radius, $R \sim U(0, R_{max})$;
    $\mathbf{x}_i(t) = \hat{\mathbf{x}}(t) + R\mathbf{r}_i / \mathcal{E}(\mathbf{r}_i, \mathbf{0})$;
**end**

---

## 13.6   Applications

Differential evolution has mostly been applied to optimize functions defined over continuous-valued landscapes [695, 811, 813, 876]. Considering an unconstrained optimization problem, such as listed in Section A.5.3, each individual, $\mathbf{x}_i$, will be represented by an $n_x$-dimensional vector where each $x_{ij} \in \mathbb{R}$. For the initial population, each individual is initialized using

$$x_{ij} \sim U(x_{min,j}, x_{max,j}) \tag{13.36}$$

The fitness function is simply the function to be optimized.

DE has also been applied to train neural networks (NN) (refer to Table 13.1 for references). In this case an individual represents a complete NN. Each element of an individual is one of the weights or biases of the NN, and the fitness function is, for example, the sum-squared error (SSE).

Table 13.1 summarizes a number of real-world applications of DE. Please note that this is not meant to be a complete list.

**Table 13.1** Applications of Differential Evolution

| Application Class | Reference |
|---|---|
| Clustering | [640, 667] |
| Controllers | [112, 124, 164, 165, 394, 429, 438, 599] |
| Filter design | [113, 810, 812, 883] |
| Image analysis | [441, 521, 522, 640, 926] |
| Integer-Programming | [390, 499, 500, 528, 530, 817] |
| Model selection | [331, 354, 749] |
| NN training | [1, 122, 550, 551, 598] |
| Scheduling | [528, 531, 699, 748] |
| System design | [36, 493, 496, 848, 839, 885] |

## 13.7   Assignments

1. Show how DE can be used to train a FFNN.

2. Discuss the influence of different values for the population diversity tolerance, $\epsilon_1$, and the gene diversity tolerance, $\epsilon_2$, as used in equations (13.11) and (13.12) for the hybrid DE.

3. Discuss the merits of the following two statements:

   (a) If the probability of recombination is very low, then DE exhibits a high probability of stagnation.

   (b) For a small population size, it is sensible to have a high probability of recombination.

4. For the DE/rand-to-best/$y/z$ strategies, suggest an approach to balance exploration and exploitation.

5. Discuss the consequences of too large and too small values of the standard deviation, $\sigma$, used in Algorithm 13.6.

6. Explain in detail why the method for adding noise to trial vectors as given in equation (13.15) may result in genetic drift.

7. With reference to the DynDE algorithm in Section 13.5.3, explain the effect of very small and very large values of the standard deviation, $\sigma$.

8. Researchers in DE have suggested that the recombination probability should be sampled from a Gaussian distribution, $N(0, 1)$, while others have suggested that $N(0.5, 0.15)$ should be used. Compare these two suggestions and provide a recommendation as to which approach is best.

9. Investigate the performance of a DE strategy if the scale factor is sampled from a Cauchy distribution.

# Chapter 14

# Cultural Algorithms

Standard evolutionary algorithms (as discussed in previous chapters) have been successful in solving diverse and complex problems in search and optimization. The search process used by standard EAs is unbiased, using little or no domain knowledge to guide the search process. However, the performance of EAs can be improved considerably if domain knowledge is used to bias the search process. Domain knowledge then serves as a mechanism to reduce the search space by pruning undesirable parts of the solution space, and by promoting desirable parts. Cultural evolution (CE) [717], based on the principles of human social evolution, was developed by Reynolds [716, 717, 724] in the early 1990s as an approach to bias the search process with prior knowledge about the domain as well as knowledge gained during the evolutionary process.

Evolutionary computation mimics biological evolution, which is based on the principle of genetic inheritance. In natural systems, genetic evolution is a slow process. Cultural evolution, on the other hand, enables societies to adapt to their changing environments at rates that exceed that of biological evolution.

The rest of this chapter is organized as follows: A compact definition of culture and artificial culture is given in Section 14.1. A general cultural algorithm (CA) framework is given in Section 14.2, outlining the different components of CA implementations. Section 14.3 describes the belief space component of CAs. A fuzzy CA approach is described in Section 14.4. Advanced topics, including constrained environments are covered in Section 14.5. Applications of CAs are summarized in Section 14.6.

## 14.1 Culture and Artificial Culture

A number of definitions of culture can be found, for example[1]:

- Culture is a system of symbolically encoded conceptual phenomena that are socially and historically transmitted within and between social groups [223].

- Culture refers to the cumulative deposit of knowledge, experience, beliefs, values, attitudes, meanings, hierarchies, religion, notions of time, roles, spatial relations, concepts of the universe, and material objects and possessions acquired

---

[1] www.tamu.edu/classes/cosc/choudhury/culture.html

*Computational Intelligence: An Introduction*, Second Edition A.P. Engelbrecht
©2007 John Wiley & Sons, Ltd

by a group of people in the course of generations through individual and group striving.

- Culture is the sum total of the learned behavior of a group of people that is generally considered to be the tradition of that people and is transmitted from generation to generation.

- Culture is a collective programming of the mind that distinguishes the members of one group or category of people from another.

In terms of evolutionary computation, culture is modeled as the source of data that influences the behavior of all individuals within that population. This differs from EP and ES where the behavioral characteristics of individuals – for the current generation only – are modeled using phenotypes. Within cultural algorithms, culture stores the general behavioral traits of the population. Cultural information is then accessible to all the individuals of a population, and over many generations.

## 14.2　Basic Cultural Algorithm

A cultural algorithm (CA) is a dual-inheritance system, which maintains two search spaces: the population space (to represent a genetic component based on Darwinian principles), and a belief space (to represent a cultural component). It is the latter that distinguishes CAs from other EAs.

The belief space models the cultural information about the population, while the population space represents the individuals on a genotypic and/or phenotypic level. Both the population and belief spaces evolve in parallel, with both influencing one another. A communication protocol therefore forms an integral part of a CA. Such a protocol defines two communication channels. One for a select group of individuals to adapt the set of beliefs, and another defining the way that the beliefs influence all of the individuals in the population space.

A pseudocode cultural algorithm is given in Algorithm 14.1, and illustrated in Figure 14.1.

---

**Algorithm 14.1** Cultural Algorithm

---

Set the generation counter, $t = 0$;
Create and initialize the population space, $\mathcal{C}(0)$;
Create and initialize the belief space, $\mathcal{B}(0)$;
**while** *stopping condition(s) not true* **do**
　　Evaluate the fitness of each $\mathbf{x}_i(t) \in \mathcal{C}(t)$;
　　Adjust $(\mathcal{B}(t), \text{Accept } (\mathcal{C}(t)))$;
　　Variate $(\mathcal{C}(t), \text{Influence } (\mathcal{B}(t)))$;
　　$t = t + 1$;
　　Select the new population;
**end**

---

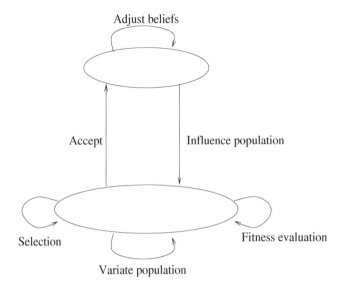

**Figure 14.1** Illustration of Population and Belief Spaces of Cultural Algorithms

At each iteration (each generation), individuals are first evaluated using the fitness function specified for the EA on the population level. An *acceptance* function is then used to determine which individuals from the current population have an influence on the current beliefs. The experience of the accepted individuals is then used to *adjust* the beliefs (to simulate evolution of culture). The adjusted beliefs are then used to *influence* the evolution of the population. The *variation* operators (crossover and mutation) use the beliefs to control the changes in individuals. This is usually achieved through self-adapting control parameters, as functions of the beliefs.

The population space is searched using any of the standard EAs, for example an EP [720] or GA [716]. Recently, particle swarm optimization (PSO) has been used on the population space level [219]. The next section discusses the belief space in more detail.

## 14.3   Belief Space

The belief space serves as a knowledge repository, where the collective behaviors (or beliefs) of the individuals in the population space are stored. The belief space is also referred to as the *meme pool*, where a *meme* is a unit of information transmitted by behavioral means. The belief space serves as a global knowledge repository of behavioral traits. The memes within the belief space are generalizations of the experience of individuals within the population space. These experiential generalizations are accumulated and shaped over several generations, and not just one generation. These generalizations express the beliefs as to what the optimal behavior of individuals constitutes.

The belief space can effectively be used to prune the population space. Each individual represents a point in the population search space: the knowledge within the belief space is used to move individuals away from undesirable areas in the population space towards more promising areas.

Some form of communication protocol is implemented to transfer information between the two search spaces. The communication protocol specifies operations that control the influence individuals have on the structure of the belief space, as well as the influence that the belief space has on the evolution process in the population level. This allows individuals to dictate their culture, causing culture to also evolve. On the other hand, the cultural information is used to direct the evolution on population level towards promising areas in the search space. It has been shown that the use of a belief space reduces computational effort substantially [717, 725].

Various CAs have been developed, which differ in the data structures used to model the belief space, the EA used on the population level, and the implementation of the communication protocol. Section 14.3.1 provides an overview of different knowledge components within the belief space. Acceptance and influence functions are discussed in Sections 14.3.2 and 14.3.4 respectively.

## 14.3.1   Knowledge Components

The belief space contains a number of knowledge components to represent the behavioral patterns of individuals from the population space. The types of knowledge components and data structures used to represent the knowledge depends on the problem being solved. The first application of CAs used version spaces represented as a lattice to store schemata [716]. For function optimization, vector representations are used (discussed below) [720]. Other representations that have been used include fuzzy systems [719, 725], ensemble structures [722], and hierarchical belief spaces [420].

In general, the belief space contains at least two knowledge components [720]:

- A **situational** knowledge component, which keeps track of the best solutions found at each generation.
- A **normative** knowledge component, which provides standards for individual behaviors, used as guidelines for mutational adjustments to individuals. In the case of function optimization, the normative knowledge component maintains a set of intervals, one for each dimension of the problem being solved. These intervals characterize the range of what is believed to be good areas to search in each dimension.

If only these two components are used, the belief space is represented as the tuple,

$$\mathcal{B}(t) = (\mathcal{S}(t), \mathcal{N}(t)) \tag{14.1}$$

where $\mathcal{S}(t)$ represents the situational knowledge component, and $\mathcal{N}(t)$ represents the normative knowledge component. The situational component is the set of best solutions,

$$\mathcal{S}(t) = \{\hat{\mathbf{y}}_l(t) : l = 1, \ldots, n_{\mathcal{S}}\} \tag{14.2}$$

and the normative component is represented as

$$\mathcal{N}(t) = (\mathcal{X}_1(t), \mathcal{X}_2(t), \ldots, \mathcal{X}_{n_x}(t)) \tag{14.3}$$

where, for each dimension, the following information is stored:

$$\mathcal{X}_j(t) = (\mathcal{I}_j(t), L_j(t), U_j(t)) \tag{14.4}$$

$\mathcal{I}_j$ denotes the closed interval, $\mathcal{I}_j(t) = [x_{min,j}(t), x_{max,j}(t)] = \{x : x_{min,j} \leq x \leq x_{max,j}\}$, $L_j(t)$ is the score for the lower bound, and $U_j(t)$ is the score for the upper bound.

In addition to the above knowledge components, the following knowledge components can be added [672, 752]:

- A **domain** knowledge component, which is similar to the situational knowledge component in that it stores the best positions found. The domain knowledge component differs from the situational knowledge component in that knowledge is not re-initialized at each generation, but contains an archive of best solutions since evolution started – very similar to the hall-of-fame used in coevolution.

- A **history** knowledge component, used in problems where search landscapes may change. This component maintains information about sequences of environmental changes. For each environmental change, the following information is stored: the current best solution, the directional change for each dimension and the current change distance.

- A **topographical** knowledge component, which maintains a multi-dimensional grid representation of the search space. Information is kept about each cell of the grid, e.g. frequency of individuals that occupy the cell. Such frequency information can be used to improve exploration by forcing mutation direction towards unexplored areas.

The type of knowledge components and the way that knowledge is represented have an influence on the acceptance and influence functions, as discussed in the following sections.

## 14.3.2 Acceptance Functions

The acceptance function determines which individuals from the current population will be used to shape the beliefs for the entire population. Static methods use absolute ranking, based on fitness values, to select the top $n\%$ individuals. Any of the selection methods for EAs (refer to Section 8.5) can be used, for example elitism, tournament selection, or roulette-wheel selection, provided that the number of individuals remains the same.

Dynamic methods do not have a fixed number of individuals that adjust the belief space. Instead, the number of individuals may change from generation to generation. Relative ranking, for example, selects individuals with above average (or median) performance [128]. Alternatively, the number of individuals is determined as

$$n_\mathcal{B}(t) = \lceil \frac{n_s \gamma}{t} \rceil \tag{14.5}$$

with $\gamma \in [0, 1]$. Using this approach, the number of individuals used to adjust the belief space is initially large, with the number decreasing exponentially over time. Other simulated-annealing based schedules can be used instead.

Adaptive methods use information about the search space and process to self-adjust the number of individuals to be selected. Reynolds and Chung [719] proposed a fuzzy acceptance function to determine the number of individuals based on generation number and individual success ratio. Membership functions are defined to implement the rules given in Algorithm 14.2.

### 14.3.3 Adjusting the Belief Space

For the purpose of this section, it is assumed that the belief space maintains a situational and normative knowledge component, and that a continuous, unconstrained function is minimized.

With the number of accepted individuals, $n_\mathcal{B}(t)$, known, the two knowledge components can be updated as follows [720]:

- **Situational knowledge**: Assuming that only one element is kept in the situational knowledge component,

$$\mathcal{S}(t+1) = \{\hat{\mathbf{y}}(t+1)\} \tag{14.6}$$

 where

$$\hat{\mathbf{y}}(t+1) = \begin{cases} \min_{l=1,\dots,n_\mathcal{B}(t)}\{\mathbf{x}_l(t)\} & \text{if } f(\min_{l=1,\dots,n_\mathcal{B}(t)}\{\mathbf{x}_l(t)\}) < f(\hat{\mathbf{y}}(t)) \\ \hat{\mathbf{y}}(t) & \text{otherwise} \end{cases}$$
$$\tag{14.7}$$

- **Normative knowledge**: In adjusting the normative knowledge component, a conservative approach is followed when narrowing intervals, thereby delaying too early exploration. Widening of intervals is applied more progressively. The interval update rule is as follows:

$$x_{min,j}(t+1) = \begin{cases} x_{lj}(t) & \text{if } x_{lj}(t) \le x_{min,j}(t) \text{ or } f(\mathbf{x}_l(t)) < L_j(t) \\ x_{min,j}(t) & \text{otherwise} \end{cases} \tag{14.8}$$

$$x_{max,j}(t+1) = \begin{cases} x_{lj}(t) & \text{if } x_{lj}(t) \ge x_{max,j}(t) \text{ or } f(\mathbf{x}_l(t)) < U_j(t) \\ x_{max,j}(t) & \text{otherwise} \end{cases} \tag{14.9}$$

$$L_j(t+1) = \begin{cases} f(\mathbf{x}_l(t)) & \text{if } x_{lj}(t) \le x_{min,j}(t) \text{ or } f(\mathbf{x}_l(t)) < L_j(t) \\ L_j(t) & \text{otherwise} \end{cases} \tag{14.10}$$

---

**Algorithm 14.2** Fuzzy Rule-base for Cultural Algorithm Acceptance Function

---

**if** *the current generation is early in the search* **then**
    **if** *the success ratio is low* **then**
        Accept a medium number of individuals (30%);
    **end**
    **if** *the success ratio is medium* **then**
        Accept a medium number of individuals;
    **end**
    **if** *the success ratio is high* **then**
        Accept a larger number of individuals (40%);
    **end**
**end**
**if** *the current generation is in the middle stages of the search* **then**
    **if** *the success ratio is low* **then**
        Accept a smaller number of individuals (20%);
    **end**
    **if** *the success ratio is medium* **then**
        Accept a medium number of individuals;
    **end**
    **if** *the success ratio is high* **then**
        Accept a medium number of individuals;
    **end**
**end**
**if** *the current generation is near the end of the search* **then**
    **if** *the success ratio is low* **then**
        Accept a smaller number of individuals;
    **end**
    **if** *the success ratio is medium* **then**
        Accept a smaller number of individuals;
    **end**
    **if** *the success ratio is high* **then**
        Accept a medium number of individuals;
    **end**
**end**

---

$$U_j(t+1) = \begin{cases} f(\mathbf{x}_{lj}(t)) & \text{if } x_{lj}(t) \geq x_{max,j}(t) \text{ or } f(\mathbf{x}_l(t)) < U_j(t) \\ U_j(t) & \text{otherwise} \end{cases} \qquad (14.11)$$

for each $\mathbf{x}_l(t), l = 1, \ldots, n_{\mathcal{B}}(t)$.

## 14.3.4 Influence Functions

Beliefs are used to adjust individuals in the population space to conform closer to the global beliefs. The adjustments are realized via influence functions. To illustrate this process, assume that an EP is used as the search algorithm in the population space. The resulting algorithm is referred to as a CAEP.

The belief space is used to determine the mutational step sizes, and the direction of changes (i.e. whether step sizes are added or subtracted). Reynolds and Chung [720] proposed four ways in which the knowledge components can be used within the influence function:

- Only the normative component is used to determine step sizes during offspring generation:

$$x'_{ij}(t) = x_{ij}(t) + \text{size}(\mathcal{I}_j(t))N_{ij}(0,1) \tag{14.12}$$

  where

$$\text{size}(\mathcal{I}_j(t)) = x_{max,j}(t) - x_{min,j}(t) \tag{14.13}$$

  is the size of the belief interval for component $j$.

- Only the situational component is used for determining change direction:

$$x'_{ij}(t) = \begin{cases} x_{ij}(t) + |\sigma_{ij}(t)N_{ij}(0,1)| & \text{if } x_{ij}(t) < \hat{y}_j(t) \in \mathcal{S}(t) \\ x_{ij}(t) - |\sigma_{ij}(t)N_{ij}(0,1)| & \text{if } x_{ij}(t) > \hat{y}_j(t) \in \mathcal{S}(t) \\ x_{ij}(t) + \sigma_{ij}(t)N_{ij}(0,1) & \text{otherwise} \end{cases} \tag{14.14}$$

  where $\sigma_{ij}$ is the strategy parameter associated with component $j$ of individual $i$.

- The normative component is used to determine change directions, and the situational component is used to determine step sizes. Equation (14.14) is used, but with

$$\sigma_{ij}(t) = \text{size}(\mathcal{I}(t)) \tag{14.15}$$

- The normative component is used for both the search directions and step sizes:

$$x_{ij}(t) = \begin{cases} x_{ij}(t) + |\text{size}(\mathcal{I}_j(t))N_{ij}(0,1)| & \text{if } x_{ij}(t) < x_{min,j}(t) \\ x_{ij}(t) - |\text{size}(\mathcal{I}_j(t))N_{ij}(0,1)| & \text{if } x_{ij}(t) > x_{max,j}(t) \\ x_{ij}(t) + \beta\text{size}(\mathcal{I}_j(t))N_{ij}(0,1) & \text{otherwise} \end{cases} \tag{14.16}$$

  where $\beta > 0$ is a scaling coefficient.

## 14.4   Fuzzy Cultural Algorithm

Section 14.3.2 discussed a fuzzy acceptance function, which initially allows many individuals to adjust the belief space, and less individuals as the generation number increases. This section discusses a completely fuzzy CA approach, where an alternative fuzzy acceptance function is used and both belief space and population space adaptations are based on fuzzy logic. This fuzzy CA was shown to improve accuracy and convergence speed [725].

## 14.4.1   Fuzzy Acceptance Function

Reynolds and Zhu [725] proposed an acceptance function that selects few individuals in the early steps of the search, but more as the generation number increases. The motivation for this approach stems from the fact that the initial population usually has only a few fit individuals that can sensibly contribute to the belief space.

As the first step, a fuzzy similarity matrix (or fuzzy equivalence relation), $\mathbf{R}$, is constructed for the population space where

$$r_{ij} = 1 - \frac{|f(\mathbf{x}_i) - f(\mathbf{x}_j)|}{\sum_{l=1}^{n_s} |f(\mathbf{x}_l)|} \tag{14.17}$$

for $i, j = 1, \ldots, n_s$, with $n_s$ the size of the population, and $\mathbf{R} = [r_{ij}]$. Before calculation of $\mathbf{R}$, the individuals are sorted from best to worst according to fitness. The equivalence relation is then used to calculate the $\alpha$-cuts matrix (refer to Section 20.4), $\mathbf{R}_\alpha$ as follows:

$$\alpha_{ij} = \begin{cases} 1 & \text{if } r_{ij} > \alpha(t) \\ 0 & \text{if } r_{ij} \leq \alpha(t) \end{cases} \tag{14.18}$$

where $\alpha(t)$ is referred to as the refinement value, and $\mathbf{R}_\alpha = [\alpha_{ij}]$ for $i, j = 1, \ldots, n_s$. A value of $\alpha_{ij} = 1$ indicates an elite individual that will be used to update the belief space, while a value of $\alpha_{ij} = 0$ indicates a non-elite individual. Because elements are sorted according to fitness, the elite elements will occupy the first elements of $\mathbf{R}_\alpha$.

The refinement value is initialized to a large value, but bounded by the interval $[0, 1]$, which will ensure a small number of elite individuals. After each generation, the value of $\alpha$ is changed according to

$$\alpha(t + 1) = \begin{cases} \gamma_1 & \text{if } t \geq \epsilon \\ 1 - \gamma_2(t - \epsilon)/n_t & \text{if } t \in [0, \epsilon) \end{cases} \tag{14.19}$$

where $\gamma_1$ and $\gamma_2$ are positive constants (suggested to be $\gamma_1 = 0.2$ and $\gamma_2 = 2.66$ [725]), and $\epsilon$ is set such that $\epsilon/t_{max} = 0.3$.

## 14.4.2   Fuzzified Belief Space

Only two knowledge components are considered. The situational component is crisp and stores the best individual. The normative component is fuzzified.

The fuzzy interval update rule occurs in the following steps for each $\mathbf{x}_l(t), l = 1, \ldots, n_B(t)$:

1. Initialization: Upper and lower bounds are initialized to the domain, as follows:

$$x_{min,lj}(0) = x_{min,j} \tag{14.20}$$
$$x_{max,lj}(0) = x_{max,j} \tag{14.21}$$

where $x_{minj}$ and $x_{max,j}$ are the domain bounds for component $j$.

2. Calculate the center of the interval for each individual as

$$c_{ij}(t) = \frac{1}{2}(x_{min,lj}(t) + x_{max,lj}(t)) \tag{14.22}$$

Interval bounds are adjusted until $x_{min,lj}(t), x_{max,lj}(t) \in [0.9c_{lj}(t), 1.1c_{lj}(t)]$.

3. Compute the fuzzy membership functions, $f_L(x_j)$ and $f_U(x_j)$ for each individual for the lower and upper bounds respectively:

$$f_L(x_j) = \frac{1}{1 + x_j^\alpha} \tag{14.23}$$

$$f_U(x_j) = \frac{1}{1 + x_j^\alpha} \tag{14.24}$$

for $\alpha \geq 0$.

4. Update the lower boundary of each component of individual $l$:

$$x_{min,lj}(t+1) = \begin{cases} x_{min,lj}(t) + \mu(x_{lj}(t))(x_{lj}(t) - x_{min,lj}(t)) \\ \quad\quad \text{if } x_{min,lj}(t) \leq x_{lj}(t) < 0.9c_{lj}(t) \\ x_{min,lj}(t) \\ \quad\quad \text{if } 0.9c_{lj}(t) \leq x_{lj}(t) < 1.1c_{lj}(t) \end{cases} \tag{14.25}$$

where $\mu(x_{lj}(t)) = f_L(x_{lj}(t))$.

5. Update the upper boundary of each component of individual $l$:

$$x_{max,lj}(t+1) = \begin{cases} x_{max,lj}(t) + \mu(x_{lj}(t))(x_{lj}(t) - x_{max,lj}(t)) \\ \quad\quad \text{if } 1.1c_{lj}(t) < x_{lj}(t) \leq x_{max,lj}(t) \\ x_{max,lj}(t) \\ \quad\quad \text{if } 0.9c_{lj}(t) \leq x_{lj}(t) \leq 1.1c_{lj}(t) \end{cases} \tag{14.26}$$

where $\mu(x_{lj}(t)) = f_U(x_{lj}(t))$.

## 14.4.3 Fuzzy Influence Function

The fuzzified normative knowledge is used to determine both the step size and the direction of change. Mutational step sizes are based on the age, $a_i(t)$, of individuals as defined in equation (11.68). For each generation that a parent performs better than its offspring, the parent's age is incremented. When the offspring replaces the parent, the corresponding age is set to 1. The influence function will effect larger mutations to individuals with a lower age value, as such individuals are still in a strong exploration phase. Assuming an EP for the population space, offspring are generated as follows for each parent, $\mathbf{x}_i(t)$:

$$x'_{ij}(t) = \begin{cases} x_{ij}(t) + \sigma_{ij}(t)|\text{size}(\mathcal{I}_j(t))N_{ij}(0,1)| & \text{if } x_{ij}(t) < x_{min,ij}(t) \\ x_{ij}(t) - \sigma_{ij}(t)|\text{size}(\mathcal{I}_j(t))N_{ij}(0,1)| & \text{if } x_{ij}(t) > x_{max,ij}(t) \\ x_{ij}(t) + \sigma_{ij}(t)\text{size}(\mathcal{I}_j(t))N_{ij}(0,1) & \text{otherwise} \end{cases} \tag{14.27}$$

where $\sigma_{ij}(t)$ is computed as

$$\sigma_{ij}(t) = \mu(a_i(t)) = \begin{cases} 1 - \frac{1}{a_i(t)} & \text{if } a_i(t) < \alpha \\ 0 & \text{if } a_i(t) \geq \alpha \end{cases} \tag{14.28}$$

where $\alpha$ is a positive constant. If a parent performs better than its offspring for more than $\alpha$ generations, then that individual will not be mutated.

## 14.5   Advanced Topics

This section discusses specific CA implementations for solving constrained optimization problems (Section 14.5.1), multi-objective optimization problems (Section 14.5.2), and dynamic environments (Section 14.5.3).

### 14.5.1   Constraint Handling

Reynolds and Peng [721], and Dos Santos Coelho and Mariani [219] used a penalty approach (refer to Section A.6.2) to solve constrained problems. The approach followed by Reynolds [721] is to simply set the fitness of any individual that violates a constraint, to an infeasible value. These penalty methods do not make use of the belief space to guide individuals to move towards feasible space. Chung and Reynolds [128] represented constraints as interval-constraint networks. These networks are then used to determine if an individual violates any of the constraints. In the case of constraint violations, information from the belief space is used to repair that individual.

Jin and Reynolds [419] expanded the interval representation of the normative knowledge component to also represent constraints. The approach fits a grid structure over the search space, where the cells are referred to as belief cells. Belief cells are then classified into one of four categories (as illustrated in Figure 14.2):

- **feasible**, which represents a part of the search space where no constraint is violated;
- **infeasible**, where at least one constraint is violated;
- **semi-feasible**, in which case the cell contains both feasible parts of the search space; and
- **unknown**

The intervals as given in the normative knowledge component are used to divide the search space as represented by these intervals into hypercubes or belief cells. The type of belief cell is determined based on the number of individuals that fall in that cell. For each cell,

- if there are no feasible or infeasible individuals, the cell is classified as unknown;
- if there are no infeasible individuals, but at least one feasible individual, the cell is classified as feasible;
- if there are no feasible individuals, but at least one infeasible individual, the cell is classified as infeasible; or
- if there are both feasible and infeasible individuals, the cell is classified as semi-feasible.

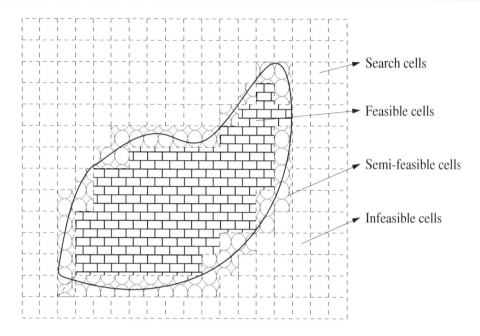

**Figure 14.2** Illustration of Belief Cells

Initially, all cells are classified as unknown.

The update of the belief space requires the counters of feasible and infeasible individuals per belief cell to be updated. At regular intervals, the normative component is updated by redefining the hypercubes based on the most recent intervals as contained in this knowledge component. After recalculation of hypercubes, all the feasible and infeasible counters have to be updated. Coello and Becerra [145] proposed that the acceptance function be adapted to use fitness as a secondary criterion in selecting individuals. Instead, feasible individuals should be given preference to update the belief space.

The influence function is changed to promote creation of feasible offspring. If $x_{ij}(t)$ is within a feasible, semi-feasible, or unknown hypercube, mutation is applied such that the individual remains in the same hypercube, or stays very close to it. If $x_{ij}(t)$ is in an infeasible cell, the direction of mutation is such that the individual moves closer to a semi-infeasible cell. If there is no semi-feasible cell in the vicinity of $x_{ij}(t)$, the move is made towards an unknown cell. Failing this, $x_{ij}(t)$ is initialized to a random position within the interval defined by the normative component.

## 14.5.2  Multi-Objective Optimization

Coello and Becerra [146] developed the first CA for solving multi-objective optimization problems (MOP). The belief space is constructed to consist of a normative component (representing intervals) and a grid. The grid is constructed over objective space, and for each cell the number of nondominated solutions that lie within that cell is recorded. This information is then used to ensure that nondominated solutions are uniformly distributed along the Pareto front.

The normative knowledge component is updated at regular intervals, while the grid is updated at each generation. Updating the grid simply involves recalculating the number of nondominated solutions per cell. An update to the normative component triggers a recalculation of the grid over objective space.

Within the population space, selection of the new population is adapted to make use of the grid information contained in the belief space. Tournament selection is used, and applied to the parents and offspring. The rules of tournament selection is updated as follows:

- If an individual dominates its competitor, then the dominating individual wins.
- If none of the two competing individuals are dominating, or if their objective values are the same, then
  - if one of the individuals lies outside of the grid, then that individual is selected.
  - if both lie in the grid, the one in the less populated cell is selected.
- If none of the cases above are satisfied, the more fit individual is selected.

This MOO approach using CAs maintains an archive of nondominated solutions [146].

## 14.5.3   Dynamic Environments

Saleem and Reynolds [752] showed that the self-adaptability of CAs, due to the dynamically changing belief space, makes CAs suitable for tracking optima in dynamically changing environments. The belief space is extended to store information from previous and current environmental states. Environmental history is stored in a table, containing the following information for each environment: the location of the best solution (i.e. the current item stored in the situational knowledge component), the fitness value of that solution, the change magnitude in each dimension, and the consequent change in fitness value. This information is then used in a dynamic influence function to introduce diversity into the population, proportional to the magnitude of change.

The step size is calculated as

$$\Delta x_{ij}(t) = \frac{2|f(\mathbf{x}_i(t)) - \tilde{f}(\mathbf{x}_i(t))|}{\hat{f}(t)} U_{ij}(0,1) \tag{14.29}$$

where $f$ represents the landscape before change, and $\tilde{f}$ represents the changed landscape; $\hat{f}(t)$ is the fitness value of the best solution as stored in the history table. This results in large step sizes for large environmental changes; thereby increasing diversity.

## 14.6  Applications

Cultural algorithms have been applied to solve a number of problems, with one of the first applications modeling the evolution of agriculture in the Valley of Oaxaca (Mexico) [715, 721]. Other applications of CAs include concept learning [826], inducing decision trees [718], real-valued function optimization [720], optimizing semantic networks [747], software testing [647], assessing the quality of GP [163], data mining [420], image segmentation [723], and robotics [287].

## 14.7  Assignments

1. Discuss how a CA can be used to train a FFNN.

2. What are the similarities and differences between CAs and ES?

3. Discuss the validity of the following statement: The belief space can be likened to a blackboard system used in multi-agent systems.

4. With reference to [24], show how CAs address the three levels of adaptation: population, individual, and component.

5. Investigate the performance of different selection operators, as discussed in Section 8.5, as acceptance function.

6. Propose membership functions for the fuzzy sets used to implement the rules as given in Algorithm 14.2 for the fuzzy acceptance function discussed in Section 14.3.2.

7. Discuss the merits of the fuzzy acceptance function of Section 14.3.2 with reference to the fuzzy rules summarized in Algorithm 14.2.

8. Implement a DE for the population space of a CA, and compare its performance with CAEP.

# Chapter 15

# Coevolution

Coevolution is the complementary evolution of closely associated species. Coevolution between two species is nicely illustrated using Holland's example of the competitive interaction between a plant and insects [377]. Consider a certain species of plant living in an environment containing insects that eat the plant. The survival "game" consists of two parts: (1) to survive, the plant needs to evolve mechanisms to defend itself from the insects, and (2) the insects need the plant as food source to survive. Both the plant and the insects evolve in complexity to obtain characteristics that will enable them to survive. For example, the plant may evolve a tough exterior, but then the insects evolve stronger jaws. Next the plant may evolve a poison to kill the insects. Next generations of the insect evolve an enzyme to digest the poison. The effect of this coevolutionary process is that, with each generation, both the plant and the insects become better at their defensive and offensive roles. In the next generation, each species change in response to the actions of the other species during the previous generation.

The biological example described above is an example of *predator-prey* competitive coevolution, 'where there is an inverse fitness interaction between the two species. A win for the one species means a failure for the other. To survive, the "losing" species adapts to counter the "winning" species in order to become the new winner. During this process the complexity of both the predator and the prey increases.

An alternative coevolutionary process is *symbiosis*, where the different species cooperate instead of compete. In this case a success in one species improves the survival strength of the other species. Symbiotic coevolution is thus achieved through a positive fitness feedback among the species that take part in this cooperating process.

In standard EAs, evolution is usually viewed as if the population attempts to adapt in a fixed physical environment. In contrast, coevolutionary (CoE) algorithms (CoEA) realize that in natural evolution the physical environment is influenced by other independently-acting biological populations. Evolution is therefore not just locally within each population, but also in response to environmental changes as caused by other populations. Another difference between standard EAs and CoEAs is that EAs define the meaning of optimality through an absolute fitness function. This fitness function then drives the evolutionary process. On the other hand, CoEAs do not define optimality using a fitness function, but attempt to evolve an optimal species where optimality is defined as defeating opponents (in the case of predator-prey CoE).

*Computational Intelligence: An Introduction*, Second Edition A.P. Engelbrecht
©2007 John Wiley & Sons, Ltd

Section 15.1 provides a summary of different types of coevolution. Competitive coevolution is discussed in Section 15.2, while some cooperative coevolution approaches are summarized in Section 15.3.

# 15.1  Coevolution Types

As indicated in the introduction of this chapter, two main classes of coevolutionary approaches can be identified, namely competitive coevolution and cooperative coevolution. For each of these classes, Fukuda and Kubota [293] identified a number of subclasses. For competitive coevolution, the following subclasses can be found:

- **Competition**, where both species are inhibited. Due to the inverse fitness interaction between the two species, success in one of the species is felt as failure by the other species.

- **Amensalism**, where one species is inhibited, and the other is not affected.

In the case of cooperative (or symbiotic) coevolution, the following subclasses can be identified:

- **Mutualism**, where both species benefit. The positive fitness interaction leads to an improvement in one species whenever the other species improves.

- **Commensalism**, where only one of the species benefits, while the other is not affected.

- **Parasitism**, where one of the species (the parasite) benefits, while the other species (the host) is harmed.

The focus of this chapter will be on competitive (predator-prey) coevolution and mutualism.

# 15.2  Competitive Coevolution

Competitive coevolution (CCE) works to produce optimal competing species through a pure bootstrapping process. Solutions to problems are found without any prior knowledge from human experts, or any other information on how to solve the problem. A CCE algorithm usually evolves two populations simultaneously. Individuals in one population represent solutions to a problem, while individuals in the other population represent test cases. Individuals in the solution population evolve to solve as many test cases as possible, while individuals in the test population evolve to present an incrementally increasing level of difficulty to the solution individuals. The fitness of individuals in the solution population is proportional to the number of test cases solved by the solution. The fitness of individuals in the test population is inversely proportional to the number of strategies that solve it.

This is the CCE approach popularized by Hillis [365]. However, Hillis was not the first to consider CCE. Miller [594, 595] and Axelrod [33] used CCE to evolve strategies for the iterated prisoner's dilemma (IPD). Holland [377] incorporated coevolution in a single population GA, where individuals compete against other individuals of the same population.

Section 15.2.1 discusses competitive fitness. The hall of fame concept is discussed in Section 15.2.1. A generic CCE algorithm is given in Section 15.2.2. Some applications of CCE are given in Section 15.2.3.

## 15.2.1 Competitive Fitness

Standard EAs use a user-defined, absolute fitness function to quantify the quality of solutions. This absolute fitness function directly represents the optimization problem. The fitness of each individual is evaluated independently from any other population, or individual, using this fitness function. In CCE, the driving force of the evolutionary process is via a relative fitness function that only expresses the performance of individuals in one population in comparison with individuals in another population. Usually, the relative fitness computes a score of how many opponents are beaten by an individual. It should be clear that the only quantification of optimality is which population's individuals perform better. No fitness function that describes the optimal point is used.

In order to calculate the relative fitness of an individual, two aspects are of importance: (1) Which individuals from the competing population are used, and (2) exactly how these competing individuals are used to compute the relative fitness. The first aspect refers to fitness sampling, and is discussed in Section 15.2.1. Relative fitness evaluation approaches are summarized in Section 15.2.1.

### Fitness Sampling

The relative fitness of individuals is evaluated against a sample of individuals from the competing population. The following sampling schemes have been developed:

- **All versus all sampling** [33], where each individual is tested against all the individuals of the other population.

- **Random sampling** [711], where the fitness of each individual is tested against a randomly selected group (consisting of one or more) individuals from the other population. The random sampling approach is computationally less complex than all versus all sampling.

- **Tournament sampling** [27], which uses relative fitness measures to select the best opponent individual.

- **All versus best sampling** [793], where all the individuals are tested against the fittest individual of the other population.

- **Shared sampling** [738, 739, 740], where the sample is selected as those opponent individuals with maximum competitive shared fitness. Shared sampling tends to select opponents that beat a large number of individuals from the competing population.

## Relative Fitness Evaluation

The following approaches can be followed to measure the relative fitness of each individual in a population. Assume that the two populations $C_1$ and $C_2$ coevolve, and the aim is to calculate the relative fitness of each individual $C_1.\mathbf{x}_i$ of population $C_1$.

- **Simple fitness**: A sample of individuals is taken from population $C_2$, and the number of individuals in population $C_2$ for which $C_1.\mathbf{x}_i$ is the winner, is counted. The relative fitness of $C_1.\mathbf{x}_i$ is simply the sum of successes for $C_1.\mathbf{x}_i$.

- **Fitness sharing** [323]: A sharing function is defined to take into consideration similarity among the individuals of population $C_1$. The simple fitness of an individual is divided by the sum of its similarities with all the other individuals in that population. Similarity can be defined as the number of individuals that also beats the individuals from the population $C_2$ sample. The consequence of the fitness sharing function is that unusual individuals are rewarded.

- **Competitive fitness sharing** [738, 739, 740]: In this case the fitness of individual $C_1.\mathbf{x}_i$ is defined as

$$f(C_1.\mathbf{x}_i) = \sum_{l=1}^{C_2.n_s} \frac{1}{C_1.n_l} \tag{15.1}$$

where $C_2.\mathbf{x}_1, \cdots, C_2.\mathbf{x}_{C_2.n_s}$ form the population $C_2$ sample, and $C_1.n_l$ is the total number of individuals in population $C_1$ that defeat individual $C_2.\mathbf{x}_l$. The competitive fitness sharing method rewards those population $C_1$ individuals that beat population $C_2$ individuals, which few other population $C_1$ individuals could beat. It is therefore not necessarily the case that the best population $C_1$ individual beats the most population $C_2$ individuals.

- **Tournament fitness** [27] holds a number of single elimination, binary tournaments to determine a relative fitness ranking. Tournament fitness results in a tournament tree, with the root element as the best individual. For each level in the tree, two opponents are randomly selected from that level, and the best of the two advances to the next level. In the case of an odd number of competitors, a single individual from the current level moves to the next level. After tournament ranking, any of the standard selection operators (refer to Section 8.5) can be used to select parents.

## Hall of Fame

Elitism is a mechanism used in standard EAs to ensure that the best parents of a current generation survive to the next generation. To be able to survive for more generations, an individual has to be highly fit in almost every population. For coevolution, Rosin and Belew [740] introduced the *hall of fame* to extend elitism in time. At each generation the best individual of a population is stored in that population's hall of fame. The hall of fame may have a limited size, in which case a new individual to be inserted in the hall of fame will replace the worst individual (or the oldest one). Individuals from one population now compete against a sample of the current opponent population and its hall of fame. The hall of fame prevents overspecialization.

## 15.2.2 Generic Competitive Coevolutionary Algorithm

The generic CCE algorithm in Algorithm 15.1 assumes that two competing populations are used. For a single population CCE algorithm, refer to Algorithm 15.2. For both these algorithms, any standard EA can be used to evolve each population for one iteration (i.e. perform reproduction and select the new population). Table 15.1 provides a summary of some publications for EAs used, including PSO.

---

**Algorithm 15.1** Competitive Coevolutionary Algorithm with Two Populations

---

Initialize two populations, $C_1$ and $C_2$;
**while** *stopping condition(s) not true* **do**
    **for** *each $C_1.\mathbf{x}_i, i = 1, \ldots, C_1.n_s$* **do**
        Select a sample of opponents from $C_2$;
        Evaluate the relative fitness of $C_1.\mathbf{x}_i$ with respect to this sample;
    **end**
    **for** *each $C_2.\mathbf{x}_i, i = 1, \ldots, C_2.n_s$* **do**
        Select a sample of opponents from $C_1$;
        Evaluate the relative fitness of $C_2.\mathbf{x}_i$ with respect to this sample;
    **end**
    Evolve population $C_1$ for one generation;
    Evolve population $C_2$ for one generation;
**end**
Select the best individual from the solution population, $S_1$;

---

---

**Algorithm 15.2** Single Population Competitive Coevolutionary Algorithm

---

Initialize one population, $C$;
**while** *stopping condition(s) not true* **do**
    **for** *each $C.\mathbf{x}_i, i = 1, \ldots, C.n_s$* **do**
        Select a sample of opponents from $C$ to exclude $C.\mathbf{x}_i$;
        Evaluate the relative fitness of $C.\mathbf{x}_i$ with respect to the sample;
    **end**
    Evolve population $C$ for one generation;
**end**
Select the best individual from $C$ as the solution;

---

Algorithm 15.1 assumes that $C_1$ is the solution population and $C_2$ is the test population.

**Table 15.1** Algorithms Used to Achieve Adaptation in CCE

| Algorithm | Reference |
|-----------|-----------|
| GA | [37, 132, 657, 750, 855] |
| EP | [119, 120, 268, 532] |
| ES | [60, 914] |
| GP | [343, 357, 358, 480] |
| PSO | [286, 580, 654] |

In CCE it is also possible to use two competing solution populations, in which case the individuals in one population serve as the test cases of individuals in the other populations. Here the solution returned by the algorithm is the best individual over both populations.

The performance of a CCE can be greatly improved if the two competing populations are as diverse as possible. Diversity can be maintained by incorporating mechanisms to facilitate niche formation. Shared sampling and the fitness sharing methods achieve this goal. Alternatively, any niching (speciation) algorithm can be implemented.

## 15.2.3    Applications of Competitive Coevolution

The first applications of coevolution were to evolve IPD strategies [33, 594, 595], and to evolve sorting algorithms [365]. Since these applications, CCE has been applied to a variety of complex real-world problems, as summarized in Table 15.2 (please note that this is not a complete list of applications).

The remainder of this section shows how CCE can be used to evolve game players for two-player, zero-sum, board games. The approach described here is based on the work of Chellapilla and Fogel [120, 121, 268] for Checkers, and further investigated by [156, 286, 580, 654] for Chess, Checkers, Tick-Tack-Toe, the IPD, and Bao. However, the model described here is not game specific.

The coevolutionary game learning  model trains neural networks  in a coevolutionary fashion to approximate the evaluation function of leaf nodes in a game tree. The learning model consists of three components:

- A game tree, expanded to a given ply-depth using game tree expansion algorithms such as minimax [629]. The root tree represents the current board state, while the other nodes in the tree represent future board states. The objective is to find the next move to take the player maximally closer to its goal, i.e. to win the game. To evaluate the desirability of future board states, an evaluation function is applied to the leaf nodes.

- A neural network evaluation function to estimate the desirability of board states represented by the leaf nodes. The NN receives a board state as its input, and

**Table 15.2** Applications of Competitive Coevolution

| Application | Reference |
|---|---|
| Game learning | [119, 120, 171, 249, 268, 286, 580, 615, 684, 739] |
| Military tactical planning | [455] |
| Controller design | [513] |
| Robot controllers | [20, 60, 132, 293, 424, 541, 631, 658, 859] |
| Evolving marketing strategies | [786] |
| Rule generation for | [415] |
| fuzzy logic controllers | |
| Constrained optimization | [144, 589, 784] |
| Autonomous vehicles | [93] |
| Scheduling | [855] |
| Neural network training | [119, 120, 286, 580, 657] |
| Drug design | [738] |
| Iterated prisoners dilemma | [33, 172, 594, 595] |

produces a scalar output as the board state desirability.

- A population of NNs, where each NN is represented by one individual, and trained in competition with other NNs. Any EA (or PSO) can be used to adapt the weights.

The objective of the above model, also summarized in Algorithm 15.3, is to evolve game-playing agents from zero knowledge about playing strategies. As is evident from Algorithm 15.3, the training process is not supervised. No target evaluation of board states is provided. The lack of desired outputs for the NN necessitates a coevolutionary training mechanism, where a NN competes against a sample of NNs in game tournaments. After each NN has played a number of games against each of its opponents, it is assigned a score based on the number of wins, losses and draws. These scores are then used as the relative fitness measure. Note that the population of NNs is randomly initialized.

# 15.3    Cooperative Coevolution

Section 15.1 referred to three different types of cooperative coevolution. This section focuses on mutualism, where individuals from different species (or subpopulations) have to cooperate in some way to solve a global task. Here, the fitness of an individual depends on that individual's ability to collaborate with individuals from other species. One of the major problems to resolve in such cooperative coevolution algorithms is that of credit assignment: How should the fitness achieved by the collective effort of all species be fairly split among the participating individuals.

De Jong and Potter [194] proposed a general framework for evolving complex solutions

---

**Algorithm 15.3** Coevolutionary Training of Game Agents

---

Create and randomly initialize a population of NNs;
**while** *stopping condition(s) not true* **do**
    **for** *each individual (or NN)* **do**
        Select a sample of competitors from the population;
        **for** *each opponent* **do**
            **for** *a specified number of times* **do**
                Play a game as first player using the NNs as board state evaluators in
                a game tree;
                Record if game was won, lost, or drawn;
                Play another game against the same opponent, but as second player;
                Record if game was won, lost, or drawn;
            **end**
        **end**
        Determine a score for the individual;
    **end**
    Evolve the population for one generation;
**end**
Return the best individual as the NN evaluation function;

---

by merging subcomponents, evolved independently from one another. A separate population is used to evolve each subcomponent using some EA. Representations from each subcomponent is then combined to form a complete solution, which is evaluated to determine a global fitness. Based on this global fitness, some credit flows back to each subcomponent reflecting how well that component collaborated with the others. This local fitness is then used within the subpopulation to evolve a better solution.

Potter and De Jong [687] applied this approach to function optimization. For an $n_x$-dimensional problem, $n_x$ subpopulations are used – one for each dimension of the problem. Each subpopulation is therefore responsible for optimizing one of the parameters of the problem, and no subpopulation can form a complete solution by itself. Collaboration is achieved by merging a representative from each subpopulation. The effectiveness of this collaboration is estimated as follows: Considering the $j$-th subpopulation, $\mathcal{C}_j$, then each individual, $\mathcal{C}_j.\mathbf{x}_i$, of $\mathcal{C}_j$ performs a single collaboration with the best individual from each of the other subpopulations by merging these best components with $\mathcal{C}_j.\mathbf{x}_i$ to form a complete solution. The credit assigned to $\mathcal{C}_j.\mathbf{x}_i$ is simply the fitness of the complete solution.

Potter and De Jong found that this approach does not perform well when problem parameters are strongly interdependent, due to the greediness of the credit assignment approach. To reduce greediness two collaboration vectors can be constructed. The first vector is constructed by considering the best individuals from each subpopulation as described above. The second vector chooses random individuals from the other subpopulations, and merges these with $\mathcal{C}_j.\mathbf{x}_i$. The best fitness of the two vectors is used as the credit for $\mathcal{C}_j.\mathbf{x}_i$.

In addition to function optimization, Potter and De Jong also applied this approach to evolve cascade neural networks [688] and robot learning [194, 690].

Other applications of cooperative coevolution include the evolution of predator-prey strategies using GP [358], evolving fuzzy membership functions [671], robot controllers [631], time series prediction [569], and neural network training [309].

# 15.4  Assignments

1. Design a CCE algorithm for playing tick-tack-toe.

2. Explain the importance of the relative fitness function in the success of a CCE algorithm.

3. Discuss the validity of the following statement: CCGA will not be successful if the genes of a chromosome are highly correlated.

4. Compare the different fitness sampling strategies with reference to computational complexity.

5. What will be the effect if fitness sampling is done only with reference to the hall of fame?

6. Why is shared sampling a good approach to calculate relative fitness?

7. Why is niche formation so important in CCE?

8. Design a CCE to evolve IPD strategies.

9. Implement a cooperative coevolutionary DE.

# Part IV

# COMPUTATIONAL SWARM INTELLIGENCE

Suppose that you and a group of friends are on a treasure finding mission. You have knowledge of the approximate area of the treasure, but do not know exactly where it is located. You want that treasure, or at least some part of it. Among your friends you have agreed on some sharing mechanism so that all who have taken part in the search will be rewarded, but with the person who found the treasure getting a higher reward than all others, and the rest being rewarded based on distance from the treasure at the time when the first one finds the treasure. Each one in the group has a metal detector and can communicate the strength of the signal and his current location to the nearest neighbors. Each person therefore knows whether one of his neighbors is nearer to the treasure than he is. What actions will you take? You basically have two choices: (1) Ignore your friends, and search for the treasure without any information that your friends may provide. In this case, if you find the treasure, it is all yours. However, if you do not find it first, you get nothing. (2) Make use of the information that you perceive from your neighboring friends, and move in the direction of your closest friend with the strongest signal. By making use of local information, and acting upon it, you increase your chances of finding the treasure, or at least maximizing your reward.

This is an extremely simple illustration of the benefits of cooperation in situations where you do not have global knowledge of an environment. Individuals within the group interact to solve the global objective by exchanging locally available information, which in the end propagates through the entire group such that the problem is solved more efficiently than can be done by a single individual.

In loose terms, the group can be referred to as a *swarm*. Formally, a swarm can be defined as a group of (generally mobile) agents that communicate with each other (either directly or indirectly), by acting on their local environment [371]. The interactions between agents result in distributive collective problem-solving strategies. Swarm intelligence (SI) refers to the problem-solving behavior that emerges from the interaction of such agents, and computational swarm intelligence (CSI) refers to algorithmic models of such behavior. More formally, swarm intelligence is the property of a system whereby the collective behaviors of unsophisticated agents interacting locally with their environment cause coherent functional global patterns to emerge [702]. Swarm intelligence has also been referred to as *collective intelligence*.

Studies of social animals and social insects have resulted in a number of computational models of swarm intelligence. Biological swarm systems that have inspired computational models include ants, termites, bees, spiders, fish schools, and bird flocks. Within these swarms, individuals are relatively simple in structure, but their collective behavior is usually very complex. The complex behavior of a swarm is a result of the pattern of interactions between the individuals of the swarm over time. This complex behavior is not a property of any single individual, and is usually not easily predicted or deduced from the simple behaviors of the individuals. This is referred to as *emergence*. More formally defined, emergence is the process of deriving some new and coherent structures, patterns and properties (or behaviors) in a complex system. These structures, patterns and behaviors come to existence without any coordinated control system, but emerge from the interactions of individuals with their local (potentially adaptive) environment.

The collective behavior of a swarm of social organisms therefore emerges in a nonlinear manner from the behaviors of the individuals of that swarm. There exists a tight coupling between individual and collective behavior: the collective behavior of individuals shapes and dictates the behavior of the swarm. On the other hand, swarm behavior has an influence on the conditions under which each individual performs its actions. These actions may change the environment, and thus the behaviors of that individual and its neighbors may also change – which again may change the collective swarm behavior. From this, the most important ingredient of swarm intelligence, and facilitator of emergent behavior, is interaction, or cooperation. Interaction among individuals aids in refining experiential knowledge about the environment. Interaction in biological swarm systems happens in a number of ways, of which social interaction is the most prominent. Here, interaction can be direct (by means of physical contact, or by means of visual, audio, or chemical perceptual inputs) or indirect (via local changes of the environment). The term *stigmergy* is used to refer to the indirect form of communication between individuals.

Examples of emergent behavior from nature are numerous:

- Termites build large nest structures with a complexity far beyond the comprehension and ability of a single termite.

- Tasks are dynamically allocated within an ant colony, without any central manager or task coordinator.

- Recruitment via waggle dances in bee species, which results in optimal foraging behavior. Foraging behavior also emerges in ant colonies as a result of simple trail-following behaviors.

- Birds in a flock and fish in a school self-organize in optimal spatial patterns. Schools of fish determine their behavior (such as swimming direction and speed) based on a small number of neighboring individuals. The spatial patterns of bird flocks result from communication by sound and visual perception.

- Predators, for example a group of lionesses, exhibit hunting strategies to outsmart their prey.

- Bacteria communicate using molecules (comparable to pheromones) to collectively keep track of changes in their environment.

- Slime moulds consist of very simple cellular organisms with limited abilities. However, in times of food shortage they aggregate to form a mobile slug with the ability to transport the assembled individuals to new feeding areas.

The objective of computational swarm intelligence models is to model the simple behaviors of individuals, and the local interactions with the environment and neighboring individuals, in order to obtain more complex behaviors that can be used to solve complex problems, mostly optimization problems. For example, particle swarm optimization (PSO) models two simple behaviors: each individual (1) moves toward its closest best neighbor, and (2) moves back to the state that the individual has experienced to be best for itself. As a result, the collective behavior that emerges is that of all individuals converging on the environment state that is best for all individuals. On the other hand, ant colony optimization models the very simple pheromone trail-following behavior of ants, where each ant perceives pheromone concentrations in its local environment and acts by probabilistically selecting the direction with the highest pheromone concentration. From this emerges the behavior of finding the best alternative (shortest path) from a collection of alternatives. Models of the local behavior of ants attending to cemeteries result in the complex behavior of grouping similar objects into clusters.

This part is devoted to the two computational swarm intelligence paradigms, particle swarm optimization (PSO), presented in Chapter 16, and ant algorithms (AA), presented in Chapter 17. The former was inspired from models of the flocking behavior of birds, while the latter was inspired from a number of models of different behaviors observed in ant and termite colonies.

# Chapter 16

# Particle Swarm Optimization

The particle swarm optimization (PSO) algorithm is a population-based search algorithm based on the simulation of the social behavior of birds within a flock. The initial intent of the particle swarm concept was to graphically simulate the graceful and unpredictable choreography of a bird flock [449], with the aim of discovering patterns that govern the ability of birds to fly synchronously, and to suddenly change direction with a regrouping in an optimal formation. From this initial objective, the concept evolved into a simple and efficient optimization algorithm.

In PSO, individuals, referred to as particles, are "flown" through hyperdimensional search space. Changes to the position of particles within the search space are based on the social-psychological tendency of individuals to emulate the success of other individuals. The changes to a particle within the swarm are therefore influenced by the experience, or knowledge, of its neighbors. The search behavior of a particle is thus affected by that of other particles within the swarm (PSO is therefore a kind of symbiotic cooperative algorithm). The consequence of modeling this social behavior is that the search process is such that particles stochastically return toward previously successful regions in the search space.

The remainder of this chapter is organized as follows: An overview of the basic PSO, i.e. the first implementations of PSO, is given in Section 16.1. The very important concepts of social interaction and social networks are discussed in Section 16.2. Basic variations of the PSO are described in Section 16.3, while more elaborate improvements are given in Section 16.5. A discussion of PSO parameters is given in Section 16.4. Some advanced topics are discussed in Section 16.6.

## 16.1   Basic Particle Swarm Optimization

Individuals in a particle swarm follow a very simple behavior: to emulate the success of neighboring individuals and their own successes. The collective behavior that emerges from this simple behavior is that of discovering optimal regions of a high dimensional search space.

A PSO algorithm maintains a swarm of particles, where each particle represents a potential solution. In analogy with evolutionary computation paradigms, a *swarm* is

*Computational Intelligence: An Introduction*, Second Edition A.P. Engelbrecht
©2007 John Wiley & Sons, Ltd

similar to a population, while a *particle* is similar to an individual. In simple terms, the particles are "flown" through a multidimensional search space, where the position of each particle is adjusted according to its own experience and that of its neighbors. Let $\mathbf{x}_i(t)$ denote the position of particle $i$ in the search space at time step $t$; unless otherwise stated, $t$ denotes discrete time steps. The position of the particle is changed by adding a velocity, $\mathbf{v}_i(t)$, to the current position, i.e.

$$\mathbf{x}_i(t+1) = \mathbf{x}_i(t) + \mathbf{v}_i(t+1) \tag{16.1}$$

with $\mathbf{x}_i(0) \sim U(\mathbf{x}_{min}, \mathbf{x}_{max})$.

It is the velocity vector that drives the optimization process, and reflects both the experiential knowledge of the particle and socially exchanged information from the particle's neighborhood. The experiential knowledge of a particle is generally referred to as the *cognitive component*, which is proportional to the distance of the particle from its own best position (referred to as the particle's *personal best* position) found since the first time step. The socially exchanged information is referred to as the *social component* of the velocity equation.

Originally, two PSO algorithms have been developed which differ in the size of their neighborhoods. These two algorithms, namely the *gbest* and *lbest* PSO, are summarized in Sections 16.1.1 and 16.1.2 respectively. A comparison between *gbest* and *lbest* PSO is given in Section 16.1.3. Velocity components are decsribed in Section 16.1.4, while an illustration of the effect of velocity updates is given in Section 16.1.5. Aspects about the implementation of a PSO algorithm are discussed in Section 16.1.6.

## 16.1.1   Global Best PSO

For the global best PSO, or *gbest* PSO, the neighborhood for each particle is the entire swarm. The social network employed by the *gbest* PSO reflects the star topology (refer to Section 16.2). For the star neighborhood topology, the social component of the particle velocity update reflects information obtained from all the particles in the swarm. In this case, the social information is the best position found by the swarm, referred to as $\hat{\mathbf{y}}(t)$.

For *gbest* PSO, the velocity of particle $i$ is calculated as

$$v_{ij}(t+1) = v_{ij}(t) + c_1 r_{1j}(t)[y_{ij}(t) - x_{ij}(t)] + c_2 r_{2j}(t)[\hat{y}_j(t) - x_{ij}(t)] \tag{16.2}$$

where $v_{ij}(t)$ is the velocity of particle $i$ in dimension $j = 1, \ldots, n_x$ at time step $t$, $x_{ij}(t)$ is the position of particle $i$ in dimension $j$ at time step $t$, $c_1$ and $c_2$ are positive acceleration constants used to scale the contribution of the cognitive and social components respectively (discussed in Section 16.4), and $r_{1j}(t), r_{2j}(t) \sim U(0, 1)$ are random values in the range $[0, 1]$, sampled from a uniform distribution. These random values introduce a stochastic element to the algorithm.

The personal best position, $\mathbf{y}_i$, associated with particle $i$ is the best position the particle has visited since the first time step. Considering minimization problems, the personal best position at the next time step, $t + 1$, is calculated as

$$\mathbf{y}_i(t+1) = \begin{cases} \mathbf{y}_i(t) & \text{if } f(\mathbf{x}_i(t+1)) \geq f(\mathbf{y}_i(t)) \\ \mathbf{x}_i(t+1) & \text{if } f(\mathbf{x}_i(t+1)) < f(\mathbf{y}_i(t)) \end{cases} \tag{16.3}$$

where $f : \mathbb{R}^{n_x} \to \mathbb{R}$ is the fitness function. As with EAs, the fitness function measures how close the corresponding solution is to the optimum, i.e. the fitness function quantifies the performance, or quality, of a particle (or solution).

The global best position, $\hat{\mathbf{y}}(t)$, at time step $t$, is defined as

$$\hat{\mathbf{y}}(t) \in \{\mathbf{y}_0(t), \ldots, \mathbf{y}_{n_s}(t)\} | f(\hat{\mathbf{y}}(t)) = \min\{f(\mathbf{y}_0(t)), \ldots, f(\mathbf{y}_{n_s}(t))\} \tag{16.4}$$

where $n_s$ is the total number of particles in the swarm. It is important to note that the definition in equation (16.4) states that $\hat{\mathbf{y}}$ is the best position discovered by any of the particles so far – it is usually calculated as the best personal best position. The global best position can also be selected from the particles of the current swarm, in which case [359]

$$\hat{\mathbf{y}}(t) = \min\{f(\mathbf{x}_0(t)), \ldots, f(\mathbf{x}_{n_s}(t))\} \tag{16.5}$$

The *gbest* PSO is summarized in Algorithm 16.1.

---

**Algorithm 16.1** *gbest* PSO

---

Create and initialize an $n_x$-dimensional swarm;
**repeat**
    **for** *each particle* $i = 1, \ldots, n_s$ **do**
        //set the personal best position
        **if** $f(\mathbf{x}_i) < f(\mathbf{y}_i)$ **then**
            $\mathbf{y}_i = \mathbf{x}_i$;
        **end**
        //set the global best position **if** $f(\mathbf{y}_i) < f(\hat{\mathbf{y}})$ **then**
            $\hat{\mathbf{y}} = \mathbf{y}_i$;
        **end**
    **end**
    **for** *each particle* $i = 1, \ldots, n_s$ **do**
        update the velocity using equation (16.2);
        update the position using equation (16.1);
    **end**
**until** *stopping condition is true*;

---

## 16.1.2 Local Best PSO

The local best PSO, or *lbest* PSO, uses a ring social network topology (refer to Section 16.2) where smaller neighborhoods are defined for each particle. The social component reflects information exchanged within the neighborhood of the particle, reflecting local knowledge of the environment. With reference to the velocity equation, the social contribution to particle velocity is proportional to the distance between a particle and the best position found by the neighborhood of particles. The velocity is calculated as

$$v_{ij}(t+1) = v_{ij}(t) + c_1 r_{1j}(t)[y_{ij}(t) - x_{ij}(t)] + c_2 r_{2j}(t)[\hat{y}_{ij}(t) - x_{ij}(t)] \tag{16.6}$$

where $\hat{y}_{ij}$ is the best position, found by the neighborhood of particle $i$ in dimension $j$. The local best particle position, $\hat{\mathbf{y}}_i$, i.e. the best position found in the neighborhood $\mathcal{N}_i$, is defined as

$$\hat{\mathbf{y}}_i(t+1) \in \{\mathcal{N}_i | f(\hat{\mathbf{y}}_i(t+1)) = \min\{f(\mathbf{x})\}, \quad \forall \mathbf{x} \in \mathcal{N}_i\} \tag{16.7}$$

with the neighborhood defined as

$$\mathcal{N}_i = \{\mathbf{y}_{i-n_{\mathcal{N}_i}}(t), \mathbf{y}_{i-n_{\mathcal{N}_i}+1}(t), \dots, \mathbf{y}_{i-1}(t), \mathbf{y}_i(t), \mathbf{y}_{i+1}(t), \dots, \mathbf{y}_{i+n_{\mathcal{N}_i}}(t)\} \tag{16.8}$$

for neighborhoods of size $n_{\mathcal{N}_i}$. The local best position will also be referred to as the neighborhood best position.

It is important to note that for the basic PSO, particles within a neighborhood have no relationship to each other. Selection of neighborhoods is done based on particle indices. However, strategies have been developed where neighborhoods are formed based on spatial similarity (refer to Section 16.2).

There are mainly two reasons why neighborhoods based on particle indices are preferred:

1. It is computationally inexpensive, since spatial ordering of particles is not required. For approaches where the distance between particles is used to form neighborhoods, it is necessary to calculate the Euclidean distance between all pairs of particles, which is of $O(n_s^2)$ complexity.

2. It helps to promote the spread of information regarding good solutions to all particles, irrespective of their current location in the search space.

It should also be noted that neighborhoods overlap. A particle takes part as a member of a number of neighborhoods. This interconnection of neighborhoods also facilitates the sharing of information among neighborhoods, and ensures that the swarm converges on a single point, namely the global best particle. The *gbest* PSO is a special case of the *lbest* PSO with $n_{\mathcal{N}_i} = n_s$.

Algorithm 16.2 summarizes the *lbest* PSO.

### 16.1.3    *gbest* versus *lbest* PSO

The two versions of PSO discussed above are similar in the sense that the social component of the velocity updates causes both to move towards the global best particle. This is possible for the *lbest* PSO due to the overlapping neighborhoods.

There are two main differences between the two approaches with respect to their convergence characteristics [229, 489]:

---

**Algorithm 16.2** *lbest* PSO

---

Create and initialize an $n_x$-dimensional swarm;
**repeat**
    **for** *each particle* $i = 1, \ldots, n_s$ **do**
        //set the personal best position
        **if** $f(\mathbf{x}_i) < f(\mathbf{y}_i)$ **then**
            $\mathbf{y}_i = \mathbf{x}_i$;
        **end**
        //set the neighborhood best position
        **if** $f(\mathbf{y}_i) < f(\hat{\mathbf{y}}_i)$ **then**
            $\hat{\mathbf{y}} = \mathbf{y}_i$;
        **end**
    **end**
    **for** *each particle* $i = 1, \ldots, n_s$ **do**
        update the velocity using equation (16.6);
        update the position using equation (16.1);
    **end**
**until** *stopping condition is true*;

---

- Due to the larger particle interconnectivity of the *gbest* PSO, it converges faster than the *lbest* PSO. However, this faster convergence comes at the cost of less diversity than the *lbest* PSO.

- As a consequence of its larger diversity (which results in larger parts of the search space being covered), the *lbest* PSO is less susceptible to being trapped in local minima. In general (depending on the problem), neighborhood structures such as the ring topology used in *lbest* PSO improves performance [452, 670].

A more in-depth discussion on neighborhoods can be found in Section 16.2.

## 16.1.4 Velocity Components

The velocity calculation as given in equations (16.2) and (16.6) consists of three terms:

- The **previous velocity**, $\mathbf{v}_i(t)$, which serves as a memory of the previous flight direction, i.e. movement in the immediate past. This memory term can be seen as a momentum, which prevents the particle from drastically changing direction, and to bias towards the current direction. This component is also referred to as the inertia component.

- The **cognitive component**, $c_1\mathbf{r}_1(\mathbf{y}_i - \mathbf{x}_i)$, which quantifies the performance of particle $i$ relative to past performances. In a sense, the cognitive component resembles individual memory of the position that was best for the particle. The effect of this term is that particles are drawn back to their own best positions, resembling the tendency of individuals to return to situations or places that satisfied them most in the past. Kennedy and Eberhart also referred to the cognitive component as the "nostalgia" of the particle [449].

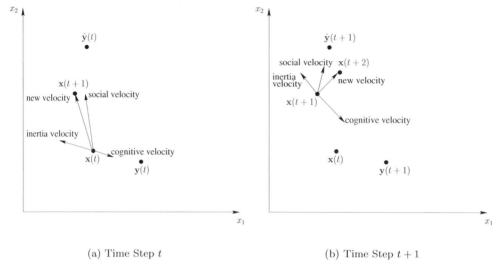

(a) Time Step $t$                                          (b) Time Step $t + 1$

Figure 16.1 Geometrical Illustration of Velocity and Position Updates for a Single Two-Dimensional Particle

- The **social component**, $c_2\mathbf{r}_2(\hat{\mathbf{y}} - \mathbf{x}_i)$, in the case of the *gbest* PSO or, $c_2\mathbf{r}_2(\hat{\mathbf{y}}_i - \mathbf{x}_i)$, in the case of the *lbest* PSO, which quantifies the performance of particle $i$ relative to a group of particles, or neighbors. Conceptually, the social component resembles a group norm or standard that individuals seek to attain. The effect of the social component is that each particle is also drawn towards the best position found by the particle's neighborhood.

The contribution of the cognitive and social components are weighed by a stochastic amount, $c_1\mathbf{r}_1$ or $c_2\mathbf{r}_2$, respectively. The effects of these weights are discussed in more detail in Section 16.4.

## 16.1.5   Geometric Illustration

The effect of the velocity equation can easily be illustrated in a two-dimensional vector space. For the sake of the illustration, consider a single particle in a two-dimensional search space.

An example movement of the particle is illustrated in Figure 16.1, where the particle subscript has been dropped for notational convenience. Figure 16.1(a) illustrates the state of the swarm at time step $t$. Note how the new position, $\mathbf{x}(t + 1)$, moves closer towards the global best $\hat{\mathbf{y}}(t)$. For time step $t + 1$, as illustrated in Figure 16.1(b), assume that the personal best position does not change. The figure shows how the three components contribute to still move the particle towards the global best particle.

It is of course possible for a particle to overshoot the global best position, mainly due to the momentum term. This results in two scenarios:

   1. The new position, as a result of overshooting the current global best, may be a

better position than the current global best. In this case the new particle position will become the new global best position, and all particles will be drawn towards it.

2. The new position is still worse than the current global best particle. In subsequent time steps the cognitive and social components will cause the particle to change direction back towards the global best.

The cumulative effect of all the position updates of a particle is that each particle converges to a point on the line that connects the global best position and the personal best position of the particle. A formal proof can be found in [863, 870].

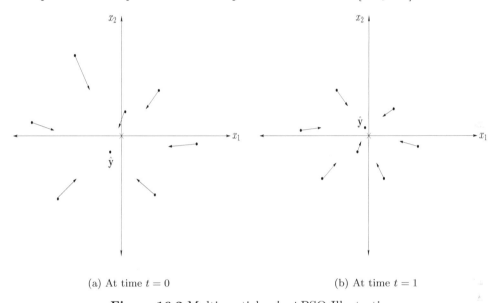

(a) At time $t = 0$          (b) At time $t = 1$

**Figure 16.2** Multi-particle *gbest* PSO Illustration

Returning to more than one particle, Figure 16.2 visualizes the position updates with reference to the task of minimizing a two-dimensional function with variables $x_1$ and $x_2$ using *gbest* PSO. The optimum is at the origin, indicated by the symbol '×'. Figure 16.2(a) illustrates the initial positions of eight particles, with the global best position as indicated. Since the contribution of the cognitive component is zero for each particle at time step $t = 0$, only the social component has an influence on the position adjustments. Note that the global best position does not change (it is assumed that $\mathbf{v}_i(0) = \mathbf{0}$, for all particles). Figure 16.2(b) shows the new positions of all the particles after the first iteration. A new global best position has been found. Figure 16.2(b) now indicates the influence of all the velocity components, with particles moving towards the new global best position.

Finally, the *lbest* PSO, as illustrated in Figure 16.3, shows how particles are influenced by their immediate neighbors. To keep the graph readable, only some of the movements are illustrated, and only the aggregate velocity direction is indicated. In neighborhood 1, both particles $a$ and $b$ move towards particle $c$, which is the best solution within that neighborhood. Considering neighborhood 2, particle $d$ moves towards $f$, so does $e$. For the next iteration, $e$ will be the best solution for neighborhood 2. Now $d$ and

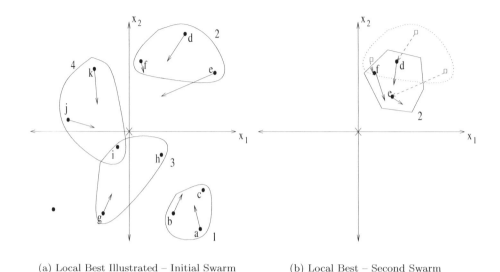

(a) Local Best Illustrated – Initial Swarm          (b) Local Best – Second Swarm

**Figure 16.3** Illustration of *lbest* PSO

$f$ move towards $e$ as illustrated in Figure 16.3(b) (only part of the solution space is illustrated). The blocks represent the previous positions. Note that $e$ remains the best solution for neighborhood 2. Also note the general movement towards the minimum.

More in-depth analyses of particle trajectories can be found in [136, 851, 863, 870].

## 16.1.6   Algorithm Aspects

A few aspects of Algorithms 16.1 and 16.2 still need to be discussed. These aspects include particle initialization, stopping conditions and defining the terms iteration and function evaluation.

With reference to Algorithms 16.1 and 16.2, the optimization process is iterative. Repeated iterations of the algorithms are executed until a stopping condition is satisfied. One such iteration consists of application of all the steps within the `repeat...until` loop, i.e. determining the personal best positions and the global best position, and adjusting the velocity of each particle. Within each iteration, a number of function evaluations (FEs) are performed. A *function evaluation* (FE) refers to one calculation of the fitness function, which characterizes the optimization problem. For the basic PSO, a total of $n_s$ function evaluations are performed per iteration, where $n_s$ is the total number of particles in the swarm.

The first step of the PSO algorithm is to initialize the swarm and control parameters. In the context of the basic PSO, the acceleration constants, $c_1$ and $c_2$, the initial velocities, particle positions and personal best positions need to be specified. In addition, the *lbest* PSO requires the size of neighborhoods to be specified. The importance of optimal values for the acceleration constants are discussed in Section 16.4.

Usually, the positions of particles are initialized to uniformly cover the search space. It is important to note that the efficiency of the PSO is influenced by the initial diversity of the swarm, i.e. how much of the search space is covered, and how well particles are distributed over the search space. If regions of the search space are not covered by the initial swarm, the PSO will have difficulty in finding the optimum if it is located within an uncovered region. The PSO will discover such an optimum only if the momentum of a particle carries the particle into the uncovered area, provided that the particle ends up on either a new personal best for itself, or a position that becomes the new global best.

Assume that an optimum needs to be located within the domain defined by the two vectors, $\mathbf{x}_{min}$ and $\mathbf{x}_{max}$, which respectively represents the minimum and maximum ranges in each dimension. Then, an efficient initialization method for the particle positions is:

$$x(0) = x_{min,j} + r_j(x_{max,j} - x_{min,j}), \quad \forall j = 1, \ldots, n_x, \ \forall i = 1, \ldots, n_s \quad (16.9)$$

where $r_j \sim U(0,1)$.

The initial velocities can be initialized to zero, i.e.

$$\mathbf{v}_i(0) = \mathbf{0} \quad (16.10)$$

While it is possible to also initialize the velocities to random values, it is not necessary, and it must be done with care. In fact, considering physical objects in their initial positions, their velocities are zero – they are stationary. If particles are initialized with nonzero velocities, this physical analogy is violated. Random initialization of position vectors already ensures random positions and moving directions. If, however, velocities are also randomly initialized, such velocities should not be too large. Large initial velocities will have large initial momentum, and consequently large initial position updates. Such large initial position updates may cause particles to leave the boundaries of the search space, and may cause the swarm to take more iterations before particles settle on a single solution.

The personal best position for each particle is initialized to the particle's position at time step $t = 0$, i.e.

$$\mathbf{y}_i(0) = \mathbf{x}_i(0) \quad (16.11)$$

Different initialization schemes have been used by researchers to initialize the particle positions to ensure that the search space is uniformly covered: Sobol sequences [661], Faure sequences [88, 89, 90], and nonlinear simplex method [662].

While it is important for application of the PSO to solve real-world problems, in that particles are uniformly distributed over the entire search space, uniformly distributed particles are not necessarily good for empirical studies of different algorithms. This typical initialization method can give false impressions of the relative performance of algorithms as shown in [312]. For many of the benchmark functions used to evaluate the performance of optimization algorithms (refer, for example, to Section A.5.3), uniform initialization will result in particles being symmetrically distributed around the optimum of the function to be optimized. In most cases it is then trivial for an optimization algorithm to find the optimum. Gehlhaar and Fogel suggest initializing in areas that do not contain the optima, in order to validate the ability of the algorithm to locate solutions outside the initialized space [312].

The last aspect of the PSO algorithms concerns the stopping conditions, i.e. criteria used to terminate the iterative search process. A number of termination criteria have been used and proposed in the literature. When selecting a termination criterion, two important aspects have to be considered:

1. The stopping condition should not cause the PSO to prematurely converge, since suboptimal solutions will be obtained.

2. The stopping condition should protect against oversampling of the fitness. If a stopping condition requires frequent calculation of the fitness function, computational complexity of the search process can be significantly increased.

The following stopping conditions have been used:

- **Terminate when a maximum number of iterations, or FEs, has been exceeded.** It is obvious to realize that if this maximum number of iterations (or FEs) is too small, termination may occur before a good solution has been found. This criterion is usually used in conjunction with convergence criteria to force termination if the algorithm fails to converge. Used on its own, this criterion is useful in studies where the objective is to evaluate the best solution found in a restricted time period.

- **Terminate when an acceptable solution has been found.** Assume that $\mathbf{x}^*$ represents the optimum of the objective function $f$. Then, this criterion will terminate the search process as soon as a particle, $\mathbf{x}_i$, is found such that $f(\mathbf{x}_i) \leq |f(\mathbf{x}^*) - \epsilon|$; that is, when an acceptable error has been reached. The value of the threshold, $\epsilon$, has to be selected with care. If $\epsilon$ is too large, the search process terminates on a bad, suboptimal solution. On the other hand, if $\epsilon$ is too small, the search may not terminate at all. This is especially true for the basic PSO, since it has difficulties in refining solutions [81, 361, 765, 782]. Furthermore, this stopping condition assumes prior knowledge of what the optimum is – which is fine for problems such as training neural networks, where the optimum is usually zero. It is, however, the case that knowledge of the optimum is usually not available.

- **Terminate when no improvement is observed over a number of iterations.** There are different ways in which improvement can be measured. For example, if the average change in particle positions is small, the swarm can be considered to have converged. Alternatively, if the average particle velocity over a number of iterations is approximately zero, only small position updates are made, and the search can be terminated. The search can also be terminated if there is no significant improvement over a number of iterations. Unfortunately, these stopping conditions introduce two parameters for which sensible values need to be found: (1) the window of iterations (or function evaluations) for which the performance is monitored, and (2) a threshold to indicate what constitutes unacceptable performance.

- **Terminate when the normalized swarm radius is close to zero.** When the normalized swarm radius, calculated as [863]

$$R_{norm} = \frac{R_{max}}{\text{diameter}(S)} \tag{16.12}$$

where $\text{diameter}(S)$ is the diameter of the initial swarm and the maximum radius, $R_{max}$, is

$$R_{max} = ||\mathbf{x}_m - \hat{\mathbf{y}}||, \quad m = 1, \ldots, n_s \tag{16.13}$$

with

$$||\mathbf{x}_m - \hat{\mathbf{y}}|| \geq ||\mathbf{x}_i - \hat{\mathbf{y}}||, \quad \forall i = 1, \ldots, n_s \tag{16.14}$$

is close to zero, the swarm has little potential for improvement, unless the global best is still moving. In the equations above, $|| \bullet ||$ is a suitable distance norm, e.g. Euclidean distance.

The algorithm is terminated when $R_{norm} < \epsilon$. If $\epsilon$ is too large, the search process may stop prematurely before a good solution is found. Alternatively, if $\epsilon$ is too small, the search may take excessively more iterations for the particles to form a compact swarm, tightly centered around the global best position.

---

**Algorithm 16.3** Particle Clustering Algorithm

---

Initialize cluster $C = \{\hat{\mathbf{y}}\}$;
**for** *about 5 times* **do**
    Calculate the centroid of cluster $C$:

$$\overline{\mathbf{x}} = \frac{\sum_{i=1,\mathbf{x}_i \in C}^{|C|} \mathbf{x}_i}{|C|} \tag{16.15}$$

    **for** $\forall \mathbf{x}_i \in S$ **do**
        **if** $||\mathbf{x}_i - \overline{\mathbf{x}}|| < \epsilon$ **then**
            $C \leftarrow C \cup \{\mathbf{x}_i\}$;
        **end**
    **endFor**
**endFor**

---

A more aggressive version of the radius method above can be used, where particles are clustered in the search space. Algorithm 16.3 provides a simple particle clustering algorithm from [863]. The result of this algorithm is a single cluster, $C$. If $|C|/n_s > \delta$, the swarm is considered to have converged. If, for example, $\delta = 0.7$, the search will terminate if at least 70% of the particles are centered around the global best position. The threshold $\epsilon$ in Algorithm 16.3 has the same importance as for the radius method above. Similarly, if $\delta$ is too small, the search may terminate prematurely.

This clustering approach is similar to the radius approach except that the clustering approach will more readily decide that the swarm has converged.

- **Terminate when the objective function slope is approximately zero.** The stopping conditions above consider the relative positions of particles in the search space, and do not take into consideration information about the slope of the objective function. To base termination on the rate of change in the objective function, consider the ratio [863],

$$f'(t) = \frac{f(\hat{\mathbf{y}}(t)) - f(\hat{\mathbf{y}}(t-1))}{f(\hat{\mathbf{y}}(t))} \qquad (16.16)$$

If $f'(t) < \epsilon$ for a number of consecutive iterations, the swarm is assumed to have converged. This approximation to the slope of the objective function is superior to the methods above, since it actually determines if the swarm is still making progress using information about the search space.

The objective function slope approach has, however, the problem that the search will be terminated if some of the particles are attracted to a local minimum, irrespective of whether other particles may still be busy exploring other parts of the search space. It may be the case that these exploring particles could have found a better solution had the search not terminated. To solve this problem, the objective function slope method can be used in conjunction with the radius or cluster methods to test if all particles have converged to the same point before terminating the search process.

In the above, convergence does not imply that the swarm has settled on an optimum (local or global). With the term *convergence* is meant that the swarm has reached an equilibrium, i.e. just that the particles converged to a point, which is not necessarily an optimum [863].

## 16.2   Social Network Structures

The feature that drives PSO is social interaction. Particles within the swarm learn from each other and, on the basis of the knowledge obtained, move to become more similar to their "better" neighbors. The social structure for PSO is determined by the formation of overlapping neighborhoods, where particles within a neighborhood influence one another. This is in analogy with observations of animal behavior, where an organism is most likely to be influenced by others in its neighborhood, and where organisms that are more successful will have a greater influence on members of the neighborhood than the less successful.

Within the PSO, particles in the same neighborhood communicate with one another by exchanging information about the success of each particle in that neighborhood. All particles then move towards some quantification of what is believed to be a better position. The performance of the PSO depends strongly on the structure of the social network. The flow of information through a social network, depends on (1) the degree of connectivity among nodes (members) of the network, (2) the amount of clustering (clustering occurs when a node's neighbors are also neighbors to one another), and (3) the average shortest distance from one node to another [892].

With a highly connected social network, most of the individuals can communicate with one another, with the consequence that information about the perceived best member quickly filters through the social network. In terms of optimization, this means faster convergence to a solution than for less connected networks. However, for the highly connected networks, the faster convergence comes at the price of susceptibility to local minima, mainly due to the fact that the extent of coverage in the search space is less than for less connected social networks. For sparsely connected networks with a large amount of clustering in neighborhoods, it can also happen that the search space is not covered sufficiently to obtain the best possible solutions. Each cluster contains individuals in a tight neighborhood covering only a part of the search space. Within these network structures there usually exist a few clusters, with a low connectivity between clusters. Consequently information on only a limited part of the search space is shared with a slow flow of information between clusters.

Different social network structures have been developed for PSO and empirically studied. This section overviews only the original structures investigated [229, 447, 452, 575]:

- The **star** social structure, where all particles are interconnected as illustrated in Figure 16.4(a). Each particle can therefore communicate with every other particle. In this case each particle is attracted towards the best solution found by the entire swarm. Each particle therefore imitates the overall best solution. The first implementation of the PSO used a star network structure, with the resulting algorithm generally being referred to as the *gbest* PSO. The *gbest* PSO has been shown to converge faster than other network structures, but with a susceptibility to be trapped in local minima. The *gbest* PSO performs best for unimodal problems.

- The **ring** social structure, where each particle communicates with its $n_\mathcal{N}$ immediate neighbors. In the case of $n_\mathcal{N} = 2$, a particle communicates with its immediately adjacent neighbors as illustrated in Figure 16.4(b). Each particle attempts to imitate its best neighbor by moving closer to the best solution found within the neighborhood. It is important to note from Figure 16.4(b) that neighborhoods overlap, which facilitates the exchange of information between neighborhoods and, in the end, convergence to a single solution. Since information flows at a slower rate through the social network, convergence is slower, but larger parts of the search space are covered compared to the star structure. This behavior allows the ring structure to provide better performance in terms of the quality of solutions found for multi-modal problems than the star structure. The resulting PSO algorithm is generally referred to as the *lbest* PSO.

- The **wheel** social structure, where individuals in a neighborhood are isolated from one another. One particle serves as the focal point, and all information is communicated through the focal particle (refer to Figure 16.4(c)). The focal particle compares the performances of all particles in the neighborhood, and adjusts its position towards the best neighbor. If the new position of the focal particle results in better performance, then the improvement is communicated to all the members of the neighborhood. The wheel social network slows down the propagation of good solutions through the swarm.

- The **pyramid** social structure, which forms a three-dimensional wire-frame as illustrated in Figure 16.4(d).

- The **four clusters** social structure, as illustrated in Figure 16.4(e). In this

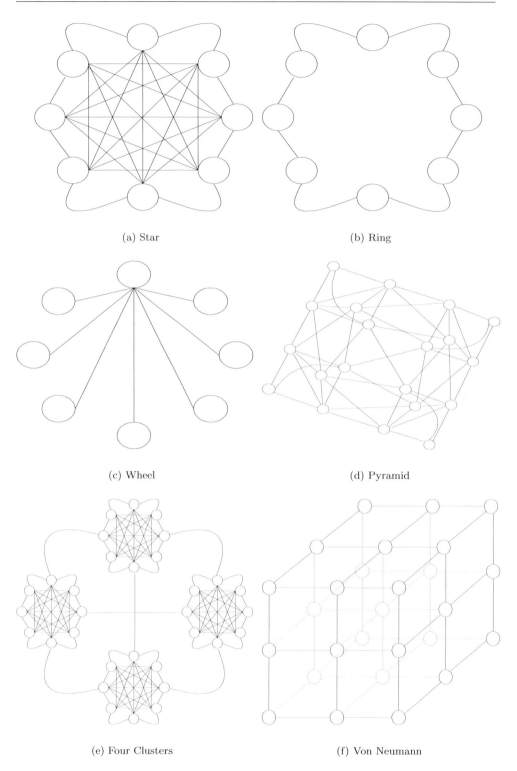

(a) Star

(b) Ring

(c) Wheel

(d) Pyramid

(e) Four Clusters

(f) Von Neumann

**Figure 16.4** Example Social Network Structures

network structure, four clusters (or cliques) are formed with two connections between clusters. Particles within a cluster are connected with five neighbors.

- The **Von Neumann** social structure, where particles are connected in a grid structure as illustrated in Figure 16.4(f). The Von Neumann social network has been shown in a number of empirical studies to outperform other social networks in a large number of problems [452, 670].

While many studies have been done using the different topologies, there is no outright best topology for all problems. In general, the fully connected structures perform best for unimodal problems, while the less connected structures perform better on multi-modal problems, depending on the degree of particle interconnection [447, 452, 575, 670].

Neighborhoods are usually determined on the basis of particle indices. For example, for the *lbest* PSO with $n_\mathcal{N} = 2$, the neighborhood of a particle with index $i$ includes particles $i - 1, i$ and $i + 1$. While indices are usually used, Suganthan based neighborhoods on the Euclidean distance between particles [820].

## 16.3 Basic Variations

The basic PSO has been applied successfully to a number of problems, including standard function optimization problems [25, 26, 229, 450, 454], solving permutation problems [753] and training multi-layer neural networks [224, 225, 229, 446, 449, 854]. While the empirical results presented in these papers illustrated the ability of the PSO to solve optimization problems, these results also showed that the basic PSO has problems with consistently converging to good solutions. A number of basic modifications to the basic PSO have been developed to improve speed of convergence and the quality of solutions found by the PSO. These modifications include the introduction of an inertia weight, velocity clamping, velocity constriction, different ways of determining the personal best and global best (or local best) positions, and different velocity models. This section discusses these basic modifications.

### 16.3.1 Velocity Clamping

An important aspect that determines the efficiency and accuracy of an optimization algorithm is the exploration–exploitation trade-off. *Exploration* is the ability of a search algorithm to explore different regions of the search space in order to locate a good optimum. *Exploitation*, on the other hand, is the ability to concentrate the search around a promising area in order to refine a candidate solution. A good optimization algorithm optimally balances these contradictory objectives. Within the PSO, these objectives are addressed by the velocity update equation.

The velocity updates in equations (16.2) and (16.6) consist of three terms that contribute to the step size of particles. In the early applications of the basic PSO, it was found that the velocity quickly explodes to large values, especially for particles far from the neighborhood best and personal best positions. Consequently, particles have large position updates, which result in particles leaving the boundaries of the search space – the particles diverge. To control the global exploration of particles, velocities are clamped to stay within boundary constraints [229]. If a particle's velocity exceeds a specified maximum velocity, the particle's velocity is set to the maximum velocity. Let $V_{max,j}$ denote the maximum allowed velocity in dimension $j$. Particle velocity is then adjusted before the position update using,

$$v_{ij}(t+1) = \begin{cases} v'_{ij}(t+1) & \text{if } v'_{ij}(t+1) < V_{max,j} \\ V_{max,j} & \text{if } v'_{ij}(t+1) \geq V_{max,j} \end{cases} \tag{16.17}$$

where $v'_{ij}$ is calculated using equation (16.2) or (16.6).

The value of $V_{max,j}$ is very important, since it controls the granularity of the search by clamping escalating velocities. Large values of $V_{max,j}$ facilitate global exploration, while smaller values encourage local exploitation. If $V_{max,j}$ is too small, the swarm may not explore sufficiently beyond locally good regions. Also, too small values for $V_{max,j}$ increase the number of time steps to reach an optimum. Furthermore, the swarm may become trapped in a local optimum, with no means of escape. On the other hand, too large values of $V_{max,j}$ risk the possibility of missing a good region. The particles may jump over good solutions, and continue to search in fruitless regions of the search space. While large values do have the disadvantage that particles may jump over optima, particles are moving faster.

This leaves the problem of finding a good value for each $V_{max,j}$ in order to balance between (1) moving too fast or too slow, and (2) exploration and exploitation. Usually, the $V_{max,j}$ values are selected to be a fraction of the domain of each dimension of the search space. That is,

$$V_{max,j} = \delta(x_{max,j} - x_{min,j}) \tag{16.18}$$

where $x_{max,j}$ and $x_{min,j}$ are respectively the maximum and minimum values of the domain of $\mathbf{x}$ in dimension $j$, and $\delta \in (0,1]$. The value of $\delta$ is problem-dependent, as was found in a number of empirical studies [638, 781]. The best value should be found for each different problem using empirical techniques such as cross-validation.

There are two important aspects of the velocity clamping approach above that the reader should be aware of:

1. Velocity clamping does not confine the positions of particles, only the step sizes as determined from the particle velocity.

2. In the above equations, explicit reference is made to the dimension, $j$. A maximum velocity is associated with each dimension, proportional to the domain of that dimension. For the sake of the argument, assume that all dimensions are clamped with the same constant $V_{max}$. Therefore if a dimension, $j$, exists such that $x_{max,j} - x_{min,j} << V_{max}$, particles may still overshoot an optimum in dimension $j$.

While velocity clamping has the advantage that explosion of velocity is controlled, it also has disadvantages that the user should be aware of (that is in addition to

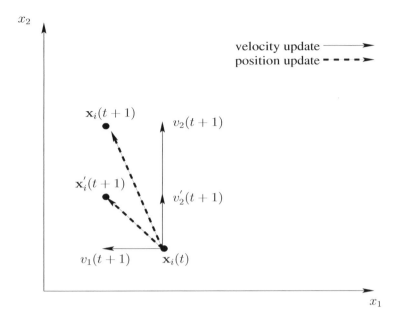

**Figure 16.5** Effects of Velocity Clamping

the problem-dependent nature of $V_{max,j}$ values). Firstly, velocity clamping changes not only the step size, but also the direction in which a particle moves. This effect is illustrated in Figure 16.5 (assuming two-dimensional particles). In this figure, $\mathbf{x}_i(t+1)$ denotes the position of particle $i$ without using velocity clamping. The position $\mathbf{x}'_i(t+1)$ is the result of velocity clamping on the second dimension. Note how the search direction and the step size have changed. It may be said that these changes in search direction allow for better exploration. However, it may also cause the optimum not to be found at all.

Another problem with velocity clamping occurs when all velocities are equal to the maximum velocity. If no measures are implemented to prevent this situation, particles remain to search on the boundaries of a hypercube defined by $[\mathbf{x}_i(t) - \mathbf{V}_{max}, \mathbf{x}_i(t) + \mathbf{V}_{max}]$. It is possible that a particle may stumble upon the optimum, but in general the swarm will have difficulty in exploiting this local area. This problem can be solved in different ways, with the introduction of an inertia weight (refer to Section 16.3.2) being one of the first solutions. The problem can also be solved by reducing $V_{max,j}$ over time. The idea is to start with large values, allowing the particles to explore the search space, and then to decrease the maximum velocity over time. The decreasing maximum velocity constrains the exploration ability in favor of local exploitation at mature stages of the optimization process. The following dynamic velocity approaches have been used:

- Change the maximum velocity when no improvement in the global best position has been seen over $\tau$ consecutive iterations [766]:

$$V_{max,j}(t+1) = \begin{cases} \gamma V_{max,j}(t) & \text{if } f(\hat{\mathbf{y}}(t)) \geq f(\hat{\mathbf{y}}(t-t')) \quad \forall t' = 1,\ldots,n_{t'} \\ V_{max,j}(t) & \text{otherwise} \end{cases}$$

(16.19)

where $\gamma$ decreases from 1 to 0.01 (the decrease can be linear or exponential using an annealing schedule similar to that given in Section 16.3.2 for the inertia weight).

- Exponentially decay the maximum velocity, using [254]

$$V_{max,j}(t+1) = (1 - (t/n_t)^{\alpha})V_{max,j}(t) \tag{16.20}$$

where $\alpha$ is a positive constant, found by trial and error, or cross-validation methods; $n_t$ is the maximum number of time steps (or iterations).

Finally, the sensitivity of PSO to the value of $\delta$ (refer to equation (16.18)) can be reduced by constraining velocities using the hyperbolic tangent function, i.e.

$$v_{ij}(t+1) = V_{max,j} \tanh\left(\frac{v'_{ij}(t+1)}{V_{max,j}}\right) \tag{16.21}$$

where $v'_{ij}(t+1)$ is calculated from equation (16.2) or (16.6).

## 16.3.2  Inertia Weight

The inertia weight  was introduced by Shi and Eberhart [780] as a mechanism to control the exploration and exploitation abilities of the swarm, and as a mechanism to eliminate the need for velocity clamping [227]. The inertia weight was successful in addressing the first objective, but could not completely eliminate the need for velocity clamping. The inertia weight, $w$, controls the momentum of the particle by weighing the contribution of the previous velocity – basically controlling how much memory of the previous flight direction will influence the new velocity. For the *gbest* PSO, the velocity equation changes from equation (16.2) to

$$v_{ij}(t+1) = wv_{ij}(t) + c_1 r_{1j}(t)[y_{ij}(t) - x_{ij}(t)] + c_2 r_{2j}(t)[\hat{y}_j(t) - x_{ij}(t)] \tag{16.22}$$

A similar change is made for the *lbest* PSO.

The value of $w$ is extremely important to ensure convergent behavior, and to optimally tradeoff exploration and exploitation. For $w \geq 1$, velocities increase over time, accelerating towards the maximum velocity (assuming velocity clamping is used), and the swarm diverges. Particles fail to change direction in order to move back towards promising areas. For $w < 1$, particles decelerate until their velocities reach zero (depending on the values of the acceleration coefficients). Large values for $w$ facilitate exploration, with increased diversity. A small $w$ promotes local exploitation. However, too small values eliminate the exploration ability of the swarm. Little momentum is then preserved from the previous time step, which enables quick changes in direction. The smaller $w$, the more do the cognitive and social components control position updates.

As with the maximum velocity, the optimal value for the inertia weight is problem-dependent [781]. Initial implementations of the inertia weight used a static value for the entire search duration, for all particles for each dimension. Later implementations made use of dynamically changing inertia values. These approaches usually start with large inertia values, which decreases over time to smaller values. In doing so, particles are allowed to explore in the initial search steps, while favoring exploitation as time increases. At this time it is crucial to mention the important relationship between the values of $w$, and the acceleration constants. The choice of value for $w$ has to be made in conjunction with the selection of the values for $c_1$ and $c_2$. Van den Bergh and Engelbrecht [863, 870] showed that

$$w > \frac{1}{2}(c_1 + c_2) - 1 \tag{16.23}$$

guarantees convergent particle trajectories. If this condition is not satisfied, divergent or cyclic behavior may occur. A similar condition was derived by Trelea [851].

Approaches to dynamically varying the inertia weight can be grouped into the following categories:

- **Random adjustments**, where a different inertia weight is randomly selected at each iteration. One approach is to sample from a Gaussian distribution, e.g.

$$w \sim N(0.72, \sigma) \tag{16.24}$$

where $\sigma$ is small enough to ensure that $w$ is not predominantly greater than one. Alternatively, Peng *et al.* used [673]

$$w = (c_1 r_1 + c_2 r_2) \tag{16.25}$$

with no random scaling of the cognitive and social components.

- **Linear decreasing**, where an initially large inertia weight (usually 0.9) is linearly decreased to a small value (usually 0.4). From Naka *et al.* [619], Ratnaweera *et al.* [706], Suganthan [820], Yoshida *et al.* [941]

$$w(t) = (w(0) - w(n_t)) \frac{(n_t - t)}{n_t} + w(n_t) \tag{16.26}$$

where $n_t$ is the maximum number of time steps for which the algorithm is executed, $w(0)$ is the initial inertia weight, $w(n_t)$ is the final inertia weight, and $w(t)$ is the inertia at time step $t$. Note that $w(0) > w(n_t)$.

- **Nonlinear decreasing**, where an initially large value decreases nonlinearly to a small value. Nonlinear decreasing methods allow a shorter exploration time than the linear decreasing methods, with more time spent on refining solutions (exploiting). Nonlinear decreasing methods will be more appropriate for smoother search spaces. The following nonlinear methods have been defined:

  - From Peram *et al.* [675],

$$w(t + 1) = \frac{(w(t) - 0.4)(n_t - t)}{n_t + 0.4} \tag{16.27}$$

  with $w(0) = 0.9$.

– From Venter and Sobieszczanski-Sobieski [874, 875],

$$w(t + 1) = \alpha w(t')$$  (16.28)

where $\alpha = 0.975$, and $t'$ is the time step when the inertia last changed. The inertia is only changed when there is no significant difference in the fitness of the swarm. Venter and Sobieszczanski-Sobieski measure the variation in particle fitness of a 20% subset of randomly selected particles. If this variation is too small, the inertia is changed. An initial inertia weight of $w(0) = 1.4$ is used with a lower bound of $w(n_t) = 0.35$. The initial $w(0) = 1.4$ ensures that a large area of the search space is covered before the swarm focuses on refining solutions.

– Clerc proposes an adaptive inertia weight approach where the amount of change in the inertia value is proportional to the relative improvement of the swarm [134]. The inertia weight is adjusted according to

$$w_i(t + 1) = w(0) + (w(n_t) - w(0)) \frac{e^{m_i(t)} - 1}{e^{m_i(t)} + 1}$$  (16.29)

where the relative improvement, $m_i$, is estimated as

$$m_i(t) = \frac{f(\hat{\mathbf{y}}_i(t)) - f(\mathbf{x}_i(t))}{f(\hat{\mathbf{y}}_i(t)) + f(\mathbf{x}_i(t))}$$  (16.30)

with $w(n_t) \approx 0.5$ and $w(0) < 1$.

Using this approach, which was developed for velocity updates without the cognitive component, each particle has its own inertia weight based on its distance from the local best (or neighborhood best) position. The local best position, $\hat{\mathbf{y}}_i(t)$ can just as well be replaced with the global best position $\hat{\mathbf{y}}(t)$. Clerc motivates his approach by considering that the more an individual improves upon his/her neighbors, the more he/she follows his/her own way, and vice versa. Clerc reported that this approach results in fewer iterations [134].

• **Fuzzy adaptive inertia**, where the inertia weight is dynamically adjusted on the basis of fuzzy sets and rules. Shi and Eberhart [783] defined a fuzzy system for the inertia adaptation to consist of the following components:

– Two inputs, one to represent the fitness of the global best position, and the other the current value of the inertia weight.

– One output to represent the change in inertia weight.

– Three fuzzy sets, namely LOW, MEDIUM and HIGH, respectively implemented as a left triangle, triangle and right triangle membership function [783].

– Nine fuzzy rules from which the change in inertia is calculated. An example rule in the fuzzy system is [229, 783]:

```
if normalized best fitness is LOW, and
   current inertia weight value is LOW
then the change in weight is MEDIUM
```

- **Increasing inertia**, where the inertia weight is linearly increased from 0.4 to 0.9 [958].

The linear and nonlinear adaptive inertia methods above are very similar to the temperature schedule of simulated annealing [467, 649] (also refer to Section A.5.2).

## 16.3.3 Constriction Coefficient

Clerc developed an approach very similar to the inertia weight to balance the exploration–exploitation trade-off, where the velocities are constricted by a constant $\chi$, referred to as the constriction coefficient [133, 136]. The velocity update equation changes to:

$$v_{ij}(t+1) = \chi[v_{ij}(t) + \phi_1(y_{ij}(t) - x_{ij}(t)) + \phi_2(\hat{y}_j(t) - x_{ij}(t))] \qquad (16.31)$$

where

$$\chi = \frac{2\kappa}{|2 - \phi - \sqrt{\phi(\phi - 4)}|} \qquad (16.32)$$

with $\phi = \phi_1 + \phi_2, \phi_1 = c_1 r_1$ and $\phi_2 = c_2 r_2$. Equation (16.32) is used under the constraints that $\phi \geq 4$ and $\kappa \in [0, 1]$. The above equations were derived from a formal eigenvalue analysis of swarm dynamics [136].

The constriction approach was developed as a natural, dynamic way to ensure convergence to a stable point, without the need for velocity clamping. Under the conditions that $\phi \geq 4$ and $\kappa \in [0, 1]$, the swarm is guaranteed to converge. The constriction coefficient, $\chi$, evaluates to a value in the range $[0, 1]$ which implies that the velocity is reduced at each time step.

The parameter, $\kappa$, in equation (16.32) controls the exploration and exploitation abilities of the swarm. For $\kappa \approx 0$, fast convergence is obtained with local exploitation. The swarm exhibits an almost hill-climbing behavior. On the other hand, $\kappa \approx 1$ results in slow convergence with a high degree of exploration. Usually, $\kappa$ is set to a constant value. However, an initial high degree of exploration with local exploitation in the later search phases can be achieved using an initial value close to one, decreasing it to zero.

The constriction approach is effectively equivalent to the inertia weight approach. Both approaches have the objective of balancing exploration and exploitation, and in doing so of improving convergence time and the quality of solutions found. Low values of $w$ and $\chi$ result in exploitation with little exploration, while large values result in exploration with difficulties in refining solutions. For a specific $\chi$, the equivalent inertia model can be obtained by simply setting $w = \chi, \phi_1 = \chi c_1 r_1$ and $\phi_2 = \chi c_2 r_2$. The differences in the two approaches are that

- velocity clamping is not necessary for the constriction model,

- the constriction model guarantees convergence under the given constraints, and

- any ability to regulate the change in direction of particles must be done via the constants $\phi_1$ and $\phi_2$ for the constriction model.

While it is not necessary to use velocity clamping with the constriction model, Eberhart and Shi showed empirically that if velocity clamping and constriction are used together, faster convergence rates can be obtained [226].

## 16.3.4    Synchronous versus Asynchronous Updates

The *gbest* and *lbest* PSO algorithms presented in Algorithms 16.1 and 16.2 perform synchronous updates of the personal best and global (or local) best positions. Synchronous updates are done separately from the particle position updates. Alternatively, asynchronous updates calculate the new best positions after each particle position update (very similar to a steady state GA, where offspring are immediately introduced into the population). Asynchronous updates have the advantage that immediate feedback is given about the best regions of the search space, while feedback with synchronous updates is only given once per iteration. Carlisle and Dozier reason that asynchronous updates are more important for *lbest* PSO where immediate feedback will be more beneficial in loosely connected swarms, while synchronous updates are more appropriate for *gbest* PSO [108].

Selection of the global (or local) best positions is usually done by selecting the absolute best position found by the swarm (or neighborhood). Kennedy proposed to select the best positions randomly from the neighborhood [448]. This is done to break the effect that one, potentially bad, solution drives the swarm. The random selection was specifically used to address the difficulties that the *gbest* PSO experience on highly multi-modal problems. The performance of the basic PSO is also strongly influenced by whether the best positions (*gbest* or *lbest*) are selected from the particle positions of the current iterations, or from the personal best positions of all particles. The difference between the two approaches is that the latter includes a memory component in the sense that the best positions are the best positions found over all iterations. The former approach neglects the temporal experience of the swarm. Selection from the personal best positions is similar to the "hall of fame" concept (refer to Sections 8.5.9 and 15.2.1) used within evolutionary computation.

## 16.3.5    Velocity Models

Kennedy [446] investigated a number of variations to the full PSO models presented in Sections 16.1.1 and 16.1.2. These models differ in the components included in the velocity equation, and how best positions are determined. This section summarizes these models.

### Cognition-Only Model

The cognition-only model excludes the social component from the original velocity equation as given in equation (16.2). For the cognition-only model, the velocity update changes to

$$v_{ij}(t+1) = v_{ij}(t) + c_1 r_{1j}(t)(y_{ij}(t) - x_{ij}(t)) \tag{16.33}$$

The above formulation excludes the inertia weight, mainly because the velocity models in this section were investigated before the introduction of the inertia weight. However, nothing prevents the inclusion of $w$ in equation (16.33) and the velocity equations that follow in this section.

The behavior of particles within the cognition-only model can be likened to nostalgia, and illustrates a stochastic tendency for particles to return toward their previous best position.

From empirical work, Kennedy reported that the cognition-only model is slightly more vulnerable to failure than the full model [446]. It tends to locally search in areas where particles are initialized. The cognition-only model is slower in the number of iterations it requires to reach a good solution, and fails when velocity clamping and the acceleration coefficient are small. The poor performance of the cognitive model is confirmed by Carlisle and Dozier [107], but with respect to dynamic changing environments (refer to Section 16.6.3). The cognition-only model was, however, successfully used within niching algorithms [89] (also refer to Section 16.6.4).

## Social-Only Model

The social-only model excludes the cognitive component from the velocity equation:

$$v_{ij}(t+1) = v_{ij}(t) + c_2 r_{2j}(t)(\hat{y}_j(t) - x_{ij}(t)) \tag{16.34}$$

for the *gbest* PSO. For the *lbest* PSO, $\hat{y}_j$ is simply replaced with $\hat{y}_{ij}$.

For the social-only model, particles have no tendency to return to previous best positions. All particles are attracted towards the best position of their neighborhood.

Kennedy empirically illustrated that the social-only model is faster and more efficient than the full and cognitive models [446], which is also confirmed by the results from Carlisle and Dozier [107] for dynamic environments.

## Selfless Model

The selfless model is basically the social model, but with the neighborhood best solution only chosen from a particle's neighbors. In other words, the particle itself is not allowed to become the neighborhood best. Kennedy showed the selfless model to be faster than the social-only model for a few problems [446]. Carlisle and Dozier's results show that the selfless model performs poorly for dynamically changing environments [107].

## 16.4   Basic PSO Parameters

The basic PSO is influenced by a number of control parameters, namely the dimension of the problem, number of particles, acceleration coefficients, inertia weight, neighborhood size, number of iterations, and the random values that scale the contribution of the cognitive and social components. Additionally, if velocity clamping or constriction is used, the maximum velocity and constriction coefficient also influence the performance of the PSO. This section discusses these parameters.

The influence of the inertia weight, velocity clamping threshold and constriction coefficient has been discussed in Section 16.3. The rest of the parameters are discussed below:

- **Swarm size**, $n_s$, i.e. the number of particles in the swarm: the more particles in the swarm, the larger the initial diversity of the swarm – provided that a good uniform initialization scheme is used to initialize the particles. A large swarm allows larger parts of the search space to be covered per iteration. However, more particles increase the per iteration computational complexity, and the search degrades to a parallel random search. It is also the case that more particles may lead to fewer iterations to reach a good solution, compared to smaller swarms. It has been shown in a number of empirical studies that the PSO has the ability to find optimal solutions with small swarm sizes of 10 to 30 particles [89, 865]. Success has even been obtained for fewer than 10 particles [863]. While empirical studies give a general heuristic of $n_s \in [10, 30]$, the optimal swarm size is problem-dependent. A smooth search space will need fewer particles than a rough surface to locate optimal solutions. Rather than using the heuristics found in publications, it is best that the value of $n_s$ be optimized for each problem using cross-validation methods.

- **Neighborhood size**: The neighborhood size defines the extent of social interaction within the swarm. The smaller the neighborhoods, the less interaction occurs. While smaller neighborhoods are slower in convergence, they have more reliable convergence to optimal solutions. Smaller neighborhood sizes are less susceptible to local minima. To capitalize on the advantages of small and large neighborhood sizes, start the search with small neighborhoods and increase the neighborhood size proportionally to the increase in number of iterations [820]. This approach ensures an initial high diversity with faster convergence as the particles move towards a promising search area.

- **Number of iterations**: The number of iterations to reach a good solution is also problem-dependent. Too few iterations may terminate the search prematurely. A too large number of iterations has the consequence of unnecessary added computational complexity (provided that the number of iterations is the only stopping condition).

- **Acceleration coefficients**: The acceleration coefficients, $c_1$ and $c_2$, together with the random vectors $\mathbf{r}_1$ and $\mathbf{r}_2$, control the stochastic influence of the cognitive and social components on the overall velocity of a particle. The constants $c_1$ and $c_2$ are also referred to as trust parameters, where $c_1$ expresses how much confidence a particle has in itself, while $c_2$ expresses how much confidence a particle has in its neighbors. With $c_1 = c_2 = 0$, particles keep flying at their current speed until they hit a boundary of the search space (assuming no inertia). If $c_1 > 0$ and $c_2 = 0$, all particles are independent hill-climbers. Each particle finds the best position in its neighborhood by replacing the current best position if the new position is better. Particles perform a local search. On the other hand, if $c_2 > 0$ and $c_1 = 0$, the entire swarm is attracted to a single point, $\hat{y}$. The swarm turns into one stochastic hill-climber.

Particles draw their strength from their cooperative nature, and are most effective when nostalgia ($c_1$) and envy ($c_2$) coexist in a good balance, i.e. $c_1 \approx c_2$. If $c_1 = c_2$, particles are attracted towards the average of $y_i$ and $\hat{y}$ [863, 870]. While most applications use $c_1 = c_2$, the ratio between these constants is problem-dependent. If $c_1 >> c_2$, each particle is much more attracted to its own personal best position, resulting in excessive wandering. On the other hand, if $c_2 >> c_1$, particles are more strongly attracted to the global best position, causing particles to rush prematurely towards optima. For unimodal problems with a smooth search space, a larger social component will be efficient, while rough multi-modal search spaces may find a larger cognitive component more advantageous.

Low values for $c_1$ and $c_2$ result in smooth particle trajectories, allowing particles to roam far from good regions to explore before being pulled back towards good regions. High values cause more acceleration, with abrupt movement towards or past good regions.

Usually, $c_1$ and $c_2$ are static, with their optimized values being found empirically. Wrong initialization of $c_1$ and $c_2$ may result in divergent or cyclic behavior [863, 870].

Clerc [134] proposed a scheme for adaptive acceleration coefficients, assuming the social velocity model (refer to Section 16.3.5):

$$c_2(t) = \frac{c_{2,min} + c_{2,max}}{2} + \frac{c_{2,max} - c_{2,min}}{2} + \frac{e^{-m_i(t)} - 1}{e^{-m_i(t)} + 1} \tag{16.35}$$

where $m_i$ is as defined in equation (16.30). The formulation of equation (16.30) implies that each particle has its own adaptive acceleration as a function of the slope of the search space at the current position of the particle.

Ratnaweera *et al.* [706] builds further on a suggestion by Suganthan [820] to linearly adapt the values of $c_1$ and $c_2$. Suganthan suggested that both acceleration coefficients be linearly decreased, but reported no improvement in performance using this scheme [820]. Ratnaweera *et al.* proposed that $c_1$ decreases linearly over time, while $c_2$ increases linearly [706]. This strategy focuses on exploration in the early stages of optimization, while encouraging convergence to a good optimum near the end of the optimization process by attracting particles more towards the neighborhood best (or global best) positions. The values of $c_1(t)$ and $c_2(t)$ at time step $t$ is calculated as

$$c_1(t) = (c_{1,min} - c_{1,max})\frac{t}{n_t} + c_{1,max} \tag{16.36}$$

$$c_2(t) = (c_{2,max} - c_{2,min})\frac{t}{n_t} + c_{2,min} \tag{16.37}$$

where $c_{1,max} = c_{2,max} = 2.5$ and $c_{1,min} = c_{2,min} = 0.5$.

A number of theoretical studies have shown that the convergence behavior of PSO is sensitive to the values of the inertia weight and the acceleration coefficients [136, 851, 863, 870]. These studies also provide guidelines to choose values for PSO parameters that will ensure convergence to an equilibrium point. The first set of guidelines are obtained from the different constriction models suggested by Clerc and Kennedy [136]. For a specific constriction model and selected $\phi$ value, the value of the constriction coefficient is calculated to ensure convergence.

For an unconstricted simplified PSO system that includes inertia, the trajectory of a particle converges if the following conditions hold [851, 863, 870, 937]:

$$1 > w > \frac{1}{2}(\phi_1 + \phi_2) - 1 \geq 0 \qquad (16.38)$$

and $0 \leq w < 1$. Since $\phi_1 = c_1 U(0,1)$ and $\phi_2 = c_2 U(0,1)$, the acceleration coefficients, $c_1$ and $c_2$ serve as upper bounds of $\phi_1$ and $\phi_2$. Equation (16.38) can then be rewritten as

$$1 > w > \frac{1}{2}(c_1 + c_2) - 1 \geq 0 \qquad (16.39)$$

Therefore, if $w$, $c_1$ and $c_2$ are selected such that the condition in equation (16.39) holds, the system has guaranteed convergence to an equilibrium state.

The heuristics above have been derived for the simplified PSO system with no stochastic component. It can happen that, for stochastic $\phi_1$ and $\phi_2$ and a $w$ that violates the condition stated in equation (16.38), the swarm may still converge. The stochastic trajectory illustrated in Figure 16.6 is an example of such behavior. The particle follows a convergent trajectory for most of the time steps, with an occasional divergent step.

Van den Bergh and Engelbrecht show in [863, 870] that convergent behavior will be observed under stochastic $\phi_1$ and $\phi_2$ if the ratio,

$$\phi_{ratio} = \frac{\phi_{crit}}{c_1 + c_2} \qquad (16.40)$$

is close to 1.0, where

$$\phi_{crit} = \sup \phi \,|\, 0.5\,\phi - 1 < w, \quad \phi \in (0, c_1 + c_2] \qquad (16.41)$$

It is even possible that parameter choices for which $\phi_{ratio} = 0.5$, may lead to convergent behavior, since particles spend 50% of their time taking a step along a convergent trajectory.

## 16.5   Single-Solution Particle Swarm Optimization

Initial empirical studies of the basic PSO and basic variations as discussed in this chapter have shown that the PSO is an efficient optimization approach – for the benchmark

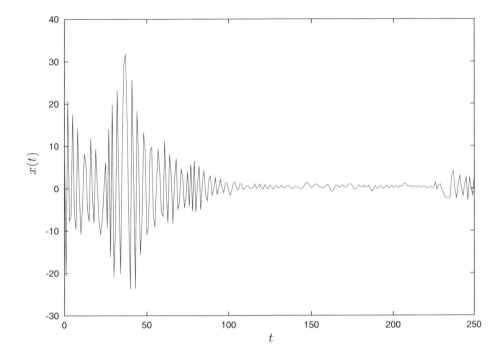

**Figure 16.6** Stochastic Particle Trajectory for $w = 0.9$ and $c_1 = c_2 = 2.0$

problems considered in these studies. Some studies have shown that the basic PSO improves on the performance of other stochastic population-based optimization algorithms such as genetic algorithms [88, 89, 106, 369, 408, 863]. While the basic PSO has shown some success, formal analysis [136, 851, 863, 870] has shown that the performance of the PSO is sensitive to the values of control parameters. It was also shown that the basic PSO has a serious defect that may cause stagnation [868].

A variety of PSO variations have been developed, mainly to improve the accuracy of solutions, diversity and convergence behavior. This section reviews some of these variations for locating a single solution to unconstrained, single-objective, static optimization problems. Section 16.5.2 considers approaches that differ in the social interaction of particles. Some hybrids with concepts from EC are discussed in Section 16.5.3. Algorithms with multiple swarms are discussed in Section 16.5.4. Multi-start methods are given in Section 16.5.5, while methods that use some form of repelling mechanism are discussed in Section 16.5.6. Section 16.5.7 shows how PSO can be changed to solve binary-valued problems.

Before these PSO variations are discussed, Section 16.5.1 outlines a problem with the basic PSO.

## 16.5.1   Guaranteed Convergence PSO

The basic PSO has a potentially dangerous property: when $\mathbf{x}_i = \mathbf{y}_i = \hat{\mathbf{y}}$, the velocity update depends only on the value of $w\mathbf{v}_i$. If this condition is true for all particles and it persists for a number of iterations, then $w\mathbf{v}_i \to 0$, which leads to stagnation of the search process. This point of stagnation may not necessarily coincide with a local minimum. All that can be said is that the particles converged to the best position found by the swarm. The PSO can, however, be pulled from this point of stagnation by forcing the global best position to change when $\mathbf{x}_i = \mathbf{y}_i = \hat{\mathbf{y}}$.

The guaranteed convergence PSO (GCPSO) forces the global best particle to search in a confined region for a better position, thereby solving the stagnation problem [863, 868]. Let $\tau$ be the index of the global best particle, so that

$$\mathbf{y}_\tau = \hat{\mathbf{y}} \tag{16.42}$$

GCPSO changes the position update to

$$x_{\tau j}(t+1) = \hat{y}_j(t) + wv_{\tau j}(t) + \rho(t)(1 - 2r_2(t)) \tag{16.43}$$

which is obtained using equation (16.1) if the velocity update of the global best particle changes to

$$v_{\tau j}(t+1) = -x_{\tau j}(t) + \hat{y}_j(t) + wv_{\tau j}(t) + \rho(t)(1 - 2r_{2j}(t)) \tag{16.44}$$

where $\rho(t)$ is a scaling factor defined in equation (16.45) below. Note that only the global best particle is adjusted according to equations (16.43) and (16.44); all other particles use the equations as given in equations (16.1) and (16.2).

The term $-x_{\tau j}(t)$ in equation (16.44) resets the global best particle's position to the position $\hat{y}_j(t)$. The current search direction, $wv_{\tau j}(t)$, is added to the velocity, and the term $\rho(t)(1 - 2r_{2j}(t))$ generates a random sample from a sample space with side lengths $2\rho(t)$. The scaling term forces the PSO to perform a random search in an area surrounding the global best position, $\hat{\mathbf{y}}(t)$. The parameter, $\rho(t)$ controls the diameter of this search area, and is adapted using

$$\rho(t+1) = \begin{cases} 2\rho(t) & \text{if } \#successes(t) > \epsilon_s \\ 0.5\rho(t) & \text{if } \#failures(t) > \epsilon_f \\ \rho(t) & \text{otherwise} \end{cases} \tag{16.45}$$

where $\#successes$ and $\#failures$ respectively denote the number of consecutive successes and failures. A failure is defined as $f(\hat{\mathbf{y}}(t)) \leq f(\hat{\mathbf{y}}(t+1))$; $\rho(0) = 1.0$ was found empirically to provide good results [863, 868]. The threshold parameters, $\epsilon_s$ and $\epsilon_f$ adhere to the following conditions:

$$\#successes(t+1) \quad > \quad \#successes(t) \Rightarrow \#failures(t+1) = 0 \tag{16.46}$$
$$\#failures(t+1) \quad > \quad \#failures(t) \Rightarrow \#successes(t+1) = 0 \tag{16.47}$$

The optimal choice of values for $\epsilon_s$ and $\epsilon_f$ is problem-dependent. Van den Bergh *et al.* [863, 868] recommends that $\epsilon_s = 15$ and $\epsilon_f = 5$ be used for high-dimensional search spaces. The algorithm is then quicker to punish poor $\rho$ settings than it is to reward successful $\rho$ values.

Instead of using static $\epsilon_s$ and $\epsilon_f$ values, these values can be learnt dynamically. For example, increase $s_c$ each time that $\#failures > \epsilon_f$, which makes it more difficult to reach the success state if failures occur frequently. Such a conservative mechanism will prevent the value of $\rho$ from oscillating rapidly. A similar strategy can be used for $\epsilon_s$.

The value of $\rho$ determines the size of the local area around $\hat{\mathbf{y}}$ where a better position for $\hat{\mathbf{y}}$ is searched. GCPSO uses an adaptive $\rho$ to find the best size of the sampling volume, given the current state of the algorithm. When the global best position is repeatedly improved for a specific value of $\rho$, the sampling volume is increased to allow step sizes in the global best position to increase. On the other hand, when $\epsilon_f$ consecutive failures are produced, the sampling volume is too large and is consequently reduced. Stagnation is prevented by ensuring that $\rho(t) > 0$ for all time steps.

## 16.5.2   Social-Based Particle Swarm Optimization

Social-based PSO implementations introduce a new social topology, or change the way in which personal best and neighborhood best positions are calculated.

### Spatial Social Networks

Neighborhoods are usually formed on the basis of particle indices. That is, assuming a ring social network, the immediate neighbors of a particle with index $i$ are particles with indices $(i - 1 \bmod n_s)$ and $(i - 1 \bmod n_s)$, where $n_s$ is the total number of particles in the swarm. Suganthan proposed that neighborhoods be formed on the basis of the Euclidean distance between particles [820]. For neighborhoods of size $n_{\mathcal{N}}$, the neighborhood of particle $i$ is defined to consist of the $n_{\mathcal{N}}$ particles closest to particle $i$. Algorithm 16.4 summarizes the spatial neighborhood selection process.

Calculation of spatial neighborhoods require that the Euclidean distance between all particles be calculated at each iteration, which significantly increases the computational complexity of the search algorithm. If $n_t$ iterations of the algorithm is executed, the spatial neighborhood calculation adds a $O(n_t n_s^2)$ computational cost. Determining neighborhoods based on distances has the advantage that neighborhoods change dynamically with each iteration.

### Fitness-Based Spatial Neighborhoods

Braendler and Hendtlass [81] proposed a variation on the spatial neighborhoods implemented by Suganthan, where particles move towards neighboring particles that have found a good solution. Assuming a minimization problem, the neighborhood of

**Algorithm 16.4** Calculation of Spatial Neighborhoods; $\mathcal{N}_i$ is the neighborhood of particle $i$

Calculate the Euclidean distance $\mathcal{E}(\mathbf{x}_{i_1}, \mathbf{x}_{i_2}), \forall i_1, i_2 = 1, \ldots, n_s$;
$S = \{i : i = 1, \ldots, n_s\}$;
**for** $i = 1, \ldots, n_s$ **do**
$\quad S' = S$;
$\quad$ **for** $i' = 1, \ldots, n_{\mathcal{N}_{i'}}$ **do**
$\quad\quad \mathcal{N}_i = \mathcal{N}_i \cup \{\mathbf{x}_{i''} : \mathcal{E}(\mathbf{x}_i, \mathbf{x}_{i''}) = \min\{\mathcal{E}(\mathbf{x}_i, \mathbf{x}_{i'''}), \forall \mathbf{x}_{i'''} \in S'\}$;
$\quad\quad S' = S' \setminus \{\mathbf{x}_{i''}\}$;
$\quad$ **end**
**end**

---

particle $i$ is defined as the $n_{\mathcal{N}}$ particles with the smallest value of

$$\mathcal{E}(\mathbf{x}_i, \mathbf{x}_{i'}) \times f(\mathbf{x}_{i'}) \tag{16.48}$$

where $\mathcal{E}(\mathbf{x}_i, \mathbf{x}_{i'})$ is the Euclidean distance between particles $i$ and $i'$, and $f(\mathbf{x}_{i'})$ is the fitness of particle $i'$. Note that this neighborhood calculation mechanism also allows for overlapping neighborhoods.

Based on this scheme of determining neighborhoods, the standard *lbest* PSO velocity equation (refer to equation (16.6)) is used, but with $\hat{\mathbf{y}}_i$ the neighborhood-best determined using equation (16.48).

### Growing Neighborhoods

As discussed in Section 16.2, social networks with a low interconnection converge slower, which allows larger parts of the search space to be explored. Convergence of the fully interconnected star topology is faster, but at the cost of neglecting parts of the search space. To combine the advantages of better exploration by neighborhood structures and the faster convergence of highly connected networks, Suganthan combined the two approaches [820]. The search is initialized with an *lbest* PSO with $n_{\mathcal{N}} = 2$ (i.e. with the smallest neighborhoods). The neighborhood sizes are then increased with increase in iteration until each neighborhood contains the entire swarm (i.e. $n_{\mathcal{N}} = n_s$).

Growing neighborhoods are obtained by adding particle position $\mathbf{x}_{i_2}(t)$ to the neighborhood of particle position $\mathbf{x}_{i_1}(t)$ if

$$\frac{||\mathbf{x}_{i_1}(t) - \mathbf{x}_{i_2}(t)||_2}{d_{max}} < \epsilon \tag{16.49}$$

where $d_{max}$ is the largest distance between any two particles, and

$$\epsilon = \frac{3t + 0.6n_t}{n_t} \tag{16.50}$$

with $n_t$ the maximum number of iterations.

This allows the search to explore more in the first iterations of the search, with faster convergence in the later stages of the search.

### Hypercube Structure

For binary-valued problems, Abdelbar and Abdelshahid [5] used a hypercube neighborhood structure. Particles are defined as neighbors if the Hamming distance between the bit representation of their indices is one. To make use of the hypercube topology, the total number of particles must be a power of two, where particles have indices from 0 to $2^{n_N} - 1$. Based on this, the hypercube has the properties [5]:

- Each neighborhood has exactly $n_N$ particles.
- The maximum distance between any two particles is exactly $n_N$.
- If particles $i_1$ and $i_2$ are neighbors, then $i_1$ and $i_2$ will have no other neighbors in common.

Abdelbar and Abdelshahid found that the hypercube network structure provides better results than the *gbest* PSO for the binary problems studied.

### Fully Informed PSO

Based on the standard velocity updates as given in equations (16.2) and (16.6), each particle's new position is influenced by the particle itself (via its personal best position) and the best position in its neighborhood. Kennedy and Mendes observed that human individuals are not influenced by a single individual, but rather by a statistical summary of the state of their neighborhood [453]. Based on this principle, the velocity equation is changed such that each particle is influenced by the successes of all its neighbors, and not on the performance of only one individual. The resulting PSO is referred to as the *fully informed* PSO (FIPS).

Two models are suggested [453, 576]:

- Each particle in the neighborhood, $\mathcal{N}_i$, of particle $i$ is regarded equally. The cognitive and social components are replaced with the term

$$\sum_{m=1}^{n_{\mathcal{N}_i}} \frac{\mathbf{r}(t)(\mathbf{y}_m(t) - \mathbf{x}_i(t))}{n_{\mathcal{N}_i}} \qquad (16.51)$$

  where $n_{\mathcal{N}_i} = |\mathcal{N}_i|$, $\mathcal{N}_i$ is the set of particles in the neighborhood of particle $i$ as defined in equation (16.8), and $\mathbf{r}(t) \sim U(0, c_1 + c_2)^{n_x}$. The velocity is then calculated as

$$\mathbf{v}_i(t+1) = \chi \left( \mathbf{v}_i(t) + \sum_{m=1}^{n_{\mathcal{N}_i}} \frac{\mathbf{r}(t)(\mathbf{y}_m(t) - \mathbf{x}_i(t))}{n_{\mathcal{N}_i}} \right) \qquad (16.52)$$

Although Kennedy and Mendes used constriction models of type $1'$ (refer to Section 16.3.3), inertia or any of the other models may be used.

Using equation (16.52), each particle is attracted towards the average behavior of its neighborhood.

Note that if $\mathcal{N}_i$ includes only particle $i$ and its best neighbor, the velocity equation becomes equivalent to that of *lbest* PSO.

- A weight is assigned to the contribution of each particle based on the performance of that particle. The cognitive and social components are replaced by

$$\frac{\sum_{m=1}^{n_{\mathcal{N}_i}} \frac{\phi_m \mathbf{p}_m(t)}{f(\mathbf{x}_m(t))}}{\sum_{m=1}^{n_{\mathcal{N}_i}} \frac{\phi_m}{f(\mathbf{x}_m(t))}} \tag{16.53}$$

where $\phi_m \sim U(0, \frac{c_1+c_2}{n_{\mathcal{N}_i}})$, and

$$\mathbf{p}_m(t) = \frac{\phi_1 \mathbf{y}_m(t) + \phi_2 \hat{\mathbf{y}}_m(t)}{\phi_1 + \phi_2} \tag{16.54}$$

The velocity equation then becomes

$$\mathbf{v}_i(t+1) = \chi \left( \mathbf{v}_i(t) + \frac{\sum_{m=1}^{n_{\mathcal{N}_i}} \left( \frac{\phi_m \mathbf{p}_m(t)}{f(\mathbf{x}_m(t))} \right)}{\sum_{m=1}^{n_{\mathcal{N}_i}} \left( \frac{\phi_m}{f(\mathbf{x}_m(t))} \right)} \right) \tag{16.55}$$

In this case a particle is attracted more to its better neighbors.

A disadvantage of the FIPS is that it does not consider the fact that the influences of multiple particles may cancel each other. For example, if two neighbors respectively contribute the amount of $a$ and $-a$ to the velocity update of a specific particle, then the sum of their influences is zero. Consider the effect when all the neighbors of a particle are organized approximately symmetrically around the particle. The change in weight due to the FIPS velocity term will then be approximately zero, causing the change in position to be determined only by $\chi \mathbf{v}_i(t)$.

## Barebones PSO

Formal proofs [851, 863, 870] have shown that each particle converges to a point that is a weighted average between the personal best and neighborhood best positions. If it is assumed that $c_1 = c_2$, then a particle converges, in each dimension to

$$\frac{y_{ij}(t) + \hat{y}_{ij}(t)}{2} \tag{16.56}$$

This behavior supports Kennedy's proposal to replace the entire velocity by random numbers sampled from a Gaussian distribution with the mean as defined in equation (16.56) and deviation,

$$\sigma = |y_{ij}(t) - \hat{y}_{ij}(t)| \tag{16.57}$$

The velocity therefore changes to

$$v_{ij}(t+1) \sim N\left(\frac{y_{ij}(t) + \hat{y}_{ij}(t)}{2}, \sigma\right) \tag{16.58}$$

In the above, $\hat{y}_{ij}$ can be the global best position (in the case of *gbest* PSO), the local best position (in the case of a neighborhood-based algorithm), a randomly selected neighbor, or the center of a FIPS neighborhood. The position update changes to

$$x_{ij}(t+1) = v_{ij}(t+1) \tag{16.59}$$

Kennedy also proposed an alternative to the barebones PSO velocity in equation (16.58), where

$$v_{ij}(t+1) = \begin{cases} y_{ij}(t) & \text{if } U(0,1) < 0.5 \\ N(\frac{y_{ij}(t)+\hat{y}_{ij}(t)}{2}, \sigma) & \text{otherwise} \end{cases} \tag{16.60}$$

Based on equation (16.60), there is a 50% chance that the $j$-th dimension of the particle dimension changes to the corresponding personal best position. This version can be viewed as a mutation of the personal best position.

## 16.5.3  Hybrid Algorithms

This section describes just some of the PSO algorithms that use one or more concepts from EC.

### Selection-Based PSO

Angeline provided the first approach to combine GA concepts with PSO [26], showing that PSO performance can be improved for certain classes of problems by adding a selection process similar to that which occurs in evolutionary algorithms. The selection procedure as summarized in Algorithm 16.5 is executed before the velocity updates are calculated.

---

**Algorithm 16.5** Selection-Based PSO

---

Calculate the fitness of all particles;
**for** *each particle* $i = 1, \ldots, n_s$ **do**
    Randomly select $n_{ts}$ particles;
    Score the performance of particle $i$ against the $n_{ts}$ randomly selected particles;
**end**
Sort the swarm based on performance scores;
Replace the worst half of the swarm with the top half, without changing the personal best positions;

---

Although the bad performers are penalized by being removed from the swarm, memory of their best found positions is not lost, and the search process continues to build on previously acquired experience. Angeline showed empirically that the selection based PSO improves on the local search capabilities of PSO [26]. However, contrary to one of the objectives of natural selection, this approach significantly reduces the diversity of the swarm [363]. Since half of the swarm is replaced by the other half, diversity is decreased by 50% at each iteration. In other words, the selection pressure is too high.

Diversity can be improved by replacing the worst individuals with mutated copies of the best individuals. Performance can also be improved by considering replacement only if the new particle improves the fitness of the particle to be deleted. This is the approach followed by Koay and Srinivasan [470], where each particle generates offspring through mutation.

**Reproduction**

Reproduction refers to the process of producing offspring from individuals of the current population. Different reproduction schemes have been used within PSO. One of the first approaches can be found in the Cheap-PSO developed by Clerc [134], where a particle is allowed to generate a new particle, kill itself, or modify the inertia and acceleration coefficient, on the basis of environment conditions. If there is no sufficient improvement in a particle's neighborhood, the particle spawns a new particle within its neighborhood. On the other hand, if a sufficient improvement is observed in the neighborhood, the worst particle of that neighborhood is culled.

Using this approach to reproduction and culling, the probability of adding a particle decreases with increasing swarm size. On the other hand, a decreasing swarm size increases the probability of spawning new particles.

The Cheap-PSO includes only the social component, where the social acceleration coefficient is adapted using equation (16.35) and the inertia is adapted using equation (16.29).

Koay and Srinivasan [470] implemented a similar approach to dynamically changing swarm sizes. The approach was developed by analogy with the natural adaptation of the amoeba to its environment: when the amoeba receives positive feedback from its environment (i.e. that sufficient food sources exist), it reproduces by releasing more spores. On the other hand, when food sources are scarce, reproduction is reduced. Taking this analogy to optimization, when a particle finds itself in a potential optimum, the number of particles is increased in that area. Koay and Srinivasan spawn only the global best particle (assuming *gbest* PSO) in order to reduce the computational complexity of the spawning process. The choice of spawning only the global best particle can be motivated by the fact that the global best particle will be the first particle to find itself in a potential optimum. Stopping conditions as given in Section 16.1.6 can be used to determine if a potential optimum has been found. The spawning process is summarized in Algorithm 16.6. It is also possible to apply the same process to the neighborhood best positions if other neighborhood networks are used, such as the ring structure.

---

**Algorithm 16.6** Global Best Spawning Algorithm

---

**if** $\hat{\mathbf{y}}(t)$ *is in a potential minimum* **then**
    **repeat**
        $\hat{\mathbf{y}} = \hat{\mathbf{y}}(t)$;
        **for** *NumberOfSpawns=1 to 10* **do**
            **for** $a = 1$ *to NumberOfSpawns* **do**
                $\hat{\mathbf{y}}_a = \hat{\mathbf{y}}(t) + N(0, \sigma)$;
                **if** $f(\hat{\mathbf{y}}_a) < f(\hat{\mathbf{y}})$ **then**
                    $\hat{\mathbf{y}} = \hat{\mathbf{y}}_a$;
                **end**
            **end**
        **end**
        **until** $f(\hat{\mathbf{y}}) \geq f(\hat{\mathbf{y}}(t))$;
        $\hat{\mathbf{y}}(t) = \hat{\mathbf{y}}$;
**end**

---

Løvberg *et al.* [534, 536] used an arithmetic crossover operator to produce offspring from two randomly selected particles. Assume that particles $a$ and $b$ are selected for crossover. The corresponding positions, $\mathbf{x}_a(t)$ and $\mathbf{x}_b(t)$ are then replaced by the offspring,

$$\mathbf{x}_{i_1}(t+1) = \mathbf{r}(t)\mathbf{x}_{i_1}(t) + (\mathbf{1} - \mathbf{r}(t))\mathbf{x}_{i_2}(t) \tag{16.61}$$

$$\mathbf{x}_{i_2}(t+1) = \mathbf{r}(t)\mathbf{x}_{i_2}(t) + (\mathbf{1} - \mathbf{r}(t))\mathbf{x}_{i_1}(t) \tag{16.62}$$

with the corresponding velocities,

$$\mathbf{v}_{i_1}(t+1) = \frac{\mathbf{v}_{i_1}(t) + \mathbf{v}_{i_2}(t)}{||\mathbf{v}_{i_1}(t) + \mathbf{v}_{i_2}(t)||}||\mathbf{v}_{i_1}(t)|| \tag{16.63}$$

$$\mathbf{v}_{i_2}(t+1) = \frac{\mathbf{v}_{i_1}(t) + \mathbf{v}_{i_2}(t)}{||\mathbf{v}_{i_1}(t) + \mathbf{v}_{i_2}(t)||}||\mathbf{v}_{i_2}(t)|| \tag{16.64}$$

where $\mathbf{r}_1(t) \sim U(0, 1)^{n_x}$. The personal best position of an offspring is initialized to its current position. That is, if particle $i_1$ was involved in the crossover, then $\mathbf{y}_{i_1}(t+1) = \mathbf{x}_{i_1}(t+1)$.

Particles are selected for breeding at a user-specified breeding probability. Given that this probability is less than one, not all of the particles will be replaced by offspring. It is also possible that the same particle will be involved in the crossover operation more than once per iteration. Particles are randomly selected as parents, not on the basis of their fitness. This prevents the best particles from dominating the breeding process. If the best particles were allowed to dominate, the diversity of the swarm would decrease significantly, causing premature convergence. It was found empirically that a low breeding probability of 0.2 provides good results [536].

The breeding process is done for each iteration after the velocity and position updates have been done.

The breeding mechanism proposed by Løvberg *et al.* has the disadvantage that parent particles are replaced even if the offspring is worse off in fitness. Also, if $f(\mathbf{x}_{i_1}(t+1)) >$

$f(\mathbf{x}_{i_1}(t))$ (assuming a minimization problem), replacement of the personal best with $\mathbf{x}_{i_1}(t+1)$ loses important information about previous personal best positions. Instead of being attracted towards the previous personal best position, which in this case will have a better fitness, the offspring is attracted to a worse solution. This problem can be addressed by replacing $\mathbf{x}_{i_1}(t)$ with its offspring only if the offspring provides a solution that improves on particle $i_1$'s personal best position, $\mathbf{y}_{i_1}(t)$.

### Gaussian Mutation

Gaussian mutation has been applied mainly to adjust position vectors after the velocity and update equations have been applied, by adding random values sampled from a Gaussian distribution to the components of particle position vectors, $\mathbf{x}_i(t+1)$.

Miranda and Fonseca [596] mutate only the global best position as follows,

$$\hat{\mathbf{y}}(t+1) = \hat{\mathbf{y}}'(t+1) + \eta' \mathbf{N}(0,1) \tag{16.65}$$

where $\hat{\mathbf{y}}'(t+1)$ represents the unmutated global best position as calculated from equation (16.4), $\eta'$ is referred to as a learning parameter, which can be a fixed value, or adapted as a strategy parameter as in evolutionary strategies.

Higashi and Iba [363] mutate the components of particle position vectors at a specified probability. Let $\mathbf{x}'_i(t+1)$ denote the new position of particle $i$ after application of the velocity and position update equations, and let $P_m$ be the probability that a component will be mutated. Then, for each component, $j = 1, \ldots, n_x$, if $U(0,1) < P_m$, then component $x'_{ij}(t+1)$ is mutated using [363]

$$x_{ij}(t+1) = x'_{ij}(t+1) + N(0,\sigma)x'_{ij}(t+1) \tag{16.66}$$

where the standard deviation, $\sigma$, is defined as

$$\sigma = 0.1(x_{max,j} - x_{min,j}) \tag{16.67}$$

Wei *et al.* [893] directly apply the original EP Gaussian mutation operator:

$$x_{ij}(t+1) = x'_{ij}(t+1) + \eta_{ij}(t)N_j(0,1) \tag{16.68}$$

where $\eta_{ij}$ controls the mutation step sizes, calculated for each dimension as

$$\eta_{ij}(t) = \eta_{ij}(t-1)\,e^{\tau' N(0,1)+\tau N_j(0,1)} \tag{16.69}$$

with $\tau$ and $\tau'$ as defined in equations (11.54) and (11.53).

The following comments can be made about this mutation operator:

- With $\eta_{ij}(0) \in (0,1]$, the mutation step sizes decrease with increase in time step, $t$. This allows for initial exploration, with the solutions being refined in later time steps. It should be noted that convergence cannot be ensured if mutation step sizes do not decrease over time.

- All the components of the same particle are mutated using the same value, $N(0, 1)$. However, for each component, an additional random number, $N_j(0, 1)$, is also used. Each component will therefore be mutated by a different amount.

- The amount of mutation depends on the dimension of the problem, by the calculation of $\tau$ and $\tau'$. The larger $n_x$, the smaller the mutation step sizes.

Secrest and Lamont [772] adjust particle positions as follows,

$$\mathbf{x}_i(t+1) = \begin{cases} \mathbf{y}_i(t) + \mathbf{v}_i(t+1) & \text{if } U(0,1) > c_1 \\ \hat{\mathbf{y}}(t) + \mathbf{v}_i(t+1) & \text{otherwise} \end{cases} \tag{16.70}$$

where

$$\mathbf{v}_i(t+1) = |\mathbf{v}_i(t+1)|\mathbf{r}_\theta \tag{16.71}$$

In the above, $\mathbf{r}_\theta$ is a random vector with magnitude of one and angle uniformly distributed from 0 to $2\pi$. $|\mathbf{v}_i(t+1)|$ is the magnitude of the new velocity, calculated as

$$|\mathbf{v}_i(t+1)| = \begin{cases} N(0, (1-c_2)||\mathbf{y}_i(t) - \hat{\mathbf{y}}(t)||_2) & \text{if } U(0,1) > c_1 \\ N(0, c_2||\mathbf{y}_i(t) - \hat{\mathbf{y}}(t)||_2) & \text{otherwise} \end{cases} \tag{16.72}$$

Coefficient $c_1 \in (0, 1)$ specifies the trust between the global and personal best positions. The larger $c_1$, the more particles are placed around the global best position; $c_2 \in (0, 1)$ establishes the point between the global best and the personal best to which the corresponding particle will converge.

## Cauchy Mutation

In the fast EP, Yao *et al.* [934, 936] showed that mutation step sizes sampled from a Cauchy distribution result in better performance than sampling from a Gaussian distribution. This is mainly due to the fact that the Cauchy distribution has more of its probability in the tails of the distribution, resulting in an increased probability of large mutation step sizes (therefore, better exploration). Stacey *et al.* [808] applied the EP Cauchy mutation operator to PSO. Given that $n_x$ is the dimension of particles, each dimension is mutated with a probability of $1/n_x$. If a component $x_{ij}(t)$ is selected for mutation, the mutation step size, $\Delta x_{ij}(t)$, is sampled from a Cauchy distribution with probability density function given by equation (11.10).

## Differential Evolution Based PSO

Differential evolution (DE) makes use of an arithmetic crossover operator that involves three randomly selected parents (refer to Chapter 13). Let $\mathbf{x}_1(t) \neq \mathbf{x}_2(t) \neq \mathbf{x}_3(t)$ be three particle positions randomly selected from the swarm. Then each dimension of particle $i$ is calculated as follows:

$$x'_{ij}(t+1) = \begin{cases} x_{1j}(t) + \beta(x_{2j}(t) - x_{3j}(t)) & \text{if } U(0,1) \leq P_c \text{ or } j = U(1, n_x) \\ x_{ij}(t) & \text{otherwise} \end{cases} \tag{16.73}$$

where $P_c \in (0, 1)$ is the probability of crossover, and $\beta > 0$ is a scaling factor.

The position of the particle is only replaced if the offspring is better. That is, $\mathbf{x}_i(t+1) = \mathbf{x}'_i(t+1)$ only if $f(\mathbf{x}'_i(t+1)) < f(\mathbf{x}_i(t))$, otherwise $\mathbf{x}_i(t+1) = \mathbf{x}_i(t)$ (assuming a minimization problem).

Hendtlass applied the above DE process to the PSO by executing the DE crossover operator from time to time [360]. That is, at specified intervals, the swarm serves as population for the DE algorithm, and the DE algorithm is executed for a number of iterations. Hendtlass reported that this hybrid produces better results than the basic PSO. Kannan *et al.* [437] applies DE to each particle for a number of iterations, and replaces the particle with the best individual obtained from the DE process.

Zhang and Xie [954] followed a somewhat different approach where only the personal best positions are changed using the following operator:

$$y'_{ij}(t+1) = \begin{cases} \hat{y}_{ij}(t) + \delta_j & \text{if } U(0,1) < P_c \text{ and } j = U(1, n_x) \\ y_{ij}(t) & \text{otherwise} \end{cases} \tag{16.74}$$

where $\delta$ is the general difference vector defined as,

$$\delta_j = \frac{y_{1j}(t) - y_{2j}(t)}{2} \tag{16.75}$$

with $y_{1j}(t)$ and $y_{2j}(t)$ randomly selected personal best positions. Then, $y_{ij}(t+1)$ is set to $y'_{ij}(t+1)$ only if the new personal best has a better fitness evaluation.

## 16.5.4 Sub-Swarm Based PSO

A number of cooperative and competitive PSO implementations that make use of multiple swarms have been developed. Some of these are described below.

Multi-phase PSO approaches divide the main swarm of particles into subgroups, where each subgroup performs a different task, or exhibits a different behavior. The behavior of a group or task performed by a group usually changes over time in response to the group's interaction with the environment. It can also happen that individuals may migrate between groups.

The breeding between sub-swarms developed by Løvberg *et al.* [534, 536] (refer to Section 16.5.3) is one form of cooperative PSO, where cooperation is implicit in the exchange of genetic material between parents of different sub-swarms.

Al-Kazemi and Mohan [16, 17, 18] explicitly divide the main swarm into two sub-swarms, of equal size. Particles are randomly assigned to one of the sub-swarms. Each sub-swarm can be in one of two phases:

- **Attraction phase**, where the particles of the corresponding sub-swarm are allowed to move towards the global best position.

- **Repulsion phase**, where the particles of the corresponding sub-swarm move away from the global best position.

For their multi-phase PSO (MPPSO), the velocity update is defined as [16, 17]:

$$v_{ij}(t + 1) = wv_{ij}(t) + c_1 x_{ij}(t) + c_2 \hat{y}_j(t) \qquad (16.76)$$

The personal best position is excluded from the velocity equation, since a hill-climbing procedure is followed where a particle's position is only updated if the new position results in improved performance.

Let the tuple $(w, c_1, c_2)$ represent the values of the inertia weight, $w$, and acceleration coefficients $c_1$ and $c_2$. Particles that find themselves in phase 1 exhibit an attraction towards the global best position, which is achieved by setting $(w, c_1, c_2) = (1, -1, 1)$. Particles in phase 2 have $(w, c_1, c_2) = (1, 1, -1)$, forcing them to move away from the global best position.

Sub-swarms switch phases either

- when the number of iterations in the current phase exceeds a user specified threshold, or

- when particles in any phase show no improvement in fitness during a user-specified number of consecutive iterations.

In addition to the velocity update as given in equation (16.76), velocity vectors are periodically initialized to new random vectors. Care should be taken with this process, not to reinitialize velocities when a good solution is found, since it may pull particles out of this good optimum. To make sure that this does not happen, particle velocities can be reinitialized based on a reinitialization probability. This probability starts with a large value that decreases over time. This approach ensures large diversity in the initial steps of the algorithm, emphasizing exploration, while exploitation is favored in the later steps of the search.

Cooperation between the subgroups is achieved through the selection of the global best particle, which is the best position found by all the particles in both sub-swarms.

Particle positions are not updated using the standard position update equation. Instead, a hill-climbing process is followed to ensure that the fitness of a particle is monotonically decreasing (increasing) in the case of a minimization (maximization) problem. The position vector is updated by randomly selecting $\varsigma$ consecutive components from the velocity vector and adding these velocity components to the corresponding position components. If no improvement is obtained for any subset of $\varsigma$ consecutive components, the position vector does not change. If an improvement is obtained, the corresponding position vector is accepted. The value of $\varsigma$ changes for each particle, since it is randomly selected, with $\varsigma \sim U(1, \varsigma_{max})$, with $\varsigma_{max}$ initially small, increasing to a maximum of $n_x$ (the dimension of particles).

The attractive and repulsive PSO (ARPSO) developed by Riget and Vesterstrøm [729, 730, 877] follows a similar process where the entire swarm alternates between an attraction and repulsion phase. The difference between the MPPSO and ARPSO lies in the velocity equation, in that there are no explicit sub-swarms in ARPSO, and ARPSO uses information from the environment to switch between phases. While

ARPSO was originally applied to one swarm, nothing prevents its application to sub-swarms.

The ARPSO is based on the diversity guided evolutionary algorithm developed by Ursem [861], where the standard Gaussian mutation operator is changed to a directed mutation in order to increase diversity. In ARPSO, if the diversity of the swarm, measured using

$$\text{diversity}(S(t)) = \frac{1}{n_s} \sum_{i=1}^{n_s} \sqrt{\sum_{j=1}^{n_x} (x_{ij}(t) - \bar{x}_j(t))^2} \qquad (16.77)$$

where $\bar{x}_j(t)$ is the average of the $j$-th dimension over all particles, i.e.

$$\bar{x}_j(t) = \frac{\sum_{i=1}^{n_s} x_{ij}(t)}{n_s} \qquad (16.78)$$

is greater than a threshold, $\varphi_{min}$, then the swarm switches to the attraction phase; otherwise the swarm switches to the repulsion phase until a threshold diversity, $\varphi_{max}$ is reached. The attraction phase uses the basic PSO velocity and position update equations. For the repulsion phase, particles repel each other using,

$$v_{ij}(t+1) = wv_{ij}(t) - c_1 r_{1j}(t)(y_{ij}(t) - x_{ij}(t)) - c_2 r_{2j}(t)(\hat{y}_{ij}(t) - x_{ij}(t)) \qquad (16.79)$$

Riget and Vesterstrøm used $\varphi_{min} = 5 \times 10^{-6}$ and $\varphi_{max} = 0.25$ as proposed by Ursem.

In the division of labor PSO (DLPSO), each particle has a response threshold, $\theta_{ik}$, for each task, $k$. Each particle receives task-related stimuli, $s_{ik}$. If the received stimuli are much lower than the response threshold, the particle engages in the corresponding task with a low probability. If $s_{ik} > \theta_{ik}$, the probability of engaging in the task is higher. Let $\vartheta_{ik}$ denote the state of particle $i$ with respect to task $k$. If $\vartheta_{ik} = 0$, then particle $i$ is not performing task $k$. On the other hand, if $\vartheta_{ik} = 1$, then task $k$ is performed by particle $i$. At each time step, each particle determines if it should become active or inactive on the basis of a given probability. An attractive particle performs task $k$ with probability

$$P(\vartheta_{ik} = 0 \rightarrow \vartheta_{ik} = 1) = P_{\theta_{ik}}(\theta_{ik}, s_{ik}) = \frac{s_{ik}^{\alpha}}{\theta_{ik}^{\alpha} + s_{ik}^{\alpha}} \qquad (16.80)$$

where $\alpha$ controls the steepness of the response function, $P_{\theta}$. For high $\alpha$, high probabilities are assigned to values of $s_{ik}$ just above $\theta_{ik}$.

The probability of an active particle to become inactive is taken as a constant, user-specified value, $P_{\vartheta}$.

The DLPSO uses only one task (the task subscript is therefore omitted), namely local search. In this case, the stimuli is the number of time steps since the last improvement in the fitness of the corresponding particle. When a particle becomes active, the local search is performed as follows: if $\vartheta_i = 1$, then $\mathbf{x}_i(t) = \hat{\mathbf{y}}_i(t)$ and $\mathbf{v}_i(t)$ is randomly assigned with length equal to the length of the velocity of the neighborhood best particle. The probability of changing to an inactive particle is $P_{\vartheta} = 1$, meaning that a particle becomes inactive immediately after its local search task is completed. Inactive particles follow the velocity and position updates of the basic PSO. The probability, $P(\vartheta_i = 1 \rightarrow \vartheta_i = 0)$ is kept high to ensure that the PSO algorithm does not degrade to a pure local search algorithm.

The local search, or exploitation, is only necessary when the swarm starts to converge to a solution. The probability of task execution should therefore be small initially, increasing over time. To achieve this, the absolute ageing process introduced by Théraulaz *et al.* [842] for ACO is used in the DLPSO to dynamically adjust the values of the response thresholds

$$\theta_i(t) = \beta \, e^{-\alpha t} \tag{16.81}$$

with $\alpha, \beta > 0$.

Empirical results showed that the DLPSO obtained significantly better results as a GA and the basic PSO [877, 878].

Krink and Løvberg [490, 534] used the life-cycle model to change the behavior of individuals. Using the life-cycle model, an individual can be in any of three phases: a PSO particle, a GA individual, or a stochastic hill-climber. The life-cycle model used here is in analogy with the biological process where an individual progresses through various phases from birth to maturity and reproduction. The transition between phases of the life-cycle is usually triggered by environmental factors.

For the life-cycle PSO (LCPSO), the decision to change from one phase to another depends on an individual's success in searching the fitness landscape. In the original model, all individuals start as particles and exhibit behavior as dictated by the PSO velocity and position update equations. The second phase in the life-cycle changes a particle to an individual in a GA, where its behavior is governed by the process of natural selection and survival of the fittest. In the last phase, an individual changes into a solitary stochastic hill-climber. An individual switches from one phase to the next if its fitness is not improved over a number of consecutive iterations. The LCPSO is summarized in Algorithm 16.7.

Application of the LCPSO results in the formation of three subgroups, one for each behavior (i.e. PSO, GA or hill-climber). Therefore, the main population may at the same time consist of individuals of different behavior.

While the original implementation initialized all individuals as PSO particles, the initial population can also be initialized to contain individuals of different behavior from the first time step. The rationale for starting with PSO particles can be motivated by the observation that the PSO has been shown to converge faster than GAs. Using hill-climbing as the third phase also makes sense, since exploitation should be emphasized in the later steps of the search process, with the initial PSO and GA phases focusing on exploration.

---

**Algorithm 16.7** Life-Cycle PSO

---

Initialize a population of individuals;
**repeat**
    **for** *all individuals* **do**
        Evaluate fitness;
        **if** *fitness did not improve* **then**
            Switch to next phase;
        **end**
    **end**
    **for** *all PSO particles* **do**
        Calculate new velocity vectors;
        Update positions;
    **end**
    **for** *all GA individuals* **do**
        Perform reproduction;
        Mutate;
        Select new population;
    **end**
    **for** *all Hill-climbers* **do**
        Find possible new neighboring solution;
        Evaluate fitness of new solution;
        Move to new solution with specified probability;
    **end**
**until** *stopping condition is true*;

---

### Cooperative Split PSO

The cooperative split PSO, first introduced in [864] by Van den Bergh and Engelbrecht, is based on the cooperative coevolutionary genetic algorithm (CCGA) developed by Potter [686] (also refer to Section 15.3). In the cooperative split PSO, denoted by CPSO-$S_K$, each particle is split into $K$ separate parts of smaller dimension [863, 864, 869]. Each part is then optimized using a separate sub-swarm. If $K = n_x$, each dimension is optimized by a separate sub-swarm, using any PSO algorithm. The number of parts, $K$, is referred to as the *split factor*.

The difficulty with the CPSO-$S_K$ algorithm is how to evaluate the fitness of the particles in the sub-swarms. The fitness of each particle in sub-swarm $S_k$ cannot be computed in isolation from other sub-swarms, since a particle in a specific sub-swarm represents only part of the complete $n_x$-dimensional solution. To solve this problem, a context vector is maintained to represent the $n_x$-dimensional solution. The simplest way to construct the context vector is to concatenate the global best positions from the $K$ sub-swarms. To evaluate the fitness of particles in sub-swarm $S_k$, all the components of the context vector are kept constant except those that correspond to the components of sub-swarm $S_k$. Particles in sub-swarm $S_k$ are then swapped into the corresponding positions of the context vector, and the original fitness function is used to evaluate the fitness of the context vector. The fitness value obtained is then assigned as the fitness of the corresponding particle of the sub-swarm.

This process has the advantage that the fitness function, $f$, is evaluated after each subpart of the context vector is updated, resulting in a much finer-grained search. One of the problems with optimizing the complete $n_x$-dimensional problem is that, even if an improvement in fitness is obtained, some of the components of the $n_x$-dimensional vector may move away from an optimum. The improved fitness could have been obtained by a sufficient move towards the optimum in the other vector components. The evaluation process of the CPSO-$S_K$ addresses this problem by tuning subparts of the solution vector separately.

---

**Algorithm 16.8** Cooperative Split PSO Algorithm

---

$K_1 = n_x \bmod K$;
$K_2 = K - (n_x \bmod K)$;
Initialize $K_1 \lceil n_x/K \rceil$-dimensional swarms;
Initialize $K_2 \lfloor n_x/K \rfloor$-dimensional swarms;
**repeat**
    **for** *each sub-swarm* $S_k, k = 1, \ldots, K$ **do**
        **for** *each particle* $i = 1, \ldots, S_k.n_s$ **do**
            **if** $f(\mathbf{b}(k, S_k.\mathbf{x}_i)) < f(\mathbf{b}(k, S_k.\mathbf{y}_i))$ **then**
                $S_k.\mathbf{y}_i = S_k.\mathbf{x}_i$;
            **end**
            **if** $f(\mathbf{b}(k, S_k.\mathbf{y}_i)) < f(\mathbf{b}(k, S_k.\hat{\mathbf{y}}))$ **then**
                $S_k.\hat{\mathbf{y}} = S_k.\mathbf{y}_i$;
            **end**
        **end**
        Apply velocity and position updates;
    **end**
**until** *stopping condition is true*;

---

The CPSO-$S_K$ algorithm is summarized in Algorithm 16.8. In this algorithm, $\mathbf{b}(k, \mathbf{z})$ returns an $n_x$-dimensional vector formed by concatenating the global best positions from all the sub-swarms, except for the $k$-th component which is replaced with $\mathbf{z}$, where $\mathbf{z}$ represents the position vector of any particle from sub-swarm $S_k$. The context vector is therefore defined as

$$\mathbf{b}(k, \mathbf{z}) = (S_1.\hat{\mathbf{y}}, \ldots, S_{k-1}.\hat{\mathbf{y}}, \mathbf{z}, S_{k+1}.\hat{\mathbf{y}}, \ldots, S_K.\hat{\mathbf{y}}) \qquad (16.82)$$

While the CPSO-$S_K$ algorithm has shown significantly better results than the basic PSO, it has to be noted that performance degrades when correlated components are split into different sub-swarms. If it is possible to identify which parameters correlate, then these parameters can be grouped into the same swarm — which will solve the problem. However, such prior knowledge is usually not available. The problem can also be addressed to a certain extent by allowing a particle to become the global best or personal best of its sub-swarm only if it improves the fitness of the context vector.

Algorithm 16.9 summarizes a hybrid search where the CPSO and GCPSO algorithms are interweaved. Additionally, a rudimentary form of cooperation is implemented between the CPSO-$S_K$ and GCPSO algorithms. Information about the best positions discovered by each algorithm is exchanged by copying the best discovered solution of the one algorithm to the swarm of the other algorithm. After completion of one iteration of the CPSO-$S_K$ algorithm, the context vector is used to replace a randomly selected particle from the GCPSO swarm (excluding the global best, $Q.\hat{\mathbf{y}}$). After completion of a GCPSO iteration, the new global best particle, $Q.\hat{\mathbf{y}}$, is split into the required subcomponents to replace a randomly selected individual of the corresponding CPSO-$S_K$ algorithm (excluding the global best positions).

## Predator-Prey PSO

The predator-prey relationship that can be found in nature is an example of a competitive environment. This behavior has been used in the PSO to balance exploration and exploitation [790]. By introducing a second swarm of predator particles, scattering of prey particles can be obtained by having prey particles being repelled by the presence of predator particles. Repelling facilitates better exploration of the search space, while the consequent regrouping promotes exploitation.

In their implementation of the predator–prey PSO, Silva *et al.* [790] use only one predator to pursue the global best prey particle. The velocity update for the predator particle is defined as

$$\mathbf{v}_p(t+1) = \mathbf{r}(\hat{\mathbf{y}}(t) - \mathbf{x}_p(t)) \tag{16.83}$$

where $\mathbf{v}_p$ and $\mathbf{x}_p$ are respectively the velocity and position vectors of the predator particle, $p$; $\mathbf{r} \sim U(0, V_{max,p})^{n_x}$. The speed at which the predator catches the best prey is controlled by $V_{max,p}$. The larger $V_{max,p}$, the larger the step sizes the predator makes towards the global best.

The prey particles update their velocity using

$$
\begin{aligned}
v_{ij}(t+1) &= wv_{ij}(t) + c_1 r_{1j}(t)(y_{ij}(t) - x_{ij}(t)) + c_2 r_{2j}(t)(\hat{y}_j(t) - x_{ij}(t)) \\
&\quad + c_3 r_{3j}(t) D(d)
\end{aligned} \tag{16.84}
$$

where $d$ is the Euclidean distance between prey particle $i$ and the predator, $r_{3j}(t) \sim U(0,1)$, and

$$D(d) = \alpha e^{-\beta d} \tag{16.85}$$

$D(d)$ quantifies the influence that the predator has on the prey. The influence grows exponentially with proximity, and is further controlled by the positive constants $\alpha$ and $\beta$.

Components of the position vector of a particle is updated using equation (16.84) based on a "fear" probability, $P_f$. For each dimension, if $U(0,1) < P_f$, then position $x_{ij}(t)$ is updated using equation (16.84); otherwise the standard velocity update is used.

---

**Algorithm 16.9** Hybrid of Cooperative Split PSO and GCPSO

---

$K_1 = n_x \bmod K$;
$K_2 = K - (n_x \bmod K)$;
Initialize $K_1 \lceil n_x/K \rceil$-dimensional swarms: $S_k, k = 1, \ldots, K$;
Initialize $K_2 \lfloor n_x/K \rfloor$-dimensional swarms: $S_k, k = K + 1, \ldots, K$;
Initialize an $n$-dimensional swarm, $Q$;
**repeat**
    **for** *each sub-swarm* $S_k, k = 1, \ldots, K$ **do**
        **for** *each particle* $i = 1, \ldots, S_k.n_s$ **do**
            **if** $f(\mathbf{b}(k, S_k.\mathbf{x}_i)) < f(\mathbf{b}(k, S_k.\mathbf{y}_i))$ **then**
                $S_k.\mathbf{y}_i = S_k.\mathbf{x}_i$;
            **end**
            **if** $f(\mathbf{b}(k, S_k.\mathbf{y}_i)) < f(\mathbf{b}(k, S_k.\hat{\mathbf{y}}))$ **then**
                $S_k.\hat{\mathbf{y}} = S_k.\mathbf{y}_i$;
            **end**
        **end**
        Apply velocity and position updates;
    **end**
    Select a random $l \sim U(1, n_s/2)|Q.\mathbf{y}_l \neq Q.\hat{\mathbf{y}}$;
    Replace $Q.\mathbf{x}_l$ with the context vector $\mathbf{b}$;
    **for** *each particle* $i = 1, \ldots, n_s$ **do**
        **if** $f(Q.\mathbf{x}_i) < f(Q.\mathbf{y}_i)$ **then**
            $Q.\mathbf{y}_i = Q.\mathbf{x}_i$;
        **end**
        **if** $f(Q.\mathbf{y}_i) < f(Q.\hat{\mathbf{y}})$ **then**
            $Q.\hat{\mathbf{y}} = Q.\mathbf{y}_i$;
        **end**
    **end**
    Apply GCPSO velocity and position updates;
    **for** *each swarm* $S_k, k = 1, \ldots, K$ **do**
        Select a random $m \sim U(1, S_k.n_s/2)|S_k.\mathbf{y}_m \neq S_k.\hat{\mathbf{y}}$;
        Replace $S_k.\mathbf{x}_m$ with the corresponding components of $Q.\hat{\mathbf{y}}$;
    **end**
**until** *stopping condition is true*;

---

## 16.5.5 Multi-Start PSO Algorithms

One of the major problems with the basic PSO is lack of diversity when particles start to converge to the same point. As an approach to prevent the premature stagnation of the basic PSO, several methods have been developed to continually inject randomness, or chaos, into the swarm. This section discusses these approaches, collectively referred to as multi-start methods.

Multi-start methods have as their main objective to increase diversity, whereby larger parts of the search space are explored. Injection of chaos into the swarm introduces a negative entropy. It is important to remember that continual injection of random positions will cause the swarm never to reach an equilibrium state. While not all the methods discussed below consider this fact, all of the methods can address the problem by reducing the amount of chaos over time.

Kennedy and Eberhart [449] were the first to mention the advantages of randomly reinitializing particles, a process referred to as *craziness*. Although Kennedy mentioned the potential advantages of a craziness operator, no evaluation of such operators was given. Since then, a number of researchers have proposed different approaches to implement a craziness operator for PSO.

When considering any method to add randomness to the swarm, a number of aspects need to be considered, including what should be randomized, when should randomization occur, how should it be done, and which members of the swarm will be affected? Additionally, thought should be given to what should be done with personal best positions of affected particles. These aspects are discussed next.

The diversity of the swarm can be increased by randomly initializing position vectors [534, 535, 863, 874, 875, 922, 923] and/or velocity vectors [765, 766, 922, 923, 924].

By initializing positions, particles are physically relocated to a different, random position in the search space. If velocity vectors are randomized and positions kept constant, particles retain their memory of their current and previous best solutions, but are forced to search in different random directions. If a better solution is not found due to random initialization of the velocity vector of a particle, the particle will again be attracted towards its personal best position.

If position vectors are initialized, thought should be given to what should be done with personal best positions and velocity vectors. Total reinitialization will have a particle's personal best also initialized to the new random position [534, 535, 923]. This effectively removes the particle's memory and prevents the particle from moving back towards its previously found best position (depending on when the global best position is updated). At the first iteration after reinitialization the "new" particle is attracted only towards the previous global best position of the swarm. Alternatively, reinitialized particles may retain their memory of previous best positions. It should be noted that the latter may have less diversity than removing particle memories, since particles are immediately moving back towards their previous personal best positions. It may, of course, happen that a new personal best position is found *en route*. When positions are reinitialized, velocities are usually initialized to zero, to have a zero momentum at the first iteration after reinitialization. Alternatively, velocities can be initialized to small random values [923]. Venter and Sobieszczanski-Sobieski [874, 875] initialize velocities to the cognitive component before reinitialization. This ensures a momentum back towards the personal best position.

The next important question to consider is when to reinitialize. If reinitialization happens too soon, the affected particles may not have had sufficient time to explore their current regions before being relocated. If the time to reinitialization is too long, it may happen that all particles have already converged. This is not really a problem, other than wasted computational time since no improvements are seen in this state. Several approaches have been identified to decide when to reinitialize:

- At fixed intervals, as is done in the mass extinction PSO developed by Xie *et al.* [923, 924]. As discussed above, fixed intervals may prematurely reinitialize a particle.

- Probabilistic approaches, where the decision to reinitialize is based on a probability. In the dissipative PSO, Xie *et al.* [922] reinitialize velocities and positions based on chaos factors that serve as probabilities of introducing chaos in the system. Let $c_v$ and $c_l$, with $c_v, c_l \in [0, 1]$, be respectively the chaos factors for velocity and location. Then, for each particle, $i$, and each dimension, $j$, if $r_{ij} \sim U(0, 1) < c_v$, then the velocity component is reinitialized to $v_{ij}(t + 1) = U(0, 1)V_{max,j}$. Also, if $r_{ij} \sim U(0, 1) < c_l$, then the position component is initialized to $x_{ij}(t + 1) \sim U(x_{min,j}, x_{max,j})$. A problem with this approach is that it will keep the swarm from reaching an equilibrium state. To ensure that an equilibrium can be reached, while still taking advantage of chaos injection, start with large chaos factors that reduce over time. The initial large chaos factors increase diversity in the first phases of the search, allowing particles to converge in the final stages. A similar probabilistic approach to reinitializing velocities is followed in [765, 766].

- Approaches based on some "convergence" condition, where certain events trigger reinitialization. Using convergence criteria, particles are allowed to first exploit their local regions before being reinitialized.

Venter and Sobieszczanski-Sobieski [874, 875] and Xie *et al.* [923] initiate reinitialization when particles do not improve over time. Venter and Sobieszczanski-Sobieski evaluate the variation in particle fitness of the current swarm. If the variation is small, then particles are centered in close proximity to the global best position. Particles that are two standard deviations away from the swarm center are reinitialized. Xie *et al.* count for each $\mathbf{x}_i \neq \hat{\mathbf{y}}$ the number of times that $f(\mathbf{x}_i) - f(\hat{\mathbf{y}}) < \epsilon$. When this count exceeds a given threshold, the corresponding particle is reinitialized. Care should be taken in setting values for $\epsilon$ and the count threshold. If $\epsilon$ is too large, particles will be reinitialized before having any chance of exploiting their current regions.

Clerc [133] defines a hope and re-hope criterion. If there is still hope that the objective can be reached, particles are allowed to continue in their current search directions. If not, particles are reinitialized around the global best position, taking into consideration the local shape of the objective function.

Van den Bergh [863] defined a number of convergence tests for a multi-start PSO, namely the normalized swarm radius condition, the particle cluster condition and the objective function slope condition (also refer to Section 16.1.6 for a discussion on these criteria).

Løvberg and Krink [534, 535] use self-organized criticality (SOC) to determine when to reinitialize particles. Each particle maintains an additional variable, $C_i$, referred to as the *criticality* of the particle. If two particles are closer than a threshold distance, $\epsilon$, from one another, then both have their criticality increased by one. The larger the criticality of all particles, the more uniform the swarm becomes. To prevent criticality from building up, each $C_i$ is reduced by a fraction in each iteration. As soon as $C_i > C$, where $C$ is the global criticality limit, particle $i$ is reinitialized. Its criticality is distributed to its immediate neighbors and $C_i = 0$. Løvberg and Krink also set the inertia weight value of each particle to $w_i = 0.2 + 0.1C_i$. This forces the particle to explore more when it is too similar to other particles.

The next issue to consider is which particles to reinitialize. Obviously, it will not be a good idea to reinitialize the global best particle! From the discussions above, a number of selection methods have already been identified. Probabilistic methods decide which particles to reinitialize based on a user-defined probability. The convergence methods use specific convergence criteria to identify particles for reinitialization. For example, SOC PSO uses criticality measures (refer to Algorithm 16.11), while others keep track of the improvement in particle fitness. Van den Bergh [863] proposed a random selection scheme, where a particle is reinitialized at each $t_r$ iteration (also refer to Algorithm 16.10). This approach allows each particle to explore its current region before being reinitialized. To ensure that the swarm will reach an equilibrium state, start with a large $t_r < n_s$, which decreases over time.

---

**Algorithm 16.10** Selection of Particles to Reinitialize; $\tau$ indicates the index of the global best particle

---

Create and initialize an $n_x$-dimensional PSO: $S$;
$s_{idx} = 0$;
**repeat**
    **if** $s_{idx} \neq \tau$ **then**
        $S.\mathbf{x}_{idx} \sim U(\mathbf{x}_{min}, \mathbf{x}_{max})$;
    **end**
    $s_{idx} = (s_{idx} + 1) \bmod t_r$;
    **for** *each particle* $i = 1, \ldots, n_s$ **do**
        **if** $f(S.\mathbf{x}_i) < f(S.\mathbf{y}_i)$ **then**
            $S.\mathbf{y}_i = S.\mathbf{x}$;
        **end**
        **if** $f(S.\mathbf{y}_i) < f(S.\hat{\mathbf{y}})$ **then**
            $S.\hat{\mathbf{y}} = S.\mathbf{y}_i$;
        **end**
    **end**
    Update velocities;
    Update position;
**until** *stopping condition is true*;

---

Finally, how are particles reinitialized? As mentioned earlier, velocities and/or positions can be reinitialized. Most approaches that reinitialize velocities set each velocity component to a random value constrained by the maximum allowed velocity. Venter and Sobieszczanski-Sobieski [874, 875] set the velocity vector to the cognitive component after reinitialization of position vectors.

Position vectors are usually initialized to a new position subject to boundary constraints; that is, $x_{ij}(t+1) \sim U(x_{min,j}, x_{max,j})$. Clerc [133] reinitializes particles on the basis of estimates of the local shape of the objective function. Clerc [135] also proposes alternatives, where a particle returns to its previous best position, and from

---

**Algorithm 16.11** Self-Organized Criticality PSO

---

Create and initialize an $n_x$-dimensional PSO: $S$;
Set $C_i = 0, \forall i = 1, \ldots, n_s$;
**repeat**
    Evaluate fitness of all particles;
    Update velocities;
    Update positions;
    Calculate criticality for all particles;
    Reduce criticality for each particle;
    **while** $\exists i = 1, \ldots, n_s$ *such that* $C_i > C$ **do**
        Disperse criticality of particle $i$;
        Reinitialize $\mathbf{x}_i$;
    **end**
**until** *stopping condition is true*;

---

there moves randomly for a fixed number of iterations.

A different approach to multi-start PSOs is followed in [863], as summarized in Algorithm 16.12. Particles are randomly initialized, and a PSO algorithm is executed until the swarm converges. When convergence is detected, the best position is recorded and all particles randomly initialized. The process is repeated until a stopping condition is satisfied, at which point the best recorded solution is returned. The best recorded solution can be refined using local search before returning the solution. The convergence criteria listed in Section 16.1.6 are used to detect convergence of the swarm.

### 16.5.6 Repelling Methods

Repelling methods have been used to improve the exploration abilities of PSO. Two of these approaches are described in this section.

### Charged PSO

Blackwell and Bentley developed the charged PSO based on an analogy of electrostatic energy with charged particles [73, 74, 75]. The idea is to introduce two opposing forces within the dynamics of the PSO: an attraction to the center of mass of the swarm and inter-particle repulsion. The attraction force facilitates convergence to a single solution, while the repulsion force preserves diversity.

The charged PSO changes the velocity equation by adding a particle acceleration, $\mathbf{a}_i$, to the standard equation. That is,

$$v_{ij}(t+1) = wv_{ij}(t) + c_1 r_1(t)[y_{ij}(t) - x_{ij}(t)] + c_2 r_2(t)[\hat{y}_j(t) - x_{ij}(t)] + a_{ij}(t) \quad (16.86)$$

---

**Algorithm 16.12** Multi-start Particle Swarm Optimization; $\hat{\mathbf{y}}$ is the best solution over all the restarts of the algorithm

---

Create and initialize an $n_x$-dimensional swarm, $S$;
**repeat**
    **if** $f(S.\hat{\mathbf{y}}) < f(\hat{\mathbf{y}})$ **then**
        $\hat{\mathbf{y}} = S.\hat{\mathbf{y}}$;
    **end**
    **if** *the swarm $S$ has converged* **then**
        Reinitialize all particles;
    **end**
    **for** *each particle $i = 1, \cdots, S.n_s$* **do**
        **if** $f(S.\mathbf{x}_i) < f(S.\mathbf{y}_i)$ **then**
            $S.\mathbf{y}_i = S.\mathbf{x}$;
        **end**
        **if** $f(S.\mathbf{y}_i) < f(S.\hat{\mathbf{y}})$ **then**
            $S.\hat{\mathbf{y}} = S.\mathbf{y}_i$;
        **end**
    **end**
    Update velocities;
    Update position;
**until** *stopping condition is true*;
Refine $\hat{\mathbf{y}}$ using local search;
Return $\hat{\mathbf{y}}$ as the solution;

---

The acceleration determines the magnitude of inter-particle repulsion, defined as [75]

$$\mathbf{a}_i(t) = \sum_{l=1, i\neq l}^{n_s} \mathbf{a}_{il}(t) \tag{16.87}$$

with the repulsion force between particles $i$ and $l$ defined as

$$\mathbf{a}_{il}(t) = \begin{cases} \left(\frac{Q_i Q_l}{||\mathbf{x}_i(t) - \mathbf{x}_l(t)||^3}\right)(\mathbf{x}_i(t) - \mathbf{x}_l(t)) & \text{if } R_c \leq ||\mathbf{x}_i(t) - \mathbf{x}_l(t)|| \leq R_p \\ \left(\frac{Q_i Q_l(\mathbf{x}_i(t) - \mathbf{x}_l(t))}{R_c^2 ||\mathbf{x}_i(t) - \mathbf{x}_l(t)||}\right) & \text{if } ||\mathbf{x}_i(t) - \mathbf{x}_l(t)|| < R_c \\ 0 & \text{if } ||\mathbf{x}_i(t) - \mathbf{x}_l(t)|| > R_p \end{cases} \tag{16.88}$$

where $Q_i$ is the charged magnitude of particle $i$, $R_c$ is referred to as the core radius, and $R_p$ is the perception limit of each particle.

Neutral particles have a zero charged magnitude, i.e. $Q_i = 0$. Only when $Q_i \neq 0$ are particles charged, and do particles repel from each other. Therefore, the standard PSO is a special case of the charged PSO with $Q_i = 0$ for all particles. Particle avoidance (inter-particle repulsion) occurs only when the separation between two particles is within the range $[R_c, R_p]$. In this case the smaller the separation, the larger the repulsion between the corresponding particles. If the separation between two particles becomes very small, the acceleration will explode to large values. The consequence will be that the particles never converge due to extremely large repulsion forces. To prevent this, acceleration is fixed at the core radius for particle separations less than $R_c$. Particles that are far from one another, i.e. further than the particle perception limit, $R_p$, do not repel one another. In this case $\mathbf{a}_{il}(t) = 0$, which allows particles to move towards the global best position. The value of $R_p$ will have to be optimized for each application.

The acceleration, $\mathbf{a}_i(t)$, is determined for each particle before the velocity update.

Blackwell and Bentley suggested as electrostatic parameters, $R_c = 1, R_p = \sqrt{3}x_{max}$ and $Q_i = 16$ [75].

Electrostatic repulsion maintains diversity, enabling the swarm to automatically detect and respond to changes in the environment. Empirical evaluations of the charged PSO in [72, 73, 75] have shown it to be very efficient in dynamic environments. Three types of swarms were defined and studied:

- **Neutral swarm**, where $Q_i = 0$ for all particles $i = 1, \ldots, n_s$.

- **Charged swarm**, where $Q_i > 0$ for all particles $i = 1, \ldots, n_s$. All particles therefore experience repulsive forces from the other particles (when the separation is less than $r_p$).

- **Atomic swarm**, where half of the swarm is charged ($Q_i > 0$) and the other half is neutral ($Q_i = 0$).

It was found that atomic swarms perform better than charged and neutral swarms [75]. As a possible explanation of why atomic swarms perform better than charged swarms, consider as worst case what will happen when the separation between particles never gets below $R_c$. If the separation between particles is always greater than or equal to $R_c$, particles repel one another, which never allows particles to converge to a single position. Inclusion of neutral particles ensures that these particles converge to an optimum, while the charged particles roam around to automatically detect and adjust to environment changes.

## Particles with Spatial Extention

Particles with spatial extension were developed to prevent the swarm from prematurely converging [489, 877]. If one particle locates an optimum, then all particles will be attracted to the optimum – causing all particles to cluster closely. The spatial extension of particles allows some particles to explore other areas of the search space, while others converge to the optimum to further refine it. The exploring particles may locate a different, more optimal solution.

The objective of spatial extension is to dynamically increase diversity when particles start to cluster. This is achieved by adding a radius to each particle. If two particles collide, i.e. their radii intersect, then the two particles bounce off. Krink *et al.* [489] and Vesterstrøm and Riget [877] investigated three strategies for spatial extension:

- **random bouncing**, where colliding particles move in a random new direction at the same speed as before the collision;

- **realistic physical bouncing**; and
- simple **velocity-line bouncing**, where particles continue to move in the same direction but at a scaled speed. With scale factor in $[0, 1]$ particles slow down, while a scale factor greater than one causes acceleration to avoid a collision. A negative scale factor causes particles to move in the opposite direction to their previous movement.

Krink *et al.* showed that random bouncing is not as efficient as the consistent bouncing methods.

To ensure convergence of the swarm, particles should bounce off on the basis of a probability. An initial large bouncing probability will ensure that most collisions result in further exploration, while a small bouncing probability for the final steps will allow the swarm to converge. At all times, some particles will be allowed to cluster together to refine the current best solution.

## 16.5.7   Binary PSO

PSO was originally developed for continuous-valued search spaces. Kennedy and Eberhart developed the first discrete PSO to operate on binary search spaces [450, 451]. Since real-valued domains can easily be transformed into binary-valued domains (using standard binary coding or Gray coding), this binary PSO can also be applied to real-valued optimization problems after such transformation (see [450, 451] for applications of the binary PSO to real-valued problems).

For the binary PSO, particles represent positions in binary space. Each element of a particle's position vector can take on the binary value 0 or 1. Formally, $\mathbf{x}_i \in \mathbb{B}^{n_x}$, or $x_{ij} \in \{0, 1\}$. Changes in a particle's position then basically implies a mutation of bits, by flipping a bit from one value to the other. A particle may then be seen to move to near and far corners of a hypercube by flipping bits.

One of the first problems to address in the development of the binary PSO, is how to interpret the velocity of a binary vector. Simply seen, velocity may be described by the number of bits that change per iteration, which is the Hamming distance between $\mathbf{x}_i(t)$ and $\mathbf{x}_i(t + 1)$, denoted by $\mathcal{H}(\mathbf{x}_i(t), \mathbf{x}_i(t + 1))$. If $\mathcal{H}(\mathbf{x}_i(t), \mathbf{x}_i(t + 1)) = 0$, zero bits are flipped and the particle does not move; $||\mathbf{v}_i(t)|| = 0$. On the other hand, $||\mathbf{v}_i(t)|| = n_x$ is the maximum velocity, meaning that all bits are flipped. That is, $\mathbf{x}_i(t + 1)$ is the complement of $\mathbf{x}_i(t)$. Now that a simple interpretation of the velocity of a bit-vector is possible, how is the velocity of a single bit (single dimension of the particle) interpreted?

In the binary PSO, velocities and particle trajectories are rather defined in terms of probabilities that a bit will be in one state or the other. Based on this probabilistic view, a velocity $v_{ij}(t) = 0.3$ implies a 30% chance to be bit 1, and a 70% chance to be bit 0. This means that velocities are restricted to be in the range $[0, 1]$ to be interpreted as a probability. Different methods can be employed to normalize velocities such that $v_{ij} \in [0, 1]$. One approach is to simply divide each $v_{ij}$ by the maximum velocity, $V_{max,j}$. While this approach will ensure velocities are in the range $[0,1]$, consider what will happen when the maximum velocities are large, with $v_{ij}(t) \ll V_{max,j}$ for all time steps, $t = 1, \ldots, n_t$. This will limit the maximum range of velocities, and thus the chances of a position to change to bit 1. For example, if $V_{max,j} = 10$, and $v_{ij}(t) = 5$, then the normalized velocity is $v'_{ij}(t) = 0.5$, with only a 50% chance that $x_{ij}(t+1) = 1$. This normalization approach may therefore cause premature convergence to bad solutions due to limited exploration abilities.

A more natural normalization of velocities is obtained by using the sigmoid function. That is,

$$v'_{ij}(t) = \text{sig}(v_{ij}(t)) = \frac{1}{1 + e^{-v_{ij}(t)}} \tag{16.89}$$

Using equation (16.89), the position update changes to

$$x_{ij}(t+1) = \begin{cases} 1 & \text{if } r_{3j}(t) < \text{sig}(v_{ij}(t+1)) \\ 0 & \text{otherwise} \end{cases} \tag{16.90}$$

with $r_{3j}(t) \sim U(0,1)$. The velocity, $v_{ij}(t)$, is now a probability for $x_{ij}(t)$ to be 0 or 1. For example, if $v_{ij}(t) = 0$, then $\text{prob}(x_{ij}(t+1) = 1) = 0.5$ (or 50%). If $v_{ij}(t) < 0$, then $\text{prob}(x_{ij}(t+1) = 1) < 0.5$, and if $v_{ij}(t) > 0$, then $\text{prob}(x_{ij}(t+1) = 1) > 0.5$. Also note that $\text{prob}(x_{ij}(t) = 0) = 1 - \text{prob}(x_{ij}(t) = 1)$. Note that $x_{ij}$ can change even if the value of $v_{ij}$ does not change, due to the random number $r_{3j}$ in the equation above.

It is only the calculation of position vectors that changes from the real-valued PSO. The velocity vectors are still real-valued, with the same velocity calculation as given in equation (16.2), but including the inertia weight. That is, $\mathbf{x}_i, \mathbf{y}_i, \hat{\mathbf{y}} \in \mathbb{B}^{n_x}$ while $\mathbf{v}_i \in \mathbb{R}^{n_x}$.

The binary PSO is summarized in Algorithm 16.13.

---

**Algorithm 16.13** *binary* PSO

---

Create and initialize an $n_x$-dimensional swarm;
**repeat**
    **for** *each particle* $i = 1, \ldots, n_s$ **do**
        **if** $f(\mathbf{x}_i) < f(\mathbf{y}_i)$ **then**
            $\mathbf{y}_i = \mathbf{x}_i$;
        **end**
        **if** $f(\mathbf{y}_i) < f(\hat{\mathbf{y}})$ **then**
            $\hat{\mathbf{y}} = \mathbf{y}_i$;
        **end**
    **end**
    **for** *each particle* $i = 1, \ldots, n_s$ **do**
        update the velocity using equation (16.2);
        update the position using equation (16.90);
    **end**
**until** *stopping condition is true*;

---

# 16.6   Advanced Topics

This section describes a few PSO variations for solving constrained problems, multi-objective optimization problems, problems with dynamically changing objective functions and to locate multiple solutions.

## 16.6.1   Constraint Handling Approaches

A very simple approach to cope with constraints is to reject infeasible particles, as follows:

- Do not allow infeasible particles to be selected as personal best or neighborhood global best solutions. In doing so, infeasible particles will never influence other particles in the swarm. Infeasible particles are, however, pulled back to feasible space due to the fact that personal best and neighborhood best positions are in feasible space. This approach can only be efficient if the ratio of number of infeasible particles to feasible particles is small. If this ratio is too large, the swarm may not have enough diversity to effectively cover the (feasible) space.

- Reject infeasible particles by replacing them with new randomly generated positions from the feasible space. Reinitialization within feasible space gives these particles an opportunity to become best solutions. However, it is also possible that these particles (or any other particle for that matter), may roam outside the boundaries of feasible space. This approach may be beneficial in cases where the feasible space is made up of a number of disjointed regions. The strategy will allow particles to cross boundaries of one feasible region to explore another feasible region.

Most applications of PSO to constrained problems make use of penalty methods to penalize those particles that violate constraints [663, 838, 951].

Similar to the approach discussed in Section 12.6.1, Shi and Krohling [784] and Laskari *et al.* [504] converted the constrained problem to an unconstrained Lagrangian.

Repair methods, which allow particles to roam into infeasible space, have also been developed. These are simple methods that apply repairing operators to change infeasible particles to represent feasible solutions. Hu and Eberhart *et al.* [388] developed an approach where particles are not allowed to be attracted by infeasible particles.

The personal best of a particle changes only if the fitness of the current position is better and if the current position violates no constraints. This ensures that the personal best positions are always feasible. Assuming that neighborhood best positions are selected from personal best positions, it is also guaranteed that neighborhood best positions are feasible. In the case where the neighborhood best positions are selected from the current particle positions, neighborhood best positions should be updated only if no constraint is violated. Starting with a feasible set of particles, if a particle moves out of feasible space, that particle will be pulled back towards feasible space. Allowing particles to roam into infeasible space, and repairing them over time by moving back to feasible best positions facilitates better exploration. If the feasible space consists of disjointed feasible regions, the chances are increased for particles to explore different feasible regions.

El-Gallad *et al.* [234] replaced infeasible particles with their feasible personal best positions. This approach assumes feasible initial particles and that personal best positions are replaced only with feasible solutions. The approach is very similar to that of Hu and Eberhart [388], but with less diversity. Replacement of an infeasible particle with its feasible personal best, forces an immediate repair. The particle is immediately brought back into feasible space. The approach of Hu and Eberhart allow an infeasible particle to explore more by pulling it back into feasible space over time. But, keep in mind that during this exploration, the infeasible particle will have no influence on the rest of the swarm while it moves within infeasible space.

Venter and Sobieszczanski-Sobieski [874, 875] proposed repair of infeasible solutions by setting

$$\mathbf{v}_i(t) = \mathbf{0} \tag{16.91}$$
$$\mathbf{v}_i(t+1) = c_1\mathbf{r}_1(t)(\mathbf{y}_i(t) - \mathbf{x}_i(t)) + c_2\mathbf{r}_2(t)(\hat{\mathbf{y}}(t) - \mathbf{x}_i(t)) \tag{16.92}$$

for all infeasible particles, $i$. In other words, the memory of previous velocity (direction of movement) is deleted for infeasible particles, and the new velocity depends only on the cognitive and social components. Removal of the momentum has the effect that infeasible particles are pulled back towards feasible space (assuming that the personal best positions are only updated if no constraints are violated).

## 16.6.2 Multi-Objective Optimization

A great number of PSO variations can be found for solving MOPs. This section describes only a few of these but provides references to other approaches.

The *dynamic neighborhood* MOPSO, developed by Hu and Eberhart [387], dynamically determines a new neighborhood for each particle in each iteration, based on distance in objective space. Neighborhoods are determined on the basis of the simplest objective. Let $f_1(\mathbf{x})$ be the simplest objective function, and let $f_2(\mathbf{x})$ be the second objective. The neighbors of a particle are determined as those particles closest to the particle with respect to the fitness values for objective $f_1(\mathbf{x})$. The neighborhood best particle is selected as the particle in the neighborhood with the best fitness according to the second objective, $f_1(\mathbf{x})$. Personal best positions are replaced only if a particle's new position dominates its current personal best solution.

The dynamic neighborhood MOPSO has a few disadvantages:

- It is not easily scalable to more than two objectives, and its usability is therefore restricted to MOPs with two objectives.

- It assumes prior knowledge about the objectives to decide which is the most simple for determination of neighborhoods.

- It is sensitive to the ordering of objectives, since optimization is biased towards improving the second objective.

Parsopoulos and Vrahatis [659, 660, 664, 665] developed the *vector evaluated* PSO (VEPSO), on the basis of the vector-evaluated genetic algorithm (VEGA) developed by Schaffer [761] (also refer to Section 9.6.3). VEPSO uses two sub-swarms, where each sub-swarm optimizes a single objective. This algorithm is therefore applicable to MOPs with only two objectives. VEPSO follows a kind of coevolutionary approach. The global best particle of the first swarm is used in the velocity equation of the second swarm, while the second swarm's global best particle is used in the velocity update of the first swarm. That is,

$$
\begin{aligned}
S_1.v_{ij}(t+1) &= wS_1.v_{ij}(t) + c_1 r_{1j}(t)(S_1.y_{ij}(t) - S_1.x_{ij}(t)) \\
&+ c_2 r_{2j}(t)(S_2.\hat{y}_i(t) - S_1.x_{ij}(t)) \\
S_2.v_{ij}(t+1) &= wS_2.v_{ij}(t) + c_1 r_{1j}(t)(S_2.y_{ij}(t) - S_2.x_{ij}(t)) \\
&+ c_2 r_{ij}(t)(S_1.\hat{y}_j(t) - S.x_{2j}(t))
\end{aligned}
$$

(16.93)

(16.94)

where sub-swarm $S_1$ evaluates individuals on the basis of objective $f_1(\mathbf{x})$, and sub-swarm $S_2$ uses objective $f_2(\mathbf{x})$.

The MOPSO algorithm developed by Coello Coello and Lechuga is one of the first PSO-based MOO algorithms that extensively uses an archive [147, 148]. This algorithm is based on the Pareto archive ES (refer to Section 12.6.2), where the objective function space is separated into a number of hypercubes.

A truncated archive is used to store non-dominated solutions. During each iteration, if the archive is not yet full, a new particle position is added to the archive if the particle represents a non-dominated solution. However, because of the size limit of the archive, priority is given to new non-dominated solutions located in less populated areas, thereby ensuring that diversity is maintained. In the case that members of the archive have to be deleted, those members in densely populated areas have the highest probability of deletion. Deletion of particles is done during the process of separating the objective function space into hypercubes. Densely populated hypercubes are truncated if the archive exceeds its size limit. After each iteration, the number of members of the archive can be reduced further by eliminating from the archive all those solutions that are now dominated by another archive member.

For each particle, a global guide is selected to guide the particle toward less dense areas of the Pareto front. To select a guide, a hypercube is first selected. Each hypercube is assigned a selective fitness value,

$$
f_{sel}(H_h) = \frac{\alpha}{f_{del}(H_h)}
$$

(16.95)

where $f_{del}(H_h) = H_h.n_s$ is the deletion fitness value of hypercube $H_h$; $\alpha = 10$ and $H_h.n_s$ represents the number of nondominated solutions in hypercube $H_h$. More densely populated hypercubes will have a lower score. Roulette wheel selection is then used to select a hypercube, $H_h$, based on the selection fitness values. The global guide for particle $i$ is selected randomly from among the members of hypercube $H_h$.

Hence, particles will have different global guides. This ensures that particles are attracted to different solutions. The local guide of each particle is simply the personal best position of the particle. Personal best positions are only updated if the new position, $\mathbf{x}_i(t+1) \prec \mathbf{y}_i(t)$. The global guide replaces the global best, and the local guide replaces the personal best in the velocity update equation.

In addition to the normal position update, a mutation operator (also referred to as a craziness operator in the context of PSO) is applied to the particle positions. The degree of mutation decreases over time, and the probability of mutation also decreases over time. That is,

$$x_{ij}(t+1) = N(0, \sigma(t))x_{ij}(t) + v_{ij}(t+1) \tag{16.96}$$

where, for example

$$\sigma(t) = \sigma(0)\,\mathrm{e}^{-t} \tag{16.97}$$

with $\sigma(0)$ an initial large variance.

The MOPSO developed by Coello Coello and Lechuga is summarized in Algorithm 16.14.

---

**Algorithm 16.14** Coello Coello and Lechuga MOPSO

---

Create and initialize an $n_x$-dimensional swarm $S$;
Let $A = \emptyset$ and $A.n_s = 0$;
Evaluate all particles in the swarm;
**for** *all non-dominated* $\mathbf{x}_i$ **do**
    $A = A \cup \{\mathbf{x}_i\}$;
**end**
Generate hypercubes;
Let $\mathbf{y}_i = \mathbf{x}_i$ for all particles;
**repeat**
    Select global guide, $\hat{\mathbf{y}}$;
    Select local guide, $\mathbf{y}_i$;
    Update velocities using equation (16.2);
    Update positions using equation (16.96);
    Check boundary constraints;
    Evaluate all particles in the swarm;
    Update the repository, $A$;
**until** *stopping condition is true*;

---

Li [517] applied a nondominated sorting approach similar to that described in Section 12.6.2 to PSO. Other MOO algorithms can be found in [259, 389, 609, 610, 940, 956].

## 16.6.3   Dynamic Environments

Early results of the application of the PSO to type I environments (refer to Section A.9) with small spatial severity showed that the PSO has an implicit ability to track changing optima [107, 228, 385]. Each particle progressively converges on a point on the line that connects its personal best position with the global best position [863, 870]. The trajectory of a particle can be described by a sinusoidal wave with diminishing amplitude around the global best position [651, 652]. If there is a small change in the location of an optimum, it is likely that one of these oscillating particles will discover the new, nearby optimum, and will pull the other particles to swarm around the new optimum.

However, if the spatial severity is large, causing the optimum to be displaced outside the radius of the contracting swarm, the PSO will fail to locate the new optimum due to loss of diversity. In such cases mechanisms need to be employed to increase the swarm diversity.

Consider spatial changes where the value of the optimum remains the same after the change, i.e. $f(\mathbf{x}^*(t)) = f(\mathbf{x}^*(t + 1))$, with $\mathbf{x}^*(t) \neq \mathbf{x}^*(t + 1)$. Since the fitness remains the same, the global best position does not change, and remains at the old optimum. Similarly, if $f(\mathbf{x}^*(t)) > f(\mathbf{x}^*(t+1))$, assuming minimization, the global best position will also not change. Consequently, the PSO will fail to track such a changing minimum. This problem can be solved by re-evaluating the fitness of particles at time $t + 1$ and updating global best and personal best positions. However, keep in mind that the same problem as discussed above may still occur if the optimum is displaced outside the radius of the swarm.

One of the goals of optimization algorithms for dynamic environments is to locate the optimum and then to track it. The self-adaptation ability of the PSO to track optima (as discussed above) assumes that the PSO did not converge to an equilibrium state in its first goal to locate the optimum. When the swarm reaches an equilibrium (i.e. converged to a solution), $\mathbf{v}_i = \mathbf{0}$. The particles have no momentum and the contributions from the cognitive and social components are zero. The particles will remain in this stable state even if the optimum does change. In the case of dynamic environments, it is possible that the swarm reaches an equilibrium if the temporal severity is low (in other words, the time between consecutive changes is large). It is therefore important to read the literature on PSO for dynamic environments with this aspect always kept in mind.

The next aspects to consider are the influence of particle memory, velocity clamping and the inertia weight. The question to answer is: to what extent do these parameters and characteristics of the PSO limit or promote tracking of changing optima? These aspects are addressed next:

- Each particle has a memory of its best position found thus far, and velocity is adjusted to also move towards this position. Similarly, each particle retains information about the global best (or local best) position. When the environment changes this information becomes stale. If, after an environment change, particles are still allowed to make use of this, now stale, information, they are drawn to the old optimum – an optimum that may no longer exist in the changed environment. It can thus be seen that particle memory (from which the global best is selected) is detrimental to the ability of PSO to track changing optima. A solution to this problem is to reinitialize or to re-evaluate the personal best positions (particle memory).

- The inertia weight, together with the acceleration coefficients balance the exploration and exploitation abilities of the swarm. The smaller $w$, the less the swarm explores. Usually, $w$ is initialized with a large value that decreases towards a small value (refer to Section 16.3.2). At the time that a change occurs, the value of $w$ may have reduced too much, thereby limiting the swarm's exploration ability and its chances of locating the new optimum. To alleviate this problem, the value of $w$ should be restored to its initial large value when a change in the environment is detected. This also needs to be done for the acceleration coefficients if adaptive accelerations are used.

- Velocity clamping also has a large influence on the exploration ability of PSO (refer to 16.3.1). Large velocity clamping (i.e. small $V_{max,j}$ values) limits the step sizes, and more rapidly results in smaller momentum contributions than lesser clamping. For large velocity clamping, the swarm will have great difficulty in escaping the old optimum in which it may find itself trapped. To facilitate tracking of changing objectives, the values of $V_{max,j}$ therefore need to be chosen carefully.

When a change in environment is detected, the optimization algorithm needs to react appropriately to adapt to these changes. To allow timeous and efficient tracking of optima, it is important to correctly and timeously detect if the environment did change. The task is easy if it is known that changes occur periodically (with the change frequency known), or at predetermined intervals. It is, however, rarely the case that prior information about when changes occur is available. An automated approach is needed to detect changes based on information received from the environment.

Carlisle and Dozier [106, 109] proposed the use of a sentry particle, or more than one sentry particle. The task of the sentry is to detect any change in its local environment. If only one sentry is used, the same particle can be used for all iterations. However, feedback from the environment will then only be from one small part of the environment. Around the sentry the environment may be static, but it may have changed elsewhere. A better strategy is to select the particle randomly from the swarm for each iteration. Feedback is then received from different areas of the environment.

A sentry particle stores a copy of its most recent fitness value. At the start of the next iteration, the fitness of the sentry particle is re-evaluated and compared with its previously stored value. If there is a difference in the fitness, a change has occurred.

More sentries allow simultaneous feedback from more parts of the environment. If multiple sentries are used detection of changes is faster and more reliable. However, the fitness of a sentry particle is evaluated twice per iteration. If fitness evaluation is costly, multiple sentries will significantly increase the computational complexity of the search algorithm.

As an alternative to using randomly selected sentry particles, Hu and Eberhart [385] proposed to monitor changes in the fitness of the global best position. By monitoring the fitness of the global best particle, change detection is based on globally provided information. To increase the accuracy of change detection, Hu and Eberhart [386] later also monitored the global second-best position. Detection of the second-best and global best positions limits the occurrence of false alarms. Monitoring of these global best positions is based on the assumption that if the optimum location changes, then the optimum value of the current location also changes.

One of the first studies in the application of PSO to dynamic environments came from Carlisle and Dozier [107], where the efficiency of different velocity models (refer to Section 16.3.5) has been evaluated. Carlisle and Dozier observed that the social-only model is faster in tracking changing objectives than the full model. However, the reliability of the social-only model deteriorates faster than the full model for larger update frequencies. The selfless and cognition-only models do not perform well on changing environments. Keep in mind that these observations were without changing the original velocity models, and should be viewed under the assumption that the swarm had not yet reached an equilibrium state. Since this study, a number of other studies have been done to investigate how the PSO should be changed to track dynamic optima. These studies are summarized in this section.

From these studies in dynamic environments, it became clear that diversity loss is the major reason for the failure of PSO to achieve more efficient tracking.

Eberhart and Shi [228] proposed using the standard PSO, but with a dynamic, randomly selected inertia coefficient. For this purpose, equation (16.24) is used to select a new $w(t)$ for each time step. In their implementation, $c_1 = c_2 = 1.494$, and velocity clamping was not done. As motivation for this change, recall from Section 16.3.1 that velocity clamping restricts the exploration abilities of the swarm. Therefore, removal of velocity clamping facilitates larger exploration, which is highly beneficial. Furthermore, from Section 16.3.2, the inertia coefficient controls the exploration–exploitation trade-off. Since it cannot be predicted in dynamically changing environments if exploration or exploitation is preferred, the randomly changing $w(t)$ ensures a good mix of focusing on both exploration and exploitation.

While this PSO implementation presented promising results, the efficiency is limited to type I environments with a low severity and type II environments where the value of the optimum is better after the change in the environment. This restriction is mainly due to the memory of particles (i.e. the personal best positions and the global best selected from the personal best positions), and that changing the inertia will not be able to kick the swarm out of the current optimum when $\mathbf{v}_i(t) \approx 0, \forall i = 1, \ldots, n_s$.

A very simple approach to increase diversity is to reinitialize the swarm, which means that all particle positions are set to new random positions, and the personal best and neighborhood best positions are recalculated. Eberhart and Shi [228] suggested the following approaches:

- Do not reinitialize the swarm, and just continue to search from the current position. This approach only works for small changes, and when the swarm has not yet reached an equilibrium state.

- Reinitialize the entire swarm. While this approach does neglect the swarm's memory and breaks the swarm out of a current optimum, it has the disadvantage that the swarm needs to search all over again. If changes are not severe, the consequence is an unnecessary increase in computational complexity due to an increase in the number of iterations to converge to the new optimum. The danger is also that the swarm may converge on a worse local optimum. Reinitialization of the entire swarm is more effective under severe environment changes.

- Reinitialize only parts of the swarm, but make sure to retain the global best position. The reinitialization of a percentage of the particle positions injects more diversity, and also preserves memory of the search space by keeping potentially good particles (existing particles may be close to the changed optimum). Hu and Eberhart experimented with a number of reinitialization percentages [386], and observed that lower reinitialization is preferred for smaller changes. Larger changes require more particles to be reinitialized for efficient tracking. It was also found empirically that total reinitialization is not efficient.

Particles' memory of previous best positions is a major cause of the inefficiency of the PSO in tracking changing optima. This statement is confirmed by the results of Carlisle and Dozier [107], where it was found that the cognition-only model showed poor performance. The cognitive component promotes the nostalgic tendency to return to previously found best positions. However, after a change in the environment, these positions become stale since they do not necessarily reflect the search space for the changed environment. The consequence is that particles are attracted to outdated positions.

Carlisle and Dozier [107], and Hu and Eberhart [385] proposed that the personal best positions of all particles be reset to the current position of the particle when a change is detected. This action effectively clears the memory of all particles. Resetting of the personal best positions is only effective when the swarm has not yet converged to a solution [386]. If this is the case, $w(t)\mathbf{v}(t) \approx 0$ and the contribution of the cognitive component will be zero for all particles, since $(\mathbf{y}_i(t) - x_i(t)) = 0, \forall i = 1, \ldots, n_s$. Furthermore, the contribution of the social component will also be approximately zero, since, at convergence, all particles orbit around the same global best position (with an approximately zero swarm radius). Consequently, velocity updates are very small, as is the case for position updates.

To address the above problem, resetting of the personal best positions can be combined with partial reinitialization of the swarm [386]. In this case, reinitialization increases diversity while resetting of the personal best positions prevents return to out-of-date positions.

Looking more carefully at the resetting of personal best positions, it may not always be a good strategy to simply reset all personal best positions. Remember that by doing so, all particles forget any experience that they have gained about the search space during the search process. It might just be that after an environment change, the personal best position of a particle is closer to the new goal than the current position. Under less severe changes, it is possible that the personal best position remains the best position for that particle. Carlisle and Dozier proposed that the personal best positions first be evaluated after the environment change and if the personal best position is worse than the current position for the new environment, only then reset the personal best position [109].

The global best position, if selected from the personal best positions, also adds to the memory of the swarm. If the environment changes, then the global best position is based on stale information. To address this problem, any of the following strategies can be followed:

- After resetting of the personal best positions, recalculate the global best position from the more up-to-date personal best positions.

- Recalculate the global best position only if it is worse under the new environment.

- Find the global best position only from the current particle positions and not from the personal best positions [142].

The same reasoning applies to neighborhood best positions.

The charged PSO discussed in Section 16.5.6 has been developed to track dynamically changing optima, facilitated by the repelling mechanism. Other PSO implementations for dynamic environments can be found in [142, 518, 955].

## 16.6.4   Niching PSO

Although the basic PSO was developed to find single solutions to optimization problems, it has been observed for PSO implementations with special parameter choices that the basic PSO has an inherent ability to find multiple solutions. Agrafiotis and Cedeño, for example, observed for a specific problem that particles coalesce into a small number of local minima [11]. However, any general conclusions about the PSO's niching abilities have to be reached with extreme care, as explained below.

The main driving forces of the PSO are the cognitive and social components. It is the social component that prevents speciation. Consider, for example, the *gbest* PSO. The social component causes all particles to be attracted towards the best position obtained by the swarm, while the cognitive component exerts an opposing force (if the global best position is not the same as a particle's personal best position) towards a particle's own best solution. The resulting effect is that particles converge to a point between the global best and personal best positions, as was formally proven in [863, 870]. If the PSO is executed until the swarm reaches an equilibrium point, then each particle will have converged to such a point between the global best position and the personal best position of that particle, with a final zero velocity.

Brits *et al.* [91] showed empirically that the *lbest* PSO succeeds in locating a small percentage of optima for a number of functions. However, keep in mind that these studies terminated the algorithms when a maximum number of function evaluations was exceeded, and not when an equilibrium state was reached. Particles may therefore still have some momentum, further exploring the search space. It is shown in [91] that the basic PSO fails miserably compared to specially designed *lbest* PSO niching algorithms. Of course, the prospect of locating more multiple solutions will improve with an increase in the number of particles, but at the expense of increased computational complexity. The basic PSO also fails another important objective of niching algorithms, namely to maintain niches. If the search process is allowed to continue, more particles converge to the global best position in the process to reach an equilibrium, mainly due to the social component (as explained above).

Emanating from the discussion in this section, it is desirable to rather adapt the basic PSO with true speciation abilities. One such algorithm, the NichePSO is described next.

The NichePSO was developed to find multiple solutions to general multi-modal problems [89, 88, 91]. The basic operating principle of NichePSO is the self-organization of particles into independent sub-swarms. Each sub-swarm locates and maintains a niche. Information exchange is only within the boundaries of a sub-swarm. No information is exchanged between sub-swarms. This independency among sub-swarms allows sub-swarms to maintain niches. To emphasize: each sub-swarm functions as a stable, individual swarm, evolving on its own, independent of individuals in other swarms.

The NichePSO starts with one swarm, referred to as the *main* swarm, containing all particles. As soon as a particle converges on a potential solution, a sub-swarm is created by grouping together particles that are in close proximity to the potential solution. These particles are then removed from the main swarm, and continue within their sub-swarm to refine (and to maintain) the solution. Over time, the main swarm shrinks as sub-swarms are spawned from it. NichePSO is considered to have converged when sub-swarms no longer improve the solutions that they represent. The global best position from each sub-swarm is then taken as an optimum (solution).

The NichePSO is summarized in Algorithm 16.15. The different steps of the algorithm are explained in more detail in the following sections.

---

**Algorithm 16.15** NichePSO Algorithm

---

Create and initialize a $n_x$-dimensional *main* swarm, $S$;
**repeat**
    Train the main swarm, $S$, for one iteration using the *cognition-only* model;
    Update the fitness of each main swarm particle, $S.\mathbf{x}_i$;
    **for** *each sub-swarm $S_k$* **do**
        Train sub-swarm particles, $S_k.\mathbf{x}_i$, using a full model PSO;
        Update each particle's fitness;
        Update the swarm radius $S_k.R$;
    **endFor**
    If possible, merge sub-swarms;
    Allow sub-swarms to absorb any particles from the main swarm that moved into the sub-swarm;
    If possible, create new sub-swarms;
**until** *stopping condition is true*;
Return $S_k.\hat{\mathbf{y}}$ for each sub-swarm $S_k$ as a solution;

---

## Main Swarm Training

The main swarm uses a cognition-only model (refer to Section 16.3.5) to promote exploration of the search space. Because the social component serves as an attractor for all particles, it is removed from the velocity update equation. This allows particles to converge on different parts of the search space.

It is important to note that velocities must be initialized to zero.

## Sub-swarm Training

The sub-swarms are independent swarms, trained using a full model PSO (refer to Section 16.1). Using a full model to train sub-swarms allows particle positions to be adjusted on the basis both of particle's own experience (the cognitive component) and of socially obtained information (the social component). While any full model PSO can be used to train the sub-swarms, the NichePSO as presented in [88, 89] uses the GCPSO (discussed in Section 16.5.1). The GCPSO is used since it has guaranteed convergence to a local minimum [863], and because the GCPSO has been shown to perform well on extremely small swarms [863]. The latter property of GCPSO is necessary because sub-swarms initially consist of only two particles. The *gbest* PSO has a tendency to stagnate with such small swarms.

## Identification of Niches

A sub-swarm is formed when a particle seems to have converged on a solution. If a particle's fitness shows little change over a number of iterations, a sub-swarm is created with that particle and its closest topological neighbor. Formally, the standard deviation, $\sigma_i$, in the fitness $f(\mathbf{x}_i)$ of each particle is tracked over a number of iterations. If $\sigma_i < \epsilon$, a sub-swarm is created. To avoid problem-dependence, $\sigma_i$ is normalized according to the domain. The closest neighbor, $l$, to the position $\mathbf{x}_i$ of particle $i$ is computed using Euclidean distance, i.e.

$$l = \arg\min_a\{||\mathbf{x}_i - \mathbf{x}_a||\} \tag{16.98}$$

with $1 \leq i, a \leq S.n_s, i \neq a$ and $S.n_s$ is the size of the main swarm.

The sub-swarm creation, or niche identification process is summarized in Algorithm 16.16. In this algorithm, $\mathcal{Q}$ represents the set of sub-swarms, $\mathcal{Q} = \{S_1, \cdots, S_K\}$, with $|\mathcal{Q}| = K$. Each sub-swarm has $S_k.n_s$ particles. During initialization of the NichePSO, $K$ is initialized to zero, and $\mathcal{Q}$ is initialized to the empty set.

## Absorption of Particles into a Sub-swarm

It is likely that particles of the main swarm move into the area of the search space covered by a sub-swarm $S_k$. Such particles are merged with the sub-swarm, for the following reasons:

---

**Algorithm 16.16** NichePSO Sub-swarm Creation Algorithm

---

**if** $\sigma_i < \epsilon$ **then**
    $k = k + 1$;
    Create sub-swarm $S_k = \{\mathbf{x}_i, \mathbf{x}_l\}$;
    Let $\mathcal{Q} \leftarrow \mathcal{Q} \cup S_k$;
    Let $S \leftarrow S \setminus S_k$;
**end**

---

- Inclusion of particles that traverse the search space of an existing sub-swarm may improve the diversity of the sub-swarm.

- Inclusion of such particles into a sub-swarm will speed up their progression towards an optimum through the addition of social information within the sub-swarm.

More formally, if for particle $i$,

$$\|\mathbf{x}_i - S_k.\hat{\mathbf{y}}\| \leq S_k.R \tag{16.99}$$

then absorb particle $i$ into sub-swarm $S_k$:

$$S_k \quad \leftarrow \quad S_k \cup \{\mathbf{x}_i\} \tag{16.100}$$
$$S \quad \leftarrow \quad S \setminus \{\mathbf{x}_i\} \tag{16.101}$$

In equation (16.99), $S_k.R$ refers to the radius of sub-swarm $S_k$, defined as

$$S_k.R = \max\{\|S_k.\hat{\mathbf{y}} - S_k.\mathbf{x}_i\|\}, \quad \forall i = 1, \ldots, S_k.n_s \tag{16.102}$$

where $S_k.\hat{\mathbf{y}}$ is the global best position of sub-swarm $S_k$.

### Merging Sub-swarms

It is possible that more than one sub-swarm form to represent the same optimum. This is due to the fact that sub-swarm radii are generally small, approximating zero as the solution represented is refined over time. It may then happen that a particle that moves toward a potential solution is not absorbed into a sub-swarm that is busy refining that solution. Consequently, a new sub-swarm is created. This leads to the redundant refinement of the same solution by multiple swarms. To solve this problem, similar sub-swarms are merged. Swarms are considered similar if the hyperspace defined by their particle positions and radii intersect. The new, larger sub-swarm then benefits from the social information and experience of both swarms. The resulting sub-swarm usually exhibits larger diversity than the original, smaller sub-swarms.

Formally stated, two sub-swarms, $S_{k_1}$ and $S_{k_2}$, intersect when

$$\|S_{k_1}.\hat{\mathbf{y}} - S_{k_2}.\hat{\mathbf{y}}\| < (S_{k_1}.R + S_{k_2}.R) \tag{16.103}$$

If $S_{k_1}.R = S_{k_2}.R = 0$, the condition in equation (16.103) fails, and the following merging test is considered:

$$||S_{k_1}.\hat{\mathbf{y}} - S_{k_2}.\hat{\mathbf{y}}|| < \mu \tag{16.104}$$

where $\mu$ is a small value close to zero, e.g. $\mu = 10^{-3}$. If $\mu$ is too large, the result may be that dissimilar swarms are merged with the consequence that a candidate solution may be lost.

To avoid tuning of $\mu$ over the domain of the search space, $||S_{k_1}.\hat{\mathbf{y}} - S_{k_2}.\hat{\mathbf{y}}||$ are normalized to the interval $[0, 1]$.

### Stopping Conditions

Any of a number of stopping conditions can be used to terminate the search for multiple solutions. It is important that the stopping conditions ensure that each sub-swarm has converged onto a unique solution.

The reader is referred to [88] for a more detailed analysis of the NichePSO.

## 16.7 Applications

PSO has been used mostly to optimize functions with continuous-valued parameters. One of the first applications of PSO was in training neural networks, as summarized in Section 16.7.1. A game learning application of PSO is summarized in Section 16.7.3. Other applications are listed in Table 16.1.

### 16.7.1 Neural Networks

The first applications of PSO was to train feedforward neural networks (FFNN) [224, 446]. These first studies in training FFNNs using PSO have shown that the PSO is an efficient alternative to NN training. Since then, numerous studies have further explored the power of PSO as a training algorithm for a number of different NN architectures. Studies have also shown for specific applications that NNs trained using PSO provide more accurate results. Section 16.7.1 and Section 16.7.1 respectively address supervised and unsupervised training. NN architecture selection approaches are discussed in Section 16.7.2.

### Supervised Learning

The main objective in supervised NN training is to adjust a set of weights such that an objective (error) function is minimized. Usually, the SSE error is used as the objective function (refer to equation (2.17)).

In order to use PSO to train an NN, a suitable representation and fitness function needs to be found. Since the objective is to minimize the error function, the fitness function is simply the given error function (e.g. the SSE given in equation (2.17)). Each particle represents a candidate solution to the optimization problem, and since the weights of a trained NN are a solution, a single particle represents one complete network. Each component of a particle's position vector represents one NN weight or bias. Using this representation, any of the PSO algorithms can be used to find the best weights for an NN to minimize the error function.

Eberhart and Kennedy [224, 446] provided the first results of applying the basic PSO to the training of FFNNs. Mendes *et al.* [575] evaluated the performance of different neighborhood topologies (refer to Section 16.2) on training FFNNs. The topologies tested included the star, ring, pyramid, square and four clusters topologies. Hirata *et al.* [367], and Gudise and Venayagamoorthy [338] respectively evaluated the ability of *lbest* and *gbest* PSO to train FFNNs. Al-Kazemi and Mohan applied the multi-phase PSO (refer to Section 16.5.4) to NN training. Van den Bergh and Engelbrecht showed that the cooperative PSO (refer to Section 16.5.4) and GCPSO (refer to Section 16.5.1) perform very well as NN training algorithms for FFNNs [863, 864]. Settles and Rylander also applied the cooperative PSO to NN training [776].

He *et al.* [359] used the basic PSO to train a fuzzy NN, after which accurate rules have been extracted from the trained network.

The real power of PSO as an NN training algorithm was illustrated by Engelbrecht and Ismail [247, 406, 407, 408] and Van den Bergh and Engelbrecht [867] in training NNs with product units. The basic PSO and cooperative PSO have been shown to outperform optimizers such as gradient-descent, LeapFrog (refer to Section 3.2.4), scaled conjugate gradient (refer to Section 3.2.3) and genetic algorithms (refer to Chapter 9). Paquet and Engelbrecht [655, 656] have further illustrated the power of the linear PSO in training support vector machines.

Salerno [753] used the basic PSO to train Elman recurrent neural networks (RNN). The PSO was successful for simple problems, but failed to train an RNN for parsing natural language phrases. Tsou and MacNish [854] also showed that the basic PSO fails to train certain RNNs, and developed a Newton-based PSO that successfully trained a RNN to learn regular language rules. Juang [430] combined the PSO as an operator in a GA to evolve RNNs. The PSO was used to enhance elitist individuals.

## Unsupervised Learning

While plenty of work has been done in using PSO algorithms to train supervised networks, not much has been done to show how PSO performs as an unsupervised training algorithm. Xiao *et al.* [921] used the *gbest* PSO to evolve weights for a self-organizing map (SOM) [476] (also refer to Section 4.5) to perform gene clustering. The training process consists of two phases. The first phase uses PSO to find an initial weight set for the SOM. The second phase initializes a PSO with the weight set obtained from the first phase. The PSO is then used to refine this weight set.

Messerschmidt and Engelbrecht [580], and Franken and Engelbrecht [283, 284, 285] used the *gbest, lbest* and Von Neumann PSO algorithms as well as the GCPSO to coevolve neural networks to approximate the evaluation function of leaf nodes in game trees. No target values were available; therefore NNs compete in game tournaments against groups of opponents in order to determine a score or fitness for each NN. During the coevolutionary training process, weights are adjusted using PSO algorithms to have NNs (particles) move towards the best game player. The coevolutionary training process has been applied successfully to the games of tick-tack-toe, checkers, bao, the iterated prisoner's dilemma, and a probabilistic version of tick-tack-toe. For more information, refer to Section 15.2.3.

## 16.7.2   Architecture Selection

Zhang and Shao [949, 950] proposed a PSO model to simultaneously optimize NN weights and architecture. Two swarms are maintained: one swarm optimizes the architecture, and the other optimizes weights. Particles in the architecture swarm are two-dimensional, with each particle representing the number of hidden units used and the connection density. The first step of the algorithm randomly initializes these architecture particles within predefined ranges.

The second swarm's particles represent actual weight vectors. For each architecture particle, a swarm of particles is created by randomly initializing weights to correspond with the number of hidden units and the connection density specified by the architecture particle. Each of these swarms is evolved using a PSO, where the fitness function is the MSE computed from the training set. After convergence of each NN weights swarm, the best weight vector is identified from each swarm (note that the selected weight vectors are of different architectures). The fitness of these NNs are then evaluated using a validation set containing patterns not used for training. The obtained fitness values are used to quantify the performance of the different architecture specifications given by the corresponding particles of the architecture swarm. Using these fitness values, the architecture swarm is further optimized using PSO.

This process continues until a termination criterion is satisfied, at which point the global best particle is one with an optimized architecture and weight values.

## 16.7.3   Game Learning

Messerschmidt and Engelbrecht [580] developed a PSO approach to train NNs in a coevolutionary mechanism to approximate the evaluation function of leaf nodes in a game tree as described in Section 15.2.3. The initial model was applied to the simple game of tick-tack-toe.

As mentioned in Section 15.2.3 the training process is not supervised. No target evaluation of board states is provided. The lack of desired outputs for the NN necessitates a coevolutionary training mechanism, where NN agents compete against other agents, and all inferior NNs strive to beat superior NNs. For the PSO coevolutionary training algorithm, summarized in Algorithm 16.17, a swarm of particles is randomly created, where each particle represents a single NN. Each NN plays in a tournament against a group of randomly selected opponents, selected from a competition pool (usually consisting of all the current particles of the swarm and all personal best positions). After each NN has played against a group of opponents, it is assigned a score based on the number of wins, losses and draws achieved. These scores are then used to determine personal best and neighborhood best solutions. Weights are adjusted using the position and velocity updates of any PSO algorithm.

---

**Algorithm 16.17** PSO Coevolutionary Game Training Algorithm

---

Create and randomly initialize a swarm of NNs;
**repeat**
    Add each personal best position to the competition pool;
    Add each particle to the competition pool;
    **for** *each particle (or NN)* **do**
        Randomly select a group of opponents from the competition pool;
        **for** *each opponent* **do**
            Play a game (using game trees to determine next moves) against the opponents, playing as first player;
            Record if game was won, lost or drawn;
            Play a game against same opponent, but as the second player;
            Record if game was won, lost or drawn;
        **end**
        Determine a score for each particle;
        Compute new personal best positions based on scores;
    **end**
    Compute neighbor best positions;
    Update particle velocities;
    Update particle positions;
**until** *stopping condition is true*;
Return global best particle as game-playing agent;

---

The basic algorithm as given in Algorithm 16.17 has been applied successfully to the zero-sum games of tick-tack-toe [283, 580], checkers [284], and bao [156]. Franken and Engelbrecht also applied the approach to the non-zero-sum game, the iterated prisoner's dilemma [285]. A variant of the approach, using two competing swarms has recently been used to train agents for a probabilistic version of tick-tac-toe [654].

# 16.8 Assignments

1. Discuss in detail the differences and similarities between PSO and EAs.

2. Discuss how PSO can be used to cluster data.

3. Why is it better to base the calculation of neighborhoods on the index assigned to particles and not on geometrical information such as Euclidean distance?

**Table 16.1** Applications of Particle Swarm Optimization

| Application | References |
|---|---|
| Clustering | [638, 639] |
| Design | [7, 316, 700, 874, 875] |
| Scheduling | [7, 470, 707, 754] |
| Planning | [771, 775, 856, 886] |
| Controllers | [143, 155, 297, 959] |
| Power systems | [6, 296, 297, 299, 437, 444] |
| Bioinformatics | [627, 705, 921] |
| Data mining | [805] |

4. Explain how PSO can be used to approximate functions using an $n$-th order polynomial.

5. Show how PSO can be used to solve a system of equations.

6. If the basic PSO is used to solve a system of equations, what problem(s) do you foresee? How can these be addressed?

7. How can PSO be used to solve problems with discrete-valued parameters?

8. For the predator-prey PSO, what will be the effect if more than one predator is used?

9. Critically discuss the following strategy applied to a dynamic inertia weight: Start with an inertia weight of 2.0, and linearly decrease it to 0.5 as a function of the iteration number.

10. The GCPSO was developed to address a specific problem with the standard PSO. What is this problem? If mutation is combined with the PSO, will this problem be addressed?

11. Consider the following adaptation of the standard *gbest* PSO algorithm: all particles, except for the *gbest* particle, use the standard PSO velocity update and position update equations. The new position of the *gbest* particle is, however, determined by using the LeapFrog algorithm. Comment on this strategy. What advantages do you see, any disadvantages? Will it solve the problem of the standard PSO in the question above?

12. Explain why the basic *gbest* PSO cannot be used to find niches (multiple solutions), neither in parallel nor sequentially (assuming that the fitness function is not allowed to be augmented).

13. Explain why velocities should be initialized to zero for the NichePSO.

14. Can it be said that PSO implements a form of
    (a) competitive coevolution?
    (b) cooperative coevolution?
    Justify your answers.

15. Discuss the validity of the following statement: *"PSO is an EA."*

# Chapter 17

# Ant Algorithms

Ants appeared on earth some 100 million years ago, and have a current total population estimated at $10^{16}$ individuals [378]. It is further estimated that the total weight of ants is in the same order of magnitude as the total weight of human beings. Most of these ants are social insects, living in colonies of 30 to millions of individuals. Ants are not the only social insects living in colonies. The complex behaviors that emerge from colonies of ants have intrigued humans, and there have been many studies of ant colonies aimed at a better understanding of these collective behaviors. Collective ant behaviors that have been studied include the foraging behavior, division of labour, cemetery organization and brood care, and construction of nests. The South African, Eugéne Marais (1872-1936) was one of the first to study termite colonies. He published his observations as early as 1927 in local magazines and newspapers. In his book, *The Soul of the Ant* [558] (first published in 1937, after his death), he described in detail his experimental procedures and observations of the workings of termite societies. The Belgian, Maurice Maeterlinck (1862–1949), published *The Life of the White Ant* [547], which was largely drawn from Marais's articles (see the discussion in [558]).

Following on form this pioneering work, the French biologist Pierre-Paul Grassé [333] postulated on the mechanics of termite communication in his studies of their nest construction behavior. Grassé determined that a form of indirect communication exists between individuals, which he termed *stigmergy*. It is through this local interaction between individuals, and between individuals and the environment, that more complex behaviors emerges. More recently, Deneubourg *et al.* [199] studied one example of stigmergy, namely pheromonal communication. From these studies, the first algorithmic models of foraging behavior have been developed and implemented [208].

While most research efforts concentrated on developing algorithmic models of foraging behavior, models have been developed for other behaviors, including division of labour, cooperative support, self-assembly, and cemetery organization. These complex behaviors emerge from the collective behavior of very unsophisticated individuals. In the context of collective behaviour, social insects are basically stimulus–response agents. Based on information perceived from the local environment, an individual performs a simple, basic action. These simple actions appear to have a large random component. Despite this simplicity in individual behavior, social insects form a highly structured social organism. To illustrate the complexity, and structured organization of ant colonies, Marais [558] pointed out the resemblance between the human body and a termite society.

*Computational Intelligence: An Introduction*, Second Edition A.P. Engelbrecht
©2007 John Wiley & Sons, Ltd

This chapter provides an introductory overview of ant algorithms. Section 17.1 considers the foraging behavior of ants and discusses the first ant algorithms developed on the basis of foraging models to solve discrete combinatorial optimization problems. Models of the cemetery organization and brood care behaviors are discussed in Section 17.2. The division of labor behavior is covered in Section 17.3. Some advanced topics are presented in Section 17.4, including application of ACO in continuous environments, MOO, dynamic environments, and handling constraints. Applications of ant algorithms (AA) are summarized in Section 17.5

# 17.1    Ant Colony Optimization Meta-Heuristic

One of the first behaviors studied by entomologists was the ability of ants to find the shortest path between their nest and a food source. From these studies and observations followed the first algorithmic models of the foraging behavior of ants, as developed by Marco Dorigo [208]. Since then, research in the development of AAs has gained momentum, resulting in a large number of algorithms and applications. Collectively, algorithms that were developed as a result of studies of ant foraging behavior are referred to as instances of the ant colony optimization meta-heuristic (ACO-MH) [211, 215]. This section provides an overview of the ACO-MH, with a focus on the basic principles of ACO algorithms, and the first algorithms that have been developed. Section 17.1.1 gives an overview of the foraging behavior of real ants, and introduces the concepts of stigmergy and artificial ants. A very simple ant algorithm implementation is discussed in Section 17.1.3 to illustrate the basic principles of AAs. Sections 17.1.4 to 17.1.11 respectively discuss the ant system (AS), ant colony system (ACS), max-min ant system (MMAS), Ant-Q, fast ant system, Antabu, AS-rank and ANTS instances of the ACO-MH. These algorithms were the first set of ant algorithms implemented, mainly with reference to the traveling salesman problem (TSP). Section 17.1.12 provides a discussion on the parameters of these algorithms.

## 17.1.1    Foraging Behavior of Ants

How do ants find the shortest path between their nest and food source, without any visible, central, active coordination mechanisms? Studies of the foraging behavior of several species of real ants revealed an initial random or chaotic activity pattern in the search for food [216, 304, 628]. As soon as a food source is located, activity patterns become more organized with more and more ants following the same path to the food source. "Auto-magically", soon all ants follow the same, shortest path. This emergent behavior is a result of a recruitment mechanism whereby ants that have located a food source influence other ants towards the food source. The recruitment mechanism differs for different species, and can either be in the form of direct contact, or indirect "communication." Most ant species use the latter form of recruitment, where communication is via pheromone trails. When an ant locates a food source, it carries a food item to the nest and lays pheromone along the trail. Forager ants decide which path to follow based on the pheromone concentrations on the different

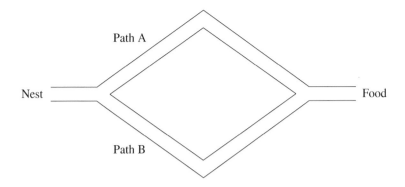

**Figure 17.1** Binary Bridge Experiment

paths. Paths with a larger pheromone concentration have a higher probability of being selected. As more ants follow a specific trail, the desirability of that path is reinforced by more pheromone being deposited by the foragers, which attracts more ants to follow that path. The collective behavior that results is a form of *autocatalytic behavior*, where *positive feedback* about a food path causes that path to be followed by more and more ants [209, 216].

The indirect communication where ants modify their environment (by laying of pheromones) to influence the behavior of other ants is referred to as *stigmergy* (refer to Section 17.1.2.

### The Bridge Experiments

Deneubourg *et al.* [199] studied the foraging behavior of the Argentine ant species *Iridomyrmex humilis* in order to develop a formal model to describe its behavior. In this laboratory experiment, as illustrated in Figure 17.1, the nest is separated from the food source by a bridge with two equally long branches. Initially, both branches were free of any pheromones. After a finite time period, one of the branches was selected, with most of the ants following the path, even with both branches being of the same length. The selection of one of the branches is due to random fluctuations in path selection, causing higher concentrations on the one path.

From this experiment, referred to as the *binary bridge* experiment (and illustrated in Figure 17.1), a simple formal model was developed to characterize the path selection process [666]. For this purpose, it is assumed that ants deposit the same amount of pheromone and that pheromone does not evaporate. Let $n_A(t)$ and $n_B(t)$ denote the number of ants on paths $A$ and $B$ respectively at time step $t$. Pasteels *et al.* [666] found empirically that the probability of the next ant to choose path $A$ at time step $t+1$ is given as,

$$P_A(t+1) = \frac{(c+n_A(t))^\alpha}{(c+n_A(t))^\alpha + (c+n_B)^\alpha} = 1 - P_B(t+1) \qquad (17.1)$$

where $c$ quantifies the degree of attraction of an unexplored branch, and $\alpha$ is the bias

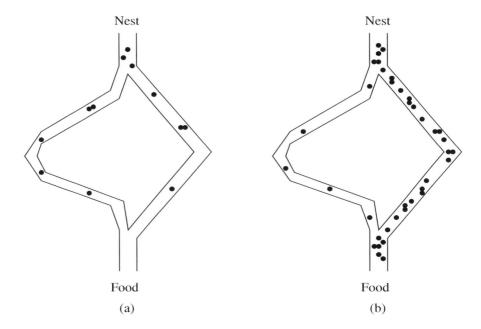

**Figure 17.2** Shortest Path Selection by Forager Ants

to using pheromone deposits in the decision process. The larger the value of $\alpha$, the higher the probability that the next ant will follow the path with a higher pheromone concentration – even if that branch has only slightly more pheromone deposits. The larger the value of $c$, the more pheromone deposits are required to make the choice of path non-random. It was found empirically that $\alpha \approx 2$ and $c \approx 20$ provide a best fit to the experimentally observed behavior.

Using the probability defined in equation (17.1), the decision rule of an ant that arrives at the binary bridge is expressed as follows: if $U(0,1) \leq P_A(t+1)$ then follow path $A$ otherwise follow path $B$.

Goss *et al.* [330] extended the binary bridge experiment, where one of the branches of the bridge was longer than the other, as illustrated in Figure 17.2. Dots in this figure indicate ants. Initially, paths are chosen randomly with approximately the same number of ants following both paths (as illustrated in Figure 17.2(a)). Over time, more and more ants follow the shorter path as illustrated in Figure 17.2(b). Selection is biased towards the shortest path, since ants that follow the shortest path return to the nest earlier than ants on the longer path. The pheromone on the shorter path is therefore reinforced sooner than that on the longer path.

Goss *et al.* [330] found that the probability of selecting the shorter path increases with the length ratio between the two paths. This has been referred to as the *differential path length* effect by Dorigo *et al.* [210, 212].

To summarize the findings of the studies of real ants, the emergence of shortest path selection behavior is explained by autocatalysis (positive feedback) and the differential path length effect [210, 212].

Although an ant colony exhibits complex adaptive behavior, a single ant follows very simple behaviors. An ant can be seen as a stimulus–response agent [629]: the ant observes pheromone concentrations and produces an action based on the pheromone-stimulus. An ant can therefore abstractly be considered as a simple computational agent. An artificial ant algorithmically models this simple behavior of real ants. The logic implemented is a simple production system with a set of production rules as illustrated in Algorithm 17.1. This algorithm is executed at each point where the ant needs to make a decision.

---

**Algorithm 17.1** Artificial Ant Decision Process

---

Let $r \sim U(0,1)$;
**for** *each potential path A* **do**
    Calculate $P_A$ using, e.g., equation (17.1);
    **if** $r \leq P_A$ **then**
        Follow path $A$;
        Break;
    **end**
**end**

---

While Algorithm 17.1 implements a simple random selection mechanism, any other probabilistic selection mechanism (as overviewed in Section 8.5) can be used, for example, roulette wheel selection.

## 17.1.2 Stigmergy and Artificial Pheromone

Generally stated, stigmergy is a class of mechanisms that mediate animal-to-animal interactions [840]. The term *stigmergy* was formally defined by Grassé [333] as a form of indirect communication mediated by modifications of the environment. This definition originated from observations of the nest-building behavior of the termite species *Bellicositermes natalensis* and *Cubitermes*. Grassé observed that coordination and regulation of nest-building activities are not on an individual level, but achieved by the current nest structure. The actions of individuals are triggered by the current configuration of the nest structure. Similar observations have been made by Marais [558], with respect to the *Termes natalensis*.

The word *stigmergy* is aptly constructed from the two Greek words [341],

- *stigma*, which means sign, and
- *ergon*, which means work.

Individuals observe signals, which trigger a specific response or action. The action may reinforce or modify signals to influence the actions of other individuals.

Two forms of stigmergy have been defined [341, 871, 902]: sematectonic and sign-based. *Sematectonic stigmergy* refers to communication via changes in the physical characteristics of the environment. Example activities that are accomplished through sematectonic stigmergy include nest building, nest cleaning, and brood sorting (refer to Section 17.2 considers algorithms based on sematectonic stigmergy). *Sign-based* stigmergy facilitates communication via a signaling mechanism, implemented via chemical compounds deposited by ants. As an example, foraging behavior emerges from ants following pheromone trails deposited by other ants.

Ant algorithms are population-based systems inspired by observations of real ant colonies. Cooperation among individuals in an ant algorithm is achieved by exploiting the stigmergic communication mechanisms observed in real ant colonies. Algorithmic modeling of the behavior of ants is thus based on the concept of *artificial stigmergy*, defined by Dorigo and Di Caro as the *"indirect communication mediated by numeric modifications of environmental states which are only locally accessible by the communicating agents"* [211]. The essence of modeling ant behavior is to find a mathematical that accurately describes the stigmergetic characteristics of the corresponding ant individuals. The main part of such a model is the definition of stigmergic variables which encapsulate the information used by artificial ants to communicate indirectly. In the contextof foraging behavior, artificial pheromone plays the role of stigmergic variable.

As discussed above, ants have the ability to always find the shortest path between their nest and the food source. As ants move from a food source to the nest, an amount of pheromone is dropped by each ant. Future ants choose paths probabilistically on on the basis of the amount of pheromone. The higher the pheromone concentration, the more the chance that the corresponding path will be selected. Some ant species have the amount of pheromone deposited proportional to the quality of the food [210].

Over time, shorter paths will have stronger pheromone concentrations, since ants return faster on those paths. Pheromone evaporates over time, with the consequence that the pheromone concentrations on the longer paths decrease more quickly than on the shorter paths.

Artificial pheromone mimics the characteristics of real pheromone, and indicates the "popularity" of a solution to the optimization problem under consideration. In effect, artificial pheromone encodes a long-term memory about the entire search process.

## 17.1.3   Simple Ant Colony Optimization

The first ant algorithm developed was the ant system (refer to Section 17.1.4)[208, 209], and since then several improvements of the ant system have been devised [300, 301, 555, 743, 815]. These algorithms have somewhat more complex decision processes than that illustrated in Algorithm 17.1. To provide a gentle, didactic introduction to ant algorithms, this section breaks the chronological order in which ant algorithms have been developed to first present the simple ACO (SACO) [212, 217]. The SACO is an algorithmic implementation of the double bridge experiment of Deneubourg *et al.* [199] (refer to Section 17.1.1), and is used in this section to illustrate the basic components and behavior of the ACO-MH.

Consider the general problem of finding the shortest path between two nodes on a

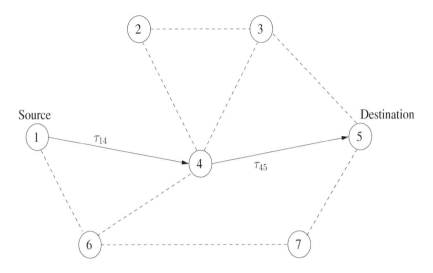

**Figure 17.3** Graph for Shortest Path Problems

graph, $G = (V, E)$, where $V$ is the set of vertices (nodes) and $E$ is a matrix representing the connections between nodes. The graph has $n_G = |V|$ nodes. The length, $L^k$, of the path constructed by ant $k$ is calculated as the number of hops in the path from the origin to the destination node. An example graph and selected path are illustrated in Figure 17.3. The length of the indicated route is 2. A pheromone concentration, $\tau_{ij}$, is associated With each edge, $(i, j)$, of the graph.

For the SACO, each edge is assigned a small random value to indicate the initial pheromone, $\tau_{ij}(0)$. Strictly speaking, edges do not have any pheromone concentrations for the first step. An ant randomly selects which edge to follow next. Using the simple decision logic summarized in Algorithm 17.1, implementation is made easier by initializing the pheromone concentration on each link to a small random value. A number of ants, $k = 1, \ldots, n_k$, are placed on the source node. For each iteration of SACO (refer to Algorithm 17.2), each ant incrementally constructs a path (solution) to the destination node. At each node, each ant executes a decision policy to determine the next link of the path. If ant $k$ is currently located at node $i$, it selects the next node $j \in \mathcal{N}_i^k$, based on the transition probability [212, 217],

$$p_{ij}^k(t) = \begin{cases} \dfrac{\tau_{ij}^\alpha(t)}{\sum_{j \in \mathcal{N}_i^k} \tau_{ij}^\alpha(t)} & \text{if } j \in \mathcal{N}_i^k \\ 0 & \text{if } j \notin \mathcal{N}_i^k \end{cases} \tag{17.2}$$

where $\mathcal{N}_i^k$ is the set of feasible nodes connected to node $i$, with respect to ant $k$. If, for any node $i$ and ant $k$, $\mathcal{N}_i^k = \emptyset$, then the predecessor to node $i$ is included in $\mathcal{N}_i^k$. Note that this may cause loops to occur within constructed paths. These loops are removed once the destination node has been reached.

In the equation above, $\alpha$ is a positive constant used to amplify the influence of pheromone concentrations. Large values of $\alpha$ give excessive importance to pheromone, especially the initial random pheromones, which may lead to rapid convergence to sub-optimal paths.

Once all ants have constructed a complete path from the origin node to the destination node, and all loops have been removed, each ant retraces its path to the source node deterministically, and deposits a pheromone amount,

$$\Delta\tau_{ij}^k(t) \propto \frac{1}{L^k(t)} \tag{17.3}$$

to each link, $(i,j)$, of the corresponding path; $L^k(t)$ is the length of the path constructed by ant $k$ at time step $t$.

That is,

$$\tau_{ij}(t+1) = \tau_{ij}(t) + \sum_{k=1}^{n_k} \Delta\tau_{ij}^k(t) \tag{17.4}$$

where $n_k$ is the number of ants.

Using equation (17.3), the total pheromone intensity of a link is proportional to the desirability of the paths in which the link occurs, based on the length of the corresponding path. The deposited pheromone, $\Delta\tau_{ij}^k$, calculated using equation (17.3), expresses the quality of the corresponding solution. For SACO, the quality of a solution (the constructed path) is simply expressed as the inverse of the length of the path in terms of the number of hops in the path. Any other measure can be used, for example the cost of traveling on the path, or the physical distance traveled. In general, if $x^k(t)$ denotes a solution at time step $t$, then $f(x^k(t))$ expresses the quality of the solution. If $\Delta\tau^k$ is not proportional to the quality of the solution and all ants deposit the same amount of pheromone (i.e. $\Delta\tau_{ij}^1 = \Delta\tau_{ij}^2 = \ldots = \Delta\tau_{ij}^{n_k}$), then it is only the differential path length effect that biases path selection towards the shortest path – very similar to the observations of Deneubourg *et al.* [199]. This discussion leads to the two main forms of solution evaluation employed by ant algorithms, namely

- **implicit** evaluation, where ants exploit the differential path length effect to bias the search of other agents, and

- **explicit** evaluation, where pheromone amounts are proportional to some quality measure of constructed solutions.

If the amount of pheromone deposited is inversely proportional to the quality of the solution (as is the case in equation (17.3)), then the larger $f(x^k(t))$ (that is, the worse the constructed solution), the smaller $1/f(x^k(t))$, hence the less the amount of pheromone deposited on the link. Thus, a long path causes all the links of that path to become less desirable as a component of the final solution. This is the case for any quality measure, $f$, that needs to be minimized.

Any of a number of termination criteria can be used in Algorithm 17.2 (and for the rest of the ant algorithms discussed later), for example,

- terminate when a maximum number of iterations, $n_t$, has been exceeded;
- terminate when an acceptable solution has been found, with $f(x^k(t)) \leq \epsilon$;

---

**Algorithm 17.2** Simple ACO Algorithm

---

Initialize $\tau_{ij}(0)$ to small random values;
Let $t = 0$;
Place $n_k$ ants on the origin node;
**repeat**
    **for** *each ant $k = 1, \ldots, n_k$* **do**
        //Construct a path $x^k(t)$;
        $x^k(t) = \emptyset$;
        **repeat**
            Select next node based on the probability defined in equation (17.2);
            Add link $(i, j)$ to path $x^k(t)$;
        **until** *destination node has been reached*;
        Remove all loops from $x^k(t)$;
        Calculate the path length $f(x^k(t))$;
    **end**
    **for** *each link $(i, j)$ of the graph* **do**
        //pheromone evaporation;
        Reduce the pheromone, $\tau_{ij}(t)$, using equation (17.5);
    **end**
    **for** *each ant $k = 1, \ldots, n_k$* **do**
        **for** *each link $(i, j)$ of $x^k(t)$* **do**
            $\Delta\tau^k = \frac{1}{f(x^k(t))}$;
            Update $\tau_{ij}$ using equation (17.4);
        **end**
    **end**
    $t = t + 1$;
**until** *stopping condition is true*;
Return the path $x^k(t)$ with smallest $f(x^k(t))$ as the solution;

---

- terminate when all ants (or most of the ants) follow the same path.

The initial experiments on the binary bridge problem [212] found that ants rapidly converge to a solution, and that little time is spent exploring alternative paths. To force ants to explore more, and to prevent premature convergence, pheromone intensities on links are allowed to "evaporate" at each iteration of the algorithm before being reinforced on the basis of the newly constructed paths. For each link, $(i, j)$, let

$$\tau_{ij}(t) \leftarrow (1 - \rho)\tau_{ij}(t) \tag{17.5}$$

with $\rho \in [0, 1]$. The constant, $\rho$, specifies the rate at which pheromones evaporate, causing ants to "forget" previous decisions. In other words, $\rho$ controls the influence of search history. For large values of $\rho$, pheromone evaporates rapidly, while small values of $\rho$ result in slower evaporation rates. The more pheromones evaporate, the more random the search becomes, facilitating better exploration. For $\rho = 1$, the search is completely random.

At this point it is important to emphasize that solution construction is the result of cooperative behavior that emerges from the simple behaviors of individual ants: each ant chooses the next link of its path based on information provided by other ants, in the form of pheromone deposits. This again refers to the autocatalytic behavior exhibited by forager ants. It is also important to note that the information used to aid in the decision making process is limited to the local environment of the ant.

In their experiments, Dorigo and Di Caro found that [212, 217]

- SACO works well for very simple graphs, with the shortest path being selected most often;

- for larger graphs, performance deteriorates with the algorithm becoming less stable and more sensitive to parameter choices;

- convergence to the shortest path is good for a small number of ants, while too many ants cause non-convergent behavior;

- evaporation becomes more important for more complex graphs. If $\rho = 0$, i.e. no evaporation, the algorithm does not converge. If pheromone evaporates too much (a large $\rho$ is used), the algorithm often converged to sub-optimal solutions for complex problems;

- for smaller $\alpha$, the algorithm generally converges to the shortest path. For complex problems, large values of $\alpha$ result in worse convergence behavior.

From these studies of the simple ACO algorithm, the importance of the exploration–exploitation trade-off becomes evident. Care should be taken to employ mechanisms to ensure that ants do not exploit pheromone concentrations such that the algorithm prematurely stagnates on sub-optimal solutions, but that ants are forced to explore alternative paths. In the sections (and chapters) that follow, different mechanisms are discussed to balance exploitation and exploration.

The simple ACO algorithm discussed in the previous section has shown some success in finding the shortest paths in graphs. The performance of the algorithm can be improved significantly by very simple changes to Algorithm 17.2. These changes include addition of heuristic information to determine the probability of selecting a link, memory to prevent cycles, and different pheromone update rules using local and/or global information about the environment. The next sections provide an overview of the early ant algorithms which are based on such changes of the simple ACO algorithm.

### 17.1.4  Ant System

The first ant algorithm was developed by Dorigo [208], referred to as ant system (AS) [4, 77, 216]. AS improves on SACO by changing the transition probability, $p_{ij}^k$, to include heuristic information, and by adding a memory capability by the inclusion of a tabu list. In AS, the probability of moving from node $i$ to node $j$ is given as

$$p_{ij}^k(t) = \begin{cases} \dfrac{\tau_{ij}^\alpha(t)\eta_{ij}^\beta(t)}{\sum_{u \in \mathcal{N}_i^k(t)} \tau_{iu}^\alpha(t)\eta_{iu}^\beta(t)} & \text{if } j \in \mathcal{N}_i^k(t) \\ 0 & \text{if } j \notin \mathcal{N}_i^k(t) \end{cases} \tag{17.6}$$

where $\tau_{ij}$ represents the *a posteriori* effectiveness of the move from node $i$ to node $j$, as expressed in the pheromone intensity of the corresponding link, $(i, j)$; $\eta_{ij}$ represents the *a priori* effectiveness of the move from $i$ to $j$ (i.e. the attractiveness, or desirability, of the move), computed using some heuristic. The pheromone concentrations, $\tau_{ij}$, indicate how profitable it has been in the past to make a move from $i$ to $j$, serving as a memory of previous best moves.

The transition probability in equation (17.6) differs from that of SACO in equation (17.2) on two aspects:

- The transition probability used by AS is a balance between pheromone intensity (i.e. history of previous successful moves), $\tau_{ij}$, and heuristic information (expressing desirability of the move), $\eta_{ij}$. This effectively balances the exploration–exploitation trade-off. The search process favors actions that it has found in the past and which proved to be effective, thereby exploiting knowledge obtained about the search space. On the other hand, in order to discover such actions, the search has to investigate previously unseen actions, thereby exploring the search space. The best balance between exploration and exploitation is achieved through proper selection of the parameters $\alpha$ and $\beta$. If $\alpha = 0$, no pheromone information is used, i.e. previous search experience is neglected. The search then degrades to a stochastic greedy search. If $\beta = 0$, the attractiveness (or potential benefit) of moves is neglected and the search algorithm is similar to SACO with its associated problems.

  The heuristic information adds an explicit bias towards the most attractive solutions, and is therefore a problem-dependent function. For example, for problems where the distance (or cost) of a path needs to be minimized,

  $$\eta_{ij} = \frac{1}{d_{ij}} \tag{17.7}$$

  where $d_{ij}$ is the distance (or cost) between the nodes $i$ and $j$.

- The set, $\mathcal{N}_i^k$, defines the set of feasible nodes for ant $k$ when located on node $i$. The set of feasible nodes may include only the immediate neighbors of node $i$. Alternatively, to prevent loops, $\mathcal{N}_i^k$ may include all nodes not yet visited by ant $k$. For this purpose, a tabu list is usually maintained for each ant. As an ant visits a new node, that node is added to the ant's tabu list. Nodes in the tabu list are removed from $\mathcal{N}_i^k$, ensuring that no node is visited more than once.

Maniezzo and Colorni used a different formulation of the probability used to determine the next node [557]:

$$p_{ij}^k(t) = \begin{cases} \frac{\alpha\tau_{ij}(t)+(1-\alpha)\eta_{ij}}{\sum_{u\in\mathcal{N}_i^k(t)}(\alpha\tau_{iu}(t)+(1-\alpha)\eta_{iu}(t))} & \text{if } j \in \mathcal{N}_i^k(t) \\ 0 & \text{otherwise} \end{cases} \tag{17.8}$$

Parameter $\alpha$ defines a relative importance of the pheromone concentration $\tau_{ij}(t)$ with respect to the desirability, $\eta_{ij}(t)$, of link $(i, j)$. The probability $p_{ij}^k$ expresses a compromise between exploiting desirability (for small $\alpha$), and pheromone intensity. This formulation removes the need for the parameter $\beta$.

Pheromone evaporation is implemented as given in equation (17.5). After completion of a path by each ant, the pheromone on each link is updated as

$$\tau_{ij}(t+1) = \tau_{ij}(t) + \Delta\tau_{ij}(t) \tag{17.9}$$

with

$$\Delta\tau_{ij}(t) = \sum_{k=1}^{n_k} \Delta\tau_{ij}^k(t) \tag{17.10}$$

where $\Delta\tau_{ij}^k(t)$ is the amount of pheromone deposited by ant $k$ on link $(i,j)$ and $k$ at time step $t$. Dorigo *et al.* [216] developed three variations of AS, each differing in the way that $\Delta\tau_{ij}^k$ is calculated (assuming a minimization problem):

- Ant-cycle AS:

$$\Delta\tau_{ij}^k(t) = \begin{cases} \frac{Q}{f(x^k(t))} & \text{if link } (i,j) \text{ occurs in path } x^k(t) \\ 0 & \text{otherwise} \end{cases} \tag{17.11}$$

  For the ant-cycle implementation, pheromone deposits are inversely proportional to the quality, $f(x^k(t))$, of the complete path constructed by the ant. Global information is therefore used to update pheromone concentrations. $Q$ is a positive constant.

  For maximization tasks,

$$\Delta\tau_{ij}^k(t) = \begin{cases} Qf(x^k(t)) & \text{if link } (i,j) \text{ occurs in path } x^k(t) \\ 0 & \text{otherwise} \end{cases} \tag{17.12}$$

- Ant-density AS:

$$\Delta\tau_{ij}^k(t) = \begin{cases} Q & \text{if link } (i,j) \text{ occurs in path } x^k(t) \\ 0 & \text{otherwise} \end{cases} \tag{17.13}$$

  Each ant deposits the same amount of pheromone on each link of its constructed path. This approach essentially counts the number of ants that followed link $(i,j)$. The higher the density of the traffic on the link, the more desirable that link becomes as component of the final solution.

- Ant-quantity AS:

$$\Delta\tau_{ij}^k(t) = \begin{cases} \frac{Q}{d_{ij}} & \text{if link } (i,j) \text{ occurs in path } x^k(t) \\ 0 & \text{otherwise} \end{cases} \tag{17.14}$$

  In this case, only local information, $d_{ij}$, is used to update pheromone concentrations. Lower cost links are made more desirable. If $d_{ij}$ represents the distance between links, then ant-quantity AS prefers selection of the shortest links.

The AS algorithm is summarized in Algorithm 17.3. During the initialization step, placement of ants is dictated by the problem being solved. If the objective is to find the shortest path between a given source and destination node, then all $n_k$ ants are placed

---

**Algorithm 17.3** Ant System Algorithm

---

$t = 0$;
Initialize all parameters, i.e. $\alpha, \beta, \rho, Q, n_k, \tau_0$;
Place all ants, $k = 1, \ldots, n_k$;
**for** *each link* $(i, j)$ **do**
    $\tau_{ij}(t) \sim U(0, \tau_0)$;
**end**
**repeat**
    **for** *each ant* $k = 1, \ldots, n_k$ **do**
        $x^k(t) = \emptyset$;
        **repeat**
            From current node $i$, select next node $j$ with probability as defined in
            equation (17.6);
            $x^k(t) = x^k(t) \cup \{(i, j)\}$;
        **until** *full path has been constructed*;
        Compute $f(x^k(t))$;
    **end**
    **for** *each link* $(i, j)$ **do**
        Apply evaporation using equation (17.5);
        Calculate $\Delta\tau_{ij}(t)$ using equation (17.10);
        Update pheromone using equation (17.4);
    **end**
    **for** *each link* $(i, j)$ **do**
        $\tau_{ij}(t + 1) = \tau_{ij}(t)$;
    **end**
    $t = t + 1$;
**until** *stopping condition is true*;
Return $x^k(t) : f(x^k(t)) = \min_{k'=1,\ldots,n_k}\{f(x^{k'}(t))\}$;

---

at the source node. On the other hand, if the objective is to construct the shortest Hamiltonian path (i.e. a path that connects all nodes, once only), then the $n_k$ ants are randomly distributed over the entire graph. By placing ants on randomly selected nodes, the exploration ability of the search algorithm is improved. Pheromones are either initialized to a constant value, $\tau_0$, or to small random values in the range $[0, \tau_0]$.

Using Algorithm 17.3 as a skeleton, Dorigo *et al.* [216] experimented with the three versions of AS. From the experiments, which focused on solving the traveling salesman problem (both the symmetric and asymmetric versions) and the quadratic assignment problem, it was concluded that the ant-cycle AS performs best, since global information about the quality of solutions is used to determine the amount of pheromones to deposit. For the ant-density and ant-quantity models, the search is not directed by any measure of quality of the solutions.

Dorigo *et al.* [216] also introduced an elitist strategy where, in addition to the reinforcement of pheromones based on equation (17.4), an amount proportional to the length of the best path is added to the pheromones on all the links that form part of the best path [216]. The pheromone update equation changes to

$$\tau_{ij}(t+1) = \tau_{ij}(t) + \Delta\tau_{ij}(t) + n_e \Delta\tau_{ij}^e(t) \tag{17.15}$$

where

$$\Delta\tau_{ij}^e(t) = \begin{cases} \frac{Q}{f(\tilde{x}(t))} & \text{if } (i,j) \in \tilde{x}(t) \\ 0 & \text{otherwise} \end{cases} \tag{17.16}$$

and $e$ is the number of elite ants. In equation (17.16), $\tilde{x}(t)$ is the current best route, with $f(\tilde{x}(t)) = min_{k=1,\ldots,n_k}\{f(x^k(t))\}$. The elitist strategy has as its objective directing the search of all ants to construct a solution to contain links of the current best route.

## 17.1.5   Ant Colony System

The ant colony system (ACS) was developed by Gambardella and Dorigo to improve the performance of AS [77, 215, 301]. ACS differs from AS in four aspects: (1) a different transition rule is used, (2) a different pheromone update rule is defined, (3) local pheromone updates are introduced, and (4) candidate lists are used to favor specific nodes. Each of these modifications is discussed next.

The ACS transition rule, also referred to as a *pseudo-random-proportional* action rule [301], was developed to explicitly balance the exploration and exploitation abilities of the algorithm. Ant $k$, currently located at node $i$, selects the next node $j$ to move to using the rule,

$$j = \begin{cases} \arg\max_{u\in\mathcal{N}_i^k(t)}\{\tau_{iu}(t)\eta_{iu}^{\beta}(t)\} & \text{if } r \leq r_0 \\ J & \text{if } r > r_0 \end{cases} \tag{17.17}$$

where $r \sim U(0,1)$, and $r_0 \in [0,1]$ is a user-specified parameter; $J \in \mathcal{N}_i^k(t)$ is a node randomly selected according to probability

$$p_{iJ}^k(t) = \frac{\tau_{iJ}(t)\eta_{iJ}^{\beta}(t)}{\sum_{u\in\mathcal{N}_i^k}\tau_{iu}(t)\eta_{iu}^{\beta}(t)} \tag{17.18}$$

$\mathcal{N}_i^k(t)$ is a set of valid nodes to visit.

The transition rule in equation (17.17) creates a bias towards nodes connected by short links and with a large amount of pheromone. The parameter $r_0$ is used to balance exploration and exploitation: if $r \leq r_0$, the algorithm exploits by favoring the best edge; if $r > r_0$, the algorithm explores. Therefore, the smaller the value of $r_0$, the less best links are exploited, while exploration is emphasized more. It is important to note that the transition rule is the same as that of AS when $r > r_0$. Also note that the ACS transition rule uses $\alpha = 1$, and is therefore omitted from equation (17.18).

Unlike AS, only the globally best ant (e.g. the ant that constructed the shortest path, $x^+(t)$) is allowed to reinforce pheromone concentrations on the links of the corresponding best path. Pheromone is updated using the global update rule,

$$\tau_{ij}(t+1) = (1-\rho_1)\tau_{ij}(t) + \rho_1\Delta\tau_{ij}(t) \tag{17.19}$$

where

$$\Delta \tau_{ij}(t) = \begin{cases} \frac{1}{f(x^+(t))} & \text{if } (i,j) \in x^+(t) \\ 0 & \text{otherwise} \end{cases} \qquad (17.20)$$

with $f(x^+)(t) = |x^+(t)|$, in the case of finding shortest paths.

The ACS global update rule causes the search to be more directed, by encouraging ants to search in the vicinity of the best solution found thus far. This strategy favors exploitation, and is applied after all ants have constructed a solution.

Gambardella and Dorigo [215, 301] implemented two methods of selecting the path, $x^+(t)$, namely

- **iteration-best**, where $x^+(t)$ represents the best path found during the current iteration, $t$, denoted as $\tilde{x}(t)$, and

- **global-best**, where $x^+(t)$ represents the best path found from the first iteration of the algorithm, denoted as $\hat{x}(t)$.

For the global-best strategy, the search process exploits more by using more global information.

Pheromone evaporation is also treated slightly differently to that of AS. Referring to equation (17.19), for small values of $\rho_1$, the existing pheromone concentrations on links evaporate slowly, while the influence of the best route is dampened. On the other hand, for large values of $\rho_1$, previous pheromone deposits evaporate rapidly, but the influence of the best path is emphasized. The effect of large $\rho_1$ is that previous experience is neglected in favor of more recent experiences. Exploration is emphasized. While the value of $\rho_1$ is usually fixed, a strategy where $\rho_1$ is adjusted dynamically from large to small values will favor exploration in the initial iterations of the search, while focusing on exploiting the best found paths in the later iterations.

In addition to the global updating rule, ACS uses the local updating rule,

$$\tau_{ij}(t) = (1 - \rho_2)\tau_{ij}(t) + \rho_2 \tau_0 \qquad (17.21)$$

with $\rho_2$ also in $(0, 1)$, and $\tau_0$ is a small positive constant. Experimental results on different TSPs showed that $\tau_0 = (n_G L)^{-1}$ provided good results [215]; $n_G$ is the number of nodes in graph $G$, and $L$ is the length of a tour produced by a nearest-neighbor heuristic for TSPs [737] ($L$ can be any rough approximation to the optimal tour length [215]).

ACS also redefines the meaning of the neighborhood set from which next nodes are selected. The set of nodes, $\mathcal{N}_i^k(t)$, is organized to contain a list of candidate nodes. These candidate nodes are preferred nodes, to be visited first. Let $n_l < |\mathcal{N}_i^k(t)|$ denote the number of nodes in the candidate list. The $n_l$ nodes closest (in distance or cost) to node $i$ are included in the candidate list and ordered by increasing distance. When a next node is selected, the best node in the candidate list is selected. If the candidate list is empty, then node $j$ is selected from the remainder of $\mathcal{N}_i^k(t)$. Selection of a non-candidate node can be based on equation (17.18), or alternatively the closest non-candidate $j \in \mathcal{N}_i^k(t)$ can be selected.

---

**Algorithm 17.4** Ant Colony System Algorithm

---

$t = 0$; Initialize parameters $\beta, \rho_1, \rho_2, r_0, \tau_0, n_k$;
Place all ants, $k = 1, \ldots, n_k$;
**for** *each link* $(i, j)$ **do**
    $\tau_{ij}(t) \sim U(0, \tau_0)$;
**end**
$\hat{x}(t) = \emptyset$;
$\bar{f}(\hat{x}(t)) = 0$;
**repeat**
    **for** *each ant* $k = 1, \ldots, n_k$ **do**
        $x^k(t) = \emptyset$;
        **repeat**
            **if** $\exists j \in$ *candidate list* **then**
                Choose $j \in \mathcal{N}_i^k(t)$ from candidate list using equations (17.17) and
                (17.18);
            **end**
            **else**
                Choose non-candidate $j \in \mathcal{N}_i^k(t)$;
            **end**
            $x^k(t) = x^k(t) \cup \{(i, j)\}$;
            Apply local update using equation (17.21);
        **until** *full path has been constructed*;
        Compute $f(x^k(t))$;
    **end**
    $x = x^k(t) : f(x^k(t)) = \min_{k'=1,\ldots,n_k} \{f(x^{k'}(t))\}$;
    Compute $f(x)$;
    **if** $f(x) < f(\hat{x}(t))$ **then**
        $\hat{x}(t) = x$;
        $f(\hat{x}(t)) = f(x)$;
    **end**
    **for** *each link* $(i, j) \in \hat{x}(t)$ **do**
        Apply global update using equation (17.19);
    **end**
    **for** *each link* $(i, j)$ **do**
        $\tau_{ij}(t + 1) = \tau_{ij}(t)$;
    **end**
    $\hat{x}(t + 1) = \hat{x}(t)$;
    $f(\hat{x}(t + 1)) = f(\hat{x}(t))$;
    $t = t + 1$;
**until** *stopping condition is true*;
Return $\hat{x}(t)$ as the solution;

---

ACS is summarized in Algorithm 17.4.

In order to apply ACS to large, complex problems, Dorigo and Gambardella added a local search procedure to iteratively improve solutions [215].

## 17.1.6 Max-Min Ant System

AS was shown to prematurely stagnate for complex problems. Here, stagnation means that all ants follow exactly the same path, and premature stagnation occurs when ants explore little and too rapidly exploit the highest pheromone concentrations. Stützle and Hoos have introduced the max-min ant system (MMAS) to address the premature stagnation problem of AS [815, 816]. The main difference between MMAS and AS is that pheromone intensities are restricted within given intervals. Additional differences are that only the best ant may reinforce pheromones, initial pheromones are set to the max allowed value, and a pheromone smoothing mechanism is used.

The pheromone global update is similar to that of ACS, in equation (17.19), where $\Delta \tau_{ij}(t)$ is calculated on the basis either of the global-best path or of the iteration-best path. The first version of MMAS focused on the iteration-best path, where $\tilde{x}(t)$ represents the best path found during the current iteration [815]. Later versions also included the global-best path, $\hat{x}(t)$, using different strategies [816]:

- Using only the global-best path to determine $\Delta \tau_{ij}(t)$, in which case $\hat{x}(t)$ is the best overall path found from the first iteration. Using only the global best path may concentrate the search too quickly around the global best solution, thus limiting exploration. This problem is reduced if the iteration-best path is used, since best paths may differ considerably from one iteration to the next, thereby allowing more exploration.

- Use mixed strategies, where both the iteration-best and the global-best paths are used to update pheromone concentrations. As default, the iteration-best path is used to favor exploration with the global-best path used periodically. Alternatively, the frequency at which the global-best path is used to update pheromones is increased over time.

- At the point of stagnation, all pheromone concentrations, $\tau_{ij}$, are reinitialized to the maximum allowed value, after which only the iteration-best path is used for a limited number of iterations.

  The $\lambda$-branching factor, with $\lambda = 0.05$, is used to determine the point of stagnation. Gambardella and Dorigo defined the mean $\lambda$-branching factor as an indication of the dimension of the search space [300]. Over time the mean $\lambda$-branching factor decreases to a small value at point of stagnation. The $\lambda_i$-branching factor is defined as the number of links leaving node $i$ with $\tau_{ij}$-values greater than $\lambda \delta_i + \tau_{i,min}$; $\delta_i = \tau_{i,max} - \tau_{i,min}$, where

$$\tau_{i,min} = \min_{j \in \mathcal{N}_i} \{\tau_{ij}\} \tag{17.22}$$

$$\tau_{i,max} = \max_{j \in \mathcal{N}_i} \{\tau_{ij}\} \tag{17.23}$$

and $\mathcal{N}_i$ is the set of all nodes connected to node $i$. If

$$\frac{\sum_{i \in V} \lambda_i}{n_G} < \epsilon \tag{17.24}$$

where $\epsilon$ is a small positive value, it is assumed that the search process stagnated.

Throughout execution of the MMAS algorithm, all $\tau_{ij}$ are restricted within predefined ranges. For the first MMAS version, $\tau_{ij} \in [\tau_{min}, \tau_{max}]$ for all links $(i, j)$, where both $\tau_{min}$ and $\tau_{max}$ were problem-dependent static parameters [815]. If after application of the global update rule $\tau_{ij}(t + 1) > \tau_{max}$, $\tau_{ij}(t + 1)$ is explicitly set equal to $\tau_{max}$. On the other hand, if $\tau_{ij}(t + 1) < \tau_{min}$, $\tau_{ij}(t + 1)$ is set to $\tau_{min}$. Using an upper bound for pheromone concentrations helps to avoid stagnation behavior: pheromone concentrations are prevented from growing excessively, which limits the probability of rapidly converging to the best path. Having a lower limit, $\tau_{min} > 0$, has the advantage that all links have a nonzero probability of being included in a solution. In fact, if $\tau_{min} > 0$ and $\eta_{ij} < \infty$, the probability of choosing a specific link as part of a solution is never zero. The positive lower bound therefore facilitates better exploration.

Stützle and Hoos [816] formally derived that the maximum pheromone concentration is asymptotically bounded. That is, for any $\tau_{ij}$ it holds that

$$\lim_{t \to \infty} \tau_{ij}(t) = \tau_{ij} \leq \frac{1}{1 - \rho} \frac{1}{f^*} \tag{17.25}$$

where $f^*$ is the cost (e.g. length) of the theoretical optimum solution.

This provides a theoretical value for the upper bound, $\tau_{max}$. Since the optimal solution is generally not known, and therefore $f^*$ is an unknown value, MMAS initializes to an estimate of $f^*$, by setting $f^* = f(\hat{x}(t))$, where $f(\hat{x}(t))$ is the cost of the global-best path. This means that the maximum value for pheromone concentrations changes as soon as a new global-best position is found. The pheromone upper bound,

$$\tau_{max}(t) = \left(\frac{1}{1 - \rho}\right) \frac{1}{f(\hat{x}(t))} \tag{17.26}$$

is therefore time-dependent.

Stützle and Hoos [816] derive formally that sensible values for the lower bound on pheromone concentrations can be calculated using

$$\tau_{min}(t) = \frac{\tau_{max}(t)(1 - \sqrt[n]{\hat{p}} n_G)}{(n_G/2 - 1)\sqrt[n]{\hat{p}} n_G} \tag{17.27}$$

where $\hat{p}$ is the probability at which the best solution is constructed. Note that $\hat{p} < 1$ to ensure that $\tau_{min}(t) > 0$. Also note that if $\hat{p}$ is too small, then $\tau_{min}(t) > \tau_{max}(t)$. In such cases, MMAS sets $\tau_{min}(t) = \tau_{max}(t)$, which has the effect that only the heuristic information, $\eta_{ij}$, is used. The value of $\hat{p}$ is a user-defined parameter that needs to be optimized for each new application.

During the initialization step of the MMAS algorithm (refer to Algorithm 17.5), all pheromone concentrations are initialized to the maximum allowed value, $\tau_{max}(0) = \tau_{max}$.

---

**Algorithm 17.5** MMAS Algorithm with Periodic Use of the Global-Best Path

---

Initialize parameters $\alpha, \beta, \rho, n_k, \hat{p}, \tau_{min}, \tau_{max}, f_\lambda$;
$t = 0, \tau_{max}(0) = \tau_{max}, \tau_{min}(0) = \tau_{min}$;
Place all ants, $k = 1, \ldots, n_k$;
$\tau_{ij}(t) = \tau_{max}(0)$, for all links $(i, j)$;
$x^+(t) = \emptyset, f(x^+(t)) = 0$;
**repeat**
    **if** *stagnation point* **then**
        **for** *each link $(i, j)$* **do**
            Calculate $\Delta\tau_{ij}(t)$ using equation (17.28);
            $\tau_{ij}(t + 1) = \tau_{ij}(t) + \Delta\tau_{ij}(t)$;
        **end**
    **end**
    **for** *each ant $k = 1, \ldots, n_k$* **do**
        $x^k(t) = \emptyset$;
        **repeat**
            Select next node $j$ with probability defined in equation (17.6);
            $x^k(t) = x^k(t) \cup \{(i, j)\}$;
        **until** *full path has been constructed*;
        Compute $f(x^k(t))$;
    **end**
    $(t \bmod f_\lambda = 0)$ ? (Iteration Best = false) : (Iteration Best = true);
    **if** *Iteration Best = true* **then**
        Find iteration-best: $x^+(t) = x^k(t) : f(x^k(t)) = \min_{k'=1,\ldots,n_k}\{f(x^{k'}(t))\}$;
        Compute $f(x^+(t))$;
    **end**
    **else**
        Find global-best: $x = x^k(t) : f(x^k(t)) = \min_{k'=1,\ldots,n_k}\{f(x^{k'}(t))\}$;
        Compute $f(x)$;
        **if** $f(x) < f(x^+(t))$ **then**
            $x^+(t) = x$;
            $f(x^+(t)) = f(x)$;
        **end**
    **end**
    **for** *each link $(i, j) \in x^+(t)$* **do**
        Apply global update using equation (17.19);
    **end**
    Constrict $\tau_{ij}(t)$ to be in $[\tau_{min}(t), \tau_{max}(t)]$ for all $(i, j)$;
    $x^+(t + 1) = x^+(t)$;
    $f(x^+(t + 1)) = f(x^+(t))$;
    $t = t + 1$;
    Update $\tau_{max}(t)$ using equation (17.26);
    Update $\tau_{min}(t)$ using equation (17.27);
**until** *stopping condition is true*;
Return $x^+(t)$ as the solution;

---

The max-min approach was also applied to ACS, referred to as MMACS. In this case, the larger the value of $r_0$ (refer to equation (17.17)), the tighter the pheromone concentration limits have to be chosen. A lower $\tau_{max}/\tau_{min}$ ratio has to be used for larger $r_0$ to prevent ants from too often preferring links with high pheromone concentrations.

Even though pheromone concentrations are clamped within a defined range, stagnation still occurred, although less than for AS. To address this problem, a pheromone smoothing strategy is used to reduce the differences between high and low pheromone concentrations. At the point of stagnation, all pheromone concentrations are increased proportional to the difference with the maximum bound, i.e.

$$\Delta\tau_{ij}(t) \propto (\tau_{max}(t) - \tau_{ij}(t)) \tag{17.28}$$

Using this smoothing strategy, stronger pheromone concentrations are proportionally less reinforced than weaker concentrations. This increases the chances of links with low pheromone intensity to be selected as part of a path, and thereby increases the exploration abilities of the algorithm.

Algorithm 17.5 summarizes MMAS, where the iteration-best path is used as default to update pheromones and the global-best is used periodically at frequency $f_\lambda$.

## 17.1.7   Ant-Q

Gambardella and Dorigo [213, 300] developed a variant of ACS in which the local update rule was inspired by Q-learning [891] (also refer to Chapter 6). In Ant-Q, the pheromone notion is dropped to be replaced by Ant-Q value (or AQ-value). The goal of Ant-Q is to learn AQ-values such that the discovery of good solutions is favored in probability.

Let $\mu_{ij}(t)$ denote the AQ-value on the link between nodes $i$ and $j$ at time step $t$. Then the transition rule (action choice rule) is given by (similarly to equation (17.17))

$$j = \begin{cases} \arg\max_{u\in\mathcal{N}_i^k(t)}\{\mu_{iu}^\alpha(t)\eta_{iu}^\beta(t)\} & \text{if } r \le r_0 \\ J & \text{otherwise} \end{cases} \tag{17.29}$$

The parameters $\alpha$ and $\beta$ weigh the importance of the learned AQ-values, $\mu_{ij}$, and the heuristic information, $\eta_{ij}$. The AQ-values express how useful it is to move to node $j$ from the current node $i$. In equation (17.29), $J$ is a random variable selected according to a probability distribution given by a function of the AQ-values, $\mu_{ij}$, and the heuristic values $\eta_{ij}$.

Three different rules have been proposed to select a value for the random variable, $J$:

- For the *pseudo-random* action choice rule, $J$ is a node randomly selected from the set $\mathcal{N}_i^k(t)$ according to the uniform distribution.

- For the *pseudo-random-proportional* action choice rule, $J \in V$ is selected according to the distribution,

$$p_{ij}^k(t) = \begin{cases} \dfrac{\mu_{ij}^\alpha(t)\eta_{ij}^\beta(t)}{\sum_{u \in \mathcal{N}_i^k(t)} \mu_{iu}^\alpha(t)\eta_{iu}^\beta(t)} & \text{if } j \in \mathcal{N}_i^k(t) \\ 0 & \text{otherwise} \end{cases} \qquad (17.30)$$

- For the *random-proportional* action choice rule, $r_0 = 0$ in equation (17.29). The next node is therefore always selected randomly based on the distribution given by equation (17.30).

Gambardella and Dorigo show that the pseudo-random-proportional action choice rule is best for Ant-Q (considering the traveling salesman problem (TSP)).

AQ-values are learned using the updating rule (very similar to that of Q-learning),

$$\mu_{ij}(t+1) = (1-\rho)\mu_{ij}(t) + \rho\left(\Delta\mu_{ij}(t) + \gamma \max_{u \in \mathcal{N}_j^k(t)} \{\mu_{ju}(t)\}\right) \qquad (17.31)$$

where $\rho$ is referred to as the discount factor (by analogy with pheromone evaporation) and $\gamma$ is the learning step size. Note that if $\gamma = 0$, then equation (17.31) reduces to the global update equation of ACS (refer to equation (17.19)).

In Ant-Q, update equation (17.31) is applied for each ant $k$ after each new node $j$ has been selected, but with $\Delta\mu_{ij}(t) = 0$. The effect is that the AQ-value associated with link $(i, j)$ is reduced by a factor of $(1 - \rho)$ each time that the link is selected to form part of a candidate solution. At the same time the AQ-value is reinforced with an amount proportional to the AQ-value of the best link, $(j, u)$, leaving node $j$ such that the future desirability of link $(j, u)$ is increased. The update according to equation (17.31) can be seen as the local update rule, similar to equation (17.21) of ACS.

In addition to the above local update, a delayed reinforcement is used, similar to the global update rule of ACS (refer to equation (17.19)). After all paths have been constructed, one for each ant, AQ-values are updated using equation (17.31), where $\Delta\mu_{ij}(t)$ is calculated using equation (17.20) with $x^+(t)$ either the global-best or the iteration-best path. Gambardella and Dorigo found from their experimental results (with respect to the TSP), that the iteration-best approach is less sensitive to changes in parameter $\gamma$, and is faster in locating solutions of the same quality as the global-best approach.

## 17.1.8   Fast Ant System

Taillard and Gambardella [831, 832] developed the fast ant system (FANT), specifically to solve the quadratic assignment problem (QAP). The main differences between FANT and the other ACO algorithms discussed so far are that (1) FANT uses only one ant, and (2) a different pheromone update rule is applied which does not make use of any evaporation.

The use of only one ant significantly reduces the computational complexity. FANT uses as transition rule equation (17.6), but with $\beta = 0$. No heuristic information is used. The pheromone update rule is defined as

$$\tau_{ij}(t+1) = \tau_{ij}(t) + w_1 \Delta \tilde{\tau}_{ij}(t) + w_2 \Delta \hat{\tau}_{ij}^+(t) \qquad (17.32)$$

where $w_1$ and $w_2$ are parameters to determine the relative reinforcement provided by the current solution at iteration $t$ and the best solution found so far. The added pheromones are calculated as

$$\Delta \tilde{\tau}_{ij}(t) = \begin{cases} 1 & \text{if } (i,j) \in \tilde{x}(t) \\ 0 & \text{otherwise} \end{cases} \qquad (17.33)$$

and

$$\Delta \hat{\tau}_{ij}(t) = \begin{cases} 1 & \text{if } (i,j) \in \hat{x}(t) \\ 0 & \text{otherwise} \end{cases} \qquad (17.34)$$

where $\tilde{x}(t)$ and $\hat{x}(t)$ are respectively the best paths found in iteration $t$ and the global-best path found from the start of the search.

Pheromones are initialized to $\tau_{ij}(0) = 1$. As soon as a new $\hat{x}(t)$ is obtained, all pheromones are reinitialized to $\tau_{ij}(0) = 1$. This step exploits the search area around the global best path, $\hat{x}(t)$. If, at time step $t$, the same solution is found as the current global best solution, the value of $w_1$ is increased by one. This step facilitates exploration by decreasing the contribution, $\Delta \hat{\tau}_{ij}(t)$, associated with the global best path.

### 17.1.9 Antabu

Roux *et al.* [743, 744] and Kaji [433] adapted AS to include a local search using tabu search (refer to Section A.5.2) to refine solutions constructed by each iteration of AS. In addition to using tabu search as local search procedure, the global update rule is changed such that each ant's pheromone deposit on each link of its constructed path is proportional to the quality of the path. Each ant, $k$, updates pheromones using

$$\tau_{ij}(t+1) = (1-\rho)\tau_{ij}(t) + \left( \frac{\rho}{f(x^k(t))} \right) \left( \frac{f(x^-(t)) - f(x^k(t))}{f(\hat{x}(t))} \right) \qquad (17.35)$$

where $f(x^-(t))$ is the cost of the worst path found so far, $f(\hat{x}(t))$ is the cost of the best path found so far, and $f(x^k(t))$ is the cost of the path found by ant $k$. Equation (17.35) is applied for each ant $k$ for each link $(i,j) \in x^k(t)$.

### 17.1.10 AS-rank

Bullnheimer *et al.* [94] proposed a modification of AS to: (1) allow only the best ant to update pheromone concentrations on the links of the global-best path, (2) to use elitist ants, and (3) to let ants update pheromone on the basis of a ranking of the ants. For AS-rank, the global update rule changes to

$$\tau_{ij}(t+1) = (1-\rho)\tau_{ij}(t) + n_e\Delta\hat{\tau}_{ij}(t) + \Delta\tau_{ij}^r(t) \tag{17.36}$$

where

$$\Delta\hat{\tau}_{ij}(t) = \frac{Q}{f(\hat{x}(t))} \tag{17.37}$$

with $\hat{x}(t)$ the best path constructed so far. If $n_e$ elite ants are used, and the $n_k$ ants are sorted such that $f(x^1(t)) \leq f(x^2(t)) \leq \ldots \leq f(x^{n_k}(t))$, then

$$\Delta\tau_{ij}^r(t) = \sum_{\sigma=1}^{n_e} \Delta\tau_{ij}^\sigma(t) \tag{17.38}$$

where

$$\Delta\tau_{ij}^\sigma(t) = \begin{cases} \frac{(n_e-\sigma)Q}{f(x^\sigma(t))} & \text{if } (i,j) \in x^\sigma(t) \\ 0 & \text{otherwise} \end{cases} \tag{17.39}$$

with $\sigma$ indicating the rank of the corresponding ant. This elitist strategy differs from that implemented in AS (refer to Section 17.1.4) in that the contribution of an elite ant to pheromone concentrations is directly proportional to its performance ranking: the better the ranking (i.e. small $\sigma$) the more the contribution.

## 17.1.11 ANTS

Approximated non-deterministic tree search (ANTS) was developed by Maniezzo and Carbonaro [555, 556] as an extension of AS. ANTS differs from AS in: (1) the transition probability calculation, (2) the global update rule, and (3) the approach to avoid stagnation.

ANTS uses the transition probability as defined in equation (17.8). The set $\mathcal{N}_i^k$ contains all feasible moves from the current node $i$. Pheromone intensities are updated after all ants have completed construction of their paths. Pheromones are updated using equations (17.5) and (17.10), but with

$$\Delta\tau_{ij}^k(t) = \tau_0\left(1 - \frac{f(x^k(t)) - \epsilon}{\overline{f}(t) - \epsilon}\right) \tag{17.40}$$

where $f(x^k(t))$ represents the cost of the corresponding path, $x^k(t)$, of ant $k$ at iteration $t$, and $\overline{f}(t)$ is a moving average on the cost of the last $\hat{n}_t$ global-best solutions found by the algorithm. If $f(\hat{x}(t))$ denotes the cost of the global-best solution at iteration $t$, then

$$\overline{f}(t) = \frac{\sum_{t'=t-\hat{n}_t}^{t} f(\hat{x}(t'))}{\hat{n}_t} \tag{17.41}$$

If $t < \hat{n}_t$, the moving average is calculated over the available $t$ best solutions. In equation (17.40), $\epsilon$ is a lower bound to the cost of the optimal solution.

---

**Algorithm 17.6** ANTS Algorithm

---

$t = 0$;
Initialize parameters $\alpha, \beta, \rho, Q, n_k, \tau_0, \hat{n}_t, \epsilon$;
Place all ants, $k = 1, \ldots, n_k$;
**for** *each link $(i, j)$* **do**
    $\tau_{ij}(t) = \tau_0$;
**end**
$\hat{x}(t) = \emptyset$;
$f(\hat{x}(t) = 0$;
**repeat**
    **for** *each ant $k = 1, \ldots, n_k$* **do**
        $x^k(t) = \emptyset$;
        **repeat**
            Select next node $j$ with probability defined in equation (17.8);
            $x^k(t) = x^k(t) \cup \{(i, j)\}$;
        **until** *full path has been constructed*;
        Compute $f(x^k(t))$;
    **end**
    $x = x^k(t) : f(x^k(t)) = \min_{k'=1,\ldots,n_k}\{f(x^{k'}(t))\}$;
    Compute $f(x)$;
    **if** $f(x) < f(\hat{x}(t))$ **then**
        $\hat{x}(t) = x$;
        $f(\hat{x}(t)) = f(x)$;
    **end**
    $\hat{n}_t = \min\{t, \hat{n}_t\}$;
    Calculate moving average, $\overline{f}(t)$, using equation (17.41);
    **for** *each link $(i, j)$* **do**
        Apply global update using equations (17.5), (17.10) and (17.40);
    **end**
    **for** *each link $(i, j)$* **do**
        $\tau_{ij}(t + 1) = \tau_{ij}(t)$;
    **end**
    $\hat{x}(t + 1) = \hat{x}(t)$;
    $f(\hat{x}(t + 1)) = f(\hat{x}(t))$;
    $t = t + 1$;
**until** *stopping condition is true*;
Return $\hat{x}(t)$ as the solution;

---

**Table 17.1** General ACO Algorithm Parameters

| Parameter | Meaning | Comment |
|-----------|---------|---------|
| $n_k$ | Number of ants | |
| $n_t$ | Maximum number of iterations | |
| $\tau_0$ | Initial pheromone amount | not for MMAS |
| $\rho$ | Pheromone persistence | $\rho_1, \rho_2$ for ACS |
| $\alpha$ | Pheromone intensification | $\alpha = 1$ for ACS |
| $\beta$ | Heuristic intensification | not for SACO, ANTS |

The calculation of the amount of pheromone, $\Delta\tau_{ij}^k$, deposited by each ant effectively avoids premature stagnation: pheromone concentrations are reduced if the cost of an ant's solution is lower than the moving average; otherwise, pheromone concentrations are increased. The update mechanism makes it possible to discriminate small achievements in the final iterations of the search algorithm, and avoids exploitation in the early iterations.

ANTS is summarized in Algorithm 17.6.

## 17.1.12   Parameter Settings

Each of the algorithms discussed thus far uses a number of control parameters that influence the performance of the algorithms (refer to Table 17.1). Here performance refers to the quality of solutions found (if any is found) and the time to reach these solutions.

The following parameters are common to most ACO algorithms:

- $n_k$, the number of ants: From the first studies of ACO algorithms, the influence of the number of ants on performance has been studied [215, 216]. An obvious influence of the number of ants relates to computational complexity. The more ants used, the more paths have to be constructed, and the more pheromone deposits calculated. As an example, the computational complexity of AS is $\mathcal{O}(n_c n_G^2 n_k)$, where $n_c = n_t n_k$ is the total number of cycles, $n_t$ is the total number of iterations, and $n_G$ is the number of nodes in the solutions (assuming that all solutions have the same number of nodes).

  The success of ACO algorithms is in the cooperative behavior of multiple ants. Through the deposited pheromones, ants communicate their experience and knowledge about the search space to other ants. The fewer ants used, the less the exploration ability of the algorithm, and consequently the less information about the search space is available to all ants. Small values of $n_k$ may then cause sub-optimal solutions to be found, or early stagnation. Too many ants are not necessarily beneficial (as has been shown, for example, with SACO). With large values of $n_k$, it may take significantly longer for pheromone intensities on good links to increase to higher levels than they do on bad links.

For solving the TSP, Dorigo *et al.* [216] found that $n_k \approx n_G$ worked well. For ACS, Dorigo and Gambardella [215] formally derived that the optimal number of ants can be calculated using

$$n_k = \frac{\log(\phi_1 - 1) - \log(\phi_2 - 2)}{r_0 \log(1 - \rho_2)} \tag{17.42}$$

where $\phi_1 \tau_0$ is the average pheromone concentration on the edges of the last best path before the global update, and $\phi_2 \tau_0$ after the global update. Unfortunately, the optimal values of $\phi_1$ and $\phi_2$ are not known. Again for the TSP, empirical analyses showed that ACS worked best when $(\phi_1 - 1)/(\phi_2 - 1) \approx 0.4$, which gives $n_k = 10$ [215].

It is important to note that the above is for a specific algorithm, used to solve a specific class of problems. The value of $n_k$ should rather be optimized for each different algorithm and problem.

- $n_t$, maximum number of iterations: It is easy to see that $n_t$ plays an important role in ensuring quality solutions. If $n_t$ is too small, ants may not have enough time to explore and to settle on a single path. If $n_t$ is too large, unnecessary computations may be done.

- $\tau_0$, initial pheromone: During the initialization step, all pheromones are either initialized to a constant value, $\tau_0$ (for MMAS, $\tau_0 = \tau_{max}$), or to random values in the range $[0, \tau_0]$. In the case of random values, $\tau_0$ is selected to be a small positive value. If a large value is selected for $\tau_0$, and random values are selected from the uniform distribution, then pheromone concentrations may differ significantly. This may cause a bias towards the links with large initial concentrations, with links that have small pheromone values being neglected as components of the final solution.

There is no easy answer to the very important question of how to select the best values for the control parameters. While many empirical studies have investigated these parameters, suggested values and heuristics to calculate values should be considered with care. These are given for specific algorithms and specific problems, and should not be accepted in general. To maximize the efficiency of any of the algorithms, the values of the relevant control parameters have to be optimized for the specific problem being solved. This can be done through elaborate empirical analyses. Alternatively, an additional algorithm can be used to "learn" the best values of the parameters for the given algorithm and problem [4, 80].

# 17.2   Cemetery Organization and Brood Care

Many ant species, for example *Lasius niger* and *Pheidole pallidula*, exhibit the behavior of clustering corpses to form cemeteries [127] (cited in [77]). Each ant seems to behave individually, moving randomly in space while picking up or depositing (dropping) corpses. The decision to pick up or drop a corpse is based on local information of the ant's current position. This very simple behavior of individual ants results in the emergence of a more complex behavior of cluster formation. It was further observed that cemeteries are sited around spatial heterogeneities [77, 563].

A similar behavior is observed in many ant species, such as *Leptothorax unifasciatus*, in the way that the colony cares for the brood [77, 912, 913]. Larvae are sorted in such a way that different brood stages are arranged in concentric rings. Smaller larvae are located in the center, with larger larvae on the periphery. The concentric clusters are organized in such a way that small larvae receive little individual space, while large larvae receive more space.

While these behaviors are still not fully understood, a number of studies have resulted in mathematical models to simulate the clustering and sorting behaviors. Based on these simulations, algorithms have been implemented to cluster data, to draw graphs, and to develop robot swarms with the ability to sort objects. This section discusses these mathematical models and algorithms. Section 17.2.1 discusses the basic ant colony clustering (ACC) model, and a generalization of this model is given in Section 17.2.2. A minimal ACC approach is summarized in Section 17.2.3.

## 17.2.1 Basic Ant Colony Clustering Model

The first algorithmic implementations that simulate cemetery formation were inspired by the studies of Chrétien of the ants *Lasius niger* [127]. Based on physical experiments, Chrétien derived the probability of an ant dropping a corpse next to an $n$-cluster as being proportional to $1 - (1 - p)^n$ for $n \leq 30$, where $p$ is a fitting parameter [78]. Ants cannot, however, precisely determine the size of clusters. Instead, the size of clusters is determined by the effort to transport the corpse (the corpse may catch on other items, making the walk more difficult). It is likely that the corpse is deposited when the effort becomes too great.

On the basis of of Chrétien's observations, Deneubourg *et al.* [200] developed a model to describe the simple behavior of ants. The main idea is that items in less dense areas should be picked up and dropped at a different location where more of the same type exist. The resulting model is referred to as the basic model.

Assuming only one type of item, all items are randomly distributed on a two-dimensional grid, or lattice. Each grid-point contains only one item. Ants are placed randomly on the lattice, and move in random directions one cell at a time. After each move, an unladen ant decides to pick up an item (if the corresponding cell has an item) based on the probability

$$P_p = \left( \frac{\gamma_1}{\gamma_1 + \lambda} \right)^2 \tag{17.43}$$

where $\lambda$ is the fraction of items the ant perceives in its neighborhood, and $\gamma_1 > 0$. When there are only a few items in the ant's neighborhood, that is $\lambda << \gamma_1$, then $P_p$ approaches 1; hence, objects have a high probability of being picked up. On the other hand, if the ant observes many objects, that is $\lambda >> \gamma_1$, $P_p$ approaches 0, and the probability that the ant will pick up an object is small.

Each loaded ant has a probability of dropping the carried object, given by

$$P_d = \left( \frac{\lambda}{\gamma_2 + \lambda} \right)^2 \tag{17.44}$$

provided that the corresponding cell is empty; $\gamma_2 > 0$. If a large number of items is observed in the neighborhood, i.e. $\lambda >> \gamma_2$, then $P_d$ approaches 1, and the probability of dropping the item is high. If $\lambda << \gamma_2$, then $P_d$ approaches 0.

The fraction of items, $\lambda$, is calculated by making use of a short-term memory for each ant. Each ant keeps track of the last $T$ time steps, and $\lambda$ is simply the number of items observed during these $T$ time steps, divided by the largest number of items that can be observed during the last $T$ time steps. If only one item can be observed during each time step, $\lambda = n_\lambda / T$, where $n_\lambda$ is the number of encountered items.

The basic model is easily expanded to more than one type of item. Let $A$ and $B$ denote two types of items. Then equations (17.43) and (17.44) are used as is, but with $\lambda$ replaced by $\lambda_A$ or $\lambda_B$ depending on the type of item encountered.

## 17.2.2   Generalized Ant Colony Clustering Model

Lumer and Faieta [540] generalized the basic model of Deneubourg *et al.* (refer to Section 17.2.1) to cluster data vectors with real-valued elements for exploratory data analysis applications. This section presents the original Lumer–Faieta algorithm and discusses extensions and modifications to improve the quality of solutions and to speed up the clustering process.

### Lumer--Faieta Algorithm

The first problem to solve, in applying the basic model to real-valued vectors, is to define a distance, or dissimilarity, $d(\mathbf{y}_a, \mathbf{y}_b)$, between data vectors $\mathbf{y}_a$ and $\mathbf{y}_b$, using any applicable norm. For real-valued vectors, the Euclidean distance between the two vectors has been used most frequently to quantify dissimilarity. The next problem is to determine how these dissimilarity measures should be used to group together similar data vectors, in such a way that

- intra-cluster distances are minimized; that is, the distances between data vectors within a cluster should be small to form a compact, condensed cluster.

- inter-cluster distances are maximized; that is, the different clusters should be well separated.

As for the basic model, data vectors are placed randomly on a two-dimensional grid. Ants move randomly around on the grid, while observing the surrounding area of $n_\mathcal{N}^2$ sites, referred to as the $n_\mathcal{N}$-patch. The surrounding area is simply a square neighborhood, $\mathcal{N}_{n_\mathcal{N} \times n_\mathcal{N}}(i)$, of the $n_\mathcal{N} \times n_\mathcal{N}$ sites surrounding the current position, $i$, of the ant. Assume that an ant is on site $i$ at time $t$, and finds data vector $\mathbf{y}_a$. The "local" density, $\lambda(\mathbf{y}_a)$, of data vector $\mathbf{y}_a$ within the ant's neighborhood is then given as

$$\lambda(\mathbf{y}_a) = \max \left\{ 0, \frac{1}{n_{\mathcal{N}}^2} \sum_{\mathbf{y}_b \in \mathcal{N}_{n_{\mathcal{N}} \times n_{\mathcal{N}}}(i)} \left( 1 - \frac{d(\mathbf{y}_a, \mathbf{y}_b)}{\gamma} \right) \right\} \qquad (17.45)$$

where $\gamma > 0$ defines the scale of dissimilarity between items $\mathbf{y}_a$ and $\mathbf{y}_b$. The constant $\gamma$ determines when two items should, or should not be located next to each other. If $\gamma$ is too large, it results in the fusion of individual clusters, clustering items together that do not belong together. If $\gamma$ is too small, many small clusters are formed. Items that belong together are not clustered together. It is thus clear that $\gamma$ has a direct influence on the number of clusters formed. The dissimilarity constant also has an influence on the speed of the clustering process. The larger $\gamma$, the faster the process is.

Using the measure of similarity, $\lambda(\mathbf{y}_a)$, the picking up and dropping probabilities are defined as [540]

$$P_p(\mathbf{y}_a) = \left( \frac{\gamma_1}{\gamma_1 + \lambda(\mathbf{y}_a)} \right)^2 \qquad (17.46)$$

$$P_d(\mathbf{y}_a) = \begin{cases} 2\lambda(\mathbf{y}_a) & \text{if } \lambda(\mathbf{y}_a) < \gamma_2 \\ 1 & \text{if } \lambda(\mathbf{y}_a) \geq \gamma_2 \end{cases} \qquad (17.47)$$

A summary of the Lumer–Faieta ant colony clustering algorithm is given in Algorithm 17.7.

The following aspects of Algorithm 17.7 need clarification:

- $n_t$ is the maximum number of iterations.

- The **grid size**: There should be more sites than data vectors, since items are not stacked. Each site may contain only one item. If the number of sites is approximately the same as the number of items, there are not enough free cells available for ants to move to. Large clusters are formed with a high classification error. On the other hand, if there are many more sites than items, many small clusters may be formed. Also, clustering takes long due to the large distances that items need to be transported.

- The **number of ants**: There should be fewer ants than data vectors. If there are too many ants, it may happen that most of the items are carried by ants. Consequently, density calculation results in close to zero values of $\lambda(\mathbf{y}_a)$. Dropping probabilities are then small, and most ants keep carrying their load without depositing it. Clusters may never be formed. On the other hand, if there are too few ants, it takes longer to form clusters.

- **Local density**: Consider that an ant carries item $\mathbf{y}_a$. If the $n_{\mathcal{N}}^2$ neighboring sites have items similar to $\mathbf{y}_a$, $\lambda(\mathbf{y}_a) \approx 1$. Item $\mathbf{y}_a$ is then dropped with high probability (assuming that the site is empty). On the other hand, if the items are very dissimilar from $\mathbf{y}_a$, $\lambda(\mathbf{y}_a) \approx 0$, and the item has a very low probability of being dropped. In the case of an unladen ant, the pick-up probability is large.

---

**Algorithm 17.7** Lumer–Faieta Ant Colony Clustering Algorithm

---

Place each data vector $\mathbf{y}_a$ randomly on a grid;
Place $n_k$ ants on randomly selected sites;
Initialize values of $\gamma_1, \gamma_2, \gamma$ and $n_t$;
**for** $t = 1$ *to* $n_t$ **do**
    **for** *each ant,* $k = 1, \cdots, n_k$ **do**
        **if** *ant* $k$ *is unladen and the site is occupied by item* $\mathbf{y}_a$ **then**
            Compute $\lambda(\mathbf{y}_a)$ using equation (17.45);
            Compute $P_p(\mathbf{y}_a)$ using equation (17.46);
            **if** $U(0,1) \leq P_p(\mathbf{y}_a)$ **then**
                Pick up item $\mathbf{y}_a$;
            **end**
        **end**
        **else**
            **if** *ant* $k$ *carries item* $\mathbf{y}_a$ *and site is empty* **then**
                Compute $\lambda(\mathbf{y}_a)$ using equation (17.45);
                Compute $P_d(\mathbf{y}_a)$ using equation (17.47);
                **if** $U(0,1) \leq P_d(\mathbf{y}_a)$ **then**
                    Drop item $\mathbf{y}_a$;
                **end**
            **end**
        **end**
        Move to a randomly selected neighboring site not occupied by another ant;
    **end**
**end**

---

- **Patch size**, $n_{\mathcal{N}}$: A larger patch size allows an ant to use more information to make its decision. Better estimates of local densities are obtained, which generally improve the quality of clusters. However, a larger patch size is computationally more expensive and inhibits quick formation of clusters during the early iterations. Smaller sizes reduce computational effort, and less information is used. This may lead to many small clusters.

## Modifications to Lumer--Faieta Algorithm

The algorithm developed by Lumer and Faieta [540] showed success in a number of applications. The algorithm does, however, have the tendency to create more clusters than necessary. A number of modifications have been made to address this problem, and to speed up the search:

- **Different Moving Speeds**: Ants are allowed to move at different speeds [540]. Fast-moving ants form coarser clusters by being less selective in their estimation of the average similarity of a data vector to its neighbors. Slower agents are more accurate in refining the cluster boundaries. Different moving speeds are easily modeled using

$$\lambda(\mathbf{y}_a) = \max\left\{0, \frac{1}{n_{\mathcal{N}}^2} \sum_{\mathbf{y}_b \in \mathcal{N}_{n_{\mathcal{N}} \times n_{\mathcal{N}}}(i)} \left(1 - \frac{d(\mathbf{y}_a, \mathbf{y}_b)}{\gamma(1 - \frac{v-1}{v_{max}})}\right)\right\} \qquad (17.48)$$

where $v \sim U(1, v_{max})$, and $v_{max}$ is the maximum moving speed.

- **Short-Term Memory**: Each ant may have a short-term memory, which allows the ant to remember a limited number of previously carried items and the position where these items have been dropped [540, 601]. If the ant picks up another item, the position of the best matching memorized data item biases the direction of the agent's random walk. This approach helps to group together similar data items.

  Handl *et al.* [344] changed the short-term memory strategy based on the observation that items stored in the memory of one ant may have been moved by some other ant, and is no longer located at the memorized position. For each memorized position, $\mathbf{y}_a$, the ant calculates the local density, $\lambda(\mathbf{y}_a)$. The ant probabilistically moves to that memorized position with highest local density. If the ant does not make the move, its memory is cleared and the dropping decision is based on the standard unbiased dropping probabilities.

- **Behavioral Switches**: Ants are not allowed to start destroying clusters if they have not performed an action for a number of time steps. This strategy allows the algorithm to escape from local optima.

- **Distance/Dissimilarity Measures**:

  For floating-point vectors, the Euclidean distance

  $$d_E(\mathbf{y}_a, \mathbf{y}_b) = \sqrt{\sum_{l=1}^{n_y}(y_{al} - y_{bl})^2} \qquad (17.49)$$

  is the most frequently used distance measure. As an alternative, Yang and Kamel [931] uses the cosine similarity function,

  $$d_C(\mathbf{y}_a, \mathbf{y}_b) = 1 - \text{sim}(\mathbf{y}_a, \mathbf{y}_b) \qquad (17.50)$$

  where

  $$\text{sim}(\mathbf{y}_a, \mathbf{y}_b) = \frac{\sum_{l=1}^{n_y} y_{al} y_{bl}}{\sqrt{\sum_{l=1}^{n_y} y_{al}^2 \sum_{l=1}^{n_y} y_{bl}^2}} \qquad (17.51)$$

  The cosine function, $\text{sim}(\mathbf{y}_a, \mathbf{y}_b)$, computes the angle between the two vectors, $\mathbf{y}_a$ and $\mathbf{y}_b$. As $\mathbf{y}_a$ and $\mathbf{y}_b$ become more similar, $\text{sim}(\mathbf{y}_a, \mathbf{y}_b)$ approaches 1.0, and $d_C(\mathbf{y}_a, \mathbf{y}_b)$ becomes 0.

- **Pick-up and Dropping Probabilities**: Pick-up and dropping probabilities different from equation (17.46) and (17.47) have been formulated specifically in attempts to speed up the clustering process:

      – Yang and Kamel [931] use a continuous, bounded, monotic increasing function to calculate the pick-up probability as

$$P_p = 1 - f_s(\lambda(\mathbf{y}_a)) \tag{17.52}$$

and the dropping probability as

$$P_d = f_s(\lambda(\mathbf{y}_a)) \tag{17.53}$$

where $f_s$ is the sigmoid function.

      – Wu and Shi [919] defines the pick-up probability as

$$P_p = \begin{cases} 1 & \text{if } \lambda(\mathbf{y}_a) \leq 0 \\ 1 - \gamma_1 \lambda(\mathbf{y}_a) & \text{if } 0 < \lambda(\mathbf{y}_a) \leq 1/\gamma_1 \\ 0 & \text{if } \lambda(\mathbf{y}_a) > 1/\gamma_1 \end{cases} \tag{17.54}$$

and the dropping probability as

$$P_d = \begin{cases} 1 & \text{if } \lambda(\mathbf{y}_a) \geq 1/\gamma_2 \\ \gamma_2 \lambda(\mathbf{y}_a) & \text{if } 0 < \lambda(\mathbf{y}_a) < 1/\gamma_2 \\ 0 & \text{if } \lambda(\mathbf{y}_a) \leq 0 \end{cases} \tag{17.55}$$

with $\gamma_1, \gamma_2 > 0$.

      – Handl *et al.* [344] used the probabilities

$$P_p = \begin{cases} 1 & \text{if } \lambda(\mathbf{y}_a) \leq 1 \\ \frac{1}{\lambda(\mathbf{y}_a)^2} & \text{otherwise} \end{cases} \tag{17.56}$$

$$P_d = \begin{cases} 1 & \text{if } \lambda(\mathbf{y}_a) > 1 \\ \lambda(\mathbf{y}_a)^4 & \text{otherwise} \end{cases} \tag{17.57}$$

The above probabilities use a different local density calculation, where an additional constraint is added:

$$\lambda(\mathbf{y}_a) = \begin{cases} \max\{0, \frac{1}{n_{\mathcal{N}}^2} \sum_{\mathbf{y}_b \in \mathcal{N}_{n_{\mathcal{N}} \times n_{\mathcal{N}}}(i)} (1 - \frac{d(\mathbf{y}_a \mathbf{y}_b)}{\gamma})\} & \text{if } \forall \mathbf{y}_b, \, (1 - \frac{d(\mathbf{y}_a, \mathbf{y}_b)}{\gamma}) > 0 \\ 0 & \text{otherwise} \end{cases} \tag{17.58}$$

- **Heterogeneous Ants:** The Lumer–Faieta clustering algorithm uses homogeneous ants. That is, all ants have the same behavior as governed by static control parameters, $n_{\mathcal{N}}$ and $\gamma$. Using static control parameters introduces the problem of finding optimal values for these parameters. Since the best values for these parameters depend on the data set being clustered, the parameters should be optimized for each new data set. Heterogeneous ants address this problem by having different values for the control parameters for each ant. Monmarché *et al.* [601] randomly selected a value for each control parameter for each ant within defined ranges. New values can be selected for each iteration of the algorithm or can remain static.

Handl *et al.* [344] used heterogeneous ants with dynamically changing values for $n_{\mathcal{N}}$ and $\gamma$. Ants start with small values of $n_{\mathcal{N}}$, which gradually increase over time. This approach saves on computational effort during the first iterations and prevents difficulties with initial cluster formation. Over time $n_\mathcal{N}$ increases to prevent a large number of small clusters form forming. The similarity coefficient, $\gamma$, is dynamically adjusted for each ant on the basis of the ant's activity. Activity is reflected by the frequency of successful pick-ups and drops of items. Initially, $\gamma^k(0) \sim U(0,1)$ for each ant $k$. Then, after each iteration,

$$
\gamma^k(t+1) = \begin{cases} \gamma^k(t) + 0.01 & \text{if } \frac{n_f^k(t)}{n_f} > 0.99 \\ \gamma^k(t) - 0.01 & \text{if } \frac{n_f^k(t)}{n_f} \leq 0.99 \end{cases} \tag{17.59}
$$

where $n_f^k(t)$ is the number of failed dropping actions of ant $k$ at time step $t$ and $n_f > 0$.

## 17.2.3   Minimal Model for Ant Clustering

As mentioned in Section 17.2.1, real ants cluster corpses around heterogeneities. The basic model proposed by Deneubourg *et al.* [199] (refer to Section 17.2.1) does not exhibit the same behavior. Martin *et al.* [563] proposed a minimal model to simulate cemetery formation where clusters are formed around heterogeneities. In this minimal model, no pick-up or dropping probabilities are used. Also, ants do not have memories. The dynamics of the minimal model consist of alternating two rules:

- The **pick-up** and **dropping rule**: Whenever an unladen ant observes a corpse (data item) in one or more of its neighboring sites, a corpse is picked up with probability one. If more than one corpse is found, one corpse is randomly selected. After moving at least one step away, the corpse is dropped but only if the ant is surrounded by at least one other corpse. The carried corpse is dropped in a randomly selected empty site.

  Ants do not walk over corpses, and can therefore potentially get trapped.

- **Random walk rule**: Ants move one site at a time, but always in the same direction for a pre-assigned random number of steps. When all these steps have been made, a new random direction is selected, as well as the number of steps. If a corpse is encountered on the ant's path, and that corpse is not picked up, then a new direction is randomly selected.

Martin *et al.* [563] showed that this very simple model provides a more accurate simulation of corpse clustering by real ants.

# 17.3   Division of Labor

The ecological success of social insects has been attributed to work efficiency achieved by division of labor among the workers of a colony, whereby each worker specializes in a subset of the required tasks [625]. Division of labor occurs in biological systems when all the individuals of that system (colony) are co-adapted through divergent specialization, in such a way that there is a fitness gain as a consequence of such specialization [774].

Division of labor has been observed in a number of *eusocial* insects [774, 850], including ants [328, 348, 558, 693, 882], wasps [298, 624], and the honey bee [393, 625, 733]. Here eusociality refers to the presence of [214]:

- cooperation in caring for the young,

- reproductive division of labor, with more or less sterile individuals working on behalf of those individuals involved with reproduction, and

- overlap of generations.

Within these insect colonies, a number of tasks are done, including reproduction, caring for the young, foraging, cemetery organization, waste disposal, and defense. Task allocation and coordination occurs mostly without any central control, especially for large colonies. Instead, individuals respond to simple local cues, for example the pattern of interactions with other individuals [328], or chemical signals. Also very interesting, is that task allocation is dynamic, based on external and internal factors. Even though certain groups of individuals may specialize in certain tasks, task switching occurs when environmental conditions demand such switches. This section provides an overview of the division of labor in insect colonies. Section 17.3 gives a summary of different types of division of labor, and gives an overview of a few mechanisms that facilitate task allocation and coordination. A simple model of division of labor based on response thresholds is discussed in Section 17.3.2. This model is generalized in Section 17.3.3 to dynamic task allocation and specialization.

## 17.3.1   Division of Labor in Insect Colonies

Social insects are characterized by one fundamental type of division of labor i.e. reproductive division of labor [216, 882]. Reproduction is carried out by a very small fraction of individuals, usually by one queen. In addition to this primary division of labor into reproductive and worker castes, a further division of labor exists among the workers. Worker division of labor can occur in the following forms [77]:

- **Temporal polyethism**, also referred to as the age subcaste [77, 216, 393, 625]. With temporal polyethism, individuals of the same age tend to perform the same tasks, and form an age caste. Evidence suggests that social interaction is a mechanism by which workers assess their relative ages [625]. As an example, in honey bee colonies, younger bees tend to the hive while older bees forage [393]. Temporal polyethism is more prominent in honey bees and eusocial wasps [625].

- **Worker polymorphism**, where workers have different morphologies [77, 558, 882]. Workers of the same morphological structure belong to the same morphological caste, and tend to perform the same tasks. For example, minor ants care for the brood, while major ants forage [77]. As another example, the termite species *Termes natalensis* has soldier ants that differ significantly in morphological structure from the worker termites [558]. Morphological castes are more prominent in ant and termite species [625].

- **Individual variability**: Even for individuals in an age or morphological caste, differences may occur among individuals in the frequency and sequence of task performance. Groups of individuals in the same caste that perform the same set of tasks in a given period are referred to as behavioral ants.

A very important aspect of division of labor is its *plasticity*. Ratios of workers that perform different tasks that maintain the colony's viability and reproductive success may vary over time. That is, workers may switch tasks. It can even happen that major ants switch over to perform tasks of minor ants. Variations in these ratios are caused by internal perturbations or external environmental conditions, e.g. climatic conditions, food availability, and predation. Eugene N. Marais reports in [558] one of the earliest observations of task switching. Very important to the survival of certain termite species is the maintenance of underground fungi gardens. During a drought, termites are forced to transport water over very large distances to feed the gardens. During such times, all worker termites, including the soldiers have been observed to perform the task of water transportation. Not even in the case of intentional harm to the nest structure do the ants deviate from the water transportation task. In less disastrous conditions, soldiers will immediately appear on the periphery of any "wounds" in the nest structure, while worker ants repair the structure.

Without a central coordinator, the question is: how are tasks allocated and coordinated, and how is it determined which individuals should switch tasks? While it is widely accepted that environmental conditions, or local cues, play this role, this behavior is still not well understood. Observations of specific species have produced some answers, for example:

- In honey bees, juvenile hormone is involved in the task coordination [393]. It was found that young and older bees have different levels of juvenile hormone: juvenile hormone blood titers typically increases with age. It is low in young bees that work in the hive, and high for foragers. Huang and Robinson observed drops in juvenile hormone when foragers switch to hive-related tasks [393].

- Gordon and Mehdiabadi [328] found for the red harvester ant, *Pogonomyrmex barbatus*, that task switches are based on recent history of brief antennal contacts between ants. They found that the time an ant spent performing midden work was positively correlated with the number of midden workers that the ant encountered while being away from the midden. It is also the case that ants busy with a different task are more likely to begin midden work when their encounters with midden workers exceed a threshold.

- Termite species maintain waste heaps, containing waste transported from fungi gardens [348]. Here division of labor is enforced by aggressive behavior directed towards workers contaminated with garbage. This behavior ensures that garbage workers rarely leave the waste disposal areas.

## 17.3.2 Task Allocation Based on Response Thresholds

Théraulaz *et al.* [841] developed a simple task allocation model based on the notion of response threshold [77, 216, 733]. Response thresholds refer to the likelihood of reacting to task-associated stimuli. Individuals with a low threshold perform a task at a lower level of stimulus than individuals with high thresholds. Individuals become engaged in a specific task when the level of task-associated stimuli exceeds their thresholds. If a task is not performed by individuals, the intensity of the corresponding stimulus increases. On the other hand, intensity decreases as more ants perform the task. Here, the task-associated stimuli serve as stigmergic variable.

This section reviews a model of response thresholds observed in ants and bees.

## Single Task Allocation

Let $s_j$ be the intensity of task-$j$-associated stimuli. The intensity can be a measure of the number of encounters, or a chemical concentration. A response threshold, $\theta_{kj}$, determines the tendency of individual $k$ to respond to the stimulus, $s_j$, associated with task $j$. For a fixed threshold model, individual $k$ engages in task $j$ with probability [77, 216, 841],

$$P_{\theta_{kj}}(s_j) = \frac{s_j^{\omega}}{s_j^{\omega} + \theta_{kj}^{\omega}} \tag{17.60}$$

where $\omega > 1$ determines the steepness of the threshold. Usually, $\omega = 2$. For $s_j << \theta_{kj}$, $P_{\theta_{kj}}(s_j)$ is close to zero, and the probability of performing task $j$ is very small. For $s_j >> \theta_{kj}$, the probability of performing task $j$ is close to one. An alternative threshold response function is [77]

$$P_{\theta_{kj}}(s_j) = 1 - e^{-s_j/\theta_{kj}} \tag{17.61}$$

Assume that there is only one task (the task subscript is therefore dropped in what follows). Also assume only two castes. If $\vartheta_k$ denotes the state of an individual, then $\vartheta_k = 0$ indicates that ant $k$ is inactive, while $\vartheta_k = 1$ indicates that the ant is performing the task. Then,

$$P(\vartheta_k = 0 \rightarrow \vartheta_k = 1) = \frac{s^2}{s^2 + \theta_k^2} \tag{17.62}$$

is the probability that an inactive ant will become active.

An active ant, busy with the task, becomes inactive with probability $P_k = p$ per time unit, i.e.

$$P(\vartheta_k = 1 \rightarrow \vartheta_k = 0) = p \tag{17.63}$$

Therefore, an active ant spends an average $1/p$ time performing the task. An individual may become engaged in the task again, immediately after releasing the task. Stimulus intensity changes over time due to increase in demand and task performance. If $\sigma$ is the increase in demand, $\gamma$ is the decrease associated with one ant performing the task, and $n_{act}$ is the number of active ants, then

$$s(t + 1) = s(t) + \sigma - \gamma n_{act} \tag{17.64}$$

The more ants engaged in the task, the smaller the intensity, $s$, becomes, and consequently, the smaller the probability that an inactive ant will take up the task. On the other hand, if all ants are inactive, or if there are not enough ants busy with the task (i.e. $\sigma > \gamma n_{act}$), the probability increases that inactive ants will participate in the task.

**Allocation of Multiple Tasks**

Let there be $n_j$ tasks, and let $n_{kj}$ be the number of workers of caste $k$ performing task $j$. Each individual has a vector of thresholds, $\theta_k$, where each $\theta_{kj}$ is the threshold allocated to the stimulus of task $j$. After $1/p$ time units of performing task $j$, the ant stops with this task, and selects another task on the basis of the probability defined in equation (17.60). It may be the case that the same task is again selected.

## 17.3.3 Adaptive Task Allocation and Specialization

The fixed response threshold model discussed in Section 17.3.2 has a number of limitations [77, 216, 841]:

- It cannot account for temporal polyethism, since it assumes that individuals are differentiated and are given pre-assigned roles.
- It cannot account for task specialization within castes.
- It is only valid over small time scales where thresholds can be considered constant.

These limitations are addressed by allowing thresholds to vary over time [77, 841]. Thresholds are adapted using a simple reinforcement mechanism. A threshold decreases when the corresponding task is performed, and increases when the task is not performed. Let $\xi$ and $\phi$ respectively represent the learning coefficient and the forgetting coefficient. Then, if ant $k$ performs task $j$ in the next time unit, then

$$\theta_{kj}(t+1) = \theta_{kj}(t) - \xi \tag{17.65}$$

If ant $k$ does not perform task $j$, then

$$\theta_{kj}(t+1) = \theta_{kj}(t) + \phi \tag{17.66}$$

If $t_{kj}$ is the fraction of time that ant $k$ spent on task $j$, then it spent $1 - t_{kj}$ time on other tasks. The change in threshold value is then given by

$$\theta_{kj}(t+1) = \theta_{kj}(t) - t_{kj}\xi + (1 - t_{kj})\phi \tag{17.67}$$

The decision to perform task $j$ is based on the response threshold function (refer to equation (17.60)).

The more ant $k$ performs task $j$, the smaller $\theta_{kj}$ becomes (obviously, $\theta_{kj}$ is bounded in the range $[\theta_{min}, \theta_{max}]$). If demand $s_j$ is high, and remains high, the probability $P_{\theta_{kj}}$ is high. This means that ant $k$ will have an increased probability of choosing task $j$.

However, if too many ants perform task $j$, $s_j$ also decreases by the factor $\gamma n_{act}$. In this case the probability increases that ants will switch tasks, and start to specialize in a new task.

# 17.4  Advanced Topics

This section shows how AAs can be used to optimize functions defined over continuous spaces, and to solve MOPs as well as dynamically changing problems.

## 17.4.1  Continuous Ant Colony Optimization

Ant colony optimization algorithms were originally developed to solve discrete optimization problems, where the values assigned to variables of the solution are constrained by a fixed finite set of discrete values. In order to apply ant colony algorithms to solve continuous optimization problems, the main problem is to determine a way to map the continuous space problem to a graph search problem. A simple solution to this problem is to encode floating-point variables using binary string representations [523]. If a floating-point variable is encoded using an $n$ bit-string, the graph representation $G = (V, E)$, contains $2n$ nodes – two nodes per bit (one for each possible value, i.e. 0 and 1). A link exists between each pair of nodes. Based on this representation, the discrete ACO algorithms can be used to solve the problem. It is, however, the case that binary representation of floating-point values loses precision. This section discusses an ACO algorithm for optimizing continuous spaces without discretizing the solution variables.

The first ACO algorithm for continuous function optimization was developed by Bilchev and Parmee [67, 68]. This approach focused on local optimization. Wodrich and Bilchev [915] extended the local search algorithm to a global search algorithm, which was further improved by Jayaraman et al. [414], Mathur et al. [564] and Rajesh et al. [701]. This global search algorithm, referred to as continuous ACO (CACO), is described next.

The CACO algorithm performs a bi-level search, with a local search component to exploit good regions of the search space, and a global search component to explore bad regions. With reference to Algorithm 17.8, the search is performed by $n_k$ ants, of which $n_l$ ants perform local searches and $n_g$ ants perform global searches. The first step of the algorithm is to create $n_r$ regions. Each region represents a point in the continuous search space. If $\mathbf{x}_i$ denotes the $i$-th region, then $x_{ij} \sim U(x_{min,j}, x_{max,j})$ for each dimension $j = 1, \ldots, n_x$ and each region $i = 1, \ldots, n_r$; $x_{min,j}$ and $x_{max,j}$ are respectively the minimum and maximum values of the domain in the $j$-th dimension. After initialization of the $n_r$ regions, the fitness of each region is evaluated, where the fitness function is simply the continuous function being optimized. Let $f(\mathbf{x}_i)$ denote the fitness of region $\mathbf{x}_i$. The pheromone, $\tau_i$, for each region is initialized to one.

The global search identifies the $n_g$ weakest regions, and uses these regions to find $n_g$

---

**Algorithm 17.8** Continuous Ant Colony Optimization Algorithm

---

Create $n_r$ regions;
$\tau_i(0) = 1$, $i = 1, \dots, n_r$;
**repeat**
    Evaluate fitness, $f(\mathbf{x}_i)$, of each region;
    Sort regions in descending order of fitness;
    Send 90% of $n_g$ global ants for crossover and mutation;
    Send 10% of $n_g$ global ants for trail diffusion;
    Update pheromone and age of $n_g$ weak regions;
    Send $n_l$ ants to probabilistically chosen good regions;
    **for** *each local ant* **do**
        **if** *region with improved fitness is found* **then**
            Move ant to better region;
            Update pheromone;
        **end**
        **else**
            Increase age of region;
            Choose new random direction;
        **end**
        Evaporate all pheromone;
    **end**
**until** *stopping condition is true*;
Return region $\mathbf{x}_i$ with best fitness as solution;

---

new regions to explore. Most of the global ants perform crossover to produce new regions, with a few ants being involved in trail diffusion. For the crossover operation, for each variable $x'_{ij}$ of the offspring, choose a random weak region $\mathbf{x}_i$ and let $x'_{ij} = x_{ij}$, with probability $P_c$ (referred to as the crossover probability). After the crossover step, the offspring $\mathbf{x}'_i$ is mutated by adding Gaussian noise to each variable $x'_{ij}$:

$$x'_{ij} \leftarrow x'_{ij} + N(0, \sigma^2) \qquad (17.68)$$

where the mutation step size, $\sigma$, is reduced at each time step:

$$\sigma \leftarrow \sigma_{max}(1 - r^{(1 - t/n_t)^{\gamma_1}}) \qquad (17.69)$$

where $r \sim U(0,1)$, $\sigma_{max}$ is the maximum step size, $t$ is the current time step, $n_t$ is the maximum number of iterations, and $\gamma_1$ controls the degree of nonlinearity. The nonlinear reduction in mutation steps limits exploration in later iterations to allow for better exploitation.

Trail diffusion implements a form of arithmetic crossover. Two weak regions are randomly selected as parents. For parents $\mathbf{x}_i$ and $\mathbf{x}_l$, the offspring $\mathbf{x}'$ is calculated as

$$x'_j = \gamma_2 x_{ij} + (1 - \gamma_2)x_{lj} \qquad (17.70)$$

where $\gamma_2 \sim U(0,1)$.

For the local search process, each ant $k$ of the $n_l$ local ants selects a region $\mathbf{x}_i$ based on the probability

$$p_i^k(t) = \frac{\tau_i^\alpha(t)\eta_i^\beta(t)}{\sum_{j \in \mathcal{N}_i^k} \tau_j^\alpha(t)\eta_i^\beta(t)} \qquad (17.71)$$

which biases towards the good regions. The ant then moves a distance away from the selected region, $\mathbf{x}_i$, to a new region, $\mathbf{x}_i'$, using

$$\mathbf{x}_i' = \mathbf{x}_i + \Delta\mathbf{x} \qquad (17.72)$$

where

$$\Delta x_{ij} = c_i - m_i a_i \qquad (17.73)$$

with $c_i$ and $m_i$ user-defined parameters, and $a_i$ the age of region $\mathbf{x}_i$. The age of a region indicates the "weakness" of the corresponding solution. If the new position, $\mathbf{x}_i'$, does not have a better fitness than $\mathbf{x}_i$, the age of $\mathbf{x}_i$ is incremented. On the other hand, if the new region has a better fitness, the position vector of the $i$-th region is replaced with the new vector $\mathbf{x}_i'$.

The direction in which an ant moves remains the same if a region of better fitness is found. However, if the new region is less fit, a new direction is randomly chosen.

Pheromone is updated by adding an amount to each $\tau_i$ proportional to the fitness of the corresponding region.

Other approaches to solving continuous optimization problems can be found in [439, 510, 511, 512, 887].

## 17.4.2   Multi-Objective Optimization

One of the first applications of multiple colony ACO algorithms was to solve multi-objective optimization problems (MOP). This section discusses such algorithms.

MOPs are solved by assigning to each colony the responsibility of optimizing one of the objectives. If $n_c$ objectives need to be optimized, a total of $n_c$ colonies are used. Colonies cooperate to find a solution that optimizes all objectives by sharing information about the solutions found by each colony.

Gambardella *et al.* [303] implemented a two-colony system to solve a variant of the vehicle routing problem. A local search heuristic is first used to obtain a feasible solution, $x(0)$, which is then improved by the two colonies, each with respect to a different objective. Each colony maintains its own pheromone matrix, initialized to have a bias towards the initial solution, $x(0)$. If at any time, $t$, one of the colonies, $C_c$, obtains a better solution, then $x(t) = \tilde{x}_c(t)$ where $\tilde{x}_c(t)$ is the iteration-best solution of colony $C_c$. At this point the colonies are reinitialized such that pheromone concentrations are biased towards the new best solution.

Ippolito *et al.* [404] use non-dominated sorting to implement sharing between colonies through local and global pheromone updates. Each colony implements an ACS algorithm to optimize one of the objectives. Separate pheromone matrices are maintained by the different colonies. Both local and global updates are based on

$$\tau_{ij}(t+1) = \tau_{ij}(t) + \gamma\tau_0 f_{ij}(t) \tag{17.74}$$

where $\gamma \in (0,1)$, and $\tau_0$ is the initial pheromone on link $(i,j)$ (all $\tau_{ij}(0) = \tau_0$); $f_{ij}(t)$ is referred to as a fitness value, whose calculation depends on whether equation (17.74) is used for local or global updates. The fitness value is calculated on the basis of a sharing mechanism, performed before applying the pheromone updates. The purpose of the sharing mechanism is to exchange information between the different colonies. Two sharing mechanisms are employed, namely local sharing and global sharing:

- **Local sharing**: After each next move has been selected (i.e. the next node is added to the current path to form a new partial solution), local sharing is applied. For this purpose, non-dominated sorting (refer to Section 9.6.3 and Algorithm 9.11) is used to rank all partial solutions in classes of non-dominance. Each non-dominance class forms a Pareto front, $\mathcal{PF}_p$ containing $n_p = |\mathcal{PF}_p|$ non-dominated solutions. The sharing mechanism assigns a fitness value $f_p^k$ to each solution in Pareto front $\mathcal{PF}_p$, as summarized in Algorithm 17.9. For the purposes of this algorithm, let $\mathbf{x}^k$ denote the solution vector that corresponds to the partial path constructed by ant $k$ at the current time step. Then, $d_{ab}$ represents the normalized Euclidean distance between solution vectors $\mathbf{x}^a$ and $\mathbf{x}^b$, calculated as

$$d_{ab} = \sqrt{\sum_{l=1}^{L}\left(\frac{x_l^a - x_l^b}{x_{max,l} - x_{min,l}}\right)^2} \tag{17.75}$$

where $L \leq n_x$ is the length of the current paths (assuming that all paths are always of the same length), and $\mathbf{x}_{max}$ and $\mathbf{x}_{min}$ respectively are the maximum and minimum values of the variables that make up a solution.

A sharing value $\sigma_{ab}$ is calculated for each distance, $d_{ab}$, as follows:

$$\sigma_{ab} = \begin{cases} 1 - \left(\frac{d_{ab}}{\sigma_{share}}\right)^2 & \text{if } d_{ab} < \sigma_{share} \\ 0 & \text{otherwise} \end{cases} \tag{17.76}$$

where

$$\sigma_{share} = \frac{0.5}{\sqrt{\frac{L}{n_p^*}}} \tag{17.77}$$

$n_p^*$ is the desired number of Pareto optimal solutions. Using the sharing values, a niche count, $\xi_a$, is calculated for each solution vector $\mathbf{x}_a$, as

$$\xi_a = \sum_{b=1}^{n_p} \sigma_{ab} \tag{17.78}$$

The fitness value of each solution $\mathbf{x}^a$ is calculated on the basis of the niche count:

$$f_p^a = \frac{f_p}{\xi_a} \tag{17.79}$$

With reference to the local update using equation (17.74), $f_{ij}(t)$ is the fitness value of the Pareto front to which the corresponding solution belongs, calculated as

$$f_{p+1} = f_{min,p} - \epsilon_p \qquad (17.80)$$

where $\epsilon_p$ is a small positive number, and

$$f_{min,p} = \begin{cases} \min_{a=1,\dots,n_p}\{f_p^a\} & \text{if } p > 1 \\ f_1 & \text{if } p = 1 \end{cases} \qquad (17.81)$$

with $f_1$ an appropriate positive constant.

- **Global sharing**: After completion of all paths, the corresponding solutions are again (as for local sharing) grouped into non-dominance classes. The fitness value, $f_{ij}(t)$ for the global update is calculated for the global-best solution similar to that of local sharing (as given in Algorithm 17.9), but this time with respect to the complete solutions.

---

**Algorithm 17.9** Multiple Colony ACO Local Sharing Mechanism

---

**for** *each Pareto front* $\mathcal{PF}_p, p = 1, \dots, n_p^*$ **do**
    **for** *each solution* $\mathbf{x}^a \in \mathcal{PF}_p$ **do**
        **for** *each solution* $\mathbf{x}^b \in \mathcal{PF}_p$ **do**
            Calculate $d_{ab}$ using equation (17.75);
            Calculate sharing value $\sigma_{ab}$ using equation (17.76);
        **end**
        Calculate niche count $\xi_a$ using equation (17.78);
        Calculate fitness value $f_p^a$ using equation (17.79);
    **end**
    Calculate $f_p$ using equation (17.80);
**end**

---

Mariano and Morales [561] adapt the Ant-Q algorithm (refer to Section 17.1.7) to use multiple colonies, referred to as families, to solve MOPs. Each colony tries to optimize one of the objectives considering the solutions found for the other objectives. The reward received depends on how the actions of a family helped to find trade-off solutions between the rest of the colonies. Each colony, $C_c$, for $c = 1, \dots, n_c$, has the same number of ants, $|C_c|$. During each iteration, colonies find solutions sequentially, with the $|C_c|$ solutions found by the one colony influencing the starting points of the next colony, by initializing AQ-values to bias toward the solutions of the preceding colony. When all colonies have constructed their solutions, the non-dominated solutions are selected. A reward is given to all ants in each of the colonies that helped in constructing a non-dominated solution. The search continues until all solutions found are non-dominated solutions, or until a predetermined number of iterations has been exceeded.

Multi-pheromone matrix methods use more than one pheromone matrix, one for each of the sub-objectives. Assuming only two objectives, Iredi *et al.* [405], uses two pheromone matrices, $\tau_1$ and $\tau_2$, one for each of the objectives. A separate heuristic information matrix is also used for each of the objectives. The AS transition rule is changed to

$$p_{ij}(t) = \begin{cases} \dfrac{\tau_{1,ij}^{\psi\alpha}(t)\tau_{2,ij}^{(1-\psi)\alpha}(t)\eta_{1_{ij}}^{\psi\beta}(t)\eta_{2_{ij}}^{(1-\psi)\beta}(t)}{\sum_{u\in\mathcal{N}_i(t)}\tau_{1,iu}^{\psi\alpha}(t)\tau_{2,iu}^{(1-\psi)\alpha}(t)\eta_{1_{i}u}^{\psi\beta}(t)\eta_{2_{i}u}^{(1-\psi)\beta}(t)} & \text{if } j \in \mathcal{N}_i(t) \\ 0 & \text{if } j \notin \mathcal{N}_i(t) \end{cases} \tag{17.82}$$

where $\psi$ is calculated for each ant as the ratio of the ant index to the total number of ants.

Every ant that generated a non-dominated solution is allowed to update both pheromone matrices, by depositing an amount of $\frac{1}{n_P}$, where $n_P$ is the number of ants that constructed a non-dominated solution. All non-dominated solutions are maintained in an archive.

Doerner *et al.* [205] modified the ACS (refer to Section 17.1.5) by using multiple pheromone matrices, one for each objective. For each ant, a weight is assigned to each objective to weight the contribution of the corresponding pheromone to determining the probability of selecting the next component in a solution. The ACS transition rule is changed to

$$j = \begin{cases} \arg\max_{u\in\mathcal{N}_i(t)}\{(\sum_{c=1}^{n_c} w_c\tau_{c,iu}(t))^\alpha\,\eta_{iu}^\beta(t)\} & \text{if } r \le r_0 \\ J & \text{if } r > r_0 \end{cases} \tag{17.83}$$

where $n_c$ is the number of objectives, $w_c$ is the weight assigned to the $c$-th objective, and $J$ is selected on the basis of the probability

$$p_{iJ}^c(t) = \frac{(\sum_{c=1}^{n_c} w_c\tau_{c,iJ}(t))^\alpha\,\eta_{iJ}^\beta(t)}{\sum_{u\in\mathcal{N}_i}(\sum_{c=1}^{n_c} w_c\tau_{c,iu}(t))^\alpha\,\eta_{iu}^\beta(t)} \tag{17.84}$$

Local pheromone updates are as for the original ACS, but done separately for each objective:

$$\tau_{c,ij} = (1-\rho)\tau_{c,ij} + \rho\tau_0 \tag{17.85}$$

The global update changes to

$$\tau_{c,ij} = (1-\rho)\tau_{c,ij} + \rho\Delta\tau_{c,ij} \tag{17.86}$$

where

$$\Delta\tau_{c,ij} = \begin{cases} 15 & \text{if } (i,j) \in \text{best and second-best solution} \\ 10 & \text{if } (i,j) \in \text{best solution} \\ 5 & \text{if } (i,j) \in \text{second-best solution} \\ 0 & \text{otherwise} \end{cases} \tag{17.87}$$

with the best solutions above referring to non-dominated solutions. An archive of non-dominated solutions is maintained.

Cardoso *et al.* [105] extended the AS to solve MOPs in dynamically changing environments. A pheromone matrix, $\tau_c$, and pheromone amplification coefficient, $\alpha_c$, is maintained for each objective function $f_c$. The AS transition rule is changed to

$$
p_{ij}(t) = \begin{cases} \dfrac{\eta_{ij}\beta(t)\prod_{c=1}^{nc}(\tau_{c,ij})^{\alpha_c}}{\sum_{u\in\mathcal{N}_i(t)}\eta_{iu}\beta(t)\prod_{c=1}^{nc}(\tau_{c,iu})^{\alpha_c}} & \text{if } j \in \mathcal{N}_i(t) \\ 0 & \text{if } j \notin \mathcal{N}_i(t) \end{cases}
\tag{17.88}
$$

All visited links are updated by each ant with an amount $\frac{Q}{f_k(\mathbf{x})}$.

The approach of Iredi *et al.* [405] discussed above makes use of a different heuristic matrix for each objective (refer to equation (17.82)). Barán and Schaerer [49] developed a different approach based on ACS, as a variation of the approach developed by Gambardella *et al.* [303]. A single colony is used, with a single pheromone matrix, but one heuristic matrix for each objective. Assuming two objectives, the ACS transition rule changes to

$$
j = \begin{cases} \arg\max_{u\in\mathcal{N}_i(t)}\{\tau_{iu}(t)\eta_{1,iu}^{\psi\beta}(t)\eta_{2,iu}^{(1-\psi)\beta}(t)\} & \text{if } r \le r_0 \\ J & \text{if } r > r_0 \end{cases}
\tag{17.89}
$$

where $\psi \in [0,1]$ and $J$ is selected on the basis of the probability,

$$
p_{iJ}^k(t) = \frac{\tau_{iJ}(t)\eta_{1,iJ}^{\psi\beta}(t)\eta_{2,iJ}^{(1-\psi)\beta}(t)}{\sum_{u\in\mathcal{N}_i}\tau_{iu}(t)\eta_{1,iu}^{\psi\beta}(t)\eta_{2,iu}^{(1-\psi)\beta}(t)}
\tag{17.90}
$$

For the local update rule, the constant $\tau_0$ is initially calculated using

$$
\tau_0 = \frac{1}{\overline{f}_1\overline{f}_2}
\tag{17.91}
$$

where $\overline{f}_1$ and $\overline{f}_2$ are the average objective values over a set of heuristically obtained solutions (prior to the execution of the ant algorithm) for the two objectives respectively. The constant is, however, changed over time. At each iteration, $\tau_0'$ is calculated using equation (17.91), but calculated over the current set of non-dominated solutions. If $\tau_0' > \tau_0$, then pheromone trails are initialized to $\tau_0 = \tau_0'$; otherwise, the global pheromone update is applied with respect to each solution $\mathbf{x} \in \mathcal{P}$, where $\mathcal{P}$ is the set on non-dominated solutions:

$$
\tau_{ij} = (1-\rho)\tau_{ij} + \frac{\rho}{f_1(\mathbf{x})f_2(\mathbf{x})}
\tag{17.92}
$$

## 17.4.3   Dynamic Environments

For dynamic optimization problems (refer to Section A.9), the search space changes over time. A current good solution may not be a good solution after a change in the environment has occurred. Ant algorithms may not be able to track changing environments due to pheromone concentrations that become too strong [77]. If most of the ants have already settled on the same solution, the high pheromone concentrations on the links representing that solution cause that solution to be constructed by all future ants with very high probability even though a change has occurred. To enable

ACO algorithms to track changing environments, mechanisms have to be employed to favor exploration.

For example, using the transition probability of ACS (refer to equation (17.18)), exploration is increased by selecting a small value for $r_0$ and increasing $\beta$. This will force more random transition decisions, where the new, updated heuristic information creates a bias towards the selection of links that are more desirable according to the changed environment.

An alternative is to use an update rule where only those links that form part of a solution have their pheromone updated, including an evaporation component similar to the local update rule of ACS (refer to Section 17.1.5). Over time the pheromone concentrations on the frequently used links decrease, and these links become less favorable. Less frequently used links will then be explored.

A very simple strategy is to reinitialize pheromone after change detection, but to keep a reference to the previous best solution found. If the location of an environment change can be identified, the pheromone of links in the neighborhood can be reinitialized to a maximum value, forcing these links to be more desirable. If these links turn out to represent bad solution components, reinforcement will be small (since it is usually proportional to the quality of the solution), and over time desirability of the links reduces due to evaporation.

Guntsch and Middendorf [340] proposed to repair solutions when a change occurred. This can be done by applying a local search procedure to all solutions. Alternatively, components affected by change are deleted from the solution, connecting the predecessor and successor of the deleted component. New components (not yet used in the solution) are then inserted on the basis of a greedy algorithm. The position where a new component is inserted is the position that causes the least cost increase, or highest cost decrease (depending on the objective).

Sim and Sun [791] used a multiple colony system, where colonies are repelled by the pheromone of other colonies to promote exploration in the case of changing environments.

Other approaches to cope with changing environments change the pheromone update rules to favor exploration: Li and Gong [520] modify both the local and global update rules. The local update rule is changed to

$$\tau_{ij}(t+1) = (1 - \rho_1(\tau_{ij}(t)))\tau_{ij}(t) + \Delta\tau_{ij}(t) \tag{17.93}$$

where $\rho_1(\tau_{ij})$ is a monotonically increasing function of $\tau_{ij}$, e.g.

$$\rho_1(\tau_{ij}) = \frac{1}{1 + e^{-(\tau_{ij} + \theta)}} \tag{17.94}$$

where $\theta > 0$.

The dynamic changing evaporation constant has the effect that high pheromone values are decreased more than low pheromone values. In the event of an environment change, and if a solution is no longer the best solution, the pheromone concentrations on the corresponding links decrease over time.

The global update is done similarly, but only with respect to the global-best and global-worst solutions, i.e.

$$\tau_{ij}(t+1) = (1 - \rho_2(\tau_{ij}(t)))\tau_{ij}(t) + \gamma_{ij}\Delta\tau_{ij}(t) \tag{17.95}$$

where

$$\gamma_{ij} = \begin{cases} +1 & \text{if } (i,j) \text{ is in the global-best solution} \\ -1 & \text{if } (i,j) \text{ is in the global-worst solution} \\ 0 & \text{otherwise} \end{cases} \tag{17.96}$$

Guntsch and Middendorf [339] proposed three pheromone update rules for dynamic environments. The objective of these update rules is to find an optimal balance between resetting enough information to allow for exploration of new solutions, while keeping enough information of the previous search process to speed up the process of finding a solution. For each of the strategies, a reset value, $\gamma_i \in [0,1]$ is calculated, and pheromone reinitialized using

$$\tau_{ij}(t+1) = (1 - \gamma_i)\tau_{ij} + \gamma_i \frac{1}{n_G - 1} \tag{17.97}$$

where $n_G$ is the number of nodes in the representation graph. The following strategies were proposed:

- **Restart strategy**: For this strategy,

$$\gamma_i = \lambda_R \tag{17.98}$$

  where $\lambda_R \in [0,1]$ is referred to as the strategy-specific parameter. This strategy does not take the location of the environment change into account.

- **$\eta$-strategy**: Heuristic information is used to decide to what degree pheromone values are equalized:

$$\gamma_i = \max\{0, d_{ij}^\eta\} \tag{17.99}$$

  where

$$d_{ij}^\eta = 1 - \frac{\bar{\eta}}{\lambda_\eta \eta_{ij}}, \ \lambda_\eta \in [0, \infty) \tag{17.100}$$

  and

$$\bar{\eta} = \frac{1}{n_G(n_G - 1)} \sum_{i=1}^{n_G} \sum_{j=1, j\neq i} \eta_{ij} \tag{17.101}$$

  Here $\gamma_i$ is proportional to the distance from the changed component, and equalization is done on all links incident to the changed component.

- **$\tau$-strategy**: Pheromone values are used to equalize links closer to the changed component more than further links:

$$\gamma_i = \min\{1, \lambda_\tau d_{ij}^\tau\}, \ \lambda_\tau \in [0, \infty) \tag{17.102}$$

  where

$$d_{ij}^\tau = \max_{\mathcal{N}_{ij}} \left\{ \prod_{(x,y)\in\mathcal{N}_{ij}} \frac{\tau_{xy}}{\tau_{max}} \right\} \tag{17.103}$$

  and $\mathcal{N}_{ij}$ is the set of all paths from $i$ to $j$.

Table **17.2** Ant Algorithm Applications

| Problem | Reference |
| --- | --- |
| Assignment problems | [29, 230, 508, 555, 608, 704, 801] |
| Bioinformatics | [573, 787] |
| Data clustering | [540, 853] |
| Robotics | [574, 912] |
| Routing | [94, 99, 131, 161, 207, 516] |
| Scheduling | [76, 282, 302, 334, 391, 572, 579, 927] |
| Sequential ordering problem | [302] |
| Set covering | [180] |
| Shortest common super-sequence | [591] |
| Text mining | [345, 370, 525, 703] |

# 17.5   Applications

Ant algorithms have been applied to a large number of real-world problems. One of the first applications is that of the ACO to solve the TSP [142,150]. ACO algorithms can, however, be applied to optimization problems for which the following problem-dependent aspects can be defined [77, 216]:

1. An appropriate **graph representation** to represent the discrete search space. The graph should accurately represent all states and transitions between states. A solution representation scheme also has to be defined.

2. An **autocatalytic (positive) feedback process**; that is, a mechanism to up-date pheromone concentrations such that current successes positively influence future solution construction.

3. **Heuristic desirability** of links in the representation graph.

4. A **constraint-satisfaction method** to ensure that only feasible solutions are constructed.

5. A solution construction method which defines the way in which solutions are built, and a state transition probability.

In addition to the above requirements, the solution construction method may specify a **local search heuristic** to refine solutions.

This section provides a detailed explanation of how ACO can be applied to solve the TSP and the QAP. Table 17.2 provides a list of other applications (this list is by no means complete).

## 17.5.1   Traveling Salesman Problem

The traveling salesman problem (TSP) is the first application to which an ACO algorithm was applied [208, 216]. It is an NP-hard combinatorial optimization problem [310], and is the most frequently used problem in ACO literature [208, 213, 214, 216, 300, 301, 339, 340, 433, 592, 814, 815, 821, 852, 900, 939]. This section shows how ACO algorithms can be used to solve the TSP.

## Problem Definition

Given a set of $n_\pi$ cities, the objective is to find a minimal length closed (Hamiltonian) tour that visits each city once. Let $\pi$ represent a solution as a permutation of the cities $\{1, \ldots, n_\pi\}$, where $\pi(i)$ indicates the $i$-th city visited. Then, $\Pi(n_\pi)$ is the set of all permutations of $\{1, \ldots, n_\pi\}$, i.e. the search space. Formally, the TSP is defined as finding the optimal permutation $\pi^*$, where

$$\pi^* = \arg \min_{\pi \in \Pi(n_\pi)} f(\pi) \qquad (17.104)$$

where

$$f(\pi) = \sum_{i,j=1}^{n_\pi} d_{ij} \qquad (17.105)$$

is the objective function, with $d_{ij}$ the distance between cities $i$ and $j$. Let $D = [d_{ij}]_{n_\pi \times n_\pi}$ denote the distance matrix.

Two versions of the TSP are defined based on the characteristics of the distance matrix. If $d_{ij} = d_{ji}$ for all $i, j = 1 \ldots, n_\pi$, then the problem is referred to as the symmetric TSP (STSP). If $d_{ij} \neq d_{ji}$, the distance matrix is asymmetric resulting in the asymmetric TSP (ATSP).

## Problem Representation

The representation graph is the 3-tuple, $G = (V, E, D)$, where $V$ is the set of nodes, each representing one city, $E$ represents the links between cities, and $D$ is the distance matrix which assigns a weight to each link $(i, j) \in E$. A solution is represented as an ordered sequence $\pi = (1, 2, \ldots, n_\pi)$ which indicates the order in which cities are visited.

## Heuristic Desirability

The desirability of adding city $j$ after city $i$ is calculated as

$$\eta_{ij}(t) = \frac{1}{d_{ij}(t)} \qquad (17.106)$$

Reference to the time step $t$ is included here to allow dynamic problems where distances may change over time.

## Constraint Satisfaction

The TSP defines two constraints:

1. All cities must be visited.
2. Each city is visited once only.

To ensure that cities are visited once only, a tabu list is maintained for each partial solution to contain all cities already visited. Let $\Upsilon^k$ denote the tabu list of the $k$-th ant. Then, $\mathcal{N}_i^k(t) = V \backslash \Upsilon^k(t)$ is the set of cities not yet visited after reaching city $i$. The first constraint is satisfied by requiring each solution to contain $n$ cities, and by result of the tabu list.

## Solution Construction

Ants are placed on random cities, and each ant incrementally constructs a solution, by selecting the next city using the transition probability of any of the previously discussed ACO algorithms.

## Local Search

The most popular search heuristics applied to TSP are the 2-opt and 3-opt heuristics [423]. Both heuristics involve exchanging links until a local minimum has been found. The 2-opt heuristic [433, 815, 900] makes two breaks in the tour and recombines nodes in the only other possible way. If the cost of the tour is improved, the modification is accepted. The 2-opt heuristic is illustrated in Figure 17.4(a). The 3-opt heuristic [215, 814, 815, 900] breaks the tour at three links, and recombines the nodes to form two alternative tours. The best of the original and two alternatives are kept. The 3-opt heuristic is illustrated in Figure 17.4(b). For the ATSP, a variation of the 3-opt is implemented to allow only exchanges that do not change the order in which cities are visited.

## 17.5.2 Quadratic Assignment Problem

The quadratic assignment problem (QAP) is possibly the second problem to which an ACO algorithm was applied [216], and is, after the TSP, the most frequently used problem for benchmarking ACO algorithms [216, 304, 340, 557, 743, 744, 835]. The QAP, introduced by Koopmans and Beckman [477], is an NP-hard problem considered as one of the hardest optimization problems. Even instances of relatively small size of

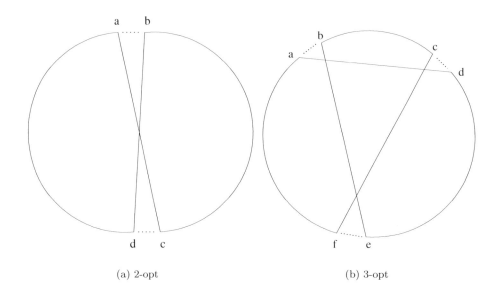

(a) 2-opt                                      (b) 3-opt

**Figure 17.4** 2-opt and 3-opt Local Search Heuristic

$n \geq 25$ cannot be solved exactly. This section discusses approaches to solve the QAP using ACO algorithms.

## Problem Definition

The main objective in QAPs is to assign facilities to locations such that the product of the flow among activities is minimized, under the constraint that all facilities must be allocated to a location and each location can have only one facility assigned to it.

The solution is a permutation $\pi$ of $\{1, \ldots, n_\pi\}$, where $n_\pi$ is the number of facilities and locations. Formally, the objective function to be minimized is defined as

$$f(\pi) = \sum_{i,j=1}^{n_\pi} d_{ij} f_{\pi(i)\pi(j)} \tag{17.107}$$

where $d_{ij}$ is the Euclidean distance between locations $i$ and $j$, $f_{hk}$ characterizes the flow between facilities $h$ and $k$, and $\pi(i)$ is the activity assigned to location $i$.

## Problem Representation

The search space is represented by a graph, $G = (V, E)$. Each node, $v_i \in V, i = 1, \ldots, n_\pi$, represents a location $i$. Associated with each link $(i, j) \in E = \{(i, j) | i, j \in V\}$ is a distance $d_{ij}$. An assignment is a permutation $\pi$ of $\{1, \ldots, n_\pi\}$.

## Solution Construction

Ants are randomly placed on the graph, and each ant constructs a permutation of facilities incrementally. When in location $i$, the ant selects a facility from the set of unallocated facilities and assigns that facility to location $i$. A complete solution is found when all locations have been visited. Different mechanisms can be used to step through the graph of locations, for example, in sequential or random order [557]. Dorigo *et al.* [216] used a different approach where the objective function is expressed as a combination of the potential vectors of distance and flow matrices. If $\mathbf{D} = [d_{ij}]_{n_\pi \times n_\pi}$ is the distance matrix, and $\mathbf{F} = [f_{hk}]_{n_\pi \times n_\pi}$ the flow matrix, define the potential vectors, $\mathbf{D}$ and $\mathbf{F}$, as

$$D_i = \sum_{j=1}^{n_\pi} D_{ij}, \ i = 1, \ldots, n_\pi \tag{17.108}$$

$$F_h = \sum_{k=1}^{n_\pi} F_{hk}, \ h = 1, \ldots, n_\pi \tag{17.109}$$

The next location is deterministically chosen from the free locations as the location with the lowest distance potential, $D_i$.

For location $i$ facility $h$ (using the potential vectors approach), can be selected deterministically as the not yet assigned facility with the highest flow potential. Alternatively, selection can be made probabilistically using an appropriate transition probability function. Stützle and Hoos [816] used the transition probability

$$p_{ih}^k(t) = \begin{cases} \frac{\tau_{ih}}{\sum_{j \in \mathcal{N}_i^k(t)} \tau_{jh}} & \text{if location } i \text{ is still free} \\ 0 & \text{otherwise} \end{cases} \tag{17.110}$$

Maniezzo and Colorni [557] defined the transition probability as

$$p_{ih}^k(t) = \begin{cases} \frac{\alpha \tau_{ih}(t) + (1-\alpha)\eta_{ih}}{\sum_{j \in \mathcal{N}_i^k(t)} \alpha \tau_{jh}(t) + (1-\alpha)\eta_{jh}} & \text{if location } i \text{ is still free} \\ 0 & \text{otherwise} \end{cases} \tag{17.111}$$

Instead of using a constructive approach to solve the QAP, Gambardella *et al.* [304] and Talbi *et al.* [835] used the hybrid AS (HAS) to modify approximate solutions. Each ant receives a complete solution obtained from a nearest-neighbor heuristic and refines the solution via a series of $n_\pi/3$ swaps. For each swap, the first location $i$ is selected randomly from $\{1, 2, \ldots, n_\pi\}$, and the second location, $j$, is selected according to the rule: if $U(0,1) < r$, then $j \sim U\{1, \ldots, n_\pi\} \backslash \{i\}$; otherwise, $j$ is selected according to the probability,

$$p_{ij}^k(t) = \frac{\tau_{i\pi^k(j)} + \tau_{j\pi^k(i)}}{\sum_{\substack{l=1 \\ l \neq i}}^{n_\pi} (\tau_{i\pi^k(l)} + \tau_{l\pi^k(i)})} \tag{17.112}$$

## Heuristic Desirability

Based on the potential vectors defined in equation (17.108) and (17.109) [216],

$$\eta_i = \frac{1}{s_{ih}} \tag{17.113}$$

where $s_{ih} = D_i F_h \in \mathbf{S} = \mathbf{DF}^T$ gives the potential goodness of assigning facility $h$ to location $i$.

## Constraint Satisfaction

The QAP defines two constraints:

- all facilities have to be assigned, and
- only one facility can be allocated per location.

To ensure that facilities are not assigned more than once, each ant maintains a facility tabu list. When a new facility has to be assigned, selection is from the facilities not yet assigned, i.e. $\{1, \ldots, n_\pi\}\backslash\Upsilon^k$. Ants terminate solution construction when all nodes (locations) have been visited, which ensures (together with the tabu list) that each location contains one facility.

## Local Search Heuristics

Gambardella *et al.* [304] consider all possible swaps in a random order. If a swap results in an improved solution, that modification is accepted. The difference in objective function value due to swapping of facilities $\pi(i)$ and $\pi(j)$ is calculated as

$$
\begin{aligned}
\Delta f(\pi, i, j, ) \;=\; & (d_{ii} - d_{jj})(f_{\pi(j)\pi(j)} - f_{\pi(i)\pi(i)}) \\
& + (d_{ij} - d_{ji})(f_{\pi(j)\pi(i)} - f_{\pi(i)\pi(j)}) \\
& + \sum_{\substack{l=1, l\neq i \\ l\neq j}}^{n_\pi} [(d_{li} - d_{lj})(f_{\pi(l)\pi(j)} - f_{\pi(l)\pi(i)}) + \\
& + (d_{il} - d_{jl})(f_{\pi(j)\pi(l)} - f_{\pi(i)\pi(l)})]
\end{aligned}
\tag{17.114}
$$

The modification is accepted if $\Delta f(\pi, i, j) < 0$.

Stützle and Hoos [816] proposed two strategies to decide if a swap is accepted:

- The **first-improvement strategy** accepts the first improving move, similar to standard hill-climbing search.
- The **best-improvement strategy** considers the entire neighborhood and accepts the move that gives the best improvement [557], similar to steepest ascent hill-climbing search.

While the first-improvement strategy is computationally less expensive than the best-improvement strategy, it may require more moves to reach a local optimum.

Roux *et al.* [743, 744] and Talbi *et al.* [835] used TS as local search method. Talbi *et al.* used a recency-based memory, with the size of the tabu list selected randomly between $n/2$ and $3n/2$.

### 17.5.3   Other Applications

Ant algorithms have been applied to many problems. Some of these are summarized in Table 17.2. Note that this table is not a complete list of applications, but just provides a flavor of different applications.

## 17.6   Assignments

1. Consider the following situation: ant $A_1$ follows the shortest of two paths to the food source, while ant $A_2$ follows the longer path. After $A_2$ reached the food source, which path back to the nest has a higher probability of being selected by $A_2$? Justify your answer.

2. Discuss the importance of the forgetting factor in the pheromone trail depositing equation.

3. Discuss the effects of the $\alpha$ and $\beta$ parameters in the transition rule of equation (17.6).

4. Show how the ACO approach to solving the TSP satisfies all the constraints of the TSP.

5. Comment on the following strategy: Let the amount of pheromone deposited be a function of the best route. That is, the ant with the best route, deposits more pheromone. Propose a pheromone update rule.

6. Comment on the similarities and differences between the ant colony approach to clustering and SOMs.

7. For the ant clustering algorithm, explain why
   (a) the 2D-grid should have more sites than number of ants;
   (b) there should be more sites than data vectors.

8. Devise a dynamic forgetting factor for pheromone evaporation.

# Part V

# ARTIFICIAL IMMUNE SYSTEMS

An artificial immune system (AIS) models the natural immune system's ability to detect cells foreign to the body. The result is a new computational paradigm with powerful pattern recognition abilities, mainly applied to anomaly detection.

Different views on how the natural immune system (NIS) functions have been developed, causing some debate among immunologists. These models include the classical view of lymphocytes that are used to distinguish between self and non-self, the clonal selection theory where stimulated B-Cells produce mutated clones, danger theory, which postulates that the NIS has the ability to distinguish between dangerous and non-dangerous foreign cells, and lastly, the network theory where it is assumed that B-Cells form a network of detectors.

Computational models have been developed for all these views, and successfully applied to solve real-world problems. This part[1] aims to provide an introduction to these computational models of the immune system. Chapter 18 provides an overview of the natural immune system, while artificial immune systems are discussed in Chapter 19.

---

[1]This part on artificial immune systems has been written by one of my PhD students, Attie Graaff. Herewith I extend my thanks to Attie for contributing this part to the book, making this a very complete book on computational intelligence.

# Chapter 18

# Natural Immune System

The body has many defense mechanisms, which among others are the skin of the body, the membrane that covers the hollow organs and vessels, and the adaptive immune system. The adaptive immune system reacts to a specific foreign body material or pathogenic material (referred to as *antigen*). During these reactions the adaptive immune system adapts to better detect the encountered antigen and a 'memory' is built up of regular encountered antigen. The obtained memory speeds up and improves the reaction of the adaptive immune system to future exposure to the same antigen. Due to this reason defense reactions are divided into three types: non-specific defense reactions, inherited defense reactions and specific defense reactions [582]. The adaptive immune system forms part of the specific defense reactions.

Different theories exist in the study of immunology regarding the functioning and organizational behavior between lymphocytes in response to encountered antigen. These theories include the classical view, clonal selection theory, network theory, and danger theory. Since the clonal selection, danger theory and network theory are based on concepts and elements within the classical view (see Section 18.1), the classical view will first be discussed in detail to form a bases onto which the other three theories will be explained in Sections 18.5, 18.6 and 18.7.

## 18.1 Classical View

The classical view of the immune system is that the immune system distinguishes between what is normal (*self*) and foreign (*non-self* or antigen) in the body. The recognition of antigens leads to the creation of specialized activated cells, which inactivate or destroy these antigens. The natural immune system mostly consists of lymphocytes and lymphoid organs. These organs are the tonsils and adenoids, thymus, lymph nodes, spleen, Peyer's patches, appendix, lymphatic vessels, and bone marrow. Lymphoid organs are responsible for the growth, development and deployment of the lymphocytes in the immune system. The lymphocytes are used to detect any antigens in the body. The immune system works on the principle of a pattern recognition system, recognizing *non-self* patterns from the *self* patterns [678].

The initial classical view was defined by Burnet [96] as B-Cells and Killer-T-Cells with antigen-specific receptors. Antigens triggered an immune response by interacting

*Computational Intelligence: An Introduction*, Second Edition A.P. Engelbrecht
©2007 John Wiley & Sons, Ltd

with these receptors. This interaction is known as stimulation (or signal 1). It was Bretscher and Cohn [87] who enhanced the initial classical view by introducing the concept of a helper T-Cell (see Section 18.3.1). This is known as the *help* signal (or signal 2). In later years, Lafferty and Cunningham added a co-stimulatory signal to the helper T-Cell model of Bretscher and Cohn. Lafferty and Cunningham [497] proposed that the helper T-Cell is co-stimulated with a signal from an antigen-presenting cell (APC). The motivation for the co-stimulated model was that T-Cells in a body had a stronger response to cells from the same species as the T-Cells in comparison to cells from different species than the T-Cells. Thus, the APC is species specific.

The rest of this chapter explains the development of the different cell types in the immune system, antigens and antibodies, immune reactions and immunity types and the detection process of foreign body material as defined by the different theories.

## 18.2   Antibodies and Antigens

Within the natural immune system, antigens are material that can trigger immune response. An immune response is the body's reaction to antigens so that the antigens are eliminated to prevent damage to the body. Antigens can be either bacteria, fungi, parasites and/or viruses [762]. An antigen must be recognized as foreign (*non-self*). Every cell has a huge variety of antigens in its surface membrane. The foreign antigen is mostly present in the cell of micro-organisms and in the cell membrane of 'donor cells'. Donor cells are transplanted blood cells obtained through transplanted organs or blood. The small segments on the surface of an antigen are called *epitopes* and the small segments on antibodies are called *paratopes* (as shown in Figure 18.1). Epitopes trigger a specific immune response and antibodies' paratopes bind to these epitopes with a certain binding strength, measured as affinity [582].

Antibodies  are chemical proteins. In contradiction to antigens, antibodies form part of *self* and are produced when lymphocytes come into contact with antigen (*non-self*). An antibody has a Y-shape (as shown Figure 18.1). Both arms of the Y consist of two identical heavy and two identical light chains. The chains are distinct into *heavy* and *light* since the heavy chain contains double the number of amino-acids than the light chain. The tips of the arms are called the variable regions and vary from one antibody to another [762]. The variable regions (paratopes) enable the antibody to match antigen and bind to the epitopes of an antigen. After a binding between an antibody and an antigen's epitope, an antigen-antibody-complex is formed, which results into the de-activation of the antigen [582]. There are five classes of antibodies: IgM, IgG, IgA, IgE, IgD [582].

## 18.3   The White Cells

All cells in the body are created in the bone marrow (as illustrated in Figure 18.2). Some of these cells develop into large cell- and particle-devouring white cells known

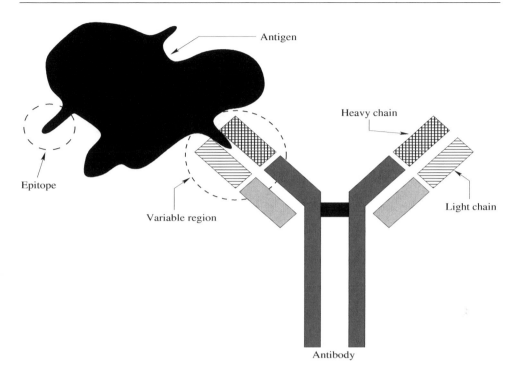

**Figure 18.1** Antigen-Antibody-Complex

as phagocytes [762]. Phagocytes include monocytes, macrophages and neutrophils. Macrophages are versatile cells that secrete powerful chemicals and play an important role in T-Cell activation. Other cells develop into small white cells known as lymphocytes.

## 18.3.1   The Lymphocytes

There are two types of lymphocytes: the T-Cell and B-Cell, both created in the bone marrow. On the surface of the T-Cells and B-Cells are receptor molecules that bind to other cells. The T-Cell binds only with molecules that are on the surface of other cells. The T-Cell first become mature in the thymus, whereas the B-Cell is already mature after creation in the bone marrow. A T-Cell becomes mature if and only if it does not have receptors that bind with molecules that represent *self* cells. It is therefore very important that the T-Cell can differentiate between *self* and *non-self* cells. Thus lymphocytes have different states: immature, mature, memory and annihilated (Figure 18.3 illustrates the life cycle of lymphocytes). These states are discussed in the subsections to follow below. Both T-Cells and B-Cells secrete lymphokines and macrophages secrete monokines. Monokines and lymphokines are known as cytokines and their function is to encourage cell growth, promote cell activation or destroy target cells [762]. These molecules on the surface of a cell are named the major histocompatibility complex molecules (MHC-molecules). Their main function is to bring to light the

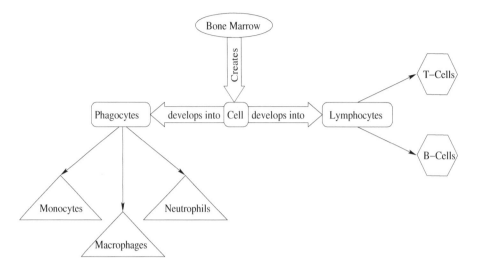

**Figure 18.2** White Cell Types

internal structure of a cell. MHC-molecules are grouped into two classes: Type I and Type II. MHC-molecules of Type I is on the surface of any cell and MHC-molecules of Type II mainly on the surface of B-Cells [678]. There are two types of T-Cells: The Helper-T-Cell and Natural-Killer-T-Cell. Each of these types of lymphocytes are described in detail below.

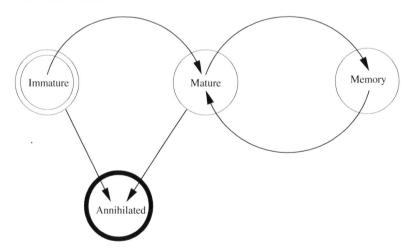

**Figure 18.3** Life Cycle of A Lymphocyte

**The B-Cell**

B-Cells are created in the bone marrow with monomeric IgM-receptors on their surfaces. A monomeric receptor is a chemical compound that can undergo a chemical

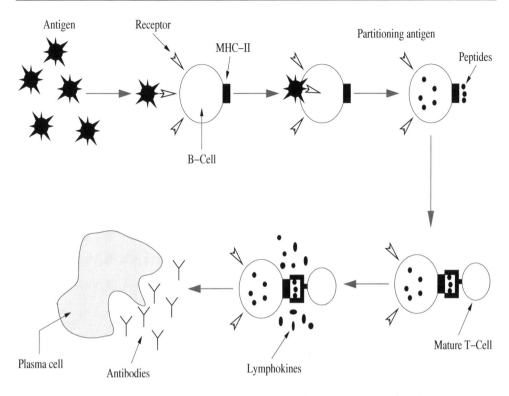

**Figure 18.4** B-Cell Develops into Plasma Cell, Producing Antibodies

reaction with other molecules to form larger molecules. In contrast to T-Cells, B-Cells leave the bone marrow as mature lymphocytes. B-Cells mostly exist in the spleen and tonsils. It is in the spleen and tonsils that the B-Cells develop into plasma cells after the B-Cells come into contact with antigens. After developing into plasma cells, the plasma cells produce antibodies that are effective against antigens [582]. The B-Cell has antigen-specific receptors and recognizes in its natural state the antigens. When contact is made between B-Cell and antigen, clonal proliferation on the B-Cell takes place and is strengthened by Helper-T-Cells (as explained the next subsection). During clonal proliferation two types of cells are formed: plasma cells and memory cells. The function of memory cells is to proliferate to plasma cells for a faster reaction to frequently encountered antigens and produce antibodies for the antigens. A plasma cell is a B-Cell that produces antibodies.

## The Helper-T-Cell (HTC)

When a B-Cell's receptor matches an antigen, the antigen is partitioned into peptides (as shown in Figure 18.4). The peptides are then brought to the surface of the B-Cell by an MHC-molecule of Type II. Macrophages also break down antigen and the broken down antigen is brought to the surface of the macrophage by an MHC-molecule of Type II. The HTC binds to the MHC-molecule on the surface of the B-Cell or macrophage and proliferates or suppresses the B-Cell response to the partitioned cell, by secreting lymphokines. This response is known as the primary response. When the HTC bounds to the MHC with a high affinity, the B-Cell is proliferated. The B-Cell then produces antibodies with the same structure or pattern as represented by the peptides. The production of antibodies is done after a *cloning process* of the B-Cell.

When the HTC does not bind with a high affinity, the B-Cell response is suppressed. Affinity is a force that causes the HTC to elect a MHC on the surface of the B-Cell with which the HTC has a stronger binding to unite, rather than with another MHC with a weaker binding. A higher affinity implies a stronger binding between the HTC and MHC. The antibodies then bind to the antigens' epitopes that have the same complementary structure or pattern. Epitopes are the portions on an antigen that are recognized by antibodies. When a B-Cell is proliferated enough, i.e. the B-Cell frequently detects antigens, it goes into a memory status, and when it is suppressed frequently it becomes annihilated and replaced by a newly created B-Cell. The immune system uses the B-Cells with memory status in a secondary response to frequently seen antigens of the same structure. The secondary response is much faster than the primary response, since no HTC signal or binding to the memory B-Cell is necessary for producing antibodies [678].

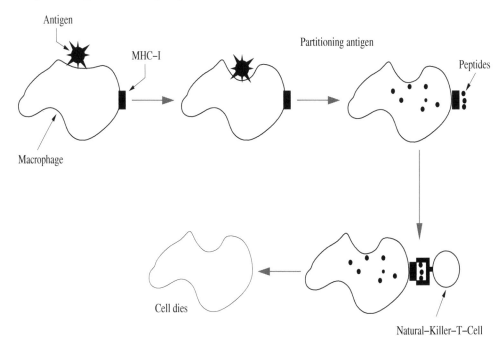

**Figure 18.5** Macrophage and NKTC

### The Natural-Killer-T-Cell (NKTC)

The NKTC binds to MHC-molecules of Type I (as illustrated in Figure 18.5). These MHC-molecules are found on all cells. Their function is to bring to light any viral proteins from a virally infected cell. The NKTC then binds to the MHC-molecule of Type I and destroys not only the virally infected cell but also the NKTC itself [678].

## 18.4   Immunity Types

Immunity can be obtained either naturally or artificially. In both cases immunity can be active or passive. This section discusses the different types of immunity.

**Active naturally-obtained immunity:**   Due to memory-cells, active naturally-obtained immunity is more or less permanent. It develops when the body gets infected or receives foreign red blood cells and actively produces antibodies to deactivate the antigen [582].

**Passive naturally-obtained immunity:**   Passive naturally-obtained immunity is short-lived since antibodies are continuously broken down without creation of new antibodies. New antibodies are not created because the antigens did not activate the *self* immune system. The immunity type develops from IgG-antibodies that are transplanted from the mother to the baby. The secreted IgA-antibodies in mothers-milk are another example of this immunity type and protect the baby from any antigens with which the mother came into contact [582].

**Active artificially-obtained immunity:**   Active artificially-obtained immunity develops when dead organisms or weakened organisms are therapeutically applied. The concept is that special treated organisms keep their antigens without provoking illness-reactions [582].

**Passive artificially-obtained immunity:**   Passive artificially-obtained immunity is obtained when a specific antibody that was produced by another human or animal, is injected into the body for an emergency treatment. Immunity is short-lived, since the immune system is not activated [582].

## 18.5   Learning the Antigen Structure

Learning in the immune system is based on increasing the population size of those lymphocytes that frequently recognize antigens. Learning by the immune system is done by a process known as affinity maturation. Affinity maturation can be broken down into two smaller processes namely, a cloning process and a somatic hyper-mutation process. The cloning process is more generally known as *clonal selection*, which is the proliferation of the lymphocytes that recognize the antigens.

The interaction of the lymphocyte with an antigen leads to an activation of the lymphocyte where upon the cell is proliferated and grown into a clone. When an antigen stimulates a lymphocyte, the lymphocyte not only secretes antibodies to bind to the antigen but also generates mutated clones of itself in an attempt to have a higher binding affinity with the detected antigen. The latter process is known as somatic hyper-mutation. Thus, through repetitive exposure to the antigen, the immune system learns and adapts to the shape of the frequently encountered antigen and moves from a random receptor creation to a repertoire that represents the antigens more precisely. Lymphocytes in a clone produce antibodies if it is a B-Cell and secrete growth factors (lymphokines) in the case of an HTC.

Since antigens determine or select the lymphocytes that need to be cloned, the process is called *clonal selection* [582]. The fittest clones are those which produce antibodies that bind to antigen best (with highest affinity). Since the total number of lymphocytes in the immune system is regulated, the increase in size of some clones decreases the size of other clones. This leads to the immune system forgetting previously learned antigens. When a familiar antigen is detected, the immune system responds with larger cloning sizes. This response is referred to as the secondary immune response [678]. Learning is also based on decreasing the population size of those lymphocytes that seldom or never detect any antigens. These lymphocytes are removed from the immune system. For the affinity maturation process to be successful, the receptor molecule repository needs to be as complete and diverse as possible to recognize any foreign shape [678].

## 18.6   The Network Theory

The network theory was first introduced by Jerne [416, 677], and states that B-Cells are interconnected to form a network of cells. When a B-Cell in the network responds to a foreign cell, the activated B-Cell stimulates all the other B-Cells to which it is connected in the network. Thus, a lymphocyte is not only stimulated by an antigen, but can also be stimulated or suppressed by neighboring lymphocytes, i.e. when a lymphocyte reacts to the stimulation of an antigen, the secretion of antibodies and generation of mutated clones (see Section 18.5) stimulate the lymphocyte's immediate neighbors. This implies that a neighbor lymphocyte can then in turn also react to the stimulation of the antigen-stimulated lymphocyte by generating mutated clones, stimulating the next group of neighbors, etc.

## 18.7   The Danger Theory

The danger theory was introduced by Matzinger [567, 568] and is based on the co-stimulated model of Lafferty and Cunningham [497]. The main idea of the *danger theory* is that the immune system distinguishes between what is dangerous and non-dangerous in the body. The *danger theory* differs from the classical view in that the immune system does not respond to all foreign cells, but only to those foreign cells

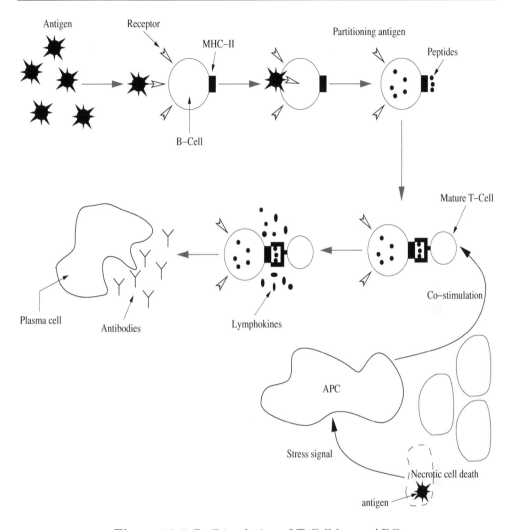

**Figure 18.6** Co-Stimulation of T-Cell by an APC

that are harmful or dangerous to the body. A foreign cell is seen to be dangerous to the body if it causes body cells to stress or die. Matzinger gives two motivational reasons for defining the new theory, which is that the immune system needs to adapt to a changing *self* and that the immune system does not always react on *foreign* or *non-self*.

Although cell death is common within the body, the immune system only reacts to those cell deaths that are not normal programmed cell death (apoptosis), i.e. non-apoptotic or necrotic deaths. When a cell is infected by a virus, the cell itself will send out a stress signal (known as signal 0) of necrotic death to activate the antigen presenting cells (APCs) (as illustrated in Figure 18.6). Thus, co-stimulation of an APC to a helper T-Cell is only possible if the APC was activated with a *danger* or stress signal. Therefore, the neighboring cells of an APC determines the APC's state. Hereon the immune reaction process is as discussed within the classical view (see Section 18.1), where now mature helper T-Cells are presented with a peptide representation of the antigen and co-stimulated by an activated APC.

The different types of signals from a dying or stressed cell is unknown. According to Matzinger these signals could either be defined as the sensing of a certain protein within a cell that leaked after the cell's death or an unexpected connection lost between connected cells after one of the cells died. Thus, if none of the above signals are fired by a cell, no immune response will be triggered by an antigen to activate the antigen presenting cells (APCs).

Thus, from a *danger* immune system perspective a T-Cell only needs to be able to differentiate APCs from any other cells. If an APC activated a T-Cell through co-stimulation, then only will the immune system respond with a clonal proliferation of the B-Cell (as discussed in Section 18.3.1). The B-Cell will then secrete antibodies to bind with the *dangerous* antigen instead of binding to all foreign harmless antigen.

## 18.8   Assignments

1. Identify the main difference between the classicial view of the NIS
   (a) and network theory,
   (b) and danger theory.
2. Discuss the merit of the following statement: *"The lymphocytes in the classical view perform a pattern matching function."*
3. At this point, discuss how the principles of a NIS can be used to solve real-world problems where anomalies need to be detected, such as fraud.
4. Discuss the merit of the following statement: *"A model of the NIS (based on the classical view) can be used as a classifier."*

# Chapter 19

# Artificial Immune Models

Chapter 18 discussed the different theories with regards to the functioning and organizational behavior of the natural immune system (NIS). These theories inspired the modeling of the NIS into an artificial immune system (AIS) for application in non-biological environments. Capabilities of the NIS within each theory, are summarised below:

- The NIS only needs to know the structure of *self*/normal cells.
- The NIS can distinguish between *self* and *foreign/non-self* cells.
- A *foreign* cell can be sensed as *dangerous* or *non-dangerous*.
- Lymphocytes are cloned and mutated to learn and adapt to the structure of the encountered *foreign* cells.
- The build-up of a *memory* on the learned structures of the *foreign* cells.
- A faster secondary response to frequently encountered *foreign* cells, due to the built-up *memory*.
- The cooperation and co-stimulation among lymphocytes to learn and react to encountered *foreign* cells.
- The formation of lymphocyte networks as a result of the cooperation and co-stimulation among lymphocytes.

This chapter discusses some of the existing AIS models. More detail is provided on the most familiar AIS models. These models are either based on or inspired by the capabilities of the NIS (as summarised above) and implement some or all of the basic AIS concepts as listed in Section 19.1. The chapter contains a section for each of the different theories in immunology as discussed in Chapter 18. In each of these sections, the AIS models that are based on or inspired by the applicable theory, are discussed. The rest of the chapter is organised as follows: Section 19.1 lists the basic concepts of an artificial immune system. A basic AIS algorithm is proposed and given in pseudo code. The different parts of the basic AIS pseudo code is briefly discussed. Section 19.2 discusses the AIS models based on or inspired by the *classical view* of the natural immune system. Section 19.3 gives an algorithm in pseudo code that is inspired by the *clonal selection* theory of the natural immune system. Other *clonal selection* inspired models are also discussed. Section 19.4 discusses the AIS models based on or inspired by the *network theory* of the natural immune system. Section 19.5 discusses the AIS models based on or inspired by *danger theory*. Section 19.6 concludes

*Computational Intelligence: An Introduction*, Second Edition A.P. Engelbrecht
©2007 John Wiley & Sons, Ltd

the chapter by giving a brief overview on some of the problem domains to which the artificial immune system has been successfully applied.

# 19.1 Artificial Immune System Algorithm

The capabilities of the NIS summarized above imply that it is mainly the inner working and cooperation between the mature T-Cells and B-Cells that is responsible for the secretion of antibodies as an immune response to antigens. The T-Cell becomes mature in the thymus. A mature T-Cell is self-tolerant, i.e. the T-Cell does not bind to self cells. The mature T-Cell's ability to discriminate between self cells and non-self cells makes the NIS capable of detecting non-self cells. When a receptor of the B-Cell binds to an antigen, the antigen is partitioned and then brought to the surface with an MHC-molecule. The receptor of the T-Cell binds with a certain affinity to the MHC-molecule on the surface of the B-Cell. The affinity can be seen as a measurement to the number of lymphokines that must be secreted by the T-Cell to clonally proliferate the B-Cell into a plasma cell that can produce antibodies. The memory of the NIS on frequently detected antigen is built-up by the B-Cells that frequently proliferate into plasma cells. Thus, to model an AIS, there are a few basic concepts that must be considered:

- There are trained detectors (artificial lymphocytes) that detect non-self patterns with a certain affinity.

- The artificial immune system may need a good repository of self patterns or self and non-self patterns to train the artificial lymphocytes (ALCs) to be self-tolerant.

- The affinity between an ALC and a pattern needs to be measured. The measured affinity indicates to what degree an ALC detects a pattern.

- To be able to measure affinity, the representation of the patterns and the ALCs need to have the same structure.

- The affinity between two ALCs needs to be measured. The measured affinity indicates to what degree an ALC links with another ALC to form a network.

- The artificial immune system has memory that is built-up by the artificial lymphocytes that frequently detect non-self patterns.

- When an ALC detects non-self patterns, it can be *cloned* and the clones can be mutated to have more diversity in the search space.

Using the above concepts as a guideline, the pseudo code in Algorithm 19.1 is a proposal of a basic AIS. Each of the algorithm's parts are briefly explained next.

1. **Initializing $\mathcal{C}$ and determining $D_T$**: The population $\mathcal{C}$ can either be populated with randomly generated ALCs or with ALCs that are initialized with a cross section of the data set to be learned. If a cross section of the data set is used to initialize the ALCs, the complement of the data set will determine the training set $D_T$. These and other initialization methods are discussed for each of the AIS models in the sections to follow.

---

**Algorithm 19.1** Basic AIS Algorithm

---

Initialize a set of ALCs as population $\mathcal{C}$;
Determine the antigen patterns as training set $D_T$;
**while** *some stopping condition(s) not true* **do**
    **for** *each antigen pattern* $\mathbf{z}_p \in D_T$ **do**
        Select a subset of ALCs for exposure to $\mathbf{z}_p$, as population $\mathcal{S} \subseteq \mathcal{C}$;
        **for** *each ALC,* $\mathbf{x}_i \in \mathcal{S}$ **do**
            Calculate the *antigen affinity* between $\mathbf{z}_p$ and $\mathbf{x}_i$;
        **end**
        Select a subset of ALCs with the highest calculated *antigen affinity* as
        population $\mathcal{H} \subseteq \mathcal{S}$;
        Adapt the ALCs in $\mathcal{H}$ with some *selection* method, based on the calculated
        *antigen affinity* and/or the *network affinity* among ALCs in $\mathcal{H}$;
        Update the *stimulation level* of each ALC in $\mathcal{H}$;
    **end**
**end**

---

2. **Stopping condition for the *while*-loop**: In most of the discussed AIS models, the stopping condition is based on convergence of the ALC population or a preset number of iterations.

3. **Selecting a subset, $\mathcal{S}$, of ALCs**: The selected subset $\mathcal{S}$ can be the entire set $\mathcal{P}$ or a number of randomly selected ALCs from $\mathcal{P}$. Selection of $\mathcal{S}$ can also be based on the stimulation level (as discussed below).

4. **Calculating the *antigen affinity***: The antigen affinity is the measurement of similarity or dissimilarity between an ALC and an antigen pattern. The most commonly used measures of affinity in existing AIS models are the Euclidean distance, $r$-continuous matching rule, hamming distance and cosine similarity.

5. **Selecting a subset, $\mathcal{H}$, of ALCs**: In some of the AIS models, the selection of *highest affinity* ALCs is based on a preset affinity threshold. Thus, the selected subset $\mathcal{H}$ can be the entire set $\mathcal{S}$, depending on the preset affinity threshold.

6. **Calculating the *network affinity***: This is the measurement of *affinity* between two ALCs. The different measures of network affinity are the same as those for *antigen affinity*. A preset network affinity threshold determines whether two or more ALCs are linked to form a network.

7. **Adapting the ALCs in subset $\mathcal{H}$**: Adaptation of ALCs can be seen as the maturation process of the ALC, supervised or unsupervised. Some of the *selection* methods that can be used are *negative selection* (or *positive selection*), *clonal selection* and/or some evolutionary technique with mutation operators. ALCs that form a *network* can influence each other to adapt to an antigen. These *selection* methods are discussed for each of the discussed AIS models in the chapter.

8. **Updating the stimulation of an ALC**: The stimulation level is calculated in different ways in existing AIS models. In some AIS models, the stimulation level is seen as the summation of antigen affinities, which determines the resource

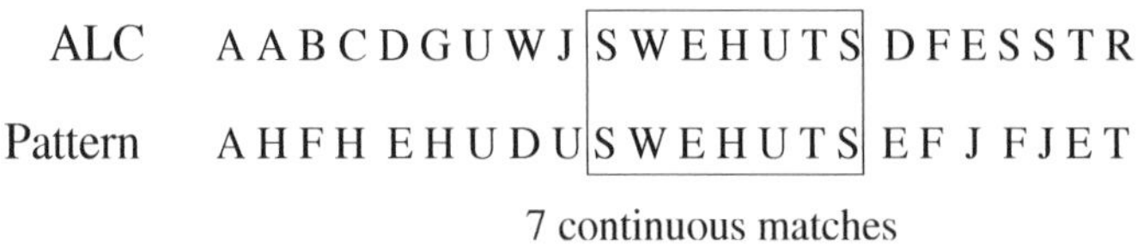

ALC      A A B C D G U W J $\boxed{\text{S W E H U T S}}$ D F E S S T R

Pattern   A H F H  E H U D U $\boxed{\text{S W E H U T S}}$ E F  J  F J E T

7 continuous matches

**Figure 19.1** $r$-Continuous Matching Rule

level of an ALC. The stimulation level can also be used to determine a selection of ALCs as the *memory* set. The *memory* set contains the ALCs that most frequently match an antigen pattern, thus *memory* status is given to these ALCs. The stimulation level is discussed for the different AIS models in the chapter.

## 19.2 Classical View Models

One of the main features in the classical view of the natural immune system is the mature T-Cells, which are self-tolerant, i.e. mature T-Cells have the ability to distinguish between *self* cells and foreign/*non-self* cells. This section discusses AIS models based on or inspired by the *classical view* of the natural immune system. Thus, the discussed AIS models train artificial lymphocytes (ALCs) on a set of *self* patterns to be *self*-tolerant, i.e. the ability to distinguish between *self* and *non-self* patterns. A training technique known as *negative selection* is discussed as well as the different measuring techniques to determine a *match* between an ALC and a *self*/*non-self* pattern.

### 19.2.1 Negative Selection

One of the AIS models based on the classical view of the natural immune system is the model introduced by Forrest *et al.* [281]. In this classical AIS, Forrest *et al.* introduced a training technique known as *negative selection*. In the model, all patterns and ALCs are represented by nominal-valued attributes or as binary strings.

The affinity between an ALC and a pattern is measured using the $r$-continuous matching rule. Figure 19.1 illustrates the $r$-continuous matching rule between an ALC and a pattern.

The $r$-continuous matching rule is a partial matching rule, i.e. an ALC detects a pattern if there are $r$-continuous or more matches in the corresponding positions. $r$ is the degree of affinity for an ALC to detect a pattern. In Figure 19.1 there are seven continuous matches between the ALC and the pattern. Thus, if $r = 4$, the ALC matches the pattern in Figure 19.1, since $7 > r$. If $r > 7$, the ALC does not match the pattern in the figure. A higher value of $r$ indicates a stronger affinity between an ALC and a pattern.

A set of ALCs in the model represents the mature T-Cells in the natural immune system. A training set of self patterns is used to train the set of ALCs using the *negative selection* technique. Algorithm 19.2 summarizes negative selection.

---

**Algorithm 19.2** Training ALCs with *Negative Selection*

---

Set counter $n_a$ as the number of self-tolerant ALCs to train;
Create an empty set of self-tolerant ALCs as $C$;
Determine the training set of self patterns as $D_T$;
**while** *size of $C$ not equal to $n_a$* **do**
    Randomly generate an ALC, $\mathbf{x}_i$;
    Matched=false;
    **for** *each self pattern $\mathbf{z}_p \in D_T$* **do**
        **if** *affinity between $\mathbf{x}_i$ and $\mathbf{z}_p$ is higher than affinity threshold $r$* **then**
            matched=true;
            break;
        **end**
    **end**
    **if** *not matched* **then**
        Add $\mathbf{x}_i$ to set $C$;
    **end**
**end**

---

For each randomly generated ALC, the affinity between the ALC and each self pattern in the training set is calculated. If the affinity between any self pattern and an ALC is higher than the affinity threshold, $r$, the ALC is discarded and a new ALC is randomly generated. The new ALC also needs to be measured against the training set of self patterns. If the affinity between all the self patterns and an ALC is lower than the affinity threshold, $r$, the ALC is added to the self-tolerant set of ALCs. Thus, the set of ALCs is negatively selected, i.e. only those ALCs with a calculated affinity less than the affinity threshold, $r$, will be included in the set of ALCs.

The trained, self-tolerant set of ALCs is then presented with a testing set of self and non-self patterns for classification. The affinity between each training pattern and the set of self-tolerant ALCs is calculated. If the calculated affinity is below the affinity threshold, $r$, the pattern is classified as a self pattern; otherwise, the pattern is classified as a non-self pattern. The training set is monitored by continually testing the ALC set against the training set for changes. A drawback of the negative selection model is that the training set needs to have a good representation of self patterns. Another drawback is the exhaustive replacement of an ALC during the monitoring of the training set until the randomly generated ALC is self-tolerant.

## 19.2.2 Evolutionary Approaches

A different approach is proposed by Kim and Bentley [457] where ALCs are not randomly generated and tested with negative selection, but an evolutionary process is used to evolve ALCs towards non-self and to maintain diversity and generality among the ALCs. The model by Potter and De Jong [689] applies a coevolutionary genetic

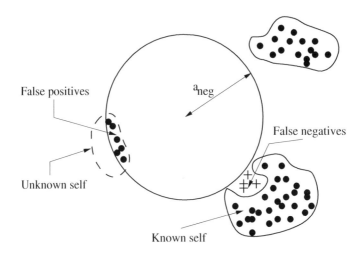

**Figure 19.2** Adapted Negative Selection

algorithm to evolve ALCs towards the selected class of non-self patterns in the training set and further away from the selected class of self patterns. Once the fitness of the ALC set evolves to a point where all the non-self patterns and none of the self patterns are detected, the ALCs represent a description of the concept. If the training set of self and non-self patterns is noisy, the ALC set will be evolved until most of the non-self patterns are detected and as few as possible self patterns are detected. The evolved ALCs can discriminate between examples and counter-examples of a given concept. Each class of patterns in the training set is selected in turn as self and all other classes as non-self to evolve the different concept in the training set.

Gonzalez *et al.* [327] present a negative selection method that is able to train ALCs with continuous-valued self patterns. The ALCs are evolved away from the training set of self patterns and well separated from one another to maximize the coverage of non-self, i.e. the least possible overlap among the evolved set of ALCs is desired. A similar approach is presented in the model of Graaff and Engelbrecht [332]. All patterns are represented as binary strings and the Hamming distance is used as affinity measure. A genetic algorithm is used to evolve ALCs away from the training set of self patterns towards a maximum non-self space coverage and a minimum overlap among existing ALCs in the set. The difference to the model of Gonzalez *et al.* [327], is that each ALC in the set has an affinity threshold. The ALCs are trained with an adapted negative selection method as illustrated in Figure 19.2.

With the adapted negative selection method the affinity threshold, $a_{neg}$, of an ALC is determined by the distance to the closest self pattern from the ALC. The affinity threshold, $a_{neg}$, is used to determine a match with a non-self pattern. Thus, if the measured affinity between a pattern and an ALC is less than the ALC's affinity threshold, $a_{neg}$, the pattern is classified as a non-self pattern. Figure 19.2 also illustrates the drawback of *false positives* and *false negatives* when the ALCs are trained with the adapted negative selection method. These drawbacks are due to an incomplete static self set. The *known self* is the incomplete static self set that is used to train the ALCs and the *unknown self* is the self patterns that are not known during training. The *unknown self* can also represent self patterns that are outliers to the set of *known self* patterns.

Surely all evolved ALCs will cover non-self space, but not all ALCs will detect non-self patterns. Therefore, Graaff and Engelbrecht [332] proposed a transition function, the life counter function, to determine an ALC's status. ALCs with annihilated status are removed in an attempt only to have mature and memory ALCs with optimum classification of non-self patterns.

# 19.3 Clonal Selection Theory Models

The process of *clonal selection* in the natural immune system was discussed in Section 18.5. *Clonal selection* in AIS is the selection of a set of ALCs with the highest calculated affinity with a *non-self* pattern. The selected ALCs are then cloned and mutated in an attempt to have a higher binding affinity with the presented *non-self* pattern. The mutated clones compete with the existing set of ALCs, based on the calculated affinity between the mutated clones and the *non-self* pattern, for survival to be exposed to the next *non-self* pattern. This section discusses some of the AIS models inspired by the *clonal selection* theory and gives the pseudo code for one of these AIS models.

## 19.3.1 CLONALG

The *selection* of a lymphocyte by a detected antigen for clonal proliferation, inspired the modeling of CLONALG. De Castro and Von Zuben [186, 190] presented CLONALG as an algorithm that performs machine-learning and pattern recognition tasks. All patterns are presented as binary strings.

The affinity between an ALC and a non-self pattern is measured as the Hamming distance between the ALC and the non-self pattern. The Hamming distance gives an indication of the similarity between two patterns, i.e. a lower Hamming distance between an ALC and a non-self pattern implies a stronger affinity.

All patterns in the training set are seen as non-self patterns. Algorithm 19.3 summarizes CLONALG for pattern recognition tasks. The different parts of the algorithm are explained next.

The set of ALCs, $\mathcal{C}$, is initialized with $n_a$ randomly generated ALCs. The ALC set is split into a memory set of ALCs, $\mathcal{M}$, and the remaining set of ALCs, $\mathcal{R}$, which are not in $\mathcal{M}$. Thus, $\mathcal{C} = \mathcal{M} \cup \mathcal{R}$ and $|\mathcal{C}| = |\mathcal{M}| + |\mathcal{R}|$ (i.e. $n_a = n_m + n_r$). The assumption in CLONALG is that there is one memory ALC for each of the patterns that needs to be recognized in $D_T$.

Each training pattern, $\mathbf{z}_p$, at random position, $p$, in $D_T$, is presented to $\mathcal{C}$. The affinity between $\mathbf{z}_p$ and each ALC in $\mathcal{C}$ is calculated. A subset of the $n_h$ highest affinity ALCs

---

**Algorithm 19.3** CLONALG Algorithm for Pattern Recognition

---

$t = t_{max}$;
Determine the antigen patterns as training set $D_T$;
Initialize a set of $n_a$ randomly generated ALCs as population $\mathcal{C}$;
Select a subset of $n_m = |D_T|$ memory ALCs, as population $\mathcal{M} \subseteq \mathcal{C}$;
Select a subset of $n_a - n_m$ ALCs, as population $\mathcal{R} \subseteq \mathcal{C}$;
**while** $t > 0$ **do**
    **for** *each antigen pattern* $\mathbf{z}_p \in D_T$ **do**
        Calculate the affinity between $\mathbf{z}_p$ and each of the ALCs in $\mathcal{C}$;
        Select $n_h$ of the highest affinity ALCs with $\mathbf{z}_p$ from $\mathcal{C}$ as subset $\mathcal{H}$;
        Sort the ALCs of $\mathcal{H}$ in ascending order, according to the ALCs affinity;
        Generate $\mathcal{W}$ as the set of clones for each ALC in $\mathcal{H}$;
        Generate $\mathcal{W}'$ as the set of mutated clones for each ALC in $\mathcal{W}$;
        Calculate the affinity between $\mathbf{z}_p$ and each of the ALCs in $\mathcal{W}'$;
        Select the ALC with the highest affinity in $\mathcal{W}'$ as $\hat{\mathbf{x}}$;
        Insert $\hat{\mathbf{x}}$ in $\mathcal{M}$ at position $p$;
        Replace $n_l$ of the lowest affinity ALCs in $\mathcal{R}$ with randomly generated ALCs;
    **end**
    $t = t - 1$;
**end**

---

is selected from $\mathcal{C}$ as subset $\mathcal{H}$. The $n_h$ selected ALCs are then sorted in ascending order of affinity with $\mathbf{z}_p$. Each ALC in the sorted $\mathcal{H}$ are cloned proportional to the calculated affinity with $\mathbf{z}_p$ and added to set $\mathcal{W}$. The number of clones, $n_{ci}$, generated for an ALC, $\mathbf{x}_i$, at position $i$ in the sorted set $\mathcal{H}$, is defined in [190] as

$$n_{ci} = round\left(\frac{\beta \times n_h}{i}\right) \qquad (19.1)$$

where $\beta$ is a multiplying factor and *round* returns the closest integer.

The ALCs in the cloned set, $\mathcal{W}$, are mutated with a mutation rate that is inversely proportional to the calculated affinity, i.e. a higher affinity implies a lower rate of mutation. The mutated clones in $\mathcal{W}$ are added to a set of mutated clones, $\mathcal{W}'$. The affinity between the mutated clones in $\mathcal{W}'$ and the selected training pattern, $\mathbf{z}_p$, is calculated.

The ALC with the highest calculated affinity in $\mathcal{W}'$, $\hat{\mathbf{x}}$, replaces the ALC at position, $p$, in set $\mathcal{M}$, if the affinity of $\hat{\mathbf{x}}$ is higher than the affinity of the ALC in set $\mathcal{M}$. Randomly generated ALCs replace $n_l$ of the lowest affinity ALCs in $\mathcal{R}$. The learning process repeats, until the maximum number of generations, $t_{max}$, has been reached. A modified version of CLONALG has been applied to multi-modal function optimization [190].

## 19.3.2 Dynamic Clonal Selection

In some cases the problem that needs to be optimized consists of self patterns that changes through time. To address these types of problems, the dynamic clonal selection algorithm (DCS) was introduced by Kim and Bentley [461]. The dynamic clonal selection algorithm is based on the AIS proposed by Hofmeyr [372]. The basic concept is to have three different populations of ALCs, categorized into immature, mature and memory ALC populations.

Kim and Bentley [461] explored the effect of three parameters on the adaptability of the model to changing self. These parameters were the tolerization period, the activation threshold and the life span. The tolerization period is a threshold on the number of generations during which ALCs can become self-tolerant. The activation threshold is used as a measure to determine if a mature ALC met the minimum number of antigen matches to be able to become a memory ALC. The life span parameter indicates the maximum number of generations that a mature ALC is allowed to be in the system.

If the mature ALC's life span meets the pre-determined life span parameter value, the mature ALC is deleted from the system. The experimental results with different parameter settings indicated that an increase in the life span with a decrease in the activation threshold resulted in the model to have an increase in detecting true non-self patterns. An increase in the tolerization period resulted in less self patterns being detected falsely than non-self patterns, but only if the self patterns were stable.

With a changing self the increase in tolerization period had no remarkable influence in the false detection of self patterns as non-self patterns. Although the DCS could incrementally learn the structure of self and non-self patterns, it lacked the ability to learn any changes in unseen self patterns. The memory ALCs in the DCS algorithm had infinite lifespan. This feature was omitted in the extended DCS by removing memory ALCs that were not self-tolerant to newly introduced self patterns [460].

A further extension to the DCS was done by introducing *hyper-mutation* on the deleted memory ALCs [459]. The deleted memory ALCs were mutated to seed the immature detector population, i.e. deleted memory ALCs form part of a gene library. Since these deleted memory ALCs contain information (which was responsible for giving them memory status), applying mutation on these ALCs will retain and fine tune the system, i.e. reinforcing the algorithm with previously trained ALCs.

## 19.3.3 Multi-Layered AIS

Knight and Timmis [468] proposed a novel clonal selection inspired model that consists of multi-layers to address some of the shortfalls of the AINE model as discussed in Section 19.4. The defined layers interact to adapt and learn the structure of the presented antigen patterns. The model follows the framework for an AIS as presented in [182]. The basic requirements in [182] for an AIS are the representation of the components (like B-Cells), the evaluation of interactions between these components or their environment (affinity) and adaptability of the model in a dynamic environment (i.e. different selection methods, e.g. negative or clonal selection).

The proposed multi-layered AIS consists of the following layers: the free-antibody layer ($\mathcal{F}$), B-Cell layer ($\mathcal{B}$) and the memory layer ($\mathcal{M}$). The set of training patterns, $D_T$, is seen as antigen. Each layer has an affinity threshold ($a_{\mathcal{F}}$, $a_{\mathcal{B}}$, $a_{\mathcal{M}}$), and a death threshold ($\epsilon_{\mathcal{F}}, \epsilon_{\mathcal{B}}, \epsilon_{\mathcal{M}}$). The death threshold is measured against the length of time since a cell was last stimulated within a specific layer. If the death threshold does not exceed this calculated length of time, the cell dies and is removed from the population in the specific layer. The affinity threshold ($a_{\mathcal{F}}$, $a_{\mathcal{B}}$, $a_{\mathcal{M}}$) determines whether an antigen binds to an entity within a specific layer. The affinity, $f_a$, between an antigen pattern and an entity in a layer is measured using Euclidean distance. Algorithm 19.4 summarizes the multi-layered AIS. The different parts of the algorithm are discussed next.

An antigen, $\mathbf{z}_p$, first enters the free-antibody layer, $\mathcal{F}$. In the free-antibody layer, the antigen pattern is then presented to $n'_f$ free-antibodies. The number of free-antibody bindings is stored in the variable $n_b$. The antigen, $\mathbf{z}_p$, then enters the B-Cell layer, $\mathcal{B}$, and is randomly presented to the B-Cells in this layer until it binds to one of the B-Cells. After binding, the stimulated B-Cell, $\mathbf{u}_k$, produces a clone, $\tilde{\mathbf{u}}_k$, if the stimulation level exceeds a predetermined stimulation threshold, $\gamma_{\mathcal{B}}$. The stimulation level is based on the number of free-antibody bindings, $n_b$, as calculated in the free-antibody layer. The clone is then mutated as $\mathbf{u}'_k$ and added to the B-Cell layer. The stimulated B-Cell, $\mathbf{u}_k$, produces free antibodies that are mutated versions of the original B-Cell. The number of free antibodies produced by a B-Cell is defined in [468] as follows:

$$f_{\mathcal{F}}\left(\mathbf{z}_p, \mathbf{u}_k\right) = \left(a_{max} - f_a\left(\mathbf{z}_p, \mathbf{u}_k\right)\right) \times \alpha \qquad (19.2)$$

where $f_{\mathcal{F}}$ is the number of antibodies that are added to the free-antibody layer, $a_{max}$ is the maximum possible distance between a B-Cell and an antigen pattern in the data space (i.e. lowest possible affinity), $f_a\left(\mathbf{z}_p, \mathbf{u}_k\right)$ is the affinity between antigen, $\mathbf{z}_p$, and B-Cell, $\mathbf{u}_k$, and $\alpha$ is some positive constant.

If an antigen does not bind to any of the B-Cells, a new B-Cell, $\mathbf{u}_{new}$, is created with the same presentation as the unbinded antigen, $\mathbf{z}_p$. The new B-Cell is added to the B-Cell layer resulting in a more diverse coverage of antigen data. The new B-Cell, $\mathbf{u}_{new}$, also produces mutated free antibodies, which are added to the free-antibody layer.

The final layer, $\mathcal{M}$, consists only of memory cells and only responds to new memory cells. The clone, $\tilde{\mathbf{u}}_k$, is presented as a new memory cell to the memory layer, $\mathcal{M}$. The memory cell with the lowest affinity to $\tilde{\mathbf{u}}_k$ is selected as $\mathbf{v}_{min}$. If the affinity between $\tilde{\mathbf{u}}_k$ and $\mathbf{v}_{min}$ is lower than the predetermined memory threshold, $a_{\mathcal{M}}$, and the affinity of $\tilde{\mathbf{u}}_k$ is less than the affinity of $\mathbf{v}_{min}$ with the antigen, $\mathbf{z}_p$ (that was responsible for the creation of the new memory cell, $\tilde{\mathbf{u}}_k$), then $\mathbf{v}_{min}$ is replaced by the new memory cell, $\tilde{\mathbf{u}}_k$. If the affinity between $\tilde{\mathbf{u}}_k$ and $\mathbf{v}_{min}$ is higher than the predetermined memory threshold, $a_{\mathcal{M}}$, the new memory cell, $\tilde{\mathbf{u}}_k$, is added to the memory layer.

The multi-layered model improved the SSAIS model [626] (discussed in Section 19.4) in that the multi-layered model obtained better compression on data while forming stable clusters.

**Algorithm 19.4** A Multi-layered AIS Algorithm

Determine the antigen patterns as training set $D_T$;
**for** *each antigen pattern,* $\mathbf{z}_p \in D_T$ **do**
    $n_b = 0$;
    *bcell_bind* $= false$;
    Randomly select $n_f'$ free antibodies from set $\mathcal{F}$ as subset $\mathcal{E}$;
    **for** *each free antibody,* $\mathbf{y}_j \in \mathcal{E}$ **do**
        Calculate the affinity, $f_a(\mathbf{z}_p, \mathbf{y}_j)$;
        **if** $f_a(\mathbf{z}_p, \mathbf{y}_j) < a_{\mathcal{F}}$ **then**
            $n_b + +$;
            Remove free antibody $\mathbf{y}_j$ from set $\mathcal{F}$;
        **end**
    **end**
    **for** *each randomly selected B-Cell,* $\mathbf{u}_k$, *at index position,* $k$, *in B-Cell set,* $\mathcal{B}$ **do**
        Calculate the affinity, $f_a(\mathbf{z}_p, \mathbf{u}_k)$;
        **if** $f_a(\mathbf{z}_p, \mathbf{u}_k) < a_{\mathcal{B}}$ **then**
            *bcell_bind* $= true$;
            break;
        **end**
    **end**
    **if** *not bcell_bind* **then**
        Initialize new B-Cell, $\mathbf{u}_{new}$, with $\mathbf{z}_p$ and add to B-Cell set $\mathcal{B}$;
        Produce $f_{\mathcal{F}}(\mathbf{z}_p, \mathbf{u}_{new})$ free antibodies and add to free-antibody set $\mathcal{F}$;
    **end**
    **else**
        Produce $f_{\mathcal{F}}(\mathbf{z}_p, \mathbf{u}_k)$ free antibodies and add to free-antibody set $\mathcal{F}$;
        Update stimulation level of $\mathbf{u}_k$ by adding $n_b$;
        **if** *stimulation level of* $\mathbf{u}_k \geq \gamma_{\mathcal{B}}$ **then**
            Clone B-Cell, $\mathbf{u}_k$, as $\tilde{\mathbf{u}}_k$;
            Mutate $\tilde{\mathbf{u}}_k$ as $\mathbf{u}_k'$;
            Add $\mathbf{u}_k'$ to B-Cell set $\mathcal{B}$;
            Select memory cell, $\mathbf{v}_{min}$, from $\mathcal{M}$ as memory cell with lowest affinity, $f_a$, to $\tilde{\mathbf{u}}_k$;
            **if** $f_a(\tilde{\mathbf{u}}_k, \mathbf{v}_{min}) < a_{\mathcal{M}}$ **then**
                Add $\tilde{\mathbf{u}}_k$ to memory set $\mathcal{M}$;
            **end**
            **else**
                **if** $f_a(\mathbf{z}_p, \tilde{\mathbf{u}}_k) < f_a(\mathbf{z}_p, \mathbf{v}_{min})$ **then**
                    Replace memory cell, $\mathbf{v}_{min}$, with clone, $\tilde{\mathbf{u}}_k$, in memory set, $\mathcal{M}$;
                **end**
            **end**
        **end**
    **end**
**end**
**for** *each cell,* $\mathbf{x}_i$, *in set* $\mathcal{F} \cup \mathcal{B} \cup \mathcal{M}$ **do**
    **if** *living time of* $\mathbf{x}_i > \epsilon_{\mathcal{F},\mathcal{B},\mathcal{M}}$ **then**
        Remove $\mathbf{x}_i$ from corresponding set;
    **end**
**end**

# 19.4 Network Theory Models

This section discusses the AIS models based on or inspired by the network theory of the natural immune system (as discussed in Section 18.6). The ALCs in a network based AIS *interact* with each other to learn the structure of a *non-self* pattern, resulting in the formation of ALC networks. The ALCs in a network co-stimulates and/or co-suppress each other to adapt to the *non-self* pattern. The stimulation of an ALC is based on the calculated affinity between the ALC and the *non-self* pattern and/or the calculated affinity between the ALC and network ALCs as co-stimulation and/or co-suppression.

## 19.4.1 Artificial Immune Network

The network theory was first modeled by Timmis and Neal [846] resulting in the artificial immune network (AINE). AINE defines the new concept of artificial recognition balls (ARBs), which are bounded by a resource limited environment. In summary, AINE consists of a population of ARBs, links between the ARBs, a set of antigen training patterns (of which a cross section is used to initialize the ARBs) and some clonal operations for learning. An ARB represents a region of antigen space that is covered by a certain type of B-Cell.

ARBs that are close to each other (similar) in antigen space are connected with weighted edges to form a number of individual network structures. The similarity (affinity) between ARBs and between the ARBs and an antigen is measured using Euclidean distance. Two ARBs are connected if the affinity between them is below the network affinity threshold, $a_n$, calculated as the average distance between all the antigen patterns in the training set, $D_T$. Algorithm 19.5 summarizes AINE.

For each iteration, all training patterns in set $D_T$ are presented to the set of ARBs, $\mathcal{A}$. After each iteration, each ARB, $\mathbf{w}_i$, calculates its stimulation level, $l_{arb,i}$, and allocates resources (i.e. B-Cells) based on its stimulation level as defined in equation (19.7). The stimulation level, $l_{arb,i}$, of an ARB, $\mathbf{w}_i$, is calculated as the summation of the antigen stimulation, $l_{a,i}$, the network stimulation, $l_{n,i}$, and the network suppression, $s_{n,i}$. The stimulation level of an ARB is defined as [626]

$$l_{arb,i}(\mathbf{w}_i) = l_{a,i}(\mathbf{w}_i) + l_{n,i}(\mathbf{w}_i) + s_{n,i}(\mathbf{w}_i) \tag{19.3}$$

with

$$l_{a,i}(\mathbf{w}_i) = \sum_{k=0}^{|\mathcal{L}_{a,i}|} 1 - \mathcal{L}_{a,ik} \tag{19.4}$$

$$l_{n,i}(\mathbf{w}_i) = \sum_{k=0}^{|\mathcal{L}_{n,i}|} 1 - \mathcal{L}_{n,ik} \tag{19.5}$$

$$s_{n,i}(\mathbf{w}_i) = - \sum_{k=0}^{|\mathcal{L}_{n,i}|} \mathcal{L}_{n,ik} \tag{19.6}$$

---

**Algorithm 19.5** Artificial Immune Network (AINE)

---

Normalise the training data;
Initialise the ARB population, $\mathcal{A}$, using a randomly selected cross section of the normalised training data;
Initialise the antigen set, $D_T$, with the remaining normalised training data;
Set the maximum number of available resources, $n_{r,max}$;
**for** *each ARB,* $\mathbf{w}_i \in \mathcal{A}$ **do**
    **for** *each ARB,* $\mathbf{w}_j \in \mathcal{A}$ **do**
        Calculate the ARB affinity, $f_a(\mathbf{w}_i, \mathbf{w}_j)$;
        **if** $f_a(\mathbf{w}_i, \mathbf{w}_j) < a_n$ *and* $i \neq j$ **then**
            Add $f_a(\mathbf{w}_i, \mathbf{w}_j)$ to the set of network stimulation levels, $\mathcal{L}_{n,i}$;
        **end**
    **end**
**end**
**while** *stopping condition not true* **do**
    **for** *each antigen,* $\mathbf{z}_p \in D_T$ **do**
        **for** *each ARB,* $\mathbf{w}_i \in \mathcal{A}$ **do**
            Calculate the antigen affinity, $f_a(\mathbf{z}_p, \mathbf{w}_i)$;
            **if** $f_a(\mathbf{z}_p, \mathbf{w}_i) < a_n$ **then**
                Add $f_a(\mathbf{z}_p, \mathbf{w}_i)$ to the set of antigen stimulation levels, $L_{a,i}$;
            **end**
        **end**
    **end**
    Allocate resources to the set of ARBs, $\mathcal{A}$, using Algorithm 19.6;
    Clone and mutate remaining ARBs in $\mathcal{A}$;
    Integrate mutated clones into $\mathcal{A}$;
**end**

---

where $L_{a,i}$ is the normalised set of affinities between an ARB, $\mathbf{w}_i$, and all antigen $\mathbf{z}_p \in D_T$ for which $f_a(\mathbf{z}_p, \mathbf{w}_i) < a_n$. The antigen stimulation, $l_{a,i}$ is therefore the sum of all antigen affinities smaller than $a_n$. Note that $0 \leq \mathcal{L}_{a,ik} \leq 1$ for all $\mathcal{L}_{a,ik} \in \mathcal{L}_{a,i}$. The network stimulation, $l_{n,i}$, and the network suppression, $s_{n,i}$, are the sum of affinities between an ARB and all its connected neighbours as defined in equations (19.5) and (19.6) respectively. Here $\mathcal{L}_{n,i}$ is the normalised set of affinities between an ARB, $\mathbf{w}_i$, and all other ARBs in set $\mathcal{A}$. Network stimulation and suppression are based on the summation of the distances to the $|\mathcal{L}_{n,i}|$ linked neighbours of the ARB, $\mathbf{w}_i$. Network suppression, $s_{n,i}$ represents the dissimilarity between an ARB and its neighbouring ARBs. Network suppression keeps the size of the ARB population under control.

The number of resources allocated to an ARB is calculated as

$$f_r(\mathbf{w}_i) = \alpha \times \left( l_{arb,i}(\mathbf{w}_i)^2 \right) \tag{19.7}$$

where $l_{arb,i}$ is the stimulation level and $\alpha$ some constant. Since the stimulation level of the ARBs in $\mathcal{A}$ are normalised, some of the ARBs will have no resources allocated. After the resource allocation step, the weakest ARBs (i.e. ARBs having zero resources) are removed from the population of ARBs. Each of the remaining ARBs in the population, $\mathcal{A}$, is then cloned and mutated if the calculated stimulation level of

---

**Algorithm 19.6** Resource Allocation in the Artificial Immune Network

---

Set the number of allocated resources, $n_r = 0$;
**for** *each ARB,* $\mathbf{w}_i \in \mathcal{A}$ **do**
    Allocate resources, $f_r(\mathbf{w}_i)$;
    $n_r = n_r + f_r(\mathbf{w}_i)$;
**end**
Sort the set of ARBs, $\mathcal{A}$, in ascending order of $f_r$;
**if** $n_r > n_{r,max}$ **then**
    $\alpha_1 = n_r - n_{r,max}$;
    **for** *each ARB,* $\mathbf{w}_i \in \mathcal{A}$ **do**
        $\alpha_2 = f_r(\mathbf{w}_i)$;
        **if** $\alpha_2 = 0$ **then**
            Remove $\mathbf{w}_i$ from set $\mathcal{A}$;
        **end**
        **else**
            $\alpha_2 = \alpha_2 - \alpha_1$;
            **if** $\alpha_2 \leq 0$ **then**
                Remove $\mathbf{w}_i$ from set $\mathcal{A}$;
                $\alpha_1 = -\alpha_2$;
            **end**
            **else**
                $f_r(\mathbf{w}_i) = \alpha_2$;
                break;
            **end**
        **end**
    **end**
**end**

---

the ARB is above a certain threshold. These mutated clones are then integrated into the population by re-calculating the network links between the ARBs in $\mathcal{A}$.

Since ARBs compete for resources based on their stimulation level, an upper limit is set to the number of resources (i.e. B-Cells) available. The stopping condition can be based on whether the maximum size of $\mathcal{A}$ has been reached.

## 19.4.2   Self Stabilizing AIS

The self stabilizing AIS (SSAIS) was developed by Neal [626] to simplify and improve AINE. The main difference between these two models is that the SSAIS does not have a shared/distributed pool with a fixed number of resources that ARBs must compete for. The resource level of an ARB is increased if the ARB has the highest stimulation for an incoming pattern. Each ARB calculates its resource level locally. After a data pattern has been presented to all of the ARBs, the resource level of the most stimulated ARB is increased by addition of the ARB's stimulation level. Algorithm 19.7 summarizes the self stabilizing AIS. The differences between Algorithm 19.5 (AINE) and Algorithm 19.7 (SSAIS) are discussed next.

---

**Algorithm 19.7** Self Stabilizing AIS

---

Normalize the training data;
Initialize the ARB population, $\mathcal{A}$, using a cross section of the normalized training data;
Initialize the antigen set, $D_T$, with the remaining normalized training data;
**for** *each antigen,* $\mathbf{z}_p \in D_T$ **do**
    Present $\mathbf{z}_p$ to each $\mathbf{w}_i \in \mathcal{A}$;
    Calculate stimulation level, $l_{arb,i}$, for each ARB, $\mathbf{w}_i$;
    Select the ARB with the highest calculated stimulation level as $\hat{\mathbf{w}}$;
    Increase the resource level of $\hat{\mathbf{w}}$;
    **for** *each ARB,* $\mathbf{w}_i \in B; \mathbf{w}_i \neq \hat{\mathbf{w}}$ **do**
        Deplete the resources of $\mathbf{w}_i$;
    **end**
    Remove ARBs with the number of allocated resources less than $n_{r,min}$;
    Generate $n_c$ clones of $\hat{\mathbf{w}}$ and mutate;
    Integrate clones (mutated or not) into $\mathcal{A}$;
**end**

---

SSAIS defines the stimulation level, $l_{arb,i}$, of an ARB as [626]

$$l_{arb,i} = l_{a,i} + \delta \tag{19.8}$$

where

$$l_{a,i}(\mathbf{w}_i, \mathbf{z}_p) = 1 - d(\mathbf{w}_i, \mathbf{z}_p) \tag{19.9}$$

and

$$\delta(\mathbf{w}_i) = \frac{l_{n,i}}{|\mathcal{L}_{n,i}|} \tag{19.10}$$

For the above, $l_{n,i}$ is defined in (19.5) and $d(\mathbf{w}, \mathbf{z}_p)$ is the Euclidean distance between an ARB, $\mathbf{w}_i$, and a training pattern, $\mathbf{z}_p$, in normalised data space (i.e. $0 \leq d(\mathbf{w}_i, \mathbf{z}_p) \leq 1$); $\delta$ is the average of the summation over the distances between ARB $\mathbf{w}_i$ to its $|\mathcal{L}_{n,i}|$ linked neighbours. Network suppression as defined in equation (19.6) is discarded to prevent premature convergence of ARBs to dominating training patterns.

For each training pattern, $\mathbf{z}_p \in D_T$, presented to the network of ARBs the resource level of each ARB that does not have the highest stimulation level is geometrically decayed using [626]

$$f_r(\mathbf{w}_i, \mathbf{z}_p) = \nu \times f_r(\mathbf{w}_i, \mathbf{x}_{p-1}) \tag{19.11}$$

where $f_r(\mathbf{w}_i, \mathbf{z}_p)$ is the number of resources for an ARB, $\mathbf{w}_i$, after being presented to $p$ training patterns; $\nu$ is the decaying rate of resources for an ARB. All ARBs with a resource level less than the fixed predefined mortality threshold, $\epsilon_{death}$, are culled from the network. Resources are only allocated by the ARB, $\hat{\mathbf{w}}$, with the highest calculated stimulation level, $l_{arb}$. The number of resources allocated to $\hat{\mathbf{w}}$ is calculated using [626]

$$f_r(\hat{\mathbf{w}}, \mathbf{z}_p) = \nu \times (f_r(\hat{\mathbf{w}}, \mathbf{z}_{p-1}) + l_{arb}(\hat{\mathbf{w}}, \mathbf{z}_p)) \tag{19.12}$$

where $l_{arb}(\hat{\mathbf{w}}, \mathbf{z}_p)$ is the stimulation level of the highest stimulated ARB after being presented to $p$ training patterns.

The highest stimulated ARB, $\hat{\mathbf{w}}$, generates $n_c$ clones, calculated as [626]

$$n_c = \frac{f_r(\hat{\mathbf{w}}, \mathbf{z}_p)}{\epsilon_{death} \times 10} \tag{19.13}$$

Thus, the number of clones generated by an ARB is proportional to the resource level of the ARB. The generated clones are mutated with a fixed mutation rate. If a clone is mutated, the clone is assigned $\epsilon_{death} \times 10$ resources from the ARB's resource level. Clones (mutated or not) are integrated with the network of ARBs, $\mathcal{A}$.

The SSAIS resulted in a model that can adapt to continuously changing data sets, and a genuinely stable AIS. A drawback to SSAIS is that the final networks that are formed have poor data compression and the SSAIS model has a time lag to adapt to the introduction of a new region of data due to the lack of diversity of the network of ARBs.

## 19.4.3   Enhanced Artificial Immune Network

Another enhancement of the AINE model was proposed by Nasraoui *et al.* [622]. The enhanced AINE was applied to the clustering (profiling) of session patterns for a specific web site. A drawback of AINE [846] is that the population of ARBs increases at a high rate. The population is also overtaken by a few ARBs with high stimulation levels that matches a small number of antigen, resulting in premature convergence of the population of ARBs. AINE [846] represents clusters of poor quality with a repertoire of ARBs that is not diverse enough to represent the antigen space due to premature convergence.

The enhanced AINE requires two passes through the antigen training set (session patterns). To avoid premature convergence of the population of ARBs, the number of resources allocated to an ARB was calculated as [622]

$$\gamma_1 \times (\log l_{arb,i}) \tag{19.14}$$

where $l_{arb,i}$ is the stimulation level of the ARB and $\gamma_1$ is some positive constant. This modification to the number of resources allocated to an ARB limits the influence of those ARBs with high stimulation to slowly overtake the population.

The number of clones, $n_c$, generated by an ARB is defined as [622]

$$n_c = \gamma_2 \times l_{arb,i} \tag{19.15}$$

where $\gamma_2$ is a positive constant. After the integration of the mutated clones, the ARBs with the same B-Cell representation are merged into one ARB. Merging identical ARBs limits the high rate of population growth. Merging of two ARBs, $\mathbf{w}_1$ and $\mathbf{w}_2$, is by applying a crossover operator to the ARBs' attributes, defined as [622]

$$w_k^* = \frac{(\mathbf{w}_{1k} + \mathbf{w}_{2k})}{2} \tag{19.16}$$

where $w_k^*$ is the value of attribute $k$ of the merged ARB.

The Euclidean distance used in AINE was replaced by the cosine similarity between two session patterns [622]. The cosine similarity, $\vartheta$, between two vectors is defined as

$$\vartheta(\mathbf{x}, \mathbf{y}) = \frac{\sum_{k=1}^{|\mathbf{x}|} \mathbf{x}_k \mathbf{y}_k}{\sqrt{\sum_{k=1}^{|\mathbf{x}|} \mathbf{x}_k \times \sum_{k=1}^{|\mathbf{y}|} \mathbf{y}_k}} \tag{19.17}$$

where $\mathbf{x}_k$ and $\mathbf{y}_k$ are the values of the $k$-th attribute in vectors $\mathbf{x}$ and $\mathbf{y}$ respectively.

## 19.4.4   Dynamic Weighted B-Cell AIS

Most of the existing network based AIS models consist of a number of interconnected (linked) B-Cells. The linked B-Cells form networks that need to be maintained. Often the required number of network B-Cells exceeds the number of training patterns that needs to be learned by the model, resulting in an increase in the number of links between these B-Cells. This makes the model unscalable and in some cases non-adaptive to dynamic environments.

The set of ARBs in [846] is an example of such an immune network. The number of ARBs could reach the same size as the training set or even exceed it, making the model unscalable. An alternative scalable network AIS, which is also adaptable in dynamic environments, was presented in [623]. This scalable model consists of dynamic weighted B-Cells (DWB-Cells). A DWB-Cell represents an influence zone, which can be defined as a weight function that decreases with the time since the antigen has been presented to the network and with the distance between the DWB-Cell and the presented antigen.

The model also proposed the incorporation of a dynamic stimulation/suppression factor into the stimulation level of a DWB-Cell to control the proliferation and redundancy of DWB-Cells in the network, thus old sub-nets die if not re-stimulated by current incoming antigen patterns. A drawback of existing immune network based learning models is that the number of interactions between the B-Cells in the network and a specific antigen are immense. The model of [623] uses K-means to cluster the DWB-Cells in order to decrease the number of interactions between an antigen pattern and the B-Cells in the network. The centroids of each of these formed clusters (or sub-nets) are used to represent the sub-nets and interact with the presented antigen pattern.

A DWB-Cell is only cloned if it reached maturity. Maturity of a DWB-Cell is reached when the cell's age exceeds a minimum time threshold. Cloning of a DWB-Cell is proportional to the cell's stimulation level. If a DWB-Cell's age exceeds a maximum time threshold, cloning of the cell is prevented, thus increasing the probability to clone newer DWB-Cells. The mechanism of somatic hyper mutation is computationally expensive and is replaced in the DWB-model by *dendritic injection*. When the immune network encounters an antigen that the network cannot react to, the specific antigen is initialized as a DWB-Cell. Thus, new information is injected into the immune network.

*Dendritic injection* models the dendritic cells, which function in the natural immune system has only recently been understood. The DWB-model has proven to be robust to noise, adaptive and scalable in learning antigen structures.

## 19.4.5    Adapted Artificial Immune Network

One of the drawbacks in the model of Timmis [845] is the difficulty to keep the network size within certain boundaries through all the iterations. This drawback makes the model unstable. The model presented in [909] adapts the model in [845] with a few changes in the implementation. Some of these changes are that the training set is used as the antigen set in the model with a set of randomly initiated antibodies. The model in [845] randomly assigns the patterns in the training set to the antigen set and antibody set.

A further change to the model is in the initialization of the network affinity threshold (NAT). In [845] the NAT was calculated as an average distance between all the patterns in the antigen set. In the adapted model, only the average distance between the $n \times k$ lowest distances in the antigen set was calculated, where $n$ is the size of the antigen set and $k$ some constant.

The stimulation level of an antibody in the adapted model is defined as the affinity between the antibody and the set of antigens. The model in [845] considered the degree of stimulation of the antibody and the degree of suppression of the antibody by other antibodies. The adapted model in [909] improves on the model in [845] in that the maximum network size is limited to the number of training patterns in the training set and stable clusters are formed with a minimal number of control parameters.

## 19.4.6    aiNet

The aiNet model developed by De Castro and Von Zuben [184, 187] implements some of the principles found in the immune network theory. A typical network in the model consists of nodes (the B-Cells or antibodies) which are connected by edges to form node pairs. A weight value (connection strength) is assigned to each edge, to indicate the similarity between two nodes. Thus, the network that is formed during training is presented by an edge-weighted graph.

Algorithm 19.8 summarizes aiNet. aiNet uses clonal selection [97] to adapt the network of antibodies, $\mathcal{B}$, to training patterns, $D_T$. Thus, the data in the training set is seen as antigens. During training the model builds a network of memory cells to determine any clusters in the data. The model uses Euclidean distance as a metric of affinity (or dissimilarity), $d(\mathbf{z}_p, \mathbf{y}_j)$, between a training pattern, $\mathbf{z}_p$, and an antibody, $\mathbf{y}_j$, in the network. A higher value of $d(\mathbf{z}_p, \mathbf{y}_j)$ implies a higher degree of dissimilarity between an antibody and an antigen training pattern.

Clonal selection is applied to $n_h$ of the antibodies with the highest affinity (dissimilarity) to a specific antigen. The number of clones, $n_c$, for each antibody, $\mathbf{y}_j$, is

---

**Algorithm 19.8** aiNet Learning Algorithm

---

Determine the antigen patterns as training set $D_T$;
**while** *stopping condition not true* **do**
    **for** *each antigen pattern,* $\mathbf{z}_p \in D_T$ **do**
        **for** *each antibody,* $\mathbf{y}_j \in \mathcal{B}$ **do**
            Calculate the antigen affinity $f_a(\mathbf{z}_p, \mathbf{y}_j)$;
        **end**
        Select $n_h$ of the highest affinity antibodies as set $\mathcal{H}$;
        **for** *each* $\mathbf{y}_j \in \mathcal{H}$ **do**
            Create $n_c$ clones of $\mathbf{y}_j$;
            Mutate the $n_c$ created clones and add to set $\mathcal{H}'$;
        **end**
        **for** *each* $\mathbf{y}'_j \in \mathcal{H}'$ **do**
            Calculate the antigen affinity, $f_a(\mathbf{z}_p, \mathbf{y}'_j)$;
        **end**
        Select $n_h\%$ of the highest affinity antibodies as set $\mathcal{M}$;
        **for** *each* $\mathbf{y}_j \in \mathcal{M}$ **do**
            **if** $f_a(\mathbf{z}_p, \mathbf{y}_j) > a_{max}$ **then**
                Remove $\mathbf{y}_j$ from $\mathcal{M}$;
            **end**
        **end**
        **for** *each* $\mathbf{y}_{j_1} \in \mathcal{M}$ **do**
            **for** *each* $\mathbf{y}_{j_2} \in \mathcal{M}$ **do**
                Calculate the network affinity $f_a(\mathbf{y}_{j_1}, \mathbf{y}_{j_2})$;
                **if** $f_a(\mathbf{y}_{j_1}, \mathbf{y}_{j_2}) < \epsilon_s$ **then**
                    Mark $\mathbf{y}_{j_1}$ and $\mathbf{y}_{j_2}$ as *elimination* from $\mathcal{M}$;
                **end**
            **end**
        **end**
        Remove all *elimination* antibodies from $\mathcal{M}$;
        $\mathcal{B} = \mathcal{B} \cup \mathcal{M}$;
    **end**
    **for** *each* $\mathbf{y}_{j_1} \in \mathcal{B}$ **do**
        **for** *each* $\mathbf{y}_{j_2} \in \mathcal{B}$ **do**
            Calculate the network affinity, $f_a(\mathbf{y}_{j_1}, \mathbf{y}_{j_2})$;
            **if** $f_a(\mathbf{y}_{j_1}, \mathbf{y}_{j_1}) < \epsilon_s$ **then**
                Mark $\mathbf{y}_{j_1}$ and $\mathbf{y}_{j_2}$ as *elimination* from $\mathcal{B}$;
            **end**
        **end**
    **end**
    Remove all *elimination* antibodies from $\mathcal{B}$;
    Replace $n_l\%$ of the lowest affinity antibodies in $\mathcal{B}$ with randomly generated antibodies;
**end**

---

proportional to the affinity, $f_a(\mathbf{z}_p, \mathbf{y}_i) = d(\mathbf{z}_p, \mathbf{y}_j)$ of the antibody. The number of clones, $n_c$, is calculated as

$$n_c = round\,(n_a(1 - f_a(\mathbf{z}_p, \mathbf{y}_j))) \tag{19.18}$$

where $n_a = |\mathcal{B}|$ is the number of antibodies. The $n_c$ clones of each antibody are mutated according to their affinity (dissimilarity) to guide the search to a locally optimized network; that is, to improve the recognition of antigens. The mutated clones are added to a set, $\mathcal{H}'$. Antibodies are mutated as follows [184],

$$\mathbf{y}_j' = \mathbf{y}_j - p_m\,(\mathbf{y}_j - \mathbf{z}_p) \tag{19.19}$$

where $\mathbf{y}_j$ is the antibody, $\mathbf{z}_p$ is the antigen training pattern, and $p_m$ is the mutation rate. The mutation rate is inverse proportional to the calculated affinity, $f_a(\mathbf{z}_p, \mathbf{y}_j)$. The higher the calculated affinity, the less $\mathbf{y}_j$ is mutated. Based on the affinity (dissimilarity) of the mutated clones with the antigen pattern $\mathbf{z}_p$, $n_h\%$ of the highest affinity mutated clones are selected as the memory set, $\mathcal{M}$. An antibody in $\mathcal{M}$ is removed if the affinity between the antibody and $\mathbf{z}_p$ is higher than the threshold, $a_{max}$.

The Euclidean distance metric is also used to determine the affinity between antibodies in the network. A smaller Euclidean distance implies a higher degree of similarity between the two antibodies. The similarity between antibodies is measured against a suppression threshold, $\epsilon_s$, resulting in elimination of the antibodies if the suppression threshold is not exceeded. This results in network pruning and elimination of *self-recognizing* antibodies.

The affinity (similarity) among antibodies in $\mathcal{M}$ is measured against $\epsilon_s$ for removal from $\mathcal{M}$. $\mathcal{M}$ is then concatenated with the set of antibodies $\mathcal{B}$. The affinity among antibodies in $\mathcal{B}$ is measured against $\epsilon_s$ for removal from $\mathcal{B}$. A percentage, $n_l\%$, of antibodies with the lowest affinity to $\mathbf{z}_p$ in $\mathcal{B}$ is replaced with randomly generated antibodies.

The *stopping condition* of the *while*-loop can be one of the following [184]:

1. **Setting an *iteration* counter**: A counter can be set to determine the maximum number of iterations.

2. **Setting the maximum size of the network**: The while-loop can be stopped when the size of the network reaches a maximum.

3. **Testing for convergence**: The algorithm terminates when the average error between the training patterns and the antibodies in $\mathcal{M}$ rises after a number of consecutive iterations.

The formed weighted-edge graph maps clusters in the data set to network clusters. The aiNet model uses the minimal spanning tree of the formed weighted-edge graph or hierarchical agglomerative clustering to determine the structure of the network clusters in the graph.

Some of the drawbacks of the aiNet model is the large number of parameters that need to be specified and that the cost of computation increases as the number of variables of a training pattern increases. The minimum spanning tree will also have difficulty in determining the network clusters if there are intersections between the clusters in the training data set. aiNet is capable of reducing data redundancy and obtaining a compressed representation of the data.

# 19.5    Danger Theory Models

In contrast to the classical view of the natural immune system, danger theory distinguishes between what is *dangerous* and *non-dangerous*, rather than what is *self* and *non-self* (as discussed in Section 18.7). A *foreign* cell is seen to be *dangerous* if the cell causes a stress *signal* of abnormal cell death in the body, i.e. necrotic cell death. One of the motivations for the danger theory is the ability of the natural immune system to adapt to a changing *self*. This motivation has inspired some of the discussed danger AIS models in this section. Each of the discussed models has a different definition of a stress *signal* and is thus problem specific. The main difference between the danger AIS models to those AIS models inspired by the classical view (as discussed in 19.2), is the inclusion of a *signal* to determine whether a *non-self* pattern is *dangerous* or not.

## 19.5.1    Mobile Ad-Hoc Networks

Dynamic source routing (DSR) is one of the communication protocols that can be used between terminal nodes in a mobile ad-hoc network. A mobile ad-hoc network consists of terminal nodes, each with a radio as communication device to transmit information to other terminal nodes in the network, i.e. no infrastructure between nodes. Thus, nodes not only function as terminals, but also as relays of the transmitted information in the network. This type of network topology requires a common routing protocol like DSR. Thus, the correct execution of the DSR protocol by each node in the ad-hoc network is crucial.

Sometimes a node *misbehaves* in the network. This *misbehaviour* can be associated with nodes that are in standby mode (not relaying information), hardware failures or malicious software (like viruses) on the node can try to overthrow the network. A danger theory inspired AIS is proposed in [757] to detect misbehaving nodes in a mobile ad-hoc network. Each node in the network applies an instance of the algorithm on observations made by the node. The nodes communicate detection information between each other.

*Self* is defined as *normal* network traffic without any packet loss. This implies that the defined *self* is dynamic, since a new node in the network might generate new unseen traffic without any packet loss, i.e *dynamic self*. Each node in the network monitors/observes the traffic for each of the neighboring nodes in the network. These observations are buffered for a specific buffering time. If the buffering time elapsed, the buffered observations are seen as self patterns and are randomly selected to generate self-tolerant detectors for the node. Existing detectors that detect a self pattern are removed and replaced by the newly generated detectors. Negative selection is used to generate these detectors.

Observations in the buffer are presented to existing detectors. If an existing detector matches an observation, the score of the detector is rewarded with a value of one; otherwise, it is penalized with a value of one. A detector can only match an observation if the observation is accompanied by a danger signal. Detectors are clustered according to their detection scores. If a source node experience a packet loss (misbehaving node), the source node will generate an observation with a danger signal along the route where the packet loss was experienced. The action taken by the neighboring nodes is to discard the observation from the buffered observations through correlation with the danger signal (also observed). This prevents the generation of detectors on *non-self* observations.

## 19.5.2 An Adaptive Mailbox

The danger theory inspired the proposal of an AIS for an adaptive mailbox. The proposed AIS in [772] classifies *interesting* from *uninteresting* emails. The algorithm is divided into two phases: an initialization phase (training) and a running phase (testing).

Algorithm 19.9 summarizes the initialization phase. The initialization phase monitors the user's actions for each new email, $\mathbf{z}$, received. If $\mathbf{z}$ is deleted by the user, an antibody, $\mathbf{y}_{new}$, is generated to detect the deleted email. After adding the new antibody to the antibody set, $\mathcal{B}$, the existing antibodies in the set are cloned and mutated to improve the generalization of the antibody set. Thus, the antibody set, $\mathcal{B}$, represents uninteresting email. The initialization phase continuous until the size of the antibody set, $\mathcal{B}$, reached a certain maximum, $n_a$.

---

**Algorithm 19.9** Initialization Phase for an Adaptive Mailbox

---

**while** $|\mathcal{B}| < n_a$ **do**
    **if** *user action = delete email*, $\mathbf{z}$, **then**
        Generate an antibody, $\mathbf{y}_{new}$, from $\mathbf{z}$;
        Add, $\mathbf{y}_{new}$, to the set of antibodies, $\mathcal{B}$;
        **for** *each antibody*, $\mathbf{y}_j \in \mathcal{B}$ **do**
            Clone and mutate antibody, $\mathbf{y}_j$, to maximize affinity with antibody, $\mathbf{y}_{new}$;
            Add, $n_h$, highest affinity clones to $\mathcal{B}$;
        **end**
    **end**
**end**

---

The running phase (see Algorithm 19.10) labels all incoming email, $\mathbf{z}$, that are deleted by the user as *uninteresting* and buffers them as antigen in $D_T$. When the buffered emails reach a specific size, $n_T$, the buffer, $D_T$, is presented to the antibody set, $\mathcal{B}$. The antibody set then adapts to the presented buffer of emails (antigens) through clonal selection. Thus, the antibody set, $\mathcal{B}$, adapts to the changing interest of the user to represent the latest general set of antibodies (uninteresting emails).

---

**Algorithm 19.10** Running Phase for an Adaptive Mailbox

---

**while** *true* **do**
    Wait for action from user;
    **if** *user action = delete email*, $\mathbf{z}$, **then**
        Assign class *uninteresting* to $\mathbf{e}$;
    **end**
    **else**
        Assign class *interesting* to $\mathbf{z}$;
    **end**
    Generate an antigen, $\mathbf{z}'$, from email, $\mathbf{z}$;
    Add $\mathbf{z}'$ to buffered set, $D_T$;
    **if** $|D_T| = n_T$ **then**
        **for** *each* $\mathbf{y}_j \in \mathcal{B}$ **do**
            $un = 0$;
            $in = 0$;
            **for** *each* $\mathbf{z}'_p \in D_T$ **do**
                Calculate $f_a(\mathbf{z}'_p, \mathbf{y}_j)$;
                **if** $\mathbf{z}'_p$ *is of class uninteresting* **then**
                    $un = un + f_a(\mathbf{z}'_p, \mathbf{y}_j)$;
                **end**
                **else**
                    $in = in + f_a(\mathbf{z}'_p, \mathbf{y}_j)$;
                **end**
            **end**
            $\alpha = un - in$;
            Clone and mutate $\mathbf{y}_j$ in proportion to $\alpha$;
        **end**
        Remove $n_l$ of the antibodies with the lowest calculated $\alpha$ from $\mathcal{B}$;
        Remove all antigens from $D_T$;
    **end**
    Calculate the degree of danger as, $\theta$;
    **while** $\theta > \theta_{max}$ **do**
        Add $\theta$ unread emails to set, $\mathcal{U}$;
        **for** *each* $\mathbf{z}'_p \in \mathcal{U}$ **do**
            **for** *each* $\mathbf{y}_j \in \mathcal{B}$ **do**
                Calculate $f_a(\mathbf{z}'_p, \mathbf{y}_j)$;
            **end**
            Set $a_h =$ highest affinity in $\mathcal{U}$;
            **if** $a_h > a_{max}$ **then**
                Move $\mathbf{z}'_p$ to temporary store or mark as deleted;
            **end**
         **end**
        **end**
    **end**
**end**

---

The number of unread emails in the inbox determines the degree of the danger signal, $\theta$. If the degree of the danger signal reaches a limit, $\theta_{max}$, the unread emails, $\mathcal{U}$, are presented to the set of antibodies, $\mathcal{B}$, for classification as *uninteresting*. An email, $\mathbf{z}'_p \in \mathcal{U}$, is classified as *uninteresting* if the highest calculated affinity, $a_h$, is higher than an affinity threshold, $a_{max}$. The *uninteresting* classified email is then moved to a temporary folder or deleted. The degree of the danger signal needs to be calculated for each new email received.

### 19.5.3   Intrusion Detection

The basic function of an intrusion detection system (IDS) is to monitor incoming traffic at a specific host connected to a network. The IDS creates a profile of *normal* user traffic and signals an alarm of intrusion for any detected *abnormal* traffic, i.e. traffic not forming part of the *normal* profile. A problem to this solution of profile creation, is that the *normal* traffic changes through time. Thus, the profile gets outdated.

The danger signal used in danger theory inspired the modeling of an adaptable IDS. The danger signal can be defined as a signal generated by the host if any incoming traffic resulted in abnormal CPU usage, memory usage or security attacks. The adaptable IDS will only signal an alarm of *abnormal* traffic if the IDS receives a danger signal from the host. If no danger signal is received from the host, the profile is adapted to accommodate the new detected *normal* traffic. Danger theory inspired AISs applied to intrusion/anomaly detection can be found in [13, 30].

## 19.6   Applications and Other AIS models

Artificial immune systems have been successfully applied to many problem domains. Some of these domains range from network intrusion and anomaly detection [13, 30, 176, 279, 280, 327, 373, 374, 457, 458, 803, 804] to data classification models [692, 890], virus detection [281], concept learning [689], data clustering [184], robotics [431, 892], pattern recognition and data mining [107, 398, 845, 847]. The AIS has also been applied to the initialization of feed-forward neural network weights [189], the initialization of centers of a radial basis function neural network [188] and the optimization of multi-modal functions [181, 294]. The interested reader is referred to [175, 183, 185] for more information on AIS applications.

## 19.7   Assignments

1. With reference to negative selection as described in Section 19.2.1, discuss the consequences of having small values for $r$. Also discuss the consequences for large values.

2. A drawback of negative selection is that the training set needs to have a good representation of self patterns. Why is this the case?

3. How can an AIS be used for classification problems where there is more than two classes?

4. How does the self stabilizing AIS improve on AINE?

5. How does the enhanced artificial immune network improve on AINE?

6. The enhanced artificial immune network calculates the number of clones generated by an ARB as $nc = l \times sl$.

   (a) Why is the number of clones a function of the stimulation level?

   (b) Explain the consequences of large and small values of $l$.

7. For the aiNet model in Algorithm 19.5, how does network suppression help to control the size of the ARB population?

8. Why should an antibody be mutated less the higher the affinity of the antibody to an antigen training pattern, considering the aiNet model?

9. Discuss the influence of different values for the danger signal threshold as applied in the adaptive mailbox problem discussed in Section 19.5.2

# Part VI

# FUZZY SYSTEMS

Two-valued, or Boolean, logic is a well-defined and used theory. Boolean logic is especially important for implementation in computing systems where information, or knowledge about a problem, is binary encoded. Boolean logic also played an important role in the development of the first AI reasoning systems, especially the inference engine of expert systems [315]. For such knowledge representation and reasoning systems, propositional and first-order predicate calculus are extensively used as representation language [539, 629]. Associated with Boolean logic is the traditional two-valued set theory, where an element either belongs to a set or not. That is, set membership is precise. Coupled with Boolean knowledge, two-valued set theory enabled the development of exact reasoning systems.

While some successes have been achieved using two-valued logic and sets, it is not possible to solve all problems by mapping the domain into two-valued variables. Most real-world problems are characterized by the ability of a representation language (or logic) to process incomplete, imprecise, vague or uncertain information. While two-valued logic and set theory fail in such environments, fuzzy logic and fuzzy sets give the formal tools to reason about such uncertain information. With fuzzy logic, domains are characterized by linguistic terms, rather than by numbers. For example, in the phrases *"it is partly cloudy"*, or *"Stephan is very tall"*, both *partly* and *very* are linguistic terms describing the *magnitude* of the fuzzy (or linguistic) variables *cloudy* and *tall*. The human brain has the ability to understand these terms, and infer from them that it will most probably not rain, and that Stephan might just be a good basket ball player (note, again, the fuzzy terms!). However, how do we use two-valued logic to represent these phrases?

Together with fuzzy logic, fuzzy set theory provides the tools to develop software products that model human reasoning (also referred to as approximate reasoning). In fuzzy sets, an element belongs to a set to a degree, indicating the certainty (or uncertainty) of membership.

The development of logic has a long and rich history, in which major philosophers played a role. The foundations of two-valued logic stemmed from the efforts of Aristotle (and other philosophers of that time), resulting in the so-called *Laws of Thought* [440]. The first version of these laws was proposed around 400 B.C., namely the *Law of the Excluded Middle*. This law states that every proposition must have only one of two outcomes: either *true* or *false*. Even in that time, immediate objections were given with examples of propositions that could be true, and simultaneously not true.

It was another great philosopher, Plato, who laid the foundations of what is today referred to as fuzzy logic. It was, however, only in the 1900s that Lejewski and Lukasiewicz [514] proposed the first alternative to the Aristotelian two-valued logic. Three-valued logic has a third value which is assigned a numeric value between *true* and *false*. Lukasiewicz later extended this to four-valued and five-valued logic. It was only recently, in 1965, that Lotfi Zadeh [944] produced the foundations of infinite-valued logic with his mathematics of fuzzy set theory.

Following the work of Zadeh, much research has been done in the theory of fuzzy systems, with applications in control, information systems, pattern recognition and decision support. Some successful real-world applications include automatic control of dam gates for hydroelectric-powerplants, camera aiming, compensation against vibrations in camcorders, cruise-control for automobiles, controlling air-conditioning systems, document archiving systems, optimized planning of bus time-tables, and many more. While fuzzy sets and logic have been used to solve real-world problems, they were also combined with other CI paradigms to form hybrid systems, for example, fuzzy neural networks and fuzzy genetic algorithms [957].

A different set theoretic approach which also uses the concept of membership functions, namely rough sets (introduced by Pawlak in 1982 [668]), is sometimes confused with fuzzy sets. While both fuzzy sets and rough sets make use of membership functions, rough sets differ in the sense that a lower and upper approximation to the rough set is determined. The lower approximation consists of all elements that belong with full certainty to the corresponding set, while the upper approximation consists of elements that may possibly belong to the set. Rough sets are frequently used in machine learning as classifier, where they are used to find the smallest number of features to discern between classes [600]. Rough sets are also used for extracting knowledge from incomplete data [600, 683]. Hybrid approaches that employ both fuzzy and rough sets have also been developed [843].

The remainder of this Part is organized as follows: Chapter 20 discusses fuzzy sets, while fuzzy logic and reasoning are covered in Chapter 21. A short overview of fuzzy controllers is given in Chapter 22. The Part is concluded with an overview of rough set theory in Chapter 23.

# Chapter 20

# Fuzzy Sets

Consider the problem of designing a set of all tall people, and assigning all the people you know to this set. Consider classical set theory where an element is either a member of the set or not. Suppose all tall people are described as those with height greater than 1.75m. Then, clearly a person of height 1.78m will be an element of the set *tall*, and someone with height 1.5m will not belong to the set of tall people. But, the same will apply to someone of height 1.73m, which implies that someone who falls only 2cm short is not considered as being tall. Also, using two-valued set theory, there is no distinction among members of the set of tall people. For example, someone of height 1.78m and one of height 2.1m belongs equally to the set! Thus, no semantics are included in the description of membership.

The alternative, fuzzy sets, has no problem with this situation. In this case all the people you know will be members of the set *tall*, but to different degrees. For example, a person of height 2.1m may be a member of the set to degree 0.95, while someone of length 1.7m may belong to the set with degree 0.4.

Fuzzy sets are an extension of crisp (two-valued) sets to handle the concept of *partial truth*, which enables the modeling of the uncertainties of natural language. The vagueness in natural language is further emphasized by linguistic terms used to describe objects or situations. For example, the phrase *when it is very cloudy, it will most probably rain*, has the linguistic terms *very* and *most probably* – which are understood by the human brain. Fuzzy sets, together with fuzzy reasoning systems, give the tools to also write software, which enables computing systems to understand such vague terms, and to reason with these terms.

This chapter formally introduces fuzzy sets. Section 20.1 defines fuzzy sets, while membership functions are discussed in Section 20.2. Operators that can be applied to fuzzy sets are covered in Section 20.3. Characteristics of fuzzy sets are summarized in Section 20.4. The chapter is concluded with a discussion of the differences between fuzziness and probability in Section 20.5.

*Computational Intelligence: An Introduction*, Second Edition A.P. Engelbrecht
©2007 John Wiley & Sons, Ltd

## 20.1    Formal Definitions

Different to classical sets, elements of a fuzzy set have membership degrees to that
set. The degree of membership to a fuzzy set indicates the certainty (or uncertainty)
that the element belongs to that set. Formally defined, suppose $X$ is the domain, or
universe of discourse, and $x \in X$ is a specific element of the domain $X$. Then, the
fuzzy set $A$ is characterized by a membership mapping function [944]

$$\mu_A : X \rightarrow [0, 1] \tag{20.1}$$

Therefore, for all $x \in X$, $\mu_A(x)$ indicates the certainty to which element $x$ belongs to
fuzzy set $A$. For two-valued sets, $\mu_A(x)$ is either 0 or 1.

Fuzzy sets can be defined for discrete (finite) or continuous (infinite) domains. The
notation used to denote fuzzy sets differ based on the type of domain over which
that set is defined. In the case of a discrete domain $X$, the fuzzy set can either
be expressed in the form of an $n_x$-dimensional vector or using the sum notation. If
$X = \{x_1, x_2, \cdots, x_{n_x}\}$, then, using set notation,

$$A = \{(\mu_A(x_i)/x_i)|x_i \in X, i = 1, \cdots, n_x\} \tag{20.2}$$

Using sum notation,

$$A = \mu_A(x_1)/x_1 + \mu_A(x_2)/x_2 + \cdots + \mu_A(x_{n_x})/x_{n_x} = \sum_{i=1}^{n_x} \mu_A(x_i)/x_i \tag{20.3}$$

where the sum should not be confused with algebraic summation. The use of sum
notation above simply serves as an indication that $A$ is a set of ordered pairs. A
continuous fuzzy set, $A$, is denoted as

$$A = \int_X \mu(x)/x \tag{20.4}$$

Again, the integral notation should not be algebraically interpreted.

## 20.2    Membership Functions

The membership function is the essence of fuzzy sets. A membership function, also
referred to as the characteristic function of the fuzzy set, defines the fuzzy set. The
function is used to associate a degree of membership of each of the elements of the
domain to the corresponding fuzzy set. Two-valued sets are also characterized by
a membership function. For example, consider the domain X of all floating-point
numbers in the range $[0, 100]$. Define the crisp set $A \subset X$ of all floating-point numbers
in the range $[10, 50]$. Then, the membership function for the crisp set $A$ is represented
in Figure 20.1. All $x \in [10, 50]$ have $\mu_A(x) = 1$, while all other floating-point numbers
have $\mu_A(x) = 0$.

Membership functions for fuzzy sets can be of any shape or type as determined by
experts in the domain over which the sets are defined. While designers of fuzzy sets

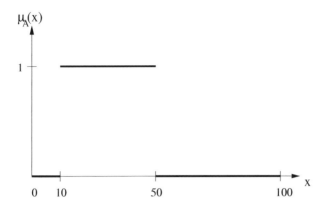

**Figure 20.1** Illustration of Membership Function for Two-Valued Sets

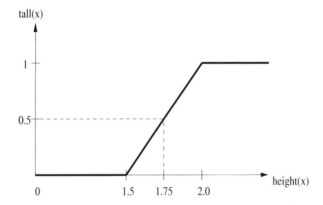

**Figure 20.2** Illustration of *tall* Membership Function

have much freedom in selecting appropriate membership functions, these functions must satisfy the following constraints:

- A membership function must be bounded from below by 0 and from above by 1.
- The range of a membership function must therefore be $[0, 1]$.
- For each $x \in X$, $\mu_A(x)$ must be unique. That is, the same element cannot map to different degrees of membership for the same fuzzy set.

Returning to the *tall* fuzzy set, a possible membership function can be defined as (also illustrated in Figure 20.2)

$$tall(x) = \begin{cases} 0 & \text{if } length(x) < 1.5m \\ (length(x) - 1.5m) \times 2.0m & \text{if } 1.5m \le length(x) \le 2.0m \\ 1 & \text{if } length(x) > 2.0m \end{cases} \quad (20.5)$$

Now, assume that a person has a length of 1.75m, then $\mu_A(1.75) = 0.5$.

While the *tall* membership function above used a discrete step function, more complex discrete and continuous functions can be used, for example:

- **Triangular** functions (refer to Figure 20.3(a)), defined as

$$\mu_A(x) = \begin{cases} 0 & \text{if } x \leq \alpha_{min} \\ \frac{x - \alpha_{min}}{\beta - \alpha_{min}} & \text{if } x \in (\alpha_{min}, \beta] \\ \frac{\alpha_{max} - x}{\alpha_{max} - \beta} & \text{if } x \in (\beta, \alpha_{max}) \\ 0 & \text{if } x \geq \alpha_{max} \end{cases} \tag{20.6}$$

- **Trapezoidal** functions (refer to Figure 20.3(b)), defined as

$$\mu_A(x) = \begin{cases} 0 & \text{if } x \leq \alpha_{min} \\ \frac{x - \alpha_{min}}{\beta_1 - \alpha_{min}} & \text{if } x \in [\alpha_{min}, \beta_1) \\ \frac{\alpha_{max} - x}{\alpha_{max} - \beta_2} & \text{if } x \in (\beta_2, \alpha_{max}) \\ 0 & \text{if } x \geq \alpha_{max} \end{cases} \tag{20.7}$$

- **Γ-membership** functions, defined as

$$\mu_A x = \begin{cases} 0 & \text{if } x \leq \alpha \\ 1 - e^{-\gamma(x - \alpha)^2} & \text{if } x > \alpha \end{cases} \tag{20.8}$$

- **S-membership** functions, defined as

$$\mu_A(x) = \begin{cases} 0 & \text{if } x \leq \alpha_{min} \\ 2\left(\frac{x - \alpha_{min}}{\alpha_{max} - \alpha_{min}}\right)^2 & \text{if } x \in (\alpha_{min}, \beta] \\ 1 - 2\left(\frac{x - \alpha_{max}}{\alpha_{max} - \alpha_{min}}\right)^2 & \text{if } x \in (\beta, \alpha_{max}) \\ 1 & \text{if } x \geq \alpha_{max} \end{cases} \tag{20.9}$$

- **Logistic** function (refer to Figure 20.3(c)), defined as

$$\mu_A(x) = \frac{1}{1 + e^{-\gamma x}} \tag{20.10}$$

- **Exponential-like** function, defined as

$$\mu_A(x) = \frac{1}{1 + \gamma(x - \beta)^2} \tag{20.11}$$

with $\gamma > 1$.

- **Gaussian** function (refer to Figure 20.3(d)), defined as

$$\mu_A(x) = e^{-\gamma(x - \beta)^2} \tag{20.12}$$

It is the task of the human expert of the domain to define the function that captures the characteristics of the fuzzy set.

## 20.3 Fuzzy Operators

As for crisp sets, relations and operators are defined for fuzzy sets. Each of these relations and operators are defined below. For this purpose let $X$ be the domain, or universe, and $A$ and $B$ are fuzzy sets defined over the domain $X$.

**Equality of fuzzy sets:** For two-valued sets, sets are equal if the two sets have exactly the same elements. For fuzzy sets, however, equality cannot be concluded if the two sets have the same elements. The degree of membership of elements to the sets must also be equal. That is, the membership functions of the two sets must be the same.

Therefore, two fuzzy sets $A$ and $B$ are equal if and only if the sets have the same domain, and $\mu_A(x) = \mu_B(x)$ for all $x \in X$. That is, $A = B$.

**Containment of fuzzy sets:** For two-valued sets, $A \subset B$ if all the elements of $A$ are also elements of $B$. For fuzzy sets, this definition is not complete, and the degrees of membership of elements to the sets have to be considered.

Fuzzy set $A$ is a subset of fuzzy set $B$ if and only if $\mu_A(x) \leq \mu_B(x)$ for all $x \in X$. That is, $A \subset B$.

Figure 20.4 shows two membership functions for which $A \subset B$.

**Complement of a fuzzy set (NOT):** The complement of a two-valued set is simply the set containing the entire domain without the elements of that set. For fuzzy sets, the complement of the set $A$ consists of all the elements of set $A$, but the membership degrees differ. Let $\overline{A}$ denote the complement of set $A$. Then, for all $x \in X$, $\mu_{\overline{A}}(x) = 1 - \mu_A(x)$. It also follows that $A \cap \overline{A} \neq \emptyset$ and $A \cup \overline{A} \neq X$.

**Intersection of fuzzy sets (AND):** The intersection of two-valued sets is the set of elements occurring in both sets. Operators that implement intersection are referred to as t-norms. The result of a t-norm is a set that contain all the elements of the two fuzzy sets, but with degree of membership that depends on the specific t-norm. A number of t-norms have been used, of which the min-operator and the product operator are the most popular. If $A$ and $B$ are two fuzzy sets, then

- **Min-operator**: $\mu_{A \cap B}(x) = \min\{\mu_A(x), \mu_B(x)\}, \ \forall x \in X$
- **Product operator**: $\mu_{A \cap B}(x) = \mu_A(x)\mu_B(x), \ \forall x \in X$

The difference between the two operations should be noted. Taking the product of membership degrees is a much stronger operator than taking the minimum, resulting in lower membership degrees for the intersection. It should also be noted that the ultimate result of a series of intersections approaches 0.0, even if the degrees of memberships to the original sets are high.

Other t-norms are [676],

- $\mu_{A \cap B}(x) = \dfrac{1}{1 + \sqrt[p]{\left(\frac{1 - \mu_A(x)}{p}\right)^p + \left(\frac{1 - \mu_B(x)}{p}\right)^p}}$, for $p > 0$.

- $\mu_{A \cap B}(x) = \max\{0, (1+p)(\mu_A(x) + \mu_B(x) - 1) - p\mu_A(x)\mu_B(x)\}$, for $p \geq -1$.

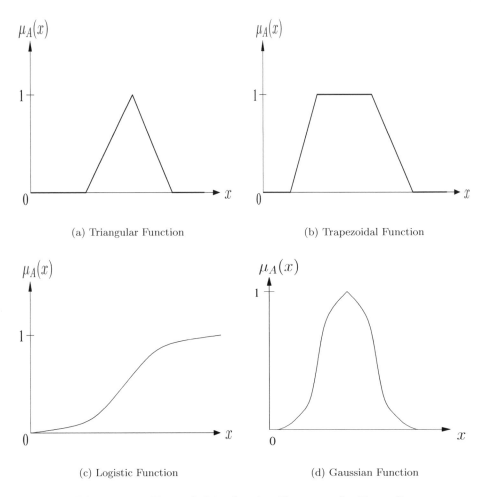

(a) Triangular Function

(b) Trapezoidal Function

(c) Logistic Function

(d) Gaussian Function

**Figure 20.3** Example Membership Functions for Fuzzy Sets

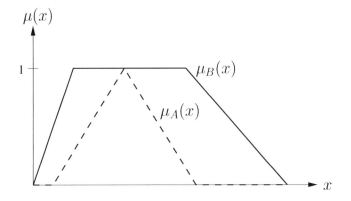

**Figure 20.4** Illustration of Fuzzy Set Containment

- $\mu_{A \cap B}(x) = \frac{\mu_A(x)\mu_B(x)}{p+(1-p)(\mu_A(x)+\mu_B(x)-\mu_A(x)\mu_B(x))}$, for $p > 0$.

- $\mu_{A \cap B}(x) = \frac{1}{\sqrt[p]{\frac{1}{(\mu_A(x))^p}+\frac{1}{(\mu_B(x))^p}-1}}$

- $\mu_{A \cap B}(x) = \frac{\mu_A(x)\mu_B(x)}{\max\{\mu_A(x)\mu_B(x),p\}}$, for $p \in [0,1]$.

- $\mu_{A \cap B}(x) = p \times \min\{\mu_A(x), \mu_B(x)\} + (1-p) \times \frac{1}{2}(\mu_A(x) + \mu_B(x))$, where $p \in [0,1]$ [930].

**Union of fuzzy sets (OR):** The union of two-valued sets contains the elements of all of the sets. The same is true for fuzzy sets, but with membership degrees that depend on the specific uninion operator used. These operators are referred to as s-norms, of which the max-operator and summation operator are most frequently used:

- **Max-operator:** $\mu_{A \cup B}(x) = \max\{\mu_A(x), \mu_B(x)\}$, $\forall x \in X$, or
- **Summation operator:** $\mu_{A \cup B}(x) = \mu_A(x) + \mu_B(x) - \mu_A(x)\mu_B(x)$, $\forall x \in X$

Again, careful consideration must be given to the differences between the two approaches above. In the limit, a series of unions will have a result that approximates 1.0, even though membership degrees are low for the original sets!

Other s-norms are [676],

- $\mu_{A \cup B}(x) = \frac{1}{1+\sqrt[p]{\left(\frac{\mu_A(x)}{1-\mu_A(x)}\right)^p + \left(\frac{\mu_B(x)}{1-\mu_B(x)}\right)^p}}$, for $p > 0$.

- $\mu_{A \cup B}(x) = \min\{1, \mu_A(x) + \mu_B(x) + p\mu_A(x)\mu_B(x)\}$, for $p \geq 0$.

- $\mu_{A \cup B}(x) = \frac{\mu_A(x)+\mu_B(x)-\mu_A(x)\mu_B(x)-(1-p)\mu_A(x)\mu_B(x)}{1-(1-p)\mu_A(x)\mu_B(x)}$, for $p \geq 0$.

- $\mu_{A \cup B}(x) = 1 - \frac{1}{\sqrt[p]{\frac{1}{(1-\mu_A(x))^p}+\frac{1}{(1-\mu_B(x))^p}-1}}$

- $1 - \frac{(1-\mu_A(x))(1-\mu_B(x))}{\max\{(1-\mu_A(x)),(1-\mu_B(x)),p\}}$ for $p \in [0,1]$.

- $\mu_{A \cup B}(x) = p \times \max\{\mu_A(x), \mu_B(x)\} + (1-p) \times \frac{1}{2}(\mu_A(x) + \mu_B(x))$, where $p \in [0,1]$ [930].

Operations on two-valued sets are easily visualized using Venn-diagrams. For fuzzy sets the effects of operations can be illustrated by graphing the resulting membership function, as illustrated in Figure 20.5. For the illustration in Figure 20.5, assume the fuzzy sets $A$, defined as floating point numbers between $[50, 80]$, and $B$, defined as numbers *about* 40 (refer to Figure 20.5(a) for definitions of the membership functions). The complement of set $A$ is illustrated in Figure 20.5(b), the intersection of the two sets are given in Figure 20.5(c) (assuming the *min* operator), and the union in Figure 20.5(d) (assuming the *max* operator).

## 20.4 Fuzzy Set Characteristics

As discussed previously, fuzzy sets are described by membership functions. In this section, characteristics of membership functions are overviewed. These characteristics include normality, height, support, core, cut, unimodality, and cardinality.

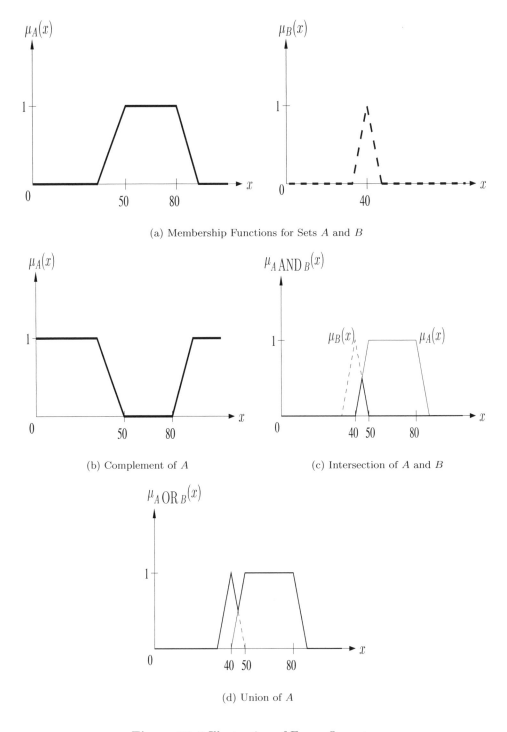

(a) Membership Functions for Sets $A$ and $B$

(b) Complement of $A$                    (c) Intersection of $A$ and $B$

(d) Union of $A$

**Figure 20.5** Illustration of Fuzzy Operators

**Normality**: A fuzzy set $A$ is normal if that set has an element that belongs to the set with degree 1. That is,

$$\exists x \in A \bullet \mu_A(x) = 1 \tag{20.13}$$

then $A$ is normal, otherwise, $A$ is subnormal. Normality can alternatively be defined as

$$\sup_x \mu_A(x) = 1 \tag{20.14}$$

**Height**: The height of a fuzzy set is defined as the supremum of the membership function, i.e.

$$height(A) = \sup_x \mu_A(x) \tag{20.15}$$

**Support**: The support of fuzzy set $A$ is the set of all elements in the universe of discourse, $X$, that belongs to $A$ with non-zero membership. That is,

$$support(A) = \{x \in X | \mu_A(x) > 0\} \tag{20.16}$$

**Core**: The core of fuzzy set $A$ is the set of all elements in the domain that belongs to $A$ with membership degree 1. That is,

$$core(A) = \{x \in X | \mu_A(x) = 1\} \tag{20.17}$$

**$\alpha$-cut**: The set of elements of $A$ with membership degree greater than $\alpha$ is referred to as the $\alpha$-cut of $A$:

$$A_\alpha = \{x \in X | \mu_A(x) \geq \alpha\} \tag{20.18}$$

**Unimodality**: A fuzzy set is unimodal if its membership function is a unimodal function, i.e. the function has just one maximum.

**Cardinality**: The cardinality of two-valued sets is simply the number of elements within the sets. This is not the same for fuzzy sets. The cardinality of fuzzy set $A$, for a finite domain, $X$, is defined as

$$card(A) = \sum_{x \in X} \mu_A(x) \tag{20.19}$$

and for an infinite domain,

$$card(A) = \int_{x \in X} \mu_A(x) dx \tag{20.20}$$

For example, if $X = \{a, b, c, d\}$, and $A = 0.3/a + 0.9/b + 0.1/c + 0.7/d$, then $card(A) = 0.3 + 0.9 + 0.1 + 0.7 = 2.0$.

**Normalization**: A fuzzy set is normalized by dividing the membership function by the height of the fuzzy set. That is,

$$normalized(A) = \frac{\mu_A(x)}{height(x)} \tag{20.21}$$

Other characteristics of fuzzy sets (i.e. concentration, dilation, contrast intensification, fuzzification) are described in Section 21.1.1.

The properties of fuzzy sets are very similar to that of two-valued sets, however, there are some differences. Fuzzy sets follow, similar to two-valued sets, the commutative, associative, distributive, transitive and idempotency properties. One of the major differences is in the properties of the cardinality of fuzzy sets, as listed below:

- $card(A) + card(B) = card(A \cap B) + card(A \cup B)$
- $card(A) + card(\overline{A}) = card(X)$

where $A$ and $B$ are fuzzy sets, and $X$ is the universe of discourse.

## 20.5   Fuzziness and Probability

There is often confusion between the concepts of fuzziness and probability. It is important that the similarities and differences between these two terms are understood. Both terms refer to degrees of certainty (or uncertainty) of events occurring. But that is where the similarities stop. Degrees of certainty as given by statistical probability are only meaningful before the associated event occurs. After that event, the probability no longer applies, since the outcome of the event is known. For example, before flipping a fair coin, there is a 50% probability that heads will be on top, and a 50% probability that it will be tails. After the event of flipping the coin, there is no uncertainty as to whether heads or tails are on top, and for that event the degree of certainty no longer applies. In contrast, membership to fuzzy sets is still relevant after an event occurred. For example, consider the fuzzy set of tall people, with Peter belonging to that set with degree 0.9. Suppose the event to execute is to determine if Peter is good at basketball. Given some membership function, the outcome of the event is a degree of membership to the set of good basketball players. After the event occurred, Peter still belongs to the set of tall people with degree 0.9.

Furthermore, probability assumes independence among events, while fuzziness is not based on this assumption. Also, probability assumes a closed world model where everything is known, and where probability is based on frequency measures of occurring events. That is, probabilities are estimated based on a repetition of a finite number of experiments carried out in a stationary environment. The probability of an event $A$ is thus estimated as

$$Prob(A) = \lim_{n \to \infty} \frac{n_A}{n} \tag{20.22}$$

where $n_A$ is the number of experiments for which event $A$ occurred, and $n$ is the total number of experiments. Fuzziness does not assume everything to be known, and is based on descriptive measures of the domain (in terms of membership functions), instead of subjective frequency measures.

Fuzziness is not probability, and probability is not fuzziness. Probability and fuzzy sets can, however, be used in a symbiotic way to express the probability of a fuzzy event.

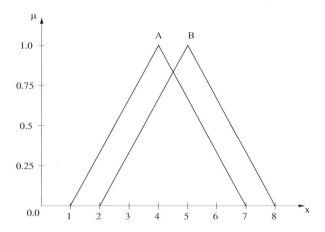

**Figure 20.6** Membership Functions for Assignments 1 and 2

# 20.6  Assignments

1. Perform intersection and union for the fuzzy sets in Figure 20.5 using the t-norms and s-norms defined in Section 20.3.

2. Give the height, support, core and normalization of the fuzzy sets in Figure 20.5.

3. Consider the two fuzzy sets:

$$
\begin{aligned}
\text{long pencils} \ &= \ \{pencil1/0.1, pencil2/0.2, pencil3/0.4, pencil4/0.6, \\
& \qquad pencil5/0.8, pencil6/1.0\} \\
\text{medium pencils} \ &= \ \{pencil1/1.0, pencil2/0.6, pencil3/0.4, pencil4/0.3, \\
& \qquad pencil5/0.1\}
\end{aligned}
$$

   (a) Determine the union of the two sets.
   (b) Determine the intersection of the two sets.

4. What is the difference between the membership function of an ordinary set and a fuzzy set?

5. Consider the membership functions of two fuzzy sets, $A$ and $B$, as given in Figure 20.6.

   (a) Draw the membership function for the fuzzy set $C = A \cap \overline{B}$, using the min-operator.
   (b) Compute $\mu_C(5)$.
   (c) Is $C$ normal? Justify your answer.

6. Consider the fuzzy sets $A$ and $B$ such that $core(A) \cap core(B) = \emptyset$. Is fuzzy set $C = A \cap B$ normal? Justify your answer.

7. Show that the min-operator is

   (a) commutative
   (b) idempotent
   (c) transitive

# Chapter 21

# Fuzzy Logic and Reasoning

The previous chapter discussed theoretical aspects of fuzzy sets. The fuzzy set operators allow rudimentary reasoning about facts. For example, consider the three fuzzy sets *tall*, *good_athlete* and *good_basketball_player*. Now assume

$$\mu_{tall}(Peter) = 0.9 \text{ and } \mu_{good\_athlete}(Peter) = 0.8$$

$$\mu_{tall}(Carl) = 0.9 \text{ and } \mu_{good\_athlete}(Carl) = 0.5$$

If it is known that a good basketball player is tall and is a good athlete, then which one of Peter or Carl will be the better basketball player? Through application of the intersection operator,

$$\mu_{good\_basketball\_player}(Peter) = \min\{0.9, 0.8\} = 0.8$$

$$\mu_{good\_basketball\_player}(Carl) = \min\{0.9, 0.5\} = 0.5$$

Using the standard fuzzy set operators, it is possible to determine that Peter will be better at the sport than Carl.

The example above is a very simplistic situation. For most real-world problems, the sought outcome is a function of a number of complex events or scenarios. For example, actions made by a controller are determined by a set of fuzzy if-then rules. The if-then rules describe situations that can occur, with a corresponding action that the controller should execute. It is, however, possible that more than one situation, as described by if-then rules, are simultaneously active, with different actions. The problem is to determine the best action to take. A mechanism is therefore needed to infer an action from a set of activated situations.

What is necessary is a formal logic system that can be used to reason about uncertainties in order to derive at plausible actions. Fuzzy logic [945] is such a system, which together with an inferencing system form a tool for approximate reasoning. Section 21.1 provides a short definition of fuzzy logic and a short overview of its main concepts. The process of fuzzy inferencing is described in Section 21.2.

## 21.1 Fuzzy Logic

Zadeh [947] defines fuzzy logic (FL) as a logical system, which is an extension of multi-valued logic that is intended to serve as a logic for approximate reasoning. The two

*Computational Intelligence: An Introduction*, Second Edition A.P. Engelbrecht
©2007 John Wiley & Sons, Ltd

most important concepts within FL is that of a linguistic variable and the fuzzy if-then rule. These concepts are discussed in the next subsections.

## 21.1.1 Linguistics Variables and Hedges

Lotfi Zadeh [946] introduced the concept of linguistic variable (or fuzzy variable) in 1973, which allows computation with words in stead of numbers. Linguistic variables are variables with values that are words or sentences from natural language. For example, referring again to the set of tall people, *tall* is a linguistic variable. Sensory inputs are linguistic variables, or nouns in a natural language, for example, temperature, pressure, displacement, etc. Linguistic variables (and hedges, explained below) allow the translation of natural language into logical, or numerical statements, which provide the tools for approximate reasoning (refer to Section 21.2).

Linguistic variables can be divided into different categories:

- Quantification variables, e.g. all, most, many, none, etc.
- Usuality variables, e.g. sometimes, frequently, always, seldom, etc.
- Likelihood variables, e.g. possible, likely, certain, etc.

In natural language, nouns are frequently combined with adjectives for quantifications of these nouns. For example, in the phrase *very tall*, the noun *tall* is quantified by the adjective *very*, indicating a person who is "taller" than tall. In fuzzy systems theory, these adjectives are referred to as hedges. A hedge serves as a modifier of fuzzy values. In other words, the hedge *very* changes the membership of elements of the set *tall* to different membership values in the set *very_tall*. Hedges are implemented through subjective definitions of mathematical functions, to transform membership values in a systematic manner.

To illustrate the implementation of hedges, consider again the set of tall people, and assume the membership function $\mu_{tall}$ characterizes the degree of membership of elements to the set *tall*. Our task is to create a new set, *very_tall* of people that are very tall. In this case, the hedge *very* can be implemented as the square function. That is, $\mu_{very\_tall}(x) = \mu_{tall}(x)^2$. Hence, if Peter belongs to the set *tall* with certainty 0.9, then he also belongs to the set *very_tall* with certainty 0.81. This makes sense according to our natural understanding of the phrase *very tall*: Degree of membership to the set *very_tall* should be less than membership to the set *tall*. Alternatively, consider the set *sort_of_tall* to represent all people that are sort of tall, i.e. people that are shorter than tall. In this case, the hedge *sort of* can be implemented as the square root function, $\mu_{sort\_of\_tall}(x) = \sqrt{\mu_{tall}(x)}$. So, if Peter belongs to the set *tall* with degree 0.81, he belongs to the set *sort_of_tall* with degree 0.9.

Different kinds of hedges can be defined, as listed below:

- **Concentration hedges** (e.g. *very*), where the membership values get relatively smaller. That is, the membership values get more concentrated around points with higher membership degrees. Concentration hedges can be defined, in general terms, as

$$\mu_{A'}(x) = \mu_A(x)^p, \quad for \ p > 1 \tag{21.1}$$

where $A'$ is the concentration of set $A$.

- **Dilation hedges** (e.g. *somewhat, sort of, generally*), where membership values increases. Dilation hedges are defined, in general, as

$$\mu_{A'}(x) = \mu_A(x)^{1/p} \quad for \ p > 1 \tag{21.2}$$

- **Contrast intensification hedges** (e.g. *extremely*), where memberships lower than $1/2$ are diminished, but memberships larger than $1/2$ are elevated. This hedge is defined as,

$$\mu_{A'}(x) = \begin{cases} 2^{p-1}\mu_A(x)^p & if \ \mu_A(x) \leq 0.5 \\ 1 - 2^{p-1}(1 - \mu_A(x))^p & if \ \mu_A(x) > 0.5 \end{cases} \tag{21.3}$$

which intensifies contrast.

- **Vague hedges** (e.g. *seldom*), are opposite to contrast intensification hedges, having membership values altered using

$$\mu_{A'}(x) = \begin{cases} \sqrt{\mu_A(x)/2} & if \ \mu_A(x) \leq 0.5 \\ 1 - \sqrt{(1 - \mu_A(x))/2} & if \ \mu_A(x) > 0.5 \end{cases} \tag{21.4}$$

Vague hedges introduce more "fuzziness" into the set.

- **Probabilistic hedges**, which express probabilities, e.g. likely, not very likely, probably, etc.

## 21.1.2 Fuzzy Rules

For fuzzy systems in general, the dynamic behavior of that system is characterized by a set of linguistic fuzzy rules. These rules are based on the knowledge and experience of a human expert within that domain. Fuzzy rules are of the general form

$$if \ antecedent(s) \ then \ consequent(s) \tag{21.5}$$

The antecedent and consequent of a fuzzy rule are propositions containing linguistic variables. In general, a fuzzy rule is expressed as

$$if \ A \ is \ a \ and \ B \ is \ b \ then \ C \ is \ c \tag{21.6}$$

where $A$ and $B$ are fuzzy sets with universe of discourse $X_1$, and $C$ is a fuzzy set with universe of discourse $X_2$. Therefore, the antecedent of a rule form a combination of fuzzy sets through application of the logic operators (i.e. complement, intersection, union). The consequent part of a rule is usually a single fuzzy set, with a corresponding membership function. Multiple fuzzy sets can also occur within the consequent, in which case they are combined using the logic operators.

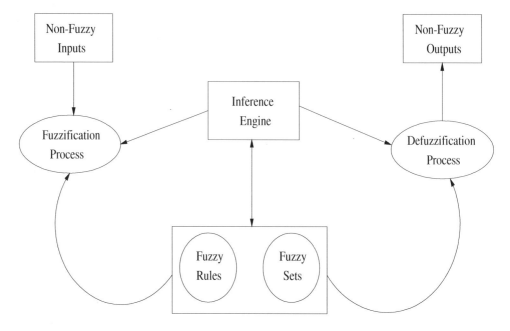

**Figure 21.1** Fuzzy Rule-Based Reasoning System

Together, the fuzzy sets and fuzzy rules form the knowledge base of a fuzzy rule-based reasoning system. In addition to the knowledge base, a fuzzy reasoning system consists of three other components, each performing a specific task in the reasoning process, i.e. fuzzification, inferencing and defuzzification. The different components of a fuzzy rule based system are illustrated in Figure 21.1.

The question is now what sense can be made from a single fuzzy rule. For example, for the rule,

$$\text{if } Age \text{ is } Old \text{ the } Speed \text{ is } Slow \tag{21.7}$$

what can be said about *Speed* if *Age* has the value of 70? Given the membership functions for *Age* and *Speed* in Figure 21.2(a), then find $\mu_{Old}(70)$, which is 0.4. For linguistic variable *Speed* find the intersection of the horizontal line with membership function *Slow*. This gives the shaded area in Figure 21.2(b). A defuzzification operator (refer to Section 21.2) is then used to find the center of gravity of the shaded area, which gives the value of *Speed* = 3.

This simple example leads to the next question: How is a plausible outcome determined if a system has more than one rule? This question is answered in the next section.

## 21.2   Fuzzy Inferencing

Together, the fuzzy sets and fuzzy rules form the knowledge base of a fuzzy rule-based reasoning system. In addition to the knowledge base, a fuzzy reasoning system consists

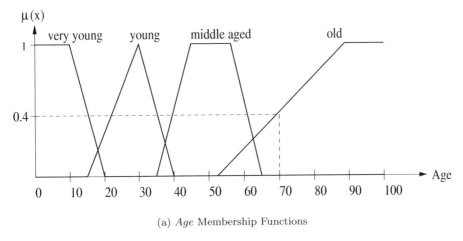

(a) *Age* Membership Functions

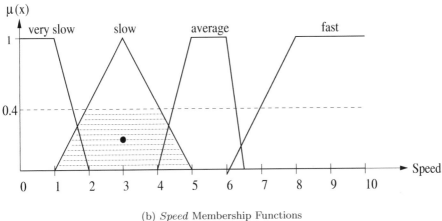

(b) *Speed* Membership Functions

**Figure 21.2** Interpreting a Fuzzy Rule

of three other components, each performing a specific task in the reasoning process, i.e. fuzzification, inferencing, and defuzzification (refer to Figure 21.1).

The remainder of this section is organized as follows: fuzzification is discussed in Section 21.2.1, fuzzy inferencing in Section 21.2.2, and defuzzification in Section 21.2.3.

## 21.2.1 Fuzzification

The antecedents of the fuzzy rules form the fuzzy "input space," while the consequents form the fuzzy "output space". The input space is defined by the combination of input fuzzy sets, while the output space is defined by the combination of output sets. The fuzzification process is concerned with finding a fuzzy representation of non-fuzzy input values. This is achieved through application of the membership functions associated with each fuzzy set in the rule input space. That is, input values from the universe of discourse are assigned membership values to fuzzy sets.

For illustration purposes, assume the fuzzy sets $A$ and $B$, and assume the corresponding membership functions have been defined already. Let $X$ denote the universe of discourse for both fuzzy sets. The fuzzification process receives the elements $a, b \in X$, and produces the membership degrees $\mu_A(a), \mu_A(b), \mu_B(a)$ and $\mu_B(b)$.

## 21.2.2   Inferencing

The task of the inferencing process is to map the fuzzified inputs (as received from the fuzzification process) to the rule base, and to produce a fuzzified output for each rule. That is, for the consequents in the rule output space, a degree of membership to the output sets are determined based on the degrees of membership in the input sets and the relationships between the input sets. The relationships between input sets are defined by the logic operators that combine the sets in the antecedent. The output fuzzy sets in the consequent are then combined to form one overall membership function for the output of the rule.

Assume input fuzzy sets $A$ and $B$ with universe of discourse $X_1$ and the output fuzzy set $C$ with $X_2$ as universe of discourse. Consider the rule

$$\text{if } A \text{ is } a \text{ and } B \text{ is } b \text{ then } C \text{ is } c \qquad (21.8)$$

From the fuzzification process, the inference engine knows $\mu_A(a)$ and $\mu_B(b)$. The first step of the inferencing process is then to calculate the firing strength of each rule in the rule base. This is achieved through combination of the antecedent sets using the operators discussed in Section 20.3. For the example above, assuming the *min*-operator, the firing strength is

$$\min\{\mu_A(a), \mu_B(b)\} \qquad (21.9)$$

For each rule $k$, the firing strength $\alpha_k$ is thus computed.

The next step is to accumulate all activated outcomes. During this step, one single fuzzy value is determined for each $c_i \in C$. Usually, the final fuzzy value, $\beta_i$, associated with each outcome $c_i$ is computed using the *max*-operator, i.e.

$$\beta_i = \max_{\forall k}\{\alpha_{k_i}\} \qquad (21.10)$$

where $\alpha_{k_i}$ is the firing strength of rule $k$ which has outcome $c_i$.

The end result of the inferencing process is a series of fuzzified output values. Rules that are not activated have a zero firing strength.

Rules can be weighted *a priori* with a factor (in the range [0,1]), representing the degree of confidence in that rule. These rule confidence degrees are determined by the human expert during the design process.

## 21.2.3 Defuzzification

The firing strengths of rules represent the degree of membership to the sets in the consequent of the corresponding rule. Given a set of activated rules and their corresponding firing strengths, the task of the defuzzification process is to convert the output of the fuzzy rules into a scalar, or non-fuzzy value.

For the sake of the argument, suppose the following hedges are defined for linguistic variable $C$ (refer to Figure 21.3(a) for the definition of the membership functions): large decrease (LD), slight increase (SI), no change (NC), slight increase (SI), and large increase (LI). Assume three rules with the following $C$ membership values: $\mu_{LI} = 0.8, \mu_{SI} = 0.6$ and $\mu_{NC} = 0.3$.

Several inference methods exist to find an approximate scalar value to represent the action to be taken:

- The **max-min method**: The rule with the largest firing strength is selected, and it is determined which consequent membership function is activated. The centroid of the area under that function is calculated and the horizontal coordinate of that centroid is taken as the output of the controller. For our example, the largest firing strength is 0.8, which corresponds to the *large_increase* membership function. Figure 21.3(b) illustrates the calculation of the output.

- The **averaging method**: For this approach, the average rule firing strength is calculated, and each membership function is clipped at the average. The centroid of the composite area is calculated and its horizontal coordinate is used as output of the controller. All rules therefore play a role in determining the action of the controller. Refer to Figure 21.3(c) for an illustration of the averaging method.

- The **root-sum-square method**: Each membership function is scaled such that the peak of the function is equal to the maximum firing strength that corresponds to that function. The centroid of the composite area under the scaled functions are computed and its horizontal coordinate is taken as output (refer to Figure 21.3(d)).

- The **clipped center of gravity method**: For this approach, each membership function is clipped at the corresponding rule firing strengths. The centroid of the composite area is calculated and the horizontal coordinate is used as the output of the controller. This approach to centroid calculation is illustrated in Figure 21.3(e).

The calculation of the centroid of the trapezoidal areas depends on whether the domain of the functions is discrete or continuous. For a discrete domain of a finite number of values, $n_x$, the output of the defuzzification process is calculated as ($\sum$ has its algebraic meaning)

$$output = \frac{\sum_{i=1}^{n_x} x_i \mu_C(x_i)}{\sum_{i=1}^{n_x} \mu_C(x_i)} \tag{21.11}$$

In the case of a continuous domain ($\int$ has its algebraic meaning),

$$output = \frac{\int_{x \in X} x \mu(x) dx}{\int_{x \in X} \mu(x) dx} \tag{21.12}$$

where $X$ is the universe of discourse.

## 21.3   Assignments

1. For the two pencil fuzzy sets in the assignments of Chapter 20, define a hedge for the set *very long pencils*, and give the resulting set.

2. Consider the following rule base:

   if $x$ is Small then $y$ is Big

   if $x$ is Medium then $y$ is Small

   if $x$ is Big then $y$ is Medium

   Given the membership functions illustrated in Figure 21.4, answer the following questions:

   (a) Using the clipped center of gravity method, draw the composite function for which the centroid needs to be calculated, for $x = 2$.

   (b) Compute the defuzzified output on the discrete domain,
       $Y = \{0, 1, 2, 3, 4, 5, 6, 7, 8\}$

3. Repeat the assignment above for the root-sum-square method.

4. Develop a set of fuzzy rules and membership functions to adapt the values of $w, c_1$ and $c_2$ of a *gbest* PSO.

5. Show how a fuzzy system can be used to adapt the learning rate of a FFNN trained using gradient descent.

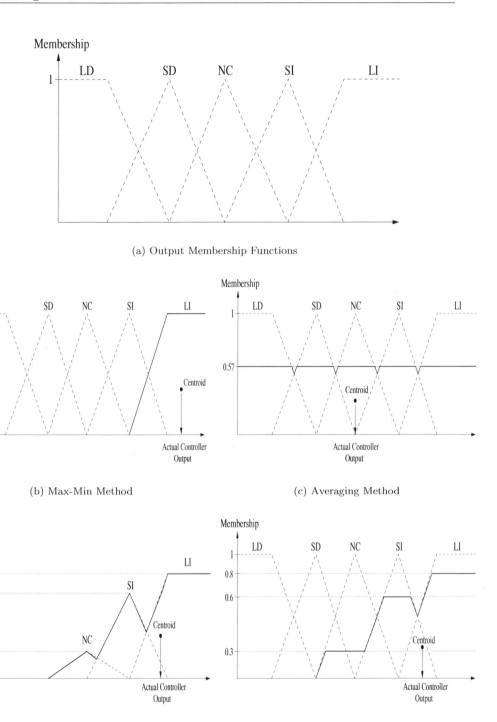

(a) Output Membership Functions

(b) Max-Min Method

(c) Averaging Method

(d) Root-Sum-Square Method

(e) Clipped Center of Gravity Method

**Figure 21.3** Defuzzification Methods for Centroid Calculation

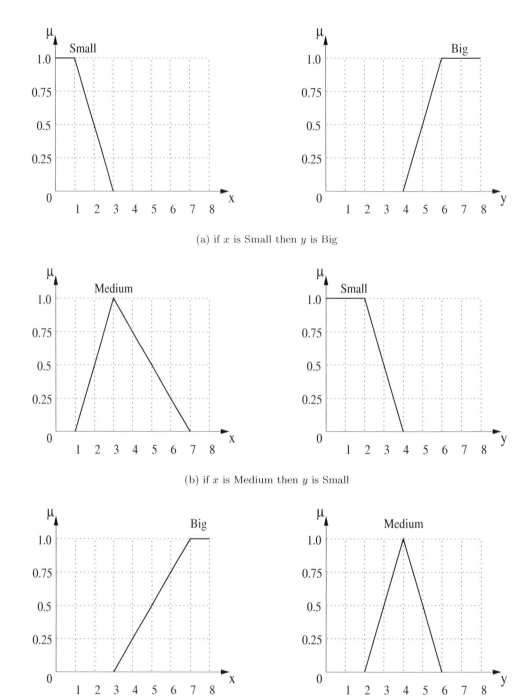

(a) if $x$ is Small then $y$ is Big

(b) if $x$ is Medium then $y$ is Small

(c) if $x$ is Big then $y$ is Medium

**Figure 21.4** Membership Functions for Assignments 2 and 3

# Chapter 22

# Fuzzy Controllers

The design of fuzzy controllers is one of the largest application areas of fuzzy logic. Where fuzzy logic is frequently described as *computing with words rather than numbers*, fuzzy control is described as *control with sentences rather than equations*. Thus, instead of describing the control strategy in terms of differential equations, control is expressed as a set of linguistic rules. These linguistic rules are easier understood by humans than systems of mathematical equations.

The first application of fuzzy control comes from the work of Mamdani and Assilian [554] in 1975, with their design of a fuzzy controller for a steam engine. The objective of the controller was to maintain a constant speed by controlling the pressure on pistons, by adjusting the heat supplied to a boiler. Since then, a vast number of fuzzy controllers have been developed for consumer products and industrial processes. For example, fuzzy controllers have been developed for washing machines, video cameras, air conditioners, etc., while industrial applications include robot control, underground trains, hydro-electrical power plants, cement kilns, etc.

This chapter gives a short overview of fuzzy controllers. Section 22.1 discusses the components of such controllers, while Section 22.2 overviews some types of fuzzy controllers.

## 22.1 Components of Fuzzy Controllers

A fuzzy controller can be regarded as a nonlinear static function that maps controller inputs onto controller outputs. A controller is used to control some system, or plant. The system has a desired response that must be maintained under whatever inputs are received. The inputs to the system can, however, change the state of the system, which causes a change in response. The task of the controller is then to take corrective action by providing a set of inputs that ensures the desired response. As illustrated in Figure 22.1, a fuzzy controller consists of four main components, which are integral to the operation of the controller:

- **Fuzzy rule base**: The rule base, or knowledge base, contains the fuzzy rules that represent the knowledge and experience of a human expert of the system. These rules express a nonlinear control strategy for the system.

*Computational Intelligence: An Introduction*, Second Edition A.P. Engelbrecht
©2007 John Wiley & Sons, Ltd

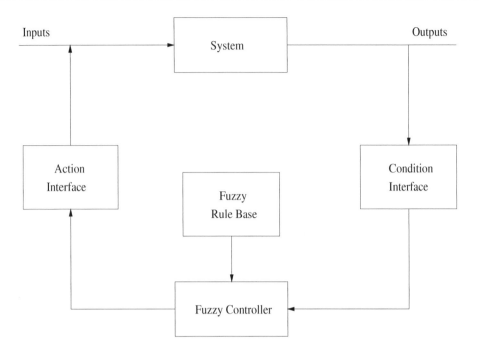

**Figure 22.1** A Fuzzy Controller

While rules are usually obtained from human experts, and are static, strategies have been developed that adapt or refine rules through learning using neural networks or evolutionary algorithms [257, 888].

- **Condition interface (fuzzifier)**: The fuzzifier receives the actual outputs of the system, and transforms these non-fuzzy values into membership degrees to the corresponding fuzzy sets. In addition to the system outputs, the fuzzification of input values to the system also occurs via the condition interface.

- **Action interface (defuzzifier)**: The action interface defuzzifies the outcome of the inference engine to produce a non-fuzzy value to represent the actual control function to be applied to the system.

- **Inference engine**: The inference engine performs inferencing upon fuzzified inputs to produce a fuzzy output (refer to Section 21.2.2).

As stated earlier, a fuzzy controller is basically a nonlinear control function. The nonlinearity in fuzzy controllers is caused by

- the fuzzification process, if nonlinear membership functions are used;
- the rule base, since rules express a nonlinear control strategy;
- the inference engine, if, for example, the *min*-operator is used for intersection and the *max*-operator is used for union; and
- the defuzzification process.

## 22.2 Fuzzy Controller Types

While there exists a number of different types of fuzzy controllers, they all have the same components and involve the same design steps. The differences between types of fuzzy controllers are mainly in the implementation of the inference engine and the defuzzifier.

The design of a fuzzy controller involves the following aspects: A universe of discourse needs to be defined, and the fuzzy sets and membership functions for both the input and output spaces have to be designed. With the help of a human expert, the linguistic rules that describe the dynamic behavior need to be defined. The designer has to decide on how the fuzzifier, inference engine and defuzzifier have to be implemented, after considering all the different options (refer to Section 21.1). Other issues that need to be considered include the preprocessing of the raw measurements as obtained from measuring equipment. Preprocessing involves the removal of noise, discretization of continuous values, scaling and transforming values into a linguistic form.

In the next sections, three controller types are discussed, namely table-based, Mamdani and Takagi-Sugeno.

### 22.2.1 Table-Based Controller

Table-based controllers are used for discrete universes, where it is feasible to calculate all combinations of inputs. The relation between all input combinations and their corresponding outputs are then arranged in a table. In cases where there are only two inputs and one output, the controller operates on a two-dimensional look-up table. The two dimensions correspond to the inputs, while the entries in the table correspond to the outputs. Finding a corresponding output involves a simple and fast look-up in the table. Table-based controllers become inefficient for situations with a large number of input and output values.

### 22.2.2 Mamdani Fuzzy Controller

Mamdani and Assilian [554] produced the first fuzzy controller. Mamdani-type controllers follow the following simple steps:

1. Identify and name input linguistic variables and define their numerical ranges.
2. Identify and name output linguistic variables and define their numerical ranges.
3. Define a set of fuzzy membership functions for each of the input variables, as well as the output variables.
4. Construct the rule base that represents the control strategy.
5. Perform fuzzification of input values.
6. Perform inferencing to determine firing strengths of activated rules.

7. Defuzzify, using centroid of gravity, to determine the corresponding action to be executed.

## 22.2.3   Takagi-Sugeno Controller

For the table-based and Mamdani controllers, the output sets are singletons (i.e. a single set), or combinations of singletons where the combinations are achieved through application of the fuzzy set operators. Output sets can, however, also be linear combinations of the inputs. Takagi and Sugeno suggested an approach to allow for such complex output sets, referred to as Takagi-Sugeno fuzzy controllers [413, 833]. In general, the rule structure for Takagi-Sugeno fuzzy controllers is

$$\text{if } f_1(A_1 \text{ is } a_1, A_2 \text{ is } a_2, \cdots, A_n \text{ is } a_n) \text{ then } C = f_2(a_1, a_2, \cdots, a_n) \tag{22.1}$$

where $f_1$ is a logical function, and $f_2$ is some mathematical function of the inputs; $C$ is the consequent, or output variable being inferred, $a_i$ is an antecedent, or input variable, and $A_i$ is a fuzzy set represented by the membership function $\mu_{A_i}$. The complete rule base is defined by $n_K$ rules.

The firing strength of each rule is computed using the *min*-operator, i.e.

$$\alpha_k = \min_{\forall i | a_i \in \mathcal{A}_k} \{\mu_{A_i}(a_i)\} \tag{22.2}$$

where $\mathcal{A}_k$ is the set of antecedents of rule $k$. Alternatively, the product can be used to calculate rule firing strengths:

$$\alpha_k = \prod_{\forall i | a_i \in \mathcal{A}_k} \mu_{A_i}(a_i) \tag{22.3}$$

The output of the controller is then determined as

$$C = \frac{\sum_{k=1}^{n_K} \alpha_k f_2(a_1, \cdots, a_n)}{\sum_{k=1}^{n_K} \alpha_k} \tag{22.4}$$

The main advantage of Takagi-Sugeno controllers is that it breaks the closed-loop approach of the Mamdani controllers. For the Mamdani controllers the system is statically described by rules. For the Takagi-Sugeno controllers, the fact that the consequent of rules is a mathematical function, provides for a more dynamic control.

## 22.3   Assignments

1. Design a Mamdani fuzzy controller to control a set of ten lifts for a building of forty storey to maximize utilization and minimize delays.

2. Design a Mamdani fuzzy controller for an automatic gearbox for motor vehicles.

3. Consider the following rule base:

if $x$ is Small then $y$ is Big

if $x$ is Medium then $y$ is Small

if $x$ is Big then $y$ is Medium

Given the membership functions illustrated in Figure 21.4, answer the following questions: using a Mamdani-type fuzzy controller, what are the firing strengths of each rule?

4. Consider the following Takagi-Sugeno rules:

if $x$ is $A_1$ and $y$ is $B_1$ then $z_1 = x + y + 1$

if $x$ is $A_2$ and $y$ is $B_1$ then $z_2 = 2x + y + 1$

if $x$ is $A_1$ and $y$ is $B_2$ then $z_3 = 2x + 3y$

if $x$ is $A_2$ and $y$ is $B_2$ then $z_4 = 2x + 5$

Compute the value of $z$ for $x = 1, y = 4$ and the antecedent fuzzy sets

$$A_1 = \{1/0.1, 2/0.6, 3/1.0\}$$

$$A_2 = \{1/0.9, 2/0.4, 3/0.0\}$$

$$B_1 = \{4/1.0, 5/1.0, 6/0.3\}$$

$$B_2 = \{4/0.1, 5/0.9, 6/1.0\}$$

# Chapter 23

# Rough Sets

Fuzzy set theory is the first to have a theoretical treatment of the problem of vagueness and uncertainty, and has had many successful implementations. Fuzzy set theory is, however, not the only theoretical logic that addresses these concepts. Pawlak [668] developed a new theoretical framework to reason with vague concepts and uncertainty. While rough set theory is somewhat related to fuzzy set theory, there are major differences.

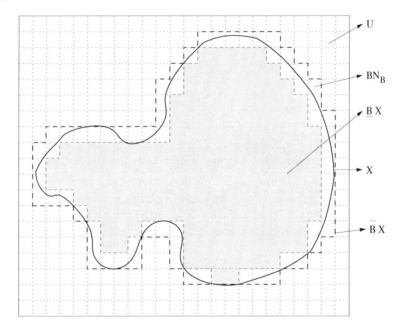

**Figure 23.1** Rough Set Illustration

Rough set theory is based on the assumption that some information, or knowledge, about the elements of the universe of discourse is initially available. This is contrary to fuzzy set theory where no such prior information is assumed. The information available about elements is used to find similar elements and indiscernible elements. Rough set theory is then based on the concepts of upper and lower approximations of sets (refer to Figure 23.1). The lower approximation contains those elements that belong to

*Computational Intelligence: An Introduction*, Second Edition A.P. Engelbrecht
©2007 John Wiley & Sons, Ltd

the set with full certainty, while the upper approximation encapsulates elements for which membership is uncertain. The boundary region of a set, which is the difference between the upper and lower approximations, thus contains all examples which cannot be classified based on the available information.

Rough sets have been shown to be fruitful in a variety of application areas, including decision support, machine learning, information retrieval and data mining. What makes rough sets so desirable for real-world applications is their robustness to noisy environments, and situations where data is incomplete. It is a supervised approach, which clarifies the set-theoretic characteristics of classes over combinatorial patterns of the attributes. In doing so, rough sets also perform automatic feature selection by finding the smallest set of input parameters necessary to discern between classes.

The idea of discernibility is defined in Section 23.1, based on the formal definition of a decision system. Section 23.2 shows how rough sets treat vagueness by forming a boundary region, while Section 23.3 discusses the treatment of uncertainty in the implementation of the rough membership function.

# 23.1 Concept of Discernibility

The discussion on rough sets will be with reference to a decision system. Firstly, an information system is formally defined as an ordered pair $\mathcal{A} = (U, A)$, where $U$ is the universe of discourse and $A$ is a non-empty set of attributes. The universe of discourse is a set of objects (or patterns, examples), while the attributes define the characteristics of a single object. Each attribute $a \in A$ is a function $a : U \to V_a$, where $V_a$ is the range of values for attribute $a$.

A decision system is an information system for which the attributes are grouped into disjoint sets of condition attributes and decision attributes. The condition attributes represent the input parameters, and the decision attributes represent the class.

The basic idea upon which rough sets rests is the discernibility between objects. If two objects are indiscernible over a set of attributes, it means that the objects have the same values for these attributes. Formally, the indiscernibility relation is defined as:

$$IND(B) = \{(x, y) \in U^2 | a(x) = a(y) \; \forall a \in B\} \tag{23.1}$$

where $B \subseteq A$. With $U/IND(B)$ is denoted the set of equivalence classes in the relation $IND(B)$. That is, $U/IND(B)$ contains one class for each set of objects that satisfy $IND(B)$ over all attributes in $B$. Objects are therefore grouped together, where the objects in different groups cannot be discerned between.

A discernibility matrix is a two-dimensional matrix where the equivalence classes form the indices, and each element is the set of attributes that can be used to discern between the corresponding classes. Formally, for a set of attributes $B \subseteq A$ in $\mathcal{A} = (U, A)$, the discernibility matrix $M_D(B)$ is defined as

$$M_D(B) = \{m_D(i, j)\}_{n \times n} \tag{23.2}$$

for $1 \le i, j \le n$, and $n = |U/IND(B)|$, with

$$m_D(i,j) = \{a \in B | a(E_i) \ne a(E_j)\} \tag{23.3}$$

for $i, j = 1, \cdots, n$; $a(E_i)$ indicates that attribute $a$ belongs to equivalence class $E_i$.

Using the discernibility matrix, discernibility functions can be defined to compute the minimal number of attributes necessary to discern equivalence classes from one another. The discernibility function $f(B)$, with $B \subseteq A$, is defined as

$$f(B) = \wedge_{i,j \in \{1 \cdots n\}} \vee \overline{m}_D(E_i, E_j) \tag{23.4}$$

where

$$\overline{m}_D(i,j) = \{\overline{a} | a \in m_D(i,j)\} \tag{23.5}$$

and $\overline{a}$ is the Boolean variable associated with $a$, and $n = |U/IND(B)|$; $\vee \overline{m}_D(E_i, E_j)$ is the disjunction over the set of Boolean variables, and $\wedge$ denotes conjunction.

The discernibility function $f(B)$ finds the minimal set of attributes required to discern any equivalence class from all others. Alternatively, the relative discernibility function $f(E, B)$ finds the minimal set of attributes required to discern a given class, $E$, from the other classes, using the set of attributes, $B$. That is,

$$f(E, B) = \wedge_{j \in \{1 \cdots n\}} \vee \overline{m}_D(E, E_j) \tag{23.6}$$

It is now possible to find all dispensible, or redundant, attributes. An attribute $a \in B \subseteq A$ is dispensible if $IND(B) = IND(B - \{a\})$. Using the definition of dispensibility, a reduct of $B \subseteq A$ is the set of attributes $B' \subseteq B$ such that all $a \in B - B'$ are dispensible, and $IND(B) = IND(B')$. The reduct of $B$ is denoted by $RED(B)$, while $RED(E, B)$ denotes the relative reduct of $B$ for equivalence class $E$. A relative reduct contains sufficient information to discern objects in one class from all other classes.

## 23.2   Vagueness in Rough Sets

Vagueness in rough set theory, where vagueness is with reference to concepts (e.g. a tall person), is based on the definition of a boundary region. The boundary region is defined in terms of an upper and lower approximation of the set under consideration.

Consider the set $X \subseteq U$, and the subset of attributes $B \subseteq A$. The lower approximation of $X$ with regard to $B$ is defined as (also refer to Figure 23.1)

$$\underline{B}X = \cup \{E \in U/IND(B) | E \subseteq X\} \tag{23.7}$$

and the upper approximation of $X$,

$$\overline{B}X = \cup \{E \in U/IND(B) | E \cap X \ne \emptyset\} \tag{23.8}$$

The lower approximation is the set of objects that can be classified with full certainty as members of $X$, while the upper approximation is the set of objects that may possibly be classified as belonging to $X$.

The region,

$$BN_B(X) = \overline{B}X - \underline{B}X \tag{23.9}$$

is defined as the $B$-boundary of $X$. If $BN_B(X) = \emptyset$, then $X$ is crisp with reference to $B$. If $BN_B(X) \neq \emptyset$, then $X$ is rough with reference to $B$.

Rough sets can thus be seen as a mathematical model of vague concepts. Vagueness can then be defined as

$$\alpha_B(X) = \frac{|\underline{B}X|}{|\overline{B}X|} \tag{23.10}$$

with $\alpha_B(X) \in [0, 1]$. If $\alpha_B(X) = 1$, the set $X$ is crisp, otherwise $X$ is rough.

## 23.3   Uncertainty in Rough Sets

A vague concept has a non-empty boundary region, where the elements of that region cannot be classified with certainty as members of the concept. All elements in the boundary region of a rough set therefore have an associated degree of membership, calculated using the rough membership function for a class $E$,

$$\mu_B^X(E, X) = \frac{|E \cap X|}{|E|} \tag{23.11}$$

with $\mu_B^X(E, X) \in [0, 1]$, $E \in U/IND(B)$ and $X \subseteq U$.

Using the rough membership function, the following definitions are valid:

$$\underline{B}X = \{x \in U | \mu_B^X(x) = 1\} \tag{23.12}$$
$$\overline{B}X = \{x \in U | \mu_B^X(x) > 0\} \tag{23.13}$$
$$BN_B(X) = \{x \in U | 0 < \mu_B^X(x) < 1\} \tag{23.14}$$

The above shows that vagueness can, in fact, be defined in terms of uncertainty.

Some properties of the rough membership function are summarized below:

- $\mu_B^X(x) = 1$ iff $x \in \underline{B}X$
- $\mu_B^X(x) = 0$ iff $x \in U - \overline{B}X$
- $0 < \mu_B^X(x) < 1$ iff $x \in BN_B(X)$
- Complement: $\mu_B^{U-X}(x) = 1 - \mu_B^X(x)$ for any $x \in U$
- Union: $\mu_B^{X \cup Y}(x) \geq \max\{\mu_B^X(x), \mu_B^Y(x)\}$ for any $x \in U$
- Intersection: $\mu_B^{X \cap Y}(x) \leq \min\{\mu_B^X(x), \mu_B^Y(x)\}$ for any $x \in U$

# 23.4  Assignments

1. Compare fuzzy sets and rough sets to show their similarities and differences.
2. Discuss the validity of the following two statements:
    (a) two-valued sets form a subset of rough sets
    (b) two-valued sets form a subset of fuzzy sets
    (c) fuzzy sets are special kinds of rough sets.
3. Discuss how rough sets can be used as classifiers.

# References

[1] H.A. Abbass. An Evolutionary Artificial Neural Networks Approach for Breast Cancer Diagnosis. *Artificial Intelligence in Medicine*, 25(3):265–281, 2002.

[2] H.A. Abbass. The Self-Adaptive Pareto Differential Evolution Algorithm. In *Proceedings of the IEEE Congress on Evolutionary Computation*, volume 1, pages 831–836, 2002.

[3] H.A. Abbass, R. Sarker, and C. Newton. PDE: A Pareto-Frontier Differential Evolution Approach for Multi-Objective Optimization Problems. In *Proceedings of the IEEE Congress on Evolutionary Computation*, volume 2, pages 971–978, 2001.

[4] F. Abbattista, N. Abbattistia, and L. Caponetti. An Evolutionary and Cooperative Agents Model for Optimization. In *Proceedings of the IEEE International Conference on Evolutionary Computation*, pages 668–671, 1995.

[5] A.M. Abdelbar and S. Abdelshahid. Swarm Optimization with Instinct-Driven Particles. In *Proceedings of the IEEE Congress on Evolutionary Computation*, pages 777–782, 2003.

[6] M.A. Abido. Particle Swarm Optimization for Multimachine Power System Stabilizer Design. In *Proceedings of the Power Engineering Society Summer Meeting*, pages 1346–1351, 2001.

[7] M.A. Abido. Optimal Power Flow using Particle Swarm Optimization. *International Journal of Electrical Power and Energy Systems*, 24(7):563–571, 2002.

[8] Y.S. Abu-Mostafa. The Vapnik-Chervonenkis Dimension: Information versus Complexity in Learning. *Neural Computation*, 1:312–317, 1989.

[9] Y.S. Abu-Mostafa. Hints and the VC Dimension. *Neural Computation*, 5:278–288, 1993.

[10] D.H. Ackley. *A Connectionist Machine for Genetic Hillclimbing*. Kluwer, Boston, M.A., 1987.

[11] D.K. Agrafiotis and W. Cedeño. Feature Selection for Structure-Activity Correlation using Binary Particle Swarms. *Journal of Medicinal Chemistry*, 45(5):1098–1107, 2002.

[12] O. Aichholzer, F. Aurenhammer, B. Brandstätter, T. Ebner, H. Krasser, C. Magele, M. Mühlmann, and W. Renhart. Evolution Strategy and Hierarchical Clustering. *IEEE Transactions on Magnetics*, 38(2):1041–1044, 2002.

[13] U. Aickelin, P.J. Bentley, S. Cayzer, J. Kim, and J. McLeod. Danger Theory: The Link between AIS and IDS? In *Proceedings of Second International Conference on Artificial Immune Systems*, pages 147–155, 2003.

[14] U. Aickelin and S. Cayzer. The Danger Theory and Its Application to Artificial Immune Systems. In *Proceedings of the First International Conference on Artificial Immune Systems*, pages 141–148, 2002.

*Computational Intelligence: An Introduction,*Second Edition A.P. Engelbrecht
©2007 John Wiley & Sons, Ltd

[15] H. Akaike. A New Look at Statistical Model Identification. *IEEE Transactions on Automatic Control*, 19(6):716–723, 1974.

[16] B. Al-Kazemi and C.K. Mohan. Multi-Phase Discrete Particle Swarm Optimization. In *Proceedings of the International Workshop on Frontiers in Evolutionary Algorithms*, pages 622–625, 2002.

[17] B. Al-Kazemi and C.K. Mohan. Multi-Phase Generalization of the Particle Swarm Optimization Algorithm. In *Proceedings of the IEEE Congress on Evolutionary Computation*, pages 489–494, 2002.

[18] B. Al-Kazemi and C.K. Mohan. Training Feedforward Neural Networks using Multi-Phase Particle Swarm Optimization. In *Proceedings of the Nineth International Conference on Neural Information Processing*, volume 5, pages 2615–2619, 2002.

[19] M.M. Ali and A. Törn. Population Set-Based Global Optimization Algorithms: Some Modifications and Numerical Studies. *Computers & Operations Research*, 31(10):1703–1725, 2004.

[20] G.N. Aly and A.M. Sameh. Evolution of Recurrent Cascade Correlation Networks with Distributed Collaborative Species. In *Proceedings of the IEEE Symposium on Combinations of Evolutionary Computation*, pages 240–249, 2000.

[21] S. Amari, N. Murata, K-R. Müller, M. Finke, and H. Yang. Asymptotic Statistical Theory of Overtraining and Cross-Validation. Technical Report METR 95-06, Department of Mathematical Engineering and Information, University of Tokyo, 1995.

[22] S. Amari, N. Murata, K-R. Müller, M. Finke, and H. Yang. Statistical Theory of Overtraining – Is Cross-Validation Asymptotically Effective? In D.S. Touretzky, M.C. Mozer, and M.E. Hasselmo, editors, *Advances in Neural Information Processing Systems*, volume 8, pages 176–182, 1996.

[23] M.R. Anderberg. *Cluster Analysis for Applications*. Academic Press, New York, 1973.

[24] P.J. Angeline. Adaptive and Self-Adaptive Evolutionary Computation. In M. Palaniswami and Y. Attikiouzel, editors, *Computational Intelligence: A Dynamic Systems Perspective*, pages 152–163, 1995.

[25] P.J. Angeline. Evolutionary Optimization versus Particle Swarm Optimization: Philosophy and Performance Differences. In *Proceedings of the Seventh Annual Conference on Evolutionary Programming*, pages 601–610, 1998.

[26] P.J. Angeline. Using Selection to Improve Particle Swarm Optimization. In *Proceedings of the IEEE Congress on Evolutionary Computation*, pages 84–89, 1998.

[27] P.J. Angeline and J.B. Pollack. Competitive Environments Evolve Better Solutions for Complex Tasks. In *Proceedings of the Fifth International Conference on Genetic Algorithms*, pages 264–270, 1993.

[28] P.J. Angeline, G.M. Saunders, and J.P. Pollack. An Evolutionary Algorithm That Constructs Recurrent Neural Networks. *IEEE Transactions on Neural Networks*, 5(1):54–65, 1994.

[29] R. Annaluru, S. Das, and A. Pahwa. Multi-Level Ant Colony Algorithm for Optimal Placement of Capacitors in Distribution Systems. In *CEC2004. Congress on Evolutionary Computation*, volume 2, pages 1932–1937, June 2004.

[30] M.J. Antunes and M.E. Correia. Towards a New Immunity-Inspired Intrusion Detection Framework. Technical Report DCC-2006-04, Departamento de Ciencia de Computadores Faculdade de Ciencias da Universidade do Porto, 2006.

[31] D.V. Arnold. Evolution Strategies with Adaptively Rescaled Mutation Vectors. Technical Report CS-2005-04, Dalhousie University, 2005.

[32] D.V. Arnold and H-G. Beyer. A General Noise Model and Its Effects on Evolution Strategy Performance. *IEEE Transactions on Evolutionary Computation*, 10(4):380–391, 2006.

[33] R. Axelrod. Evolution of Strategies in The Iterated Prisoner's Dilemma. In L. Davis, editor, *Genetic Algorithms and Simulated Annealing*, pages 32–41. Morgan Kaufmann, 1987.

[34] S. Aydin and H. Temeltas. Time Optimal Trajectory Planning for Mobile Robots by Differential Evolution Algorithm and Neural Networks. In *Proceedings of the IEEE International Symposium on Industrial Electronics*, volume 1, pages 352–357, 2003.

[35] B.V. Babu and M.M.L. Jehan. Differential Evolution for Multi-Objective Optimization. In *Proceedings of the IEEE Congress on Evolutionary Computation*, volume 4, pages 2696–2703, 2003.

[36] B.V. Babu and K.K.N. Sastry. Estimation of Heat Transfer Parameters in A Trickle-Bed Reactor using Differential Evolution and Orthogonal Collocation. *Computers & Chemical Engineering*, 23(3):327–339, 1999.

[37] V. Bachelet and E-G. Talbi. COSEARCH: A Co-Evolutionary Metaheuristic. In *Proceedings of the IEEE Congress on Evolutionary Computation*, volume 2, pages 1550–1557, 2000.

[38] T. Bäck. Selective Pressure in Evolutionary Algorithms: A Characterization of Selection Mechanisms. In *Proceedings of the First IEEE Conference on Evolutionary Computation*, pages 57–62, 1994.

[39] T. Bäck. *Evolutionary Algorithms in Theory and Practice*. Oxford University Press, New York, 1996.

[40] T. Bäck. On The Behavior of Evolutionary Algorithms in Dynamic Environments. In *IEEE World Congress on Evolutionary Computation, Proceedings of the IEEE Congress on Evolutionary Computation*, pages 446–451, 1998.

[41] T. Bäck, D.B. Fogel, and Z. Michalewicz. *Evolutionary Computation 2: Advanced Algorithms and Operators*. IOP Press, 2000.

[42] T. Bäck and U. Hammel. Evolution Strategies Applied to Perturbed Objective Functions. In *IEEE World Congress on Computational Intelligence, Proceedings of the IEEE Conference on Evolutionary Computation*, volume 1, pages 40–45, 1994.

[43] T. Bäck, F. Hoffmeister, and H-P. Schwefel. A Survey of Evolution Strategies. In L.B. Belew and R.K. Booker, editors, *Proceedings of the Fourth International Conference on Genetic Algorithms*, pages 2–9, 1991.

[44] T. Bäck and M. Schütz. Evolution Strategies for Mixed-Integer Optimization of Optical Multilayer Systems. In J.R. McDonnell, R.G. Reynolds, and D.B. Fogel, editors, *Proceedings of the Fourth Annual Conference on Evolutionary Programming*, pages 35–51, Cambridge, M.A., 1995.

[45] T. Bäck and H.-P. Schwefel. An Overview of Evolutionary Algorithms for Parameter Optimization. *Evolutionary Computation*, 1(1):1–23, 1993.

[46] J.E. Baker. Reducing Bias and Inefficiency in The Selection Algorithm. In J. Grefenstette, editor, *Proceedings of the Second International Conference of Genetic Algorithms*, pages 14–21, Hillsdale, N.J., 1987. Erlbaum.

[47] P. Baldi. Computing with Arrays of Bell-Shaped and Sigmoid Functions. In R.P. Lippmann, J.E. Moody, and D.S. Touretzky, editors, *Neural Information Processing Systems*, volume 3, pages 735–742, San Mateo, C.A., 1991. Morgan Kaufmann.

[48] W. Banzhaf. Interactive Evolution. In *Handbook of Evolutionary Computation*, pages C2.10:1–C2.10:6. IOP Press, 1997.

[49] B. Barán and M. Schaerer. A Multiobjective Ant Colony System for Vehicle Routing Problem with Time Windows. In *Proceedings of the Twenty First IASTED International Conference on Applied Informatics*, pages 97–102, 2003.

[50] E. Barnard. Performance and Generalization of the Classification Figure of Merit Criterion Function. *IEEE Transactions on Neural Networks*, 2(2):322–325, 1991.

[51] R. Battiti. First- and Second-Order Methods for Learning: Between Steepest Descent and Newton's Method. *Neural Computation*, 4:141–166, 1992.

[52] D.C. Bauer, J. Cannady, and R.C. Garcia. Detecting Anomalous Behavior: Optimization of Network Traffic Parameters via An Evolution Strategy. In *Proceedings of the IEEE Southeast Conference*, pages 34–39, 2001.

[53] E.B. Baum. On the Capabilities of Multilayer Perceptrons. *Journal of Complexity*, 4:193–215, 1988.

[54] E.B. Baum and D. Haussler. What Size Net Gives Valid Generalization? In D.S. Touretzky, editor, *Advances in Neural Information Processing Systems*, volume 1, pages 81–90, 1989.

[55] D. Beasley, D.R. Bull, and R.R. Martin. A Sequential Niching Technique for Multimodal Function Optimization. *Evolutionary Computation*, 1(2):101–125, 1993.

[56] R.L. Becerra and C.A. Coello Coello. Culturizing Differential Evolution for Constrained Optimization. In *Proceedings of the Fifth Mexican International Conference in Computer Science*, pages 304–311, 2004.

[57] S. Becker and Y. Le Cun. Improving The Convergence of Back-Propagation Learning with Second Order Methods. In D.S. Touretzky, G.E. Hinton, and T.J. Sejnowski, editors, *Proceedings of the 1988 Connectionist Summer School*, pages 29–37. Morgan Kaufmann, 1988.

[58] L.M. Belue and K.W. Bauer. Determining Input Features for Multilayer Perceptrons. *Neurocomputing*, 7:111–121, 1995.

[59] A. Berlanga, P. Isasi, A. Sanchis, and J.M. Molina. Neural Networks Robot Controller Trained with Evolution Strategies. In *Proceedings of the IEEE Congress on Evolutionary Computation*, pages 413–419, 1999.

[60] A. Berlanga, A. Sanchis, P. Isasi, and J.M. Molina. A General Learning Co-Evolution Method to Generalize Autonomous Robot Navigation Behavior. In *Proceedings of the IEEE Congress on Evolutionary Computation*, volume 1, pages 769–776, 2000.

[61] H-G. Beyer. Toward a Theory of 'Evolution Strategies'. Some Asymptotical Results from The $(1 \dotplus \lambda)$-Theory. *Evolutionary Computation*, 1(2):165–188, 1993.

[62] H-G. Beyer. Toward a Theory of Evolution Strategies: On The Benefits of Sex – The$(\mu/\mu, \lambda)$ Theory. *Evolutionary Computation*, 3(1):81–111, 1995.

[63] H-G. Beyer. Toward a Theory of Evolution Strategies: The $(\mu, \lambda)$-Theory. *Evolutionary Computation*, 2(4):381–407, 1995.

[64] H-G. Beyer. Toward a Theory of Evolution Strategies: Self-Adaptation. *Evolutionary Computation*, 3(3):311–347, 1996.

[65] H-G. Beyer. Mutate Large, but Inherit Small! On the Analysis of Rescaled Mutations in $(1, \lambda)$-ES with Noisy Fitness Data. In A.E. Eiben, T. Bäck, M. Schoenauer, and H-P. Schwefel, editors, *Proceedings of the Parallel Problem Solving from Nature Conference*, pages 109–118. Springer Verlag, 1998.

[66] Z. Bian, Y. Yu, B. Zheng, M. Wang, and H. Mao. A Novel Evolution Strategy Algorithm Based on the Selected Direction by The Polar Coordinates. In *International Symposium on Systems and Control in Aerospace and Astronautics*, pages 907–911, 2006.

[67] G. Bilchev and I.C. Parmee. The Ant Colony Metaphor for Searching Continuous Design Spaces. In T. Fogarty, editor, *Proceedings of the AISB Workshop on Evolutionary Computation, Lecture Notes in Computer Science*, volume 993, pages 25–39. Springer-Verlag, 1995.

[68] G. Bilchev and I.C. Parmee. Constrained Optimisation with an Ant Colony Search Model. In *Proceedings of Adaptive Computing in Engineering Design and Control*, pages 145–151, 1996.

[69] H.K. Birru. Empirical Study of Two Classes of Bit Variation Operators in Evolutionary Computation. In *Proceedings of the IEEE Congress on Evolutionary Computation*, volume 3, 1999.

[70] H.K. Birru, K. Chellapilla, and S.S. Rao. Local Search Operators in Fast Evolutionary Programming. In *Proceedings of the IEEE Congress on Evolutionary Computation*, volume 2, 1999.

[71] C. Bishop. Exact Calculation of the Hessian Matrix for the Multilayer Perceptron. *Neural Computation*, 4:494–501, 1992.

[72] T.M. Blackwell. Particle Swarms and Population Diversity II: Experiments. In *Genetic and Evolutionary Computation Conference, Workshop on Evolutionary Algorithms for Dynamic Optimization Problems*, pages 14–18, 2003.

[73] T.M. Blackwell. Swarms in Dynamic Environments. In *Proceedings of the Genetic and Evolutionary Computation Conference, Lecture Notes in Computer Science*, volume 2723, pages 1–12, 2003.

[74] T.M. Blackwell and P.J. Bentley. Don't Push Me! Collision-Avoiding Swarms. In *Proceedings of the IEEE Congress on Evolutionary Computation*, volume 2, pages 1691–1696, 2002.

[75] T.M. Blackwell and P.J. Bentley. Dynamic Search with Charged Swarms. In *Proceedings of the Genetic and Evolutionary Computation Conference*, pages 19–26, 2002.

[76] C. Blum. Beam-ACO – Hybridizing Ant Colony Optimization with Beam Search: An Application to Open Shop Scheduling. *Computers and Operations Research*, 32(6):1565–1591, 2004.

[77] E. Bonabeau, M. Dorigo, and G. Theraulaz. *Swarm Intelligence: From Natural to Artificial Systems*. Oxford University Press, 1999.

[78] E. Bonabeau, G. Theraulaz, V. Fourcassié, and J-L. Deneubourg. The Phase-Ordering Kinetics of Cemetery Organization in Ants. *Physical Review E*, 57:4568–4571, 1998.

[79] S. Bös. Optimal Weight Decay in a Perceptron. In *Proceedings of the International Conference on Artificial Neural Networks*, pages 551–556, 1996.

[80] H.M. Botee and E. Bonabeau. Evolving Ant Colony Optimization. *Advanced in Complex Systems*, 1:149–159, 1998.

[81] D. Braendler and T. Hendtlass. The Suitability of Particle Swarm Optimization for Training Neural Hardware. In *Proceedings of the Fifteenth International Conference on Industrial and Engineering, Applications of Artificial Intelligence and Expert Systems, Lecture Notes in Computer Science*, volume 2358, pages 190–199. Springer-Verlag, 2002.

[82] J. Branke. Memory Enhanced Evolutionary Algorithm for Changing Optimization Problems. In *Proceedings of the IEEE Congress on Evolutionary Computation*, volume 3, pages 1875–1882, 1999.

[83] J. Branke. *Evolutionary Optimization in Dynamic Environments*. Springer, 2001.

[84] L. Breiman. Bagging Predictors. *Machine Learning*, 24(2):123–140, 1996.

[85] H. Bremermann, M. Rogson, and S. Salaff. Global Properties of Evolution Processess. In H. Pattee, E. Edlsack, L. Fein, and A. Callahan, editors, *Natural Automata and Useful Simulations*, pages 3–41, Washington, D.C., 1966. Spartan Books.

[86] H.J. Bremermann. Optimization through Evolution and Recombination. In M.C. Yovits, G.T. Jacobi, and G.D. Goldstine, editors, *Self-Organization Systems*, pages 93–106. Spartan Books, 1962.

[87] P. Bretscher and M. Cohn. A Theory of Self-Nonself Discrimination. *Science*, 169:1042–1049, 1970.

[88] R. Brits. Niching Strategies for Particle Swarm Optimization. Master's thesis, Department of Computer Science, University of Pretoria, 2002.

[89] R. Brits, A.P. Engelbrecht, and F. van den Bergh. A Niching Particle Swarm Optimizer. In *Proceedings of the Fourth Asia-Pacific Conference on Simulated Evolution and Learning*, pages 692–696, 2002.

[90] R. Brits, A.P. Engelbrecht, and F. van den Bergh. Solving Systems of Unconstrained Equations using Particle Swarm Optimization. In *Proceedings of the IEEE Conference on Systems, Man, and Cybernetics*, volume 3, pages 102–107, Oct 2002.

[91] R. Brits, A.P. Engelbrecht, and F. van den Bergh. Locating Multiple Optima using Particle Swarm Optimization. *Applied Mathematics and Computation*, 2007.

[92] D.S. Broomhead and D. Lowe. Multivariate Functional Interpolation and Adaptive Networks. *Complex Systems*, 2:321–355, 1988.

[93] M.D. Bugajska and A.C. Schultz. Anytime Coevolution of Form and Function. In *Proceedings of the IEEE Congress on Evolutionary Computation*, volume 1, pages 359–366, 2003.

[94] B. Bullnheimer, G. Kotsis, and C. Strauss. Parallelization Strategies for The Ant System. In G. Toraldo A. Murli, P. Pardalos, editor, *Kluwer Series on Applied Optimization*, pages 87–100, 1997.

[95] W.L. Buntine and A.S. Weigend. Computing Second Order Derivatives in Feed-Forward Networks: A Review. *IEEE Transactions on Neural Networks*, 5(3):480–488, 1994.

[96] F.M. Burnet. *The Clonal Selection Theory of Acquired Immunity*. Vanderbilt University Press, Nashville, T.N., 1959.

[97] F.M. Burnet. Clonal Selection and After. In G.I. Bell, A.S. Perelson, and G.H. Pimbley Jr., editors, *Theoretical Immunology*, pages 63–85. Marcel Dekker Inc., New York, 1978.

[98] P. Burrascano. A Pruning Technique Maximizing Generalization. In *Proceedings of the International Joint Conference on Neural Networks*, volume 1, pages 347–350, 1993.

[99] D. Camara and A.A.F. Loureiro. A GPS/Ant-Like Routing Algorithm for Ad Hoc Networks. In *Proceedings of the IEEE Wireless Communications and Networking Conference*, pages 1232–1236, 2000.

[100] E. Cantú-Paz. A Survey of Parallel Genetic Algorithms. *Calculateurs Paralleles, Reseaux et Systems Repartis*, 10(2):141–171, 1998.

[101] E. Cantú-Paz. Migration Policies and Takeover Times in Parallel Genetic Algorithms. In W. Banzhaf, J. Daida, A.E. Eiben, M.H. Garzon, V. Honavar, M. Jakiela, and R.E. Smith, editors, *Proceedings of the Genetic and Evolutionary Computation Conference*, page 775, San Francisco, C.A., 1999. Morgan Kaufmann.

[102] E. Cantú-Paz. Migration Policies, Selection Pressure, and Parallel Evolutionary Algorithms. Technical Report IlliGAL Report No. 99015, University of Illinois at Urbana-Champaign, 1999.

[103] E. Cantú-Paz. Parallel Genetic Algorithms with Distributed Panmictic Populations. Technical Report IlliGAL Report No. 99006, University of Illinois at Urbana-Champaign, 1999.

[104] E. Cantú-Paz. Topologies, Migration Rates, and Multi-Population Parallel Genetic Algorithms. In W. Banzhaf, J. Daida, A.E. Eiben, M.H. Garzon, V. Honavar, M. Jakiela, and R.E. Smith, editors, *Proceedings of the Genetic and Evolutionary Computation Conference*, pages 91–98, San Francisco, C.A., 1999. Morgan Kaufmann.

[105] P. Cardoso, M. Jesús, and A. Márquez. MONACO – Multi-Objective Network Optimization Based on an ACO. In *Proceedings of Encuentros de Geometría Computacional*, 2003.

[106] A. Carlisle. *Applying the Particle Swarm Optimizer to Non-Stationary Environments*. PhD thesis, Auburn University, 2002.

[107] A. Carlisle and G. Dozier. Adapting Particle Swarm Optimization to Dynamic Environments. In *Proceedings of the International Conference on Artificial Intelligence*, pages 429–434, 2000.

[108] A. Carlisle and G. Dozier. An Off-the-Shelf PSO. In *Proceedings of the Workshop on Particle Swarm Optimization*, pages 1–6, 2001.

[109] A. Carlisle and G. Dozier. Tracking Changing Extrema with Adaptive Particle Swarm Optimizer. In *Proceedings of the Fifth Biannual World Automation Congress*, pages 265–270, 2002.

[110] T.D.H. Cau and R.J. Kaye. Multiple Distributed Energy Storage Scheduling using Constructive Evolutionary Programming. In *Proceedings of the Twenty-Secondth IEEE Power Engineering Society International Conference on Power Industry Computer Applications*, pages 402–407, 2001.

[111] J.L. Ceciliano and R. Bieva. Transmission Network Planning using Evolutionary Programming. In *Proceedings of the IEEE Congress on Evolutionary Computation*, volume 3, 1999.

[112] CS. Chang and D. Du. Differential Evolution Based Tuning of Fuzzy Automatic Train Operation for Mass Rapid Transit System. *IEE Proceedings of Electric Power Applications*, 147(3):206–212, 2000.

[113] T-T. Chang and H-C. Chang. Application of differential evolution to passive shunt harmonic filter planning. In *Proceedings of the Eigth International Conference on Harmonics and Quality of Power*, volume 1, pages 149–153, 1999.

[114] Y-P. Chang and C-J. Wu. Design of Harmonic Filters using Combined Feasible Direction Method and Differential Evolution. In *Proceedings of the International Conference on Power System Technology*, volume 1, pages 812–817, 2004.

[115] D. Chaturvedi, K. Deb, and S.K. Chakraborty. Structural Optimization using Real-Coded Genetic Algorithms. In P.K. Roy and S.D. Mehta, editors, *Proceedings of the Symposium on Genetic Algorithms*, pages 73–82, Dehradun, 1995. Mahendra Pal Singh.

[116] Y. Chauvin. A Back-Propagation Algorithm with Optimal use of Hidden Units. In D.S. Touretzky, editor, *Advances in Neural Information Processing Systems*, volume 1, pages 519–526, 1989.

[117] Y. Chauvin. Dynamic Behavior of Constrained Back-Propagation Networks. In D.S. Touretzky, editor, *Advances in Neural Information Processing Systems*, volume 2, pages 642–649, 1990.

[118] K. Chellapilla. Combining Mutation Operators in Evolutionary Programming. *IEEE Transactions on Evolutionary Computation*, 2(3):91–96, 1998.

[119] K. Chellapilla and D.B. Fogel. Evolution, Neural Networks, Games, and Intelligence. In *Proceedings of the IEEE*, pages 1471–1496, 1999.

[120] K. Chellapilla and D.B. Fogel. Evolving Neural Networks to Play Checkers without Expert Knowledge. *IEEE Transactions on Neural Networks*, 10(6):1382–1391, 1999.

[121] K. Chellapilla and D.B. Fogel. Anaconda Defeats Hoyle 6-0: A Case Study Competing an Evolved Checkers Program against Commercially Available Software. In *Proceedings of the IEEE Congress on Evolutionary Computation*, pages 857–863, 2000.

[122] C-W. Chen, D-Z. Chen, and G-Z. Cao. An Improved Differential Evolution Algorithm in Training and Encoding Prior Knowledge into Feedforward Networks with Application in Chemistry. *Chemometrics and Intelligent Laboratory Systems*, 64(1):27–43, 2002.

[123] S. Chen, S.A. Billings, C.F.N. Cowan, and P.P. Grant. Practical Identification of NARMAX Models using Radial Basis Functions. *International Journal of Control*, 52:1327–1350, 1990.

[124] J-P. Chiou and F-S. Wang. A Hybrid Method of Differential Evolution with Application to Optimal Control Problems of A Bioprocess System. In *IEEE World Congress on Computational Intelligence, Proceedings of the IEEE International Conference on Evolutionary Computation*, pages 627–632, 1998.

[125] J-P. Chiou and F-S. Wang. Hybrid Method of Evolutionary Algorithms for Static and Dynamic Optimization Problems with Application to A Fed-Batch Fermentation Process. *Computers & Chemical Engineering*, 23(9):1277–1291, 1999.

[126] D-H. Choi and S-Y. Oh. A New Mutation Rule for Evolutionary Programming Motivated from Backpropagation Learning. *IEEE Transactions on Evolutionary Computation*, 4(2):188–190, 2000.

[127] L. Chrétien. *Organisation Spatiale du Matériel Provenant de l'excavation du nid chez Messor Barbarus et des Cadavres d'ouvrières chez "Lasius niger" (Hymenopterae: Formicidae)*. PhD thesis, Université Libre de Bruxelles, 1996.

[128] C-J. Chung and R.G. Reynolds. A Testbed for Solving Optimization Problems using Cultural Algorithms. In L.J Fogel, P.J. Angeline, and T. Bäck, editors, *Proceedings of the Fifth Annual Conference on Evolutionary Programming*, pages 225–236, Cambridge, M.A., 1996. MIT Press.

[129] T. Cibas, F. Fogelman Soulié, P. Gallinari, and S. Raudys. Variable Selection with Neural Networks. *Neurocomputing*, 12:223–248, 1996.

[130] A. Cichocki and R. Unbehauen. *Neural Networks for Optimization and Signal Processing*. Wiley, New York, 1993.

[131] V.A. Cicirello and S.F. Smith. Ant Colony Control for Autonomous Decentralized Shop Floor Routing. In *Proceedings of the Fifth International Symposium on Autonomous Decentralized Systems*, pages 383–390, 2001.

[132] J.M. Claverie, K. de Jong, and A.F. Sheta. Robust Nonlinear Control Design using Competitive Coevolution. In *Proceedings of the IEEE Congress on Evolutionary Computation*, volume 1, pages 403–409, 2000.

[133] M. Clerc. The Swarm and the Queen: Towards a Deterministic and Adaptive Particle Swarm Optimization. In *Proceedings of the IEEE Congress on Evolutionary Computation*, volume 3, pages 1951–1957, 1999.

[134] M. Clerc. Think Locally, Act Locally: The Way of Life of Cheap-PSO, An Adaptive PSO. Technical report, http://clerc.maurice.free.fr/pso/, 2001.

[135] M. Clerc. Discrete Particle Swarm Optimization. In *New Optimization Techniques in Engineering, Lecture Notes in Computer Science*, volume 3612. Springer-Verlag, 2004.

[136] M. Clerc and J. Kennedy. The Particle Swarm-Explosion, Stability, and Convergence in a Multidimensional Complex Space. *IEEE Transactions on Evolutionary Computation*, 6(1):58–73, 2002.

[137] I. Cloete and J. Ludik. Increased Complexity Training. In J. Mira, J. Cabestany, and A. Prieto, editors, *International Workshop on Artificial Neural Networks, Lecture Notes in Computer Science*, pages 267–271, Berlin, 1993. Springer-Verlag.

[138] I. Cloete and J. Ludik. Delta Training Strategies. In *IEEE World Congress on Computational Intelligence, Proceedings of the International Joint Conference on Neural Networks*, volume 1, pages 295–298, 1994.

[139] I. Cloete and J. Ludik. Incremental Training Strategies. In *Proceedings of the International Conference on Artificial Neural Networks*, volume 2, pages 743–746, 1994.

[140] G. Coath and S.K. Halgamuge. A Comparison of Constraint-Handling Methods for The Application of Particle Swarm Optimization to Constrained Nonlinear Optimization Problems. In *Proceedings of the IEEE Congress on Evolutionary Computation*, volume 4, pages 2419–2425, 2003.

[141] H.G. Cobb. An Investigation into The Use of Hypermutation as An Adaptive Operator in Genetic Algorithms having Continuous, Time-Dependent Nonstationary Environments. Technical Report AIC-90-001, Naval Research Laboratory, Washington, D.C., 1990.

[142] J.P. Coelho, P.B. De Moura Oliveira, and J. Boa Ventura Cunha. Non-Linear Concentration Control System Design using A New Adaptive PSO. In *Proceedings of the 5th Portuguese Conference on Automatic Control*, 2002.

[143] J.P. Coelho, P.M. Oliveira, and J.B. Cunha. Greenhouse Air Temperature Control using the Particle Swarm Optimisation Algorithm. In *Proceedings of the Fifteenth Triennial World Congress of the International Federation of Automatic Control*, 2002.

[144] C.A. Coello Coello. Self-Adaptive Penalties for GA-Based Optimization. In *Proceedings of the IEEE Congress on Evolutionary Computation*, volume 1, 1999.

[145] C.A. Coello Coello and R.L. Becerra. Constrained Optimization using an Evolutionary Programming-based Cultural Algorithm. In I.C. Parmee, editor, *Proceedings of the Fifth International Conference on Adaptive Computing in Design and Manufacture*, volume 5, pages 317–328. Springer-Verlag, 2002.

[146] C.A. Coello Coello and R.L. Becerra. Evolutionary Multiobjective Optimization using a Cultural Algorithm. In *Proceedings of the IEEE Swarm Intelligence Symposium*, pages 6–13, 2003.

[147] C.A. Coello Coello and M.S. Lechuga. MOPSO: A Proposal for Multiple Objective Particle Swarm Optimization. In *Proceedings of the IEEE Congress on Evolutionary Computation*, volume 2, pages 1051–1056, 2002.

[148] C.A. Coello Coello, G. Toscano Pulido, and M. Salazar Lechuga. An Extension of Particle Swarm Optimization that can Handle Multiple Objectives. In *Workshop on Multiple Objective Metaheuristics*, 2002.

[149] C.A. Coello Coello, D.A. van Veldhuizen, and G.B. Lamont. *Evolutionary Algorithms for Solving Multi-Objective Problems*. Plenum US, 2002.

[150] C.A. Coello Coello, D.A. Van Veldhuizen, and G.B. Lamont. *Evolutionary Algorithms for Solving Multi-Objective Problems*. Kluwer Academic Publishers, 2002.

[151] D. Cohn, L. Atlas, and R. Ladner. Improving Generalization with Active Learning. *Machine Learning*, 15:201–221, 1994.

[152] D. Cohn and G. Tesauro. Can Neural Networks do Better than the Vapnik-Chervonenkis Bounds? In R. Lippmann, J. Moody, and D.S. Touretzky, editors, *Advances in Neural Information Processing Systems*, volume 3, pages 911–917, 1991.

[153] D.A. Cohn. Neural Network Exploration using Optimal Experiment Design. Technical Report AI Memo No 1491, Artificial Intelligence Laboratory, Massachusetts Institute of Technology, 1994.

[154] D.A. Cohn, Z. Ghahramani, and M.I. Jordan. Active Learning with Statistical Models. *Journal of Artificial Intelligence Research*, 4:129–145, 1996.

[155] A. Conradie, R. Miikkulainen, and C. Aldrich. Adaptive Control Utilizing Neural Swarming. In *Proceedings of the Genetic and Evolutionary Computation Conference*, pages 60–67, 2002.

[156] J. Conradie and A.P. Engelbrecht. Training Bao Game-Playing Agents using Coevolutionary Particle Swarm Optimization. In *Proceedings of the IEEE Symposium on Computational Intelligence in Games*, pages 67–74, 2006.

[157] D. Corne, M. Dorigo, and F. Glover. *New Ideas in Optimization*. McGraw-Hill, 1999.

[158] M. Cosnard, P. Koiran, and H. Paugam-Moisy. Complexity Issues in Neural Network Computations. In I. Simon, editor, *Proceedings of the First Latin American Symposium on Theoretical Informatics, Lecture Notes in Computer Science*, volume 583, pages 530–543. Springer-Verlag, 1992.

[159] L. Costa and P. Oliveira. An Evolution Strategy for Multiobjective Optimization. In *Proceedings of the IEEE Congress on Evolutionary Computation*, volume 1, pages 97–102, 2002.

[160] M. Cottrell, B. Girard, Y. Girard, M. Mangeas, and C. Muller. SSM: A Statistical Stepwise Method for Weight Elimination. In *Proceedings of the International Conference on Artificial Neural Networks*, volume 1, pages 681–684, 1994.

[161] T. Coudert, P. Berruet, and J-L. Philippe. Integration of Reconfiguration in Transitic Systems: An Agent-Based Approach. In *Proceedings of the IEEE International Conference on Systems, Man and Cybernetics*, volume 4, pages 4008–4014, 2003.

[162] R. Coulom. Feedforward Neural Networks in Reinforcement Learning Applied to High-Dimensional Motor Control. In N. Cesa-Bianchi *et. al.*, editor, *Lecture Notes in Artificial Intelligence*, pages 403–413, Berlin Heidelberg, 2002. Springer-Verlag.

[163] G.S. Cowan and R.G. Reynolds. Learning to Access the Quality of Genetic Programs using Cultural Algorithms. In *Proceedings of the IEEE Congress on Evolutionary Computation*, volume 3, pages 1679–1686, 1999.

[164] I.L. López Cruz, L.G. van Willigenburg, and G. van Straten. Efficient Differential Evolution algorithms for multimodal optimal control problems. *Applied Soft Computing*, 3(2):97–122, 2003.

[165] I.L. López Cruz, L.G. van Willigenburg, and G. van Straten. Optimal Control of Nitrate in Lettuce by a Hybrid Approach: Differential Evolution and Adjustable Control Weight Gradient Algorithms. *Computers and Electronics in Agriculture*, 40(1-3):179–197, 2003.

[166] Y. Le Cun, J.S. Denker, and S.A. Solla. Optimal Brain Damage. In D. Touretzky, editor, *Advances in Neural Information Processing Systems*, volume 2, pages 598–605, 1990.

[167] Y. Le Cun, I. Kanter, and S.A. Solla. Second Order Properties of Error Surfaces: Learning Time and Generalization. In R.P. Lippmann, J.E. Moody, and D.S. Touretzky, editors, *Advances in Neural Information Processing Systems*, volume 3, pages 918–924, 1990.

[168] T. Czernichow. Architecture Selection through Statistical Sensitivity Analysis. In *Proceedings of the International Conference on Artificial Neural Networks*, pages 179–184, 1996.

[169] N. Damavandi and S. Safavi-Nacini. A Hybrid Evolutionary Programming Method for Circuit Optimization. *IEEE Transactions on Circuits and Systems - I: Regular Papers*, 52(5):902–910, 2005.

[170] C. Darken and J. Moody. Note on Learning Rate Schedules for Stochastic Optimization. In R. Lippmann, J. Moody, and D.S. Touretzky, editors, *Advances in Neural Information Processing Systems*, volume 3, 1991.

[171] P.J. Darwen and J.B. Pollack. Co-Evolutionary Learning on Noisy Tasks. In P.J. Angeline, Z. Michalewicz, M. Schoenauer, X. Yao, and A. Zalzala, editors, *Proceedings of the IEEE Congress on Evolutionary Computation*, volume 3, pages 1724–1731, 6-9 1999.

[172] P.J. Darwen and X. Yao. Speciation as Automatic Categorical Modularization. *IEEE Transactions on Evolutionary Computation*, 1(2):101–108, 1997.

[173] C.R. Darwin. *On the Origin of Species by Means of Natural Selection or Preservation of Favoured Races in the Struggle for Life*. Murray, London, 1859.

[174] I. Das and J. Dennis. A Closer Look at Drawbacks of Minimizing Weighted Sums of Objectives for Pareto Set Generation in Multicriteria Optimization Problems. *Structural Optimization*, 14(1):63–69, 1997.

[175] D. Dasgupta. *Artificial Immune Systems and their Applications*. Springer: Berlin, 1998.

[176] D. Dasgupta and S. Forrest. An Anomaly Detection Algorithm Inspired by the Immune System. In D. Dasgupta, editor, *Artificial Immune Systems and Their Applications*, pages 262–277. Springer-Verlag, 1999.

[177] J. Davidson. *Stochastic Limit Theory*. Oxford Scholarship Online Monographs, 1994.

[178] L. Davis. Hybridization and Numerical Representation. In L. Davis, editor, *The Handbook of Genetic Algorithms*, pages 61–71. Van Nostrand Reinhold, 1991.

[179] R. Dawkins. *The Blind Whatchmaker*. Norton, New York, 1986.

[180] R.M. de A Silva and G.L. Ramalho. Ant System for the Set Covering Problem. In *Proceedings of the IEEE International Conference on Systems, Man, and Cybernetics*, pages 3129–3133, 2001.

[181] L.N. de Castro and J. Timmis. An Artificial Immune Network for Multimodal Function Optimization. In *Proceedings of the IEEE Congress on Evolutionary Computation*, pages 699–704, 2002.

[182] L.N. de Castro and J. Timmis. *Artificial Immune Systems: A New Computational Approach*. Springer-Verlag, London, UK, 2002.

[183] L.N. de Castro and F.J. Von Zuben. Artificial Immune Systems: Part I - Basic Theory and Applications. Technical Report DCA-RT 01/99, Department of Computer Engineering and Industrial Automation, School of Electrical and Computer Engineering, State University of Campinas, Brazil, 1999.

[184] L.N. de Castro and F.J. Von Zuben. An Evolutionary Immune Network for Data Clustering. In *Proceedings of the IEEE Brazilian Symposium on Artificial Neural Networks*, pages 84–89, 2000.

[185] L.N. de Castro and F.J. Von Zuben. Artificial Immune Systems: Part II - A Survey Of Applications. Technical Report DCA-RT 02/00, Department of Computer Engineering and Industrial Automation, School of Electrical and Computer Engineering, State University of Campinas, Brazil, February 2000.

[186] L.N. de Castro and F.J. Von Zuben. The Clonal Selection Algorithm with Engineering Applications. In *Proceedings of the Genetic and Evolutionary Computational Conference*, pages 36–37, 2000.

[187] L.N. de Castro and F.J. Von Zuben. AiNet: An Artificial Immune Network for Data Analysis. In Hussein A. Abbass, Ruhul A. Sarker, and Charles S. Newton, editors, *Data Mining: A Heuristic Approach*. Idea Group Publishing, USA, 2001.

[188] L.N. de Castro and F.J. Von Zuben. An Immunological Approach to Initialize Centers of Radial Basis Function Neural Networks. In *Proceedings of the Fifth Brazilian Conference on Neural Networks*, pages 79–84, 2001.

[189] L.N. de Castro and F.J. Von Zuben. An Immunological Approach to Initialize Feedforward Neural Network Weights. In *Proceedings of the International Conference on Artificial Neural Networks and Genetic Algorithms*, pages 126–129, 2001.

[190] L.N. de Castro and F.J. Von Zuben. Learning and Optimization Using the Clonal Selection Principle. *IEEE Transactions on Evolutionary Computation, Special Issue on Artificial Immune Systems*, 6(3):239–251, 2002.

[191] K. de Jong. *An Analysis of the Behavior of a Class of Genetic Adaptive Systems*. PhD thesis, University of Michigan, 1975.

[192] K. de Jong and J. Sarma. Generation Gaps Revisited. In *Foundations of Genetic Algorithms*, volume 2, pages 19–28. Morgan Kaufmann, 1992.

[193] K.A. de Jong and R.W. Morrison. A Test Problem Generator for Non-Stationary Environments. In *Proceedings of the IEEE Congress on Evolutionary Computation*, pages 2047–2053, 1999.

[194] K.A. de Jong and M.A. Potter. Evolving Complex Structures via Cooperative Coevolution. In *Proceedings of the Fourth Annual Conference on Evolutionary Programming*, pages 307–317, Cambridge, MA, 1995. MIT Press.

[195] K. Deb. *Multi-Objective Optimization using Evolutionary Algorithms*. Wiley & Sons, 2002.

[196] K. Deb and R.B. Agrawal. Simulated Binary Crossover for Continuous Space. *Complex Systems*, 9:115–148, 1995.

[197] K. Deb, S. Agrawal, A. Patrap, and T. Meyarivan. A Fast Elitist Non-dominated Sorting Genetic Algorithm for Multi-Objective Optimization: NSGA-II. In *Proceedings of the Sixth Parallel Problem Solving in Nature Conference*, pages 849–858, 2000.

[198] K. Deb, D. Joshi, and A. Anand. Real-Coded Evolutionary Algorithms with Parent-Centric Recombination. In *Proceedings of the IEEE Congress on Evolutionary Computation*, pages 61–66, 2002.

[199] J-L. Deneubourg, S. Aron, S. Goss, and J-M. Pasteels. The Self-Organizing Exploratory Pattern of the Argentine Ant. *Journal of Insect Behavior*, 3:159–168, 1990.

[200] J-L. Deneubourg, S. Goss, N. Franks, A. Sendova-Franks, C. Detrain, and L. Chrétien. The Dynamics of Collective Sorting: Robot-Like Ant and Ant-Like Robot. In J.A. Meyer and S.W. Wilson, editors, *Proceedings of the First Conference on Simulation of Adaptive Behavior: From Animals to Animats*, pages 356–365. MIT Press, 1991.

[201] J.E. Dennis and R.B. Schnabel. *Numerical Methods for Unconstrained Optimization and Nonlinear Equations*. Prentice-Hall, 1983.

[202] J. Depenau and M. Møller. Aspects of Generalization and Pruning. In *IEEE World Congress on Computational Intelligence, Proceedings of the International Joint Conference on Neural Networks*, volume 3, pages 504–509, 1994.

[203] K.I. Diamantaras and S.Y. Kung. *Principal Component Neural Networks: Theory and Applications*. Wiley, New York, 1996.

[204] E. Diaz-Dorado, J.C. Pidre, and E.M. Garcia. Planning of Large Rural Low-Voltage Networks using Evolution Strategies. *IEEE Transactions on Power Systems*, 18(2):1594–1600, 2003.

[205] K. Doerner, W.J. Gutjahr, R.F. Hartl, C. Strauss, and C. Stummer. Pareto Ant Colony Optimization: A Metaheuristic Approach to Multiobjective Portfolio Selection. *Annals of Operations Research*, 131:79–99, 2004.

[206] J.U. Dolinsky, I.D. Jenkinson, and G.J. Colquhoun. Application of Genetic Programming to the Calibration of Industrial Robots. *Computers in Industry*, 58(3):255–264, 2007.

[207] A.V. Donati, R. Montemanni, L.M. Gambardella, and A.E. Rizzoli. Integration of a Robust Shortest Path Algorithm with a Time Dependent Vehicle Routing Model and Applications. In *Proceedings of the IEEE International Symposium on Computational Intelligence for Measurement Systems and Applications*, pages 26–31, 2003.

[208] M. Dorigo. *Optimization, Learning and Natural Algorithms*. PhD thesis, Politecnico di Milano, 1992.

[209] M. Dorigo. Learning by Probabilistic Boolean Networks. In *Proceedings of the IEEE International Conference on Neural Networks*, pages 887–891, 1994.

[210] M. Dorigo, E. Bonabeau, and G. Theraulaz. Ant Algorithms and Stigmergy. *Future Generation Computer Systems*, 16(9):851–871, 2000.

[211] M. Dorigo and G. Di Caro. Ant Colony Optimization: A New Meta-Heuristic. In *Proceedings of the IEEE Congress on Evolutionary Computation*, volume 2, page 1477, July 1999.

[212] M. Dorigo and G. Di Caro. The Ant Colony Optimization Meta-Heuristic. In D. Corne, M. Dorigo, and F. Glover, editors, *New Ideas in Optimization*, pages 11–32. McGraw-Hill, 1999.

[213] M. Dorigo and L. Gambardella. A Study of Some Properties of Ant-Q. In *Proceedings of the Fourth International Conference on Parallel Problem Solving From Nature*, pages 656–665, 1996.

[214] M. Dorigo and L.M. Gambardella. Ant Colonies for the Travelling Salesman Problem. *Biosystems*, 43(2):73–81, 1997.

[215] M. Dorigo and L.M. Gambardella. Ant Colony System: A Cooperative Learning Approach to the Traveling Salesman Problem. *IEEE Transactions on Evolutionary Computation*, 1(1):53–66, 1997.

[216] M. Dorigo, V. Maniezzo, and A. Colorni. Ant System: Optimization by a Colony of Cooperating Agents. *IEEE Transactions on Systems, Man, and Cybernetics-Part B*, 26(1):29–41, 1996.

[217] M. Dorigo and T. Stützle. An Experimental Study of the Simple Ant Colony Optimization Algorithm. In *Proceedings of the WSES International Conference on Evolutionary Computation*, pages 253–258, 2001.

[218] B. Dorizzi, G. Pellieux, F. Jacquet, T. Czernichow, and A. Muñoz. Variable Selection using Generalized RBF Networks: Application to the Forecast of the French T-Bonds. In *Proceedings of Computational Engineering in Systems Applications*, pages 122–127, 1996.

[219] L. dos Santos Coelho and V.C. Mariani. An Efficient Particle Swarm Optimization Approach Based on Cultural Algorithm Applied to Mechanical Design. In *Proceedings of the IEEE Congress on Evolutionary Computation*, pages 1099–1104, 2006.

[220] H. Drucker. Boosting using Neural Networks. In A. Sharkey, editor, *Combining Artificial Neural Nets, Perspectives in Neural Computing*, pages 51–78. Springer, 1999.

[221] W. Duch and J. Korczak. Optimization and Global Minimization Methods Suitable for Neural Networks. *Neural Computing Surveys*, 2:163–212, 1998.

[222] R. Durbin and D.E. Rumelhart. Product Units: A Computationally Powerful and Biologically Plausible Extension to Backpropagation Networks. *Neural Computation*, 1:133–142, 1989.

[223] W. Durham. *Co-Evolution: Genes, Culture and Human Diversity*. Stanford University Press, 1994.

[224] R.C. Eberhart and J. Kennedy. A New Optimizer using Particle Swarm Theory. In *Proceedings of the Sixth International Symposium on Micromachine and Human Science*, pages 39–43, 1995.

[225] R.C. Eberhart and Y. Shi. Evolving Artificial Neural Networks. In *Proceedings of the International Conference on Neural Networks and Brain*, pages PL5–PL13, 1998.

[226] R.C. Eberhart and Y. Shi. Comparing Inertia Weights and Constriction Factors in Particle Swarm Optimization. In *Proceedings of the IEEE Congress on Evolutionary Computation*, volume 1, pages 84–88, 2000.

[227] R.C. Eberhart and Y. Shi. Particle Swarm Optimization: Developments, Applications and Resources. In *Proceedings of the IEEE Congress on Evolutionary Computation*, volume 1, pages 27–30, May 2001.

[228] R.C. Eberhart and Y. Shi. Tracking and Optimizing Dynamic Systems with Particle Swarms. In *Proceedings of the IEEE Congress on Evolutionary Computation*, volume 1, pages 94–100, 2001.

[229] R.C. Eberhart, P.K. Simpson, and R.W. Dobbins. *Computational Intelligence PC Tools*. Academic Press Professional, first edition, 1996.

[230] J. Eggers, D. Feillet, S. Kehl, M.O. Wagner, and B. Yannou. Optimization of the Keyboard Arrangement Problem using an Ant Colony Algorithm. *European Journal of Operational Research*, 148(3):672–686, 2003.

[231] A. E. Eiben and C. A. Schippers. On Evolutionary Exploration and Exploitation. *Fundamenta Informaticae*, 35(1-4):35–50, 1998.

[232] A.E. Eiben, P-E. Raué, and Z. Ruttkay. Genetic Algorithms with Multi-parent Recombination. In Y. Davidor, H-P. Schwefel, and R. Männer, editors, *Proceedings of the Parallel Problem Solving from Nature Conference*, pages 78–87, Berlin, 1994. Springer.

[233] A.E. Eiben, C.H.M. van Kemenade, and J.N. Kok. Orgy in the Computer: Multi-Parent Reproduction in Genetic Algorithms. Technical Report CS-R9548, Centrum voor Wiskunde en Informatica, 1995.

[234] A.I. El-Gallad, M.E. El-Hawary, A.A. Sallam, and A. Kalas. Enhancing the Particle Swarm Optimizer via Proper Parameters Selection. In *Proceedings of the Canadian Conference on Electrical and Computer Engineering*, pages 792–797, 2002.

[235] M.Y. El-Sharkh and A.A. El-Keib. Maintenance Scheduling of Generation and Transmission Systems using Fuzzy Evolutionary Programming. *IEEE Transactions on Power Systems*, 18(2):862–866, 2003.

[236] J.L. Elman. Distributed Representations, Simple Recurrent Networks, and Grammatical Structure. *Machine Learning*, 7(2/3):195–226, 1991.

[237] A.P. Engelbrecht. Data Generation using Sensitivity Analysis. In *Proceedings of the International Symposium on Computational Intelligence*, 2000.

[238] A.P. Engelbrecht. A New Pruning Heuristic Based on Variance Analysis of Sensitivity Information. *IEEE Transactions on Neural Networks*, 12(6), 2001.

[239] A.P. Engelbrecht. Sensitivity Analysis for Selective Learning by Feedforward Neural Networks. *Fundamenta Informaticae*, 45(1):295–328, 2001.

[240] A.P. Engelbrecht and I. Cloete. A Sensitivity Analysis Algorithm for Pruning Feedforward Neural Networks. In *Proceedings of the IEEE International Conference in Neural Networks*, volume 2, pages 1274–1277, 1996.

[241] A.P. Engelbrecht and I. Cloete. Feature Extraction from Feedforward Neural Networks using Sensitivity Analysis. In *Proceedings of the International Conference on Systems, Signals, Control, Computers*, volume 2, pages 221–225, 1998.

[242] A.P. Engelbrecht and I. Cloete. Selective Learning using Sensitivity Analysis. In *IEEE World Congress on Computational Intelligence, Proceedings of the International Joint Conference on Neural Networks*, pages 1150–1155, 1998.

[243] A.P. Engelbrecht and I. Cloete. Incremental Learning using Sensitivity Analysis. In *Proceedings of the IEEE International Joint Conference on Neural Networks*, volume 2, pages 1350–1355, 1999.

[244] A.P. Engelbrecht, I. Cloete, J. Geldenhuys, and J.M. Zurada. Automatic Scaling using Gamma Learning for Feedforward Neural Networks. In J. Mira and F. Sandoval, editors, *Proceedings of the International Workshop on Artificial Neural Networks, Lecture Notes in Computer Science*, volume 930, pages 374–381, 1995.

[245] A.P. Engelbrecht, I. Cloete, and J.M. Zurada. Determining the Significance of Input Parameters using Sensitivity Analysis. In J. Mira and F. Sandoval, editors, *International Workshop on Artificial Neural Networks, Lecture Notes in Computer Science*, volume 930, pages 382–388, 1995.

[246] A.P. Engelbrecht, L. Fletcher, and I. Cloete. Variance Analysis of Sensitivity Information for Pruning Feedforward Neural Networks. In *Proceedings of the IEEE International Joint Conference on Neural Networks*, 1999.

[247] A.P. Engelbrecht and A. Ismail. Training Product Unit Neural Networks. *Stability and Control: Theory and Applications*, 2(1-2):59–74, 1999.

[248] A.P. Engelbrecht, S. Rouwhorst, and L. Schoeman. A Building Block Approach to Genetic Programming for Rule Discovery. In H.A. Abbass, R.A. Sarker, and C.S. Newton, editors, *Data Mining: A Heuristic Approach*, pages 174–189. Idea Group Publishing, 2002.

[249] T.M. English. Learning to Focus Selectively on Possible Lines of Play in Checkers. In *Proceedings of the IEEE Congress on Evolutionary Computation*, volume 2, pages 1019–1024, 2001.

[250] L.J. Eshelman, R.A. Caruana, and J.D. Schaffer. Biases in the Crossover Landscape. In J.D. Schaffer, editor, *Proceedings of the Third International Conference on Genetic Algorithms*, pages 10–19, 1989.

[251] L.J. Eshelman and J.D. Schaffer. Real-Coded Genetic Algorithms and Interval Schemata. In D. Whitley, editor, *Foundations of Genetic Algorithms*, volume 2, pages 187–202, San Mateo, 1993. Morgan Kaufmann.

[252] S. Fahlman and C. Lebiere. The Cascade-Correlation Learning Architecture. Technical Report CMU-CS-90-100, Carnegie Mellon University, 1990.

[253] S.E. Fahlman. Fast Learning Variations on Back-Propagation: An Empirical Study. In D.S. Touretzky, G.E. Hinton, and T.J. Sejnowski, editors, *Proceedings of the 1988 Connectionist Summer School*, pages 38–51. Morgan Kaufmann, 1988.

[254] H-Y. Fan. A Modification to Particle Swarm Optimization Algorithm. *Engineering Computations*, 19(7-8):970–989, 2002.

[255] J. Farmer, N. Packard, and A. Perelson. The Immune System, Adaptation and Machine Learning. *Physica D*, 22:187–204, 1986.

[256] M. Fathi-Torbaghan and L. Hildebrand. Model-Free Optimization of Fuzzy Rule-based System using Evolution Strategies. *IEEE Transactions on Systems, Man, and Cybernetics*, 27(2):270–277, 1997.

[257] J. Favilla, A. Machion, and F. Gomide. Fuzzy Traffic Control: Adaptive Strategies. In *Proceedings of the IEEE Symposium on Fuzzy Systems*, 1993.

[258] V. Feoktistov and S. Janaqi. Generalization of The Strategies in Differential Evolution. In *Proceedings of the Eighteenth Parallel and Distributed Processing Symposium*, page 165, 2004.

[259] J.E. Fieldsend and S. Singh. A Multi-Objective Algorithm Based upon Particle Swarm Optimisation. In *Proceedings of the UK Workshop on Computational Intelligence*, pages 37–44, 2003.

[260] W. Finnoff, F. Hergert, and H.G. Zimmermann. Improving Model Selection by Nonconvergent Methods. *Neural Networks*, 6:771–783, 1993.

[261] L. Fletcher, V. Katkovnik, F.E. Steffens, and A.P. Engelbrecht. Optimizing the Number of Hidden Nodes of a Feedforward Artificial Neural Network. In *IEEE World Congress on Computational Intelligence, Proceedings of the International Joint Conference on Neural Networks*, pages 1608–1612, 1998.

[262] R. Fletcher. *Practical Methods of Optimization*. John Wiley & Sons, 1987.

[263] C.A. Floudas and P.M. Pardalos. *Recent Advances in Global Optimization*. Princeton Series in Computer Science, Princeton University Press, 1991.

[264] T.C. Fogarty. Varying the Probability of Mutation in the Genetic Algorithm. In J.D. Schaffer, editor, *Proceedings of the Third International Conference on Genetic Algorithms*, pages 104–109, San Mateo, C.A., 1989. Morgan Kaufmann.

[265] D.B. Fogel. *System Identification through Simulated Evolution: A Machine Learning Approach to Modeling*. Ginn Press, Needham Heights, MA, 1991.

[266] D.B. Fogel. *Evolving Artificial Intelligence*. PhD thesis, University of California, 1992.

[267] D.B. Fogel. Applying Fogel and Burgin's 'Competitive Goal-Seeking through Evolutionary Programming' to Coordination, Trust, and Bargaining Games. In *Proceedings of the IEEE Congress on Evolutionary Computation*, volume 2, pages 1210–1216, 2000.

[268] D.B. Fogel. *Blondie24: Playing at the edge of A.I.* Morgan Kaufmann, 2001.

[269] D.B. Fogel, G.B. Fogel, and K. Ohkura. Multiple-Vector Self-Adaptation in Evolutionary Algorithms. *BioSystems*, 61(2-3):155–162, 2001.

[270] D.B. Fogel and L.J. Fogel. Optimal Routing of Multiple Autonomous Underwater Vehicles through Evolutionary Programming. In *Proceedings of the Symposium on Autonomous Underwater Vehicle Technology*, pages 44–47, 1990.

[271] D.B. Fogel, L.J. Fogel, and J.W. Atmar. Meta-Evolutionary Programming. In *Proceedings of the Twenty-Fifth Conference on Signals, Systems and Computers*, volume 1, pages 540–545, 1991.

[272] D.B. Fogel, L.J. Fogel, and V.W. Porto. Evolutionary Programming for Training Neural Networks. In *Proceedings of the IEEE International Joint Conference on Neural Networks*, volume 1, pages 601–605, 1990.

[273] D.B. Fogel, T.J. Hays, and D.R. Johnson. A Platform for Evolving Characters in Competitive Games. In *Proceedings of the IEEE Congress on Evolutionary Computation*, volume 2, pages 1420–1426, 2004.

[274] G.B. Fogel, G.W. Greenwood, and K. Chellapilla. Evolutionary Computation with Extinction: Experiments and Analysis. In *Proceedings of the IEEE Congress on Evolutionary Computation*, volume 2, pages 1415–1420, 2000.

[275] L.J. Fogel. Autonomous Automata. *Industrial Research*, 4:14–19, 1962.

[276] L.J. Fogel. *On the Organization of Intellect*. PhD thesis, University of California, Los Angeles, 1964.

[277] L.J. Fogel, P.J. Angeline, and D.B. Fogel. An Evolutionary Programming Approach to Self-Adaptation on Finite State Machines. In J. McDonnell, R. Reynolds, and D.B. Fogel, editors, *Proceedings of the Fourth Annual Conference on Evolutionary Programming*, pages 355–365. MIT Press, 1995.

[278] L.J. Fogel, A. Owens, and M. Walsh. *Artificial Intelligence through Simulated Evolution*. John Wiley & Sons, 1966.

[279] S. Forrest and S. Hofmeyr. Immunology as Information Processing. In L.A. Segel and I. Cohen, editors, *Design Principles for the Immune System and Other Distributed Autonomous Systems*. Oxford University Press, Santa Fe Institute Studies in the Sciences of Complexity. New York, 2001.

[280] S. Forrest, S. Hofmeyr, and A. Somayaji. Computer Immunology. *Communications of the ACM*, 40(10):88–96, 1997.

[281] S. Forrest, A.S. Perelson, L. Allen, and R. Cherukuri. Self-Nonself Discrimination in a Computer. In *Proceedings of the IEEE Symposium on Research in Security and Privacy*, pages 202–212, 1994.

[282] O. Fournier, P. Lopez, and J-D. Lan Sun Luk. Cyclic Scheduling Following the Social Behavior of Ant Colonies. In *Proceedings of the IEEE International Conference on Systems, Man and Cybernetics*, volume 3, page 5, October 2002.

[283] N. Franken. PSO-Based Coevolutionary Game Learning. Master's thesis, Department of Computer Science, University of Pretoria, 2004.

[284] N. Franken and A.P. Engelbrecht. Comparing PSO Structures to Learn the Game of Checkers from Zero Knowledge. In *Proceedings of the IEEE Congress on Evolutionary Computation*, pages 234–241, 2003.

[285] N. Franken and A.P. Engelbrecht. PSO Approaches to Co-Evolve IPD Strategies. In *Proceedings of the IEEE Congress on Evolutionary Computation*, pages 356–363, 2004.

[286] N. Franken and A.P. Engelbrecht. Particle Swarm Optimisation Approaches to Co-evolve Strategies for the Iterated Prisoner's Dilemma. *IEEE Transactions on Evolutionary Computation*, 9(6):562–579, 2005.

[287] B. Franklin and M. Bergerman. Cultural Algorithms: Concepts and Experiments. In *Proceedings of the IEEE Congress on Evolutionary Computation*, volume 2, pages 1245–1251, 2000.

[288] A.S. Fraser. Simulation of Genetic Systems by Automatic Digital Computers I: Introduction. *Australian Journal of Biological Science*, 10:484–491, 1957.

[289] A.S. Fraser. Simulation of Genetic Systems by Automatic Digital Computers II: Effects of Linkage on Rates of Advance Under Selection. *Australian Journal of Biological Science*, 10:492–499, 1957.

[290] Y. Freund and R.E. Schapire. A Short Introduction to Boosting. *Journal of Japanese Society for Artificial Intelligence*, 14(5):771–780, 1999.

[291] B. Fritzke. Incremental Learning of Local Linear Mappings. In *Proceedings of the International Conference on Artificial Neural Networks*, pages 217–222, 1995.

[292] O. Fujita. Optimization of the Hidden Unit Function in Feedforward Neural Networks. *Neural Networks*, 5:755–764, 1992.

[293] T. Fukuda and N. Kubota. Learning, Adaptation and Evolution of Intelligent Robotic System. In *Proceedings of the IEEE International Symposium on Intelligent Control*, pages 2–7, 1998.

[294] T. Fukuda, K. Mori, and M. Tsukiyama. Parallel Search for Multi-Modal Function Optimization with Diversity and Learning of Immune Algorithm. In D. Dasgupta, editor, *Artificial Immune Systems and their Applications*, pages 210–220. Springer, 1998.

[295] K. Fukumizu. Active Learning in Multilayer Perceptrons. In D.S. Touretzky, M.C. Mozer, and M.E. Hasselmo, editors, *Advances in Neural Information Processing Systems*, volume 8, pages 295–301, 1996.

[296] Y. Fukuyama, S. Takayama, Y. Nakanishi, and H. Yoshida. A Particle Swarm Optimization for Reactive Power and Voltage Control in Electric Power Systems. In *Proceedings of the Genetic and Evolutionary Computation Conference*, pages 1523–1528, 1999.

[297] Y. Fukuyama and H. Yoshida. A Particle Swarm Optimization for Reactive Power and Voltage Control in Electric Power Systems. In *Proceedings of the IEEE Congress on Evolutionary Computation*, volume 1, pages 87–93, 2001.

[298] R. Gadagkar and N.V. Joshi. Quantitative Ethology of Social Wasps: Time-Activity Budgets and Caste Differences in *Ropalidia Marginata (L)*. *Animal Behavior*, 31:26–31, 1983.

[299] Z-L. Gaing. Particle Swarm Optimization to Solving the Economic Dispatch Considering the Generator Constraints. *IEEE Transactions on Power Systems*, 18(3):1187–1195, 2003.

[300] L.M. Gambardella and M. Dorigo. Ant-Q: A Reinforcement Learning Approach to the TSP. In *Proceedings of Twelfth International Conference on Machine Learning*, pages 252–260, 1995.

[301] L.M. Gambardella and M. Dorigo. Solving Symmetric and Asymmetric TSPs by Ant Colonies. In *Proceedings of IEEE International Conference on Evolutionary Computation*, pages 622–627, 1996.

[302] L.M. Gambardella and M. Dorigo. An Ant Colony System Hybridized with a New Local Search for the Sequential Ordering Problem. *Informs Journal of Computing*, 12(3):237–255, 2000.

[303] L.M. Gambardella, E. Taillard, and G. Agazzi. MACS-VRPTW: A Multiple Ant Colony System for Vehicle Routing Problems with Time Windows. Technical Report IDSIA-06-99, IDSIA, Lugano, Switzerland, 1999.

[304] L.M. Gambardella, E.D. Taillard, and M. Dorigo. Ant Colonies for the QAP. *Journal of the Operational Research Society*, 50:167–176, 1999.

[305] J. Gan and K. Warwick. A Variable Radius Niche Technique for Speciation in Genetic Algorithms. In *Proceedings of the Genetic and Evolutionary Computation Conference*, pages 96–103. Morgan-Kaufmann, 2000.

[306] J. Gan and K. Warwick. Dynamic Niche Clustering: A Fuzzy Variable Radius Niching Technique for Multimodal Optimization in GAs. In *Proceedings of the IEEE Congress on Evolutionary Computation*, volume I, pages 215–222, 2001.

[307] J. Gan and K. Warwick. Modelling Niches of Arbitrary Shape in Genetic Algorithms using Niche Linkage in the Dynamic Niche Clustering Framework. In *Proceedings of the IEEE World Congress on Evolutionary Computation*, pages 43–48, 2002.

[308] W. Gao. Fast Immunized Evolutionary Programming. In *Proceedings of the IEEE Congress on Evolutionary Computation*, volume 1, pages 666–670, 2004.

[309] N. Garcia-Pedrajas, C. Hervas-Martinez, and J. Munoz-Perez. COVNET: A Cooperative Coevolutionary Model for Evolving Artificial Neural Networks. *IEEE Transactions on Neural Networks*, 25(3):575–596, 2003.

[310] M.R. Garey and D.S. Johnson. *Computers and Intractability: A Guide to the Theory of NP-Completeness*. W.H. Freeman, San Francisco, 1979.

[311] T.D. Gedeon, P.M. Wong, and D. Harris. Balancing Bias and Variance: Network Topology and Pattern Set Reduction Techniques. In J. Mira and F. Sandoval, editors, *Proceedings of the International Workshop on Artificial Neural Networks, Lecture Notes in Computer Science*, volume 930, pages 551–558, 1995.

[312] D. Gehlhaar and D. Fogel. Tuning Evolutionary Programming for Conformationally Flexible Molecular Docking. In L. Fogel, P. Angeline, and T. Bäck, editors, *Proceedings of the Fifth Annual Conference on Evolutionary Programming*, pages 419–429. MIT Press, 1996.

[313] S. Geman, E. Bienenstock, and R. Dousart. Neural Networks and the Bias/Variance Dilemma. *Neural Computation*, 4:1–58, 1992.

[314] J. Ghosh and Y. Shin. Efficient Higher-Order Neural Networks for Classification and Function Approximation. *International Journal of Neural Systems*, 3(4):323–350, 1992.

[315] J.C. Giarratano. *Expert Systems: Principles and Programming*. PWS Publishing, third edition, 1998.

[316] D. Gies and Y. Rahmat-Samii. Reconfigurable Antenna Array Design using Parallel PSO. In *Proceedings of the IEEE Society International Conference on Antennas and Propagation*, pages 177–180, 2003.

[317] P.E. Gill, W. Murray, and M.H. Wright. *Practical Optimization*. Academic Press, 1983.

[318] F. Girosi, M. Jones, and T. Poggio. Regularization Theory and Neural Network Architectures. *Neural Computation*, 7:219–269, 1995.

[319] F. Glover. Future Paths for Integer Programming and Links to Artificial Intelligence. *Computers and Operations Research*, 13:533–549, 1986.

[320] D.E. Goldberg and K. Deb. A Comparison of Selection Schemes used in Genetic Algorithms. In G.J.E. Rawlins, editor, *Foundations of Genetic Algorithms*, pages 69–93. Morgan Kaufmann, 1991.

[321] D.E. Goldberg, K. Deb, and J.H. Clark. Don't Worry, Be Messy. In *Proceedings of the Fourth International Conference on Genetic Algorithms and Their Applications*, pages 24–30, 1991.

[322] D.E. Goldberg, H. Kargupta, K. Deb, and G. Harik. Rapid, Accurate Optimization of Difficult Problems using Fast Messy Genetic Algorithms. In *Proceedings of the Fifth International Conference on Genetic Algorithms*, pages 56–64. Morgan Kaufmann, 1993.

[323] D.E. Goldberg, B. Korb, and K. Deb. Messy Genetic Algorithms: Motivation, Analysis, and First Results. *Complex Systems*, 3:493–530, 1989.

[324] D.E. Goldberg, B. Korb, and K. Deb. Messy Genetic Algorithms Revisited: Studies in Mixed Size and Scale. *Complex Systems*, 3:415–444, 1990.

[325] D.E. Goldberg and J. Richardson. Genetic Algorithm with Sharing for Multimodal Function Optimization. In *Proceedings of the Second International Conference on Genetic Algorithms*, pages 41–49, 1987.

[326] D.E. Goldberg and L. Wang. Adaptive Niching via Coevolutionary Sharing. In D. Quagliarella, J. Périaux, C. Poloni, and G. Winter, editors, *Genetic Algorithms and Evolution Strategy in Engineering and Computer Science*, pages 21–38. John Wiley and Sons, Chichester, 1998.

[327] F. Gonzalez, D. Dasgupta, and R. Kozma. Combining Negative Selection and Classification Techniques for Anomaly Detection. In *Proceedings of the Congress on Evolutionary Computation*, volume 1, pages 705–710, 2002.

[328] D.M. Gordon and N.J. Mehdiabadi. Encounter Rate and Task Allocation in Harvester Ants. *Behavioral Ecololgy and Sociobiology*, 45:370–377, 1999.

[329] V.S. Gordon and D. Whitley. Serial and Parallel Genetic Algorithms as Function Optimizers. In S. Forrest, editor, *Proceedings of the Fifth International Conference on Genetic Algorithms*, pages 177–183. Morgan Kaufmann, 1993.

[330] S. Goss, S. Aron, J.L. Deneubourg, and J.M. Pasteels. Self-Organized Shortcuts in the Argentine Ant. *Naturwissenschaften*, 76:579–581, 1989.

[331] J.C. Goswami, R. Mydur, and P. Wu. Application of Differential Evolution Algorithm to Model-Based Well Log-Data Inversion. In *Proceedings of the International Symposium of the Antennas and Propagation Society*, volume 1, pages 318–321, 2002.

[332] A.J. Graaff and A.P. Engelbrecht. Optimised Coverage of Non-self with Evolved Lymphocytes in an Artificial Immune System. *International Journal of Computational Intelligence Research*, 2(2):127–150, 2006.

[333] P-P. Grassé. La Reconstruction du nid et les Coordinations Individuelles chez *Bellicositermes Natalensis* et *Cubitermes sp.* la Théorie de la Stigmergie: Essai d'interprétation du Comportement des Termites Constructeurs. *Insectes Sociaux*, 6:41–80, 1959.

[334] M. Gravel, W.L. Price, and C. Gagné. Scheduling Continuous Casting of Aluminum using a Multiple Objective Ant Colony Optimization Metaheuristic. *European Journal of Operational Research*, 143(1):218–229, 2002.

[335] J.J. Grefenstette. Parallel Adaptive Algorithms for Function Optimization. Technical Report CS-81-19, Vanderbilt University, Computer Science Department, Nashville, 1981.

[336] J.J. Grefenstette. Genetic Algorithms for Changing Environments. In R. Maenner and B. Manderick, editors, *Proceedings of the Parallel Problem Solving from Nature Conference*, volume 2, pages 137–144, 1992.

[337] H. Gu and H. Takahashi. Estimating Learning Curves of Concept Learning. *Neural Networks*, 10(6):1089–1102, 1997.

[338] V.G. Gudise and G.K. Venayagamoorthy. Comparison of Particle Swarm Optimization and Backpropagation as Training Algorithms for Neural Networks. In *Proceedings of the IEEE Swarm Intelligence Symposium*, pages 110–117, 2003.

[339] M. Guntsch and M. Middendorf. Pheromone Modification Strategies for Ant Algorithms Applied to Dynamic TSP. In *Proceedings of the Workshop on Applications of Evolutionary Computing*, pages 213–222, 2001.

[340] M. Guntsch and M. Middendorf. Applying Population Based ACO to Dynamic Optimization Problems. In *Proceedings of Third International Workshop on Ant Colony Optimization and Swarm Intelligence*, pages 111–122, 2003.

[341] K. Hadeli, P. Valckenaers, M. Kollingbau, and H. Van Brussel. Multi-Agent Coordination and Control using Stigmergy. *Computers in Industry*, 53(1):75–96, 2004.

[342] M. Hagiwara. Removal of Hidden Units and Weights for Back Propagation Networks. In *Proceedings of the International Joint Conference on Neural Networks*, volume 1, pages 351–354, 1993.

[343] H. Handa, N. Baba, O. Katai, T. Sawaragi, and T. Horiuchi. Genetic Algorithm Involving Coevolution Mechanism to Search for Effective Genetic Information. In *Proceedings of the IEEE International Conference on Evolutionary Computation*, pages 709–714, 1997.

[344] J. Handl, J. Knowles, and M. Dorigo. Ant-Based Clustering: A Comparative Study of Its Relative Performance with Respect to $k$-Means, Average Link and 1D-SOM. Technical Report TR/IRIDIA/2003-24, Université Libre de Bruxelles, 2003.

[345] J. Handl and B. Meyer. Improved Ant-Based Clustering and Sorting in a Document Retrieval Interface. In *Proceedings of the Seventh International Conference on Parallel Problem Solving from Nature, Lecture Notes in Computer Science*, volume 2439, pages 913–923. Springer-Verlag, 2002.

[346] S.J. Hanson and L.Y. Pratt. Comparing Biases for Minimal Network Construction with Back-Propagation. In D.S. Touretzky, editor, *Advances in Neural Information Processing Systems*, volume 1, pages 177–185, 1989.

[347] S. Harding and J.F. Miller. Evolution of Robot Controller Using Cartesian Genetic Programming. In *Lecture Notes in Computer Science*, volume 3447, pages 62–73, 2005.

[348] A.G. Hart and F.L.W. Ratnieks. Task Partitioning, Division of Labour and Nest Compartmentalisation Collectively Isolate Hazardous Waste in the Leafcutting Ant *Atta Cephalotes. Behavioral Ecology and Sociobiology*, 49:387–392, 2001.

[349] E.F. Hartman, J.D. Keeler, and J.M. Kowalski. Layered Neural Networks with Gaussian Hidden Units as Universal Approximators. *Neural Computation*, 2(2):210–215, 1990.

[350] Y Hasegawa, K. Mase, and T. Fukuda. Re-Grasping Behavior Acquisition by Evolutionary Programming. In *Proceedings of the IEEE Congress on Evolutionary Computation*, volume 1, 1999.

[351] B. Hassibi and D.G. Stork. Second Order Derivatives for Network Pruning: Optimal Brain Surgeon. In C. Lee Giles, S.J. Hanson, and J.D. Cowan, editors, *Advances in Neural Information Processing Systems*, volume 5, pages 164–171, 1993.

[352] B. Hassibi, D.G. Stork, and G. Wolff. Optimal Brain Surgeon: Extensions and Performance Comparisons. In J.D. Cowan, G. Tesauro, and J. Alspector, editors, *Advances in Neural Information Processing Systems*, volume 6, pages 263–270, 1994.

[353] D. Haussler, M. Kearns, M. Opper, and R. Schapire. Estimating Average-Case Learning Curves using Bayesian, Statistical Physics and VC Dimension Method. In J. Moody, S.J. Hanson, and R. Lippmann, editors, *Advances in Neural Information Processing Systems*, volume 4, pages 855–862, 1992.

[354] D.S. Hawkins, D.M. Allen, and A.J. Stromberg. Determining the number of components in mixtures of linear models. *Computational Statistics & Data Analysis*, 38(1):15–48, 2001.

[355] M. Hayashi. A Fast Algorithm for the Hidden Units in a Multilayer Perceptron. In *Proceedings of the International Joint Conference on Neural Networks*, volume 1, pages 339–342, 1993.

[356] S. Haykin. *Neural Networks: A Comprehensive Foundation*. MacMillan, 1994.

[357] T. Haynes and S. Sen. Evolving Behavioral Strategies in Predators and Prey. In S. Sen, editor, *International Joint Conference on Artificial Intelligence, Workshop on Adaptation and Learning in Multiagent Systems*, pages 32–37, Montreal, Quebec, Canada, 1995. Morgan Kaufmann.

[358] T. Haynes and S. Sen. Cooperation of the Fittest. In J.R. Koza, editor, *Late Breaking Papers at the Genetic Programming Conference*, pages 47–55, Stanford University, C.A., 1996. Stanford Bookstore.

[359] Z. He, C. Wei, L. Yang, X. Gao, S. Yao, R.C. Eberhart, and Y. Shi. Extracting Rules from Fuzzy Neural Network by Particle Swarm Optimization. In *Proceedings of the IEEE Congress on Evolutionary Computation*, pages 74–77, 1998.

[360] T. Hendtlass. A Combined Swarm Differential Evolution Algorithm for Optimization Problems. In *Proceedings of the Fourteenth International Conference on Industrial and Engineering Applications of Artificial Intelligence and Expert Systems, Lecture Notes in Computer Science*, volume 2070, pages 11–18. Springer-Verlag, 2001.

[361] T. Hendtlass and M. Randall. A Survey of Ant Colony and Particle Swarm Meta-Heuristics and Their Application to Discrete Optimization Problems. In *Proceedings of the Inaugural Workshop on Artificial Life*, pages 15–25, 2001.

[362] A. Hertz, E. Taillard, and R. de Werra. A Tutorial on Tabu Search. In *Proceedings of Giornate di Lavoro (Enterprise Systems: Management of Technical and Organizational Changes)*, pages 13–24, 1995.

[363] H. Higashi and H. Iba. Particle Swarm Optimization with Gaussian Mutation. In *Proceedings of the IEEE Swarm Intelligence Symposium*, pages 72–79, 2003.

[364] L. Hildebrand, B. Reusch, and M. Fathi. Directed Mutation – A New Self-Adaptation for Evolution Strategies. In *Proceedings of the IEEE Congress on Evolutionary Computation*, pages 1550–1557, 1999.

[365] W.D. Hillis. Co-Evolving Parasites Improve Simulated Evolution as an Optimization Procedure. In S. Forrest, editor, *Emergent Computation: Self-Organizing, Collective, and Cooperative Computing Networks*, pages 228–234. MIT Press, 1990.

[366] R. Hinterding. Gaussian Mutation and Self-Adaption for Numeric Genetic Algorithms. In *Proceedings of the International Conference on Evolutionary Computation*, volume 1, page 384, 1995.

[367] N. Hirata, A. Ishigame, and H. Nishigaito. Neuro Stabilizing Control Based on Lyapunov Method for Power System. In *Proceedings of the Fourty-First SICE Annual Conference*, volume 5, pages 3169–3171, 2002.

[368] Y. Hirose, K. Yamashita, and S. Hijiya. Back-Propagation Algorithm which Varies the Number of Hidden Units. *Neural Networks*, 4:61–66, 1991.

[369] R.J.W Hodgson. Particle Swarm Optimization Applied to the Atomic Cluster Optimization Problem. In *Proceedings of the Genetic and Evolutionary Computation Conference*, pages 68–73, 2002.

[370] K. Hoe, W. Lai, and T. Tai. Homogeneous Ants for Web document Similarity Modeling and Categorization. In *Proceedings of the Third International Workshop on Ant Algorithms, Lecture Notes in Computer Science*, volume 2463, pages 256–261. Springer-Verlag, 2002.

[371] J. Hoffmeyer. The Swarming Body. In I. Rauch and G.F. Carr, editors, *Semiotics Around the World, Proceedings of the Fifth Congress of the International Association for Semiotic Studies*, pages 937–940, 1994.

[372] S. Hofmeyr. *An Immunological Model of Distributed Detection and Its Application to Computer Security*. PhD thesis, University of New Mexico, 1999.

[373] S. Hofmeyr and S. Forrest. Immunity by Design: An Artificial Immune System. In *Proceedings of the Genetic and Evolutionary Computation Conference*, pages 1289–1296, 1999.

[374] S. Hofmeyr and S. Forrest. Architecture for an Artificial Immune System. *Evolutionary Computation*, 8(4):443–473, 2000.

[375] A. Hole. Vapnik-Chervonenkis Generalization Bounds for Real Valued Neural Networks. *Neural Computation*, 8:1277–1299, 1996.

[376] J.H. Holland. *Adaptation in Natural and Artificial Systems*. University of Michigan Press, Ann Arbor, 1975.

[377] J.H. Holland. ECHO: Explorations of Evolution in a Miniature World. In J.D. Farmer and J. Doyne, editors, *Proceedings of the Second Conference on Artificial Life*, 1990.

[378] B. Hölldobler and E.O. Wilson. *Journey of the Ants: A Story of Scientific Exploration*. Harvard University Press, 1994.

[379] L. Holmström and P. Koistinen. Using Additive Noise in Back-Propagation Training. *IEEE Transactions on Neural Networks*, 3:24–38, 1992.

[380] A. Homaifar, A.H-Y. Lai, and X. Qi. Constrained Optimization via Genetic Algorithms. *Simulation*, 2(4):242–254, 1994.

[381] A. Hoorfar. Mutation-Based Evolutionary Algorithms and their Applications to Optimization of Antennas in Layered Media. In *Proceedings of the Antennas and Propagation Society International Symposium*, volume 4, pages 2876–2879, 1999.

[382] J. Horn, N. Nafpliotis, and D.E. Goldberg. A Niched Pareto Genetic Algorithm for Multiobjective Optimization. In *Proceedings of the IEEE Symposium on Circuits and Systems*, pages 2264–2267, 1991.

[383] K. Hornik. Multilayer Feedforward Networks are Universal Approximators. *Neural Networks*, 2:359–366, 1989.

[384] O. Hrstka and A. Kucerová. Improvements of Real Coded Genetic Algorithms Based on Differential Operators Preventing Premature Convergence. *Advances in Engineering Software*, 35(3-4):237–246, 2004.

[385] X. Hu and R.C. Eberhart. Tracking Dynamic Systems with PSO: Where's the Cheese? In *Proceedings of the Workshop on Particle Swarm Optimization*, pages 80–83, 2001.

[386] X. Hu and R.C. Eberhart. Adaptive Particle Swarm Optimization: Detection and Response to Dynamic Systems. In *Proceedings of the IEEE Congress on Evolutionary Computation*, volume 2, pages 1666–1670, 2002.

[387] X. Hu and R.C. Eberhart. Multiobjective Optimization using Dynamic Neighborhood Particle Swarm Optimization. In *Proceedings of the IEEE Congress on Evolutionary Computation*, volume 2, pages 1677–1681, 2002.

[388] X. Hu and R.C. Eberhart. Solving Constrained Nonlinear Optimization Problems with Particle Swarm Optimization. In *Proceedings of the Sixth World Multiconference on Systemics, Cybernetics and Informatics*, 2002.

[389] X. Hu, R.C. Eberhart, and Y. Shi. Particle Swarm with Extended Memory for Multiobjective Optimization. In *Proceedings of the IEEE Swarm Intelligence Symposium*, pages 193–197, 2003.

[390] H-J. Huang and F-S. Wang. Fuzzy Decision-Making Design of Chemical Plant using Mixed-Integer Hybrid Differential Evolution. *Computers & Chemical Engineering*, 26(12):1649–1660, 2002.

[391] S-J. Huang. Enhancement of Hydroelectric Generation Scheduling using Ant Colony System Based Optimization Approaches. *IEEE Transactions on Energy Conversion*, 3:296–301, September 2001.

[392] T-Y. Huang and Y-Y. Chen. Modified Evolution Strategies with a Diversity-Based Parent-Inclusion Scheme. In *Proceedings of the IEEE International Conference on Control Applications*, pages 379–384, 2000.

[393] Z-Y. Huang and G.E. Robinson. Regulation of Honey Bee Division of Labor by Colony Age Demography. *Behavioral Ecology and Sociobiology*, 39:147–158, 1996.

[394] W. Huapeng and H. Handroos. Utilization of Differential Evolution in Inverse Kinematics Solution of a Parallel Redundant Manipulator. In *Proceedings of the Fourth International Conference on Knowledge-Based Intelligent Engineering Systems and Allied Technologies*, volume 2, pages 812–815, 2000.

[395] S. Huband, P. Hingston, L. While, and L. Barone. An Evolution Strategy with Probabilistic Mutation for Multi-Objective Optimisation. In *Proceedings of the IEEE Congress on Evolutionary Computation*, volume 4, pages 2284–2291, 2003.

[396] P.J. Huber. *Robust Statistics*. John Wiley & Sons, 1981.

[397] H. Hüning. A Node Splitting Algorithm that Reduces the Number of Connections in a Hamming Distance Classifying Network. In J. Mira, J. Cabestany, and A. Prieto, editors, *International Workshop on Artificial Neural Networks, Lecture Notes in Computer Science*, volume 686, pages 102–107, Berlin, 1993. Springer-Verlag.

[398] J.E. Hunt and D.E. Cooke. Learning using an Artificial Immune System. *Journal of Network and Computer Applications*, 19(2):189–212, 1996.

[399] S.D. Hunt and J.R. Deller (Jr). Selective Training of Feedforward Artificial Neural Networks using Matrix Perturbation Theory. *Neural Networks*, 8(6):931–944, 1995.

[400] D.R. Hush, J.M. Salas, and B. Horne. Error Surfaces for Multi-Layer Perceptrons. In *International Joint Conference on Neural Networks*, volume 1, pages 759–764, 1991.

[401] A. Hussain, J.J. Soraghan, and T.S. Durbani. A New Neural Network for Non-linear Time-Series Modelling. *NeuroVest Journal*, pages 16–26, 1997.

[402] J-N. Hwang, J.J. Choi, S. Oh, and R.J. Marks II. Query-Based Learning Applied to Partially Trained Multilayer Perceptrons. *IEEE Transactions on Neural Networks*, 2(1):131–136, 1991.

[403] J. Iivarinen, T. Kohonen, J. Kangas, and S. Kaski. Visualizing the Clusters on the Self-Organizing Map. In *Proceedings of the Conference on AI Research in Finland*, pages 122–126, 1994.

[404] M.G. Ippolito, E. Riva Sanseverino, and F. Vuinovich. Multi-Objective Ant Colony Search Algorithm for Optimal Electrical Distribution System Strategical Planning. In *Proceedings of the IEEE Congress on Evolutionary Computation*, pages 1924–1931, 2004.

[405] S. Iredi, D. Merkle, and M. Middendorf. Bi-Criterion Optimization with Multi Colony Ant Algorithms. In *Proceedings of the First International Conference on Evolutionary Multicriterion Optimization, Lecture Notes in Computer Science*, volume 1993, pages 359–372. Springer-Verlag, 2001.

[406] A. Ismail. Training and Optimization of Product Unit Neural Networks. Master's thesis, Department of Computer Science, University of Pretoria, 2001.

[407] A. Ismail and A.P. Engelbrecht. Training Product Units in Feedforward Neural Networks using Particle Swarm Optimization. In *Proceedings of the International Conference on Artificial Intelligence*, pages 36–40, 1999.

[408] A. Ismail and A.P. Engelbrecht. Global Optimization Algorithms for Training Product Unit Neural Networks. In *Proceedings of the IEEE International Joint Conference on Neural Networks*, volume 1, pages 132–137, 2000.

[409] K. Izumi, M.M.A. Hashem, and K. Watanabe. An Evolution Strategy with Competing Subpopulations. In *Proceedings of the IEEE International Symposium on Computational Intelligence in Robotics and Automation*, pages 306–311, 1997.

[410] R.A. Jacobs. Increased Rates of Convergence Through Learning Rate Adaption. *Neural Networks*, 1(4):295–308, 1988.

[411] C.Z. Janikow and Z. Michalewicz. An Experimental Comparison of Binary and Floating Point Representations in Genetic Algorithms. In R.K. Belew and L.B. Booker, editors, *Proceedings of the Fourth International Conference in Genetic Algorithms*, pages 31–36. Morgan Kaufmann, 1991.

[412] D.J. Janson and J.F. Frenzel. Training Product Unit Neural Networks with Genetic Algorithms. *IEEE Expert*, 8(5):26–33, 1993.

[413] J. Jantzen. Design of Fuzzy Controllers. Technical Report 98-E864, Department of Automation, Technical University of Denmark, 1998.

[414] V.K. Jayaraman, B.D. Kulkarni, S. Karale, and P. Shelokar. Ant Colony Framework for Optimal Design and Scheduling of Batch Plants. *Computers and Chemical Engineering*, 24(8):1901–1912, 2000.

[415] J. Jeong and S-Y. Oh. Automatic Rule Generation for Fuzzy Logic Controllers using Rule-Level Co-Evolution of Subpopulations. In *Proceedings of the IEEE Congress on Evolutionary Computation*, volume 3, 1999.

[416] N.K. Jerne. Towards a Network Theory of the Immune System. *Annals of Immunology*, 125C(1-2):373–389, 1974.

[417] C. Jiang and C. Wang. Improved Evolutionary Programming with Dynamic Mutation and Metropolis Criteria for Multi-Objective Reactive Power Optimisation. *IEE Proceedings*, 152(2):291–294, 2005.

[418] C. Jiang and C. Wang. Improved evolutionary programming with dynamic mutation and metropolis criteria for multi-objective reactive power optimisation. *IEE Proceedings: Generation, Transmission and Distribution*, 152(2), 2005.

[419] X. Jin and R.G. Reynolds. Using Knowledge-Based Evolutionary Computation to Solve Nonlinear Constraint Optimization Problems: A Cultural Algorithm Approach. In *Proceedings of the IEEE Congress on Evolutionary Computation*, volume 3, pages 1672–1678, 1999.

[420] X. Jin and R.G. Reynolds. Mining Knowledge in Large Scale Databases using Cultural Algorithms with Constraint Handling Mechanisms. In *Proceedings of the IEEE Congress on Evolutionary Computation*, volume 2, pages 1498–1506, 2000.

[421] Y. Jin, M. Olhofer, and B. Sendhoff. Dynamic Weighted Aggregation for Evolutionary Multi-Objective Optimization: Why Does It Work and How? In *Proceedings of the Genetic and Evolutionary Computation Conference*, pages 1042–1049, 2001.

[422] Y. Jin, M. Olhofer, and B. Sendhoff. Dynamic Weighted Aggregation for Evolutionary Multi-Objective Optimization: Why does it Work and How? In *Proceedings of the Genetic and Evolutionary Computation Conference*, pages 1042–1049, 2001.

[423] D.S. Johnson and L.A. McGeoch. The Traveling Salesman Problem: A Case Study in Local Optimization. In J.K. Lenstra E.H.L. Aarts, editor, *Local Search in Combinatorial Optimization*, pages 215–310. John Wiley & Sons, 1997.

[424] J. Johnson and M. Sugisaka. Complexity Science for the Design of Swarm Robot Control Systems. In *Proceedings of the Twenty-Sixth Annual Conference of the IEEE Industrial Electronics Society*, volume 1, pages 695–700, 2000.

[425] J.A. Joines and C.R. Houck. On the Use of Non-Stationary Penalty Functions to Solve Nonlinear Constrained Optimization Problems with Genetic Algorithms. In *Proceedings of the IEEE Congress on Evolutionary Computation*, pages 579–584, 1994.

[426] I.T. Jolliffe. *Principal Component Analysis*. Springer-Verlag, New York, USA, 1986.

[427] T. Jones. Crossover, Macromutation, and Population-based Search. In L. Eshelman, editor, *Proceedings of the Sixth International Conference on Genetic Algorithms*, pages 73–80, San Francisco, C.A., 1995. Morgan Kaufmann.

[428] M.I. Jordan. Attractor Dynamics and Parallelism in a Connectionst Sequential Machine. In *Proceedings of the Cognitive Science Conference*, pages 531–546, 1986.

[429] R. Joshi and A.C. Sanderson. Minimal Representation Multisensor Fusion using Differential Evolution. In *Proceedings of the International Symposium on Computational Intelligence in Robotics and Automation*, pages 255–273, 1997.

[430] C-F. Juang. A Hybrid of Genetic Algorithm and Particle Swarm Optimization for Recurrent Network Design. *IEEE Transactions on Systems, Man, and Cybernetics - Part B: Cybernetics*, 34(2):997–1006, 2003.

[431] J.H. Jun, D.W. Lee, and K.B. Sim. Realization of Cooperative Swarm Behavior in Distributed Autonomous Robotic Systems using Artificial Immune System. In *Proceedings of IEEE International Conference on Systems, Man and Cybernetics*, volume 6, pages 614–619, 1999.

[432] L.P. Kaelbling, M.I. Littman, and A.W. Moore. Reinforcement Learning: A Survey. *Journal of Artificial Intelligence Research*, 4:237–285, 1996.

[433] T. Kaji. Approach by Ant Tabu Agents for Traveling Salesman Problem. In *Proceedings of the IEEE International Conference on Systems, Man, and Cybernetics*, volume 5, pages 3429–3434, 2001.

[434] R. Kamimura. Principal Hidden Unit Analysis: Generation of Simple Networks by Minimum Entropy Method. In *Proceedings of the International Joint Conference on Neural Networks*, volume 1, pages 317–320, 1993.

[435] R. Kamimura and S. Nakanishi. Weight Decay as a Process of Redundancy Reduction. In *IEEE World Congress on Computational Intelligence, Proceedings of the International Joint Conference on Neural Networks*, volume 3, pages 486–491, 1994.

[436] J. Kamruzzaman, Y. Kumagai, and H. Hikita. Study on Minimal Net Size, Convergence Behavior and Generalization Ability of Heterogeneous Backpropagation Network. In I. Aleksander and J. Taylor, editors, *Artificial Neural Networks*, volume 2, pages 203–206, 1992.

[437] S. Kannan, S.M.R. Slochanal, P. Subbaraj, and N.P. Padhy. Application of Particle Swarm Optimization Technique and its Variants to Generation Expansion Planning. *Electric Power Systems Research*, 70(3):203–210, 2004.

[438] M.D. Kapadi and R.D. Gudi. Optimal Control of Fed-Batch Fermentation Involving Multiple Feeds using Differential Evolution. *Process Biochemistry*, 39(11):1709–1721, 2004.

[439] N. Karaboga, A. Kalinli, and D. Karaboga. Designing Digital IIR Filters using Ant Colony Optimisation Algorithm. *Engineering Applications of Artificial Intelligence*, 17:301–309, 2004.

[440] S. Karner. Laws of Thought. *Encyclopedia of Philosophy*, 4:414–417, 1967.

[441] K-U. Kasemir and K. Betzler. Detecting Ellipses of Limited Eccentricity in Images with High Noise Levels. *Image and Vision Computing*, 21(10):221–227, 2003.

[442] S. Kaski, J. Venna, and T. Kohonen. Coloring that Reveals Cluster Structures in Multivariate Data. *Australian Journal of Intelligent Information Processing Systems*, 6:82–88, 2000.

[443] M. Kasper. Shape Optimization by Evolution Strategies. *IEEE Transactions on Magnetics*, 28(2):1556–1560, 1992.

[444] I.N. Kassabalidis, M.A. El-Shurkawi, R.J. Marks, L.S. Moulin, and A.P. Alves da Silva. Dynamic Security Border Identification using Enhanced Particle Swarm Optimization. *IEEE Transactions on Power Systems*, 17(3):723–729, 2002.

[445] Y. Katada, M. Svinin, Y. Matsumura, K. Ohkura, and K. Ueda. Stable Grasp Planning by Evolutionary Programming. *IEEE Transactions on Industrial Electronics*, 48(4):749–756, 2001.

[446] J. Kennedy. The Particle Swarm: Social Adaptation of Knowledge. In *Proceedings of the IEEE International Conference on Evolutionary Computation*, pages 303–308, 1997.

[447] J. Kennedy. Small Worlds and Mega-Minds: Effects of Neighborhood Topology on Particle Swarm Performance. In *Proceedings of the IEEE Congress on Evolutionary Computation*, volume 3, pages 1931–1938, 1999.

[448] J. Kennedy. Bare Bones Particle Swarms. In *Proceedings of the IEEE Swarm Intelligence Symposium*, pages 80–87, 2003.

[449] J. Kennedy and R.C. Eberhart. Particle Swarm Optimization. In *Proceedings of the IEEE International Joint Conference on Neural Networks*, pages 1942–1948, 1995.

[450] J. Kennedy and R.C. Eberhart. A Discrete Binary Version of the Particle Swarm Algorithm. In *Proceedings of the World Multiconference on Systemics, Cybernetics and Informatics*, pages 4104–4109, 1997.

[451] J. Kennedy, R.C. Eberhart, and Y. Shi. *Swarm Intelligence*. Morgan Kaufmann, 2001.

[452] J. Kennedy and R. Mendes. Population Structure and Particle Performance. In *Proceedings of the IEEE Congress on Evolutionary Computation*, pages 1671–1676, 2002.

[453] J. Kennedy and R. Mendes. Neighborhood Topologies in Fully-Informed and Best-of-Neighborhood Particle Swarms. In *Proceedings of the IEEE International Workshop on Soft Computing in Industrial Applications*, pages 45–50, 2003.

[454] J. Kennedy and W. Spears. Matching Algorithms to Problems: An Experimental Test of the Particle Swarm and Some Genetic Algorithms on the Multimodal Problem Generator. In *Proceedings of the IEEE Congress on Evolutionary Computation*, pages 78–83, May 1998.

[455] R.H. Kewley and M.J. Embrechts. Computational Military Tactical Planning System. *IEEE Transactions on Systems, Man and Cybernetics*, 32(2):161–171, 2003.

[456] H-S. Kim, J-H. Park, and Y-K. Choi. Variable Structure Control of Brushless DC Motor using Evolution Strategy with Varying Search Space. In *Proceedings of the IEEE International Conference on Evolutionary Computation*, pages 764–769, 1996.

[457] J. Kim and P.J. Bentley. Negative Selection and Niching by an Artificial Immune System for Network Intrusion Detection. In *Genetic and Evolutionary Computation Conference*, pages 149–158, 1999.

[458] J. Kim and P.J. Bentley. An Evaluation of Negative Selection in an Artificial Immune System for Network Intrusion Detection. In *Proceedings of the Genetic and Evolutionary Computation Conference*, pages 1330–1337, 2001.

[459] J. Kim and P.J. Bentley. A Model of Gene Library Evolution in the Dynamic Clonal Selection Algorithm. In *Proceedings of the First International Conference on Artificial Immune Systems*, volume 1, pages 182–189, 2002.

[460] J. Kim and P.J. Bentley. Immune Memory in the Dynamic Clonal Selection Algorithm. In *Proceedings of the First International Conference on Artificial Immune Systems*, volume 1, pages 59–67, 2002.

[461] J. Kim and P.J. Bentley. Towards an Artificial Immune System for Network Intrusion Detection: An Investigation of Dynamic Clonal Selection. In *Proceedings of Congress on Evolutionary Computation*, pages 1015–1020, 2002.

[462] J-H. Kim, H-K. Chae, J-Y. Jeon, and S-W. Lee. Identification and Control of Systems with Friction using Accelerated Evolutionary Programming. *IEEE Control Systems Magazine*, 16(4):38–47, 1996.

[463] J-H. Kim and H. Myung. Evolutionary Programming Techniques for Constrained Optimization Problems. *IEEE Transactions on Evolutionary Computation*, 1(2):129–140, 1997.

[464] M-K. Kim, C-G. Lee, and H-K. Jung. Multiobjective Optimal Design of Three-phase Induction Motor using Improved Evolution Strategy. *IEEE Transactions on Magnetics*, 34(5):2980–2983, 1998.

[465] S. Kim and J-H. Kim. Optimal Trajectory Planning of a Redundant Manipulator using Evolutionary Programming. In *Proceedings of the IEEE International Conference on Evolutionary Computation*, pages 738–743, 1996.

[466] S. Kirkpatrick. Optimization by Simulated Annealing – Quantitative Studies. *Journal of Statistical Physics*, 34:975–986, 1984.

[467] S. Kirkpatrick, C.D. Gelatt, and M.P. Vecchi. Optimization by Simulated Annealing. *Science*, 220:671–680, 1983.

[468] T. Knight and J. Timmis. A Multi-Layered Immune Inspired Approach to Data Mining. In *Proceedings of the Fourth International Conference on Recent Advances in Soft Computing*, pages 266–271, 2002.

[469] J.D. Knowles and D.W. Corne. The Pareto Archived Evolution Strategy: A New Baseline Algorithm for Pareto Multiobjective Optimisation. In *Proceedings of the IEEE Congress on Evolutionary Computation*, pages 98–105, 1999.

[470] C.A. Koay and D. Srinivasan. Particle Swarm Optimization-Based Approach for Generator Maintenance Scheduling. In *Proceedings of the IEEE Swarm Intelligence Symposium*, pages 167–173, 2003.

[471] K. Kohara. Selective Presentation Learning for Forecasting by Neural Networks. In *International Workshop on Applications of Neural Networks in Telecommunications*, pages 316–323, 1995.

[472] T. Kohonen. Self-Organization and Associative Memory. In *Springer Series in Information Sciences*, volume 8. Springer-Verlag, 1984.

[473] T. Kohonen. *Context-Addressable Memories*. Springer-Verlag, second edition, 1987.

[474] T. Kohonen. *Self-Organizing Maps*. Springer Series in Information Sciences, 1995.

[475] T. Kohonen. *Self-Organizing Maps*. Springer, second edition, 1997.

[476] T. Kohonen. *Self-Organizing Maps*. Springer, 2000.

[477] T.C. Koopmans and M.J. Beckman. Assignment Problems and the Location of Economic Activities. *Econometrica*, 25:53–76, 1957.

[478] J.R. Koza. Hierarchical Genetic Algorithms Operating on Populations of Computer Programs. In N.S. Sridharan, editor, *Proceedings of the Eleventh International Joint Conference on Artificial Intelligence*, volume 1, pages 768–774, 1989.

[479] J.R. Koza. Genetic programming: A Paradigm for Genetically Breeding Populations of Computer Programs to Solve Problems. Technical Report STAN-CS-90-1314, Department of Computer Science, Stanford University, 1990.

[480] J.R. Koza. Genetic Evolution and Co-Evolution of Computer Programs. In C. Taylor, C. Langton, J.D. Farmer, and S. Rasmussen, editors, *Artificial Life II*, volume X, pages 603–629. Addison-Wesley, Santa Fe Institute, New Mexico, USA, 1991.

[481] J.R. Koza. Genetic Evolution and Co-evolution of Game Strategies. In *Proceedings of the International Conference on Game Theory and Its Applications*. Stony Brook, 1992.

[482] J.R. Koza. *Genetic Programming: On the Programming of Computers by Means of Natural Selection.* MIT Press, 1992.

[483] J.R. Koza. *Genetic Programming II: Automatic Discovery of Reusable Programs.* MIT Press, 1994.

[484] J.R. Koza and D. Andre. Automatic Discovery of Protein Motifs using Genetic Programming. In X. Yao, editor, *Evolutionary Computation: Theory and Applications*, Singapore, 1996. World Scientific.

[485] J.R. Koza and D. Andre. Classifying Protein Segments as Transmembrane Domains Using Architecture-Altering Operations in Genetic Programming. In P.J. Angeline and K.E. Kinnear (Jr), editors, *Advances in Genetic Programming, chapter 8*, volume 2, pages 155–176, Cambridge, M.A., 1996. MIT Press.

[486] J.R. Koza and J.P. Rice. Automatic Programming of Robots using Genetic Programming. In *Proceedings of Tenth National Conference on Artificial Intelligence*, pages 194–201. AAAI Press/MIT Press, 1992.

[487] S. Koziel and Z. Michalewicz. Evolutionary Algorithms, Homomorphous Mappings, and Constrained Optimization. *Evolutionary Computation*, 7(1):19–44, 1999.

[488] O. Kramer, C-K. Ting, and H.K. Büning. A New Mutation Operator for Evolution Strategies for Constrained Problems. In *Proceedings of the IEEE Congress on Evolutionary Computation*, volume 3, pages 2600–2606, 2005.

[489] T. Krink, J.S. Vesterstrøm, and J. Riget. Particle Swarm Optimisation with Spatial Particle Extension. In *Proceedings of the Fourth Congress on Evolutionary Computation*, volume 2, pages 1474–1479, 2002.

[490] T. Krink and M. Løvberg. The Life Cycle Model: Combining Particle Swarm Optimisation, Genetic Algorithms and Hill Climbers. In *Proceedings of the Parallel Problem Solving from Nature Conference, Lecture Notes in Computer Science*, volume 2439, pages 621–630. Springer-Verlag, 2002.

[491] A. Krogh and J.A. Hertz. A Simple Weight Decay can Improve Generalization. In J. Moody, S.J Hanson, and R. Lippmann, editors, *Advances in Neural Information Processing Systemsxi*, volume 4, pages 950–957, 1992.

[492] F. Kursawe. Towards Self-Adapting Evolution Strategies. In *Proceedings of the Second IEEE Conference on Evolutionary Computation*, pages 283–288, 1995.

[493] D.G. Kurup, M. Himdi, and A. Rydberg. Synthesis of Uniform Amplitude Unequally Spaced Antenna Arrays using the Differential Evolution Algorithm. *IEEE Transactions on Antennas and Propagation*, 51(9):2210–2217, 2003.

[494] I. Kuscu and C. Thornton. Design of Artificial Neural Networks Using Genetic Algorithms: Review and Prospect. Technical report, Cognitive and Computing Sciences, University of Sussex, 1994.

[495] T-Y. Kwok and D-Y. Yeung. Constructive Feedforward Neural Networks for Regression Problems: A Survey. Technical Report HKUST-CS95-43, Department of Computer Science, The Hong Kong University of Science & Technology, 1995.

[496] A. Kyprianou, K. Worden, and M. Panet. Identification of Hysteretic Systems using the Differential Evolution Algorithm. *Journal of Sound and Vibration*, 248(2):289–314, 2001.

[497] K.J. Lafferty and A.J. Cunningham. A New Analysis of Allogeneic Interactions. In *The Australian Journal of Experimental Biology and Medical Science*, volume 53, pages 27–42, 1975.

[498] J. Lampinen. A Constraint Handling Approach for the Differential Evolution Algorithm. In *Proceedings of the IEEE Congress on Evolutionary Computation*, volume 2, pages 1468–1473, 2002.

[499] J. Lampinen and I. Zelinka. Mixed Integer-Discrete-Continuous Optimization by Differential Evolution, Part I: The Optimization Method. In *Proceedings of the Fifth International Mendel Conference on Soft Computing*, pages 71–76, 1999.

[500] J. Lampinen and I. Zelinka. Mixed Variable Non-Linear Optimization by Differential Evolution. In *Proceedings of the Second International Prediction Conference*, pages 45–55, 1999.

[501] K.J. Lang, A.H. Waibel, and G.E. Hinton. A Time-Delay Neural Network Architecture for Isolated Word Recognition. *Neural Networks*, 3:33–43, 1990.

[502] R. Lange and R. Männer. Quantifying a Critical Training Set Size for Generalization and Overfitting using Teacher Neural Networks. In *International Conference on Artificial Neural Networks*, volume 1, pages 497–500, 1994.

[503] S. Lange and T. Zeugmann. Incremental Learning from Positive Data. *Journal of Computer and System Sciences*, 53:88–103, 1996.

[504] E.C. Laskari, K.E. Parsopoulos, and M.N. Vrahatis. Particle Swarm Optimization for Integer programming. In *Proceedings of the IEEE Congress on Evolutionary Computation*, volume 2, pages 1582–1587, 2002.

[505] C-Y. Lee and X. Yao. Evolutionary Programming using Mutations Based on the Levy Probability Distribution. *IEEE Transactions on Evolutionary Computation*, 8(2):1–13, 2004.

[506] S. Lee and R. Kill. Multilayer Feedforward Potential Funcion Networks. In *Proceedings of the IEEE Second International Conference on Neural Networks*, volume 1, pages 161–171, 1988.

[507] S-W. Lee, H-Byung Jun, and K-B. Sim. Performance Improvement of Evolution Strategies using Reinforcement Learning. In *Proceedings of the IEEE International Fuzzy Systems Conference*, volume 2, pages 639–644, 1999.

[508] Z-J. Lee, C-Y. Lee, and F. Su. An Immunity-Based Ant Colony Optimization Algorithm for Solving Weapon-Target Assignment Problem. *Applied Soft Computing*, 2(1):39–47, 2002.

[509] L.R. Leerink, C. Lee Giles, B.G. Horne, and M.A. Jabri. Learning with Product Units. *Advances in Neural Information Processing Systems*, 7:537–544, 1995.

[510] W. Lei and W. Qidi. Ant System Algorithm for Optimization in Continuous Space. In *Proceedings of the IEEE International Conference on Control Applications*, pages 395–400, 2001.

[511] W. Lei and W. Qidi. Further Example Study on Ant System Algorithm based Continuous Space Optimization. In *Proceedings of the Fourth World Congress on Intelligent Control and Automation*, pages 2541–2545, 2002.

[512] W. Lei and W. Qidi. Performance Evaluation of Ant System Optimization Process. In *Proceedings of the Fourth World Congress on Intelligent Control and Automation*, pages 2546–2550, 2002.

[513] D. Leitch and P.J. Probert. New Techniques for Genetic Development of a Class of Fuzzy Controllers. *IEEE Transactions on Systems, Man and Cybernetics*, 28(1):112–123, 1998.

[514] C. Lejewski and J. Lukasiewicz. *Encyclopedia of Philosophy*, 5:104–107, 1967.

[515] A.U. Levin, T.K. Leen, and J.E. Moody. Fast Pruning using Principal Components. In J.D. Cowan, G. Tesauro, and J. Alspector, editors, *Advances in Neural Information Processing Systems*, volume 6, pages 35–42, 1994.

[516] J. Li, H-Z. Liu, B. Yang, J-B. Yu, N. Xu, and C-H. Li. Application of an EACS Algorithm to Obstacle Detour Routing in VLSI Physical Design. In *Proceedings of the International Conference on Machine Learning and Cybernetics*, pages 1553–1558, 2003.

[517] X. Li. A Non-Dominated Sorting Particle Swarm Optimizer for Multiobjective Optimization. In *Proceedings of the Genetic and Evolutionary Computation Conference, Lecture Notes in Computer Science*, volume 2723, pages 37–48. Springer-Verlag, 2003.

[518] X. Li and K.H. Dam. Comparing Particle Swarms for Tracking Extrema in Dynamic Environments. In *Proceedings of the IEEE Congress on Evolutionary Computation*, volume 3, pages 1772–1779, 2003.

[519] X-L. Li, X-D. He, and S-M. Yaun. Learning Bayesian Networks Structures from Incomplete Data Based on Extending Evolutionary Programming. In *Proceedings of the Fourth International Conference on Machine Learning and Cybernetics*, pages 2039–2043, 2005.

[520] Y. Li and S. Gong. Dynamic Ant Colony Optimisation for TSP. *International Journal of Advanced Manufacturing Technology*, 22(7-8):528–533, 2003.

[521] Y. Li, L. Rao, R. He, G. Xu, X. Gou, W. Yan, L. Wang, and S. Yang. Three EIT Approaches for Static Imaging of Head. In *Proceedings of the Twenty-Sixth Annual International Conference of the Engineering in Medicine and Biology Sociery*, volume 1, pages 578–581, 2004.

[522] Y. Li, L. Rao, R. He, G. Xu, Q. Wu, M. Ge, and W. Tan. Image Reconstruction of EIT using Differential Evolution Algorithm. In *Proceedings of the Twenty-Fifth IEEE Annual International Conference on Engineering in Medicine and Biology Society*, volume 2, pages 1011–1014, 2003.

[523] Y. Li, T-J. Wu, and D.J. Hill. An Accelerated Ant Colony Algorithm for Complex Nonlinear System Optimization. In *Proceedings of the IEEE International Symposium on Intelligent Control*, pages 709–713, 2003.

[524] K-H. Liang, X. Yao, and C. Newton. Dynamic Control of Adaptive Parameters in Evolutionary Programming. In *Proceedings of the Second Asia-Pacific Conference on Simulated Evolution and Learning, Lecture Notes in Computer Science*, volume 1585, pages 42–49, 1998.

[525] Y-C. Liang, S. Kultural-Konak, and A.E. Smith. Meta Heuristics for the Orienteering Problem. In *Proceedings of the IEEE Congress on Evolutionary Computation*, volume 1, pages 384–389, May 2002.

[526] S-F. Lim and S-B. Ho. Dynamic Creation of Hidden Units with Selective Pruning in Backpropagation. In *IEEE World Congress on Computational Intelligence, Proceedings of the International Joint Conference on Neural Networks*, volume 3, pages 492–497, 1994.

[527] L. Lin. Self-Improving Reactive Agents Based on Reinforcement Learning, Planning and Teaching. *Machine Learning*, 8:293–321, 1992.

[528] Y-C. Lin, K-S. Hwang, and F-S. Wang. Plant Scheduling and Planning using Mixed-Integer Hybrid Differential Evolution with Multiplier Updating. In *Proceedings of the IEEE Congress on Evolutionary Computation*, volume 1, pages 593–600, 2000.

[529] Y-C. Lin, K-S. Hwang, and F-S. Wang. Hybrid Differential Evolution with Multiplier Updating Method for Nonlinear Constrained Optimization Problems. In *Proceedings of the IEEE Congress on Evolutionary Computation*, volume 1, pages 872–877, 2002.

[530] Y-C. Lin, K-S. Hwang, and F-S. Wang. A Mixed-Coding Scheme of Evolutionary Algorithms to Solve Mixed-Integer Nonlinear Programming Problems. *Computers & Mathematics with Applications*, 47(8-9):237–246, 2004.

[531] Y-C. Lin, F-S. Wang, and K-S. Hwang. A hybrid Method of Evolutionary Algorithms For Mixed-Integer Nonlinear Optimization Problems. In *Proceedings of the IEEE Congress on Evolutionary Computation*, volume 3, 1999.

[532] Y. Liu, X. Yao, Q. Zhao, and T. Higuchi. Scaling Up Fast Evolutionary Programming with Cooperative Coevolution. In *Proceedings of the IEEE Congress on Evolutionary Computation*, volume 2, pages 1101–1108, 2001.

[533] M.F. Møller. A Scaled Conjugate Gradient Algorithm for Fast Supervised Learning. *Neural Networks*, 6:525–533, 1993.

[534] M. Løvberg. Improving Particle Swarm Optimization by Hybridization of Stochastic Search Heuristics and Self-Organized Criticality. Master's thesis, Department of Computer Science, University of Aarhus, Denmark, 2002.

[535] M. Løvberg and T. Krink. Extending Particle Swarm Optimisers with Self-Organized Criticality. In *Proceedings of the IEEE Congress on Evolutionary Computation*, volume 2, pages 1588–1593, 2002.

[536] M. Løvberg, T.K. Rasmussen, and T. Krink. Hybrid Particle Swarm Optimiser with Breeding and Subpopulations. In *Proceedings of the Genetic and Evolutionary Computation Conference*, pages 469–476, 2001.

[537] J. Ludik and I. Cloete. Training Schedules for Improved Convergence. In *IEEE International Joint Conference on Neural Networks*, volume 1, pages 561–564, 1993.

[538] J. Ludik and I. Cloete. Incremental Increased Complexity Training. In *European Symposium on Artificial Neural Networks*, pages 161–165, 1994.

[539] G.F. Luger and W.A. Stubblefield. *Artificial Intelligence, Structures and Strategies for Complex Problem Solving*. Addison-Wesley, third edition, 1997.

[540] E. Lumer and B. Faieta. Diversity and Adaptation in Populations of Clustering Ants. In *Proceedings of the Third International Conference on Simulation of Adaptive Behavior: From Animals to Animats*, volume 3, pages 499–508. MIT Press, 1994.

[541] H.H. Lund, J. Hallam, and W-P. Lee. Evolving Robot Morphology. In *Proceedings of the IEEE International Conference on Evolutionary Computation*, pages 197–202, 1997.

[542] J.T. Ma and L.L. Lai. Determination of Operational Parameters of Electrical Machines using Evolutionary Programming. In *Proceedings of the Seventh International Conference on Electrical Machines and Drives*, pages 116–120, 1995.

[543] I.F. MacGill and R.J. Kaye. Decentralised Coordination of Power System Operation using Dual Evolutionary Programming. *IEEE Transactions on Power Systems*, 14(1):112–119, 1999.

[544] D.J.C. MacKay. *Bayesian Methods for Adaptive Models*. PhD thesis, California Institute of Technology, 1992.

[545] N.K. Madavan. Multiobjective Optimization using a Pareto Differential Evolution Approach. In *Proceedings of the IEEE Congress on Evolutionary Computation*, volume 2, pages 1145–1150, 2002.

[546] M.T. Madsen, R. Uppaluri, E.A. Hoffman, and G. McLennan. Pulmonary CT Image Classification using Evolutionary Programming. In *Proceedings of the IEEE Nuclear Science Symposium*, volume 2, pages 1179–1182, 1997.

[547] M. Maeterlinck. *The Life of the White Ant*. Dodd-Mead, New York, 1927.

[548] C.A. Magele, K. Preis, W. Renhart, R. Dyczij-Edlinger, and K.R. Richter. Higher Order Evolution Strategies for the Global Optimization of Electromagnetic Devices. *IEEE Transactions on Magnetics*, 29(2):1775–1778, 1993.

[549] G.D. Magoulas, V.P. Plagianakos, and M.N. Vrahatis. Adaptive Stepsize Algorithms for On-line Training of Neural Networks. *Nonlinear Analysis: Theory, Methods and Applications (in press)*, 2001.

[550] G.D. Magoulas, V.P. Plagianakos, and M.N. Vrahatis. Hybrid methods using evolutionary algorithms for on-line training. In *Proceedings of the International Joint Conference on Neural Networks*, volume 3, pages 2218–2223, 2001.

[551] G.D. Magoulas, V.P. Plagianakos, and M.N. Vrahatis. Neural Network-Based Colonoscopic Diagnosis using On-Line Learning and Differential Evolution. *Applied Soft Computing*, 4(4):369–379, 2004.

[552] G.D. Magoulas, M.N. Vrahatis, and G.S. Androulakis. Effective Backpropagation Training with Variable Stepsize. *Neural Networks*, 10(1):69–82, 1997.

[553] S.W. Mahfoud. *Niching Methods for Genetic Algorithms*. PhD thesis, University of Illinois, Illinois, 1995.

[554] E.H. Mamdani and S. Assilian. An Experiment in Linguistic Synthesis with a Fuzzy Logic Controller. *International Journal of Man-Machine Studies*, 7:1–13, 1975.

[555] V. Maniezzo and A Carbonaro. Ant Colony Optimization: An Overview. In C. Ribeiro, editor, *Essays and Surveys in Metaheuristics*, pages 21–44. Kluwer, 1999.

[556] V. Maniezzo and A. Carbonaro. An ANTS Heuristic for the Frequency Assignment Problem. *Future Generation Computer Systems*, 16(9):927–935, 2000.

[557] V. Maniezzo and A. Colorni. The Ant System Applied to the Quadratic Assignment Problem. *IEEE Transactions on Knowledge and Data Engineering*, 11(5):769–778, 1999.

[558] E.N. Marais. *Die Siel van die Mier (The Soul of the Ant)*. J.L. van Schaik, Pretoria, South Africa, fifth edition, 1948. (first published in 1937).

[559] E.N. Marais. *The Soul of the Ape*. Hammersworth, London, second edition edition, 1969.

[560] E.N. Marais. *The Soul of the White Ant*. Human and Rousseau Publishers, Cape Town, 1970.

[561] C.E. Mariano and E. Morales. A Multiple Objective Ant-Q Algorithm for the Design of Water Distribution Irrigation Networks. Technical Report HC-9904, Instituto Mexicano de Technologiía del Agua, 1999.

[562] S. Markon, D.V. Arnold, T. Bäck, T. Beielstein, and H-G. Beyer. Thresholding – A Selection Operator for Noisy ES. In *Proceedings of the IEEE Congress on Evolutionary Computation*, pages 465–472, 2001.

[563] M. Martin, B. Chopard, and P. Albuquerque. Formation of an Ant Cemetery: Swarm Intelligence or Statistical Accident? *Future Generation Computer Systems*, 18(7):951–959, 2002.

[564] M. Mathur, S.B. Karale, S. Priye, V.K. Jayaraman, and B.D. Kulkarni. Ant Colony Approach to Continuous Function Optimization. *Industrial Engineering Chemistry Research*, 39(10):3814–3822, 2000.

[565] Y. Matsumura, K. Ohkura, and K. Ueda. Evolutionary Programming with Noncoding Segments for Realvalued Function Optimization. In *Proceedings of the IEEE International Conference on Systems, Man and Cybernetics*, volume 4, pages 242–247, 1999.

[566] Y. Matsumura, K. Ohkura, and K. Ueda. Evolutionary Dynamics of Evolutionary Programming in Noisy Environment. In *Proceedings of the IEEE Congress on Evolutionary Computation*, pages 17–24, 2001.

[567] P. Matzinger. The Danger Model in its Historical Context. *Scandinavian Journal of Immunology*, 54:4–9, 2001.

[568] P. Matzinger. The Real Function of the Immune System. *http://cmmg.biosci.wayne.edu/asg/polly.html*, 2004.

[569] H.A. Mayer and R. Schwaiger. Evolutionary and Coevolutionary Approaches to Time Series Prediction using Generalized Multi-Layer Perceptrons. In *Proceedings of the IEEE Congress on Evolutionary Computation*, volume 1, 1999.

[570] E. Mayr. *Animal Species and Evolution*. Belknap, Cambridge, MA, 1963.

[571] J.R. McDonnell and D. Waagen. Evolving Recurrent Perceptrons for Time-Series Modeling. *IEEE Transactions on Neural Networks*, 5(1):24–38, 1994.

[572] P.R. McMullen. An Ant Colony Optimization Approach to Addressing a JIT Sequencing Problem with Multiple Objectives. *Artificial Intelligence in Engineering*, 15(3):309–317, 2001.

[573] P. Meksangsouy and N. Chaiyaratana. DNA Fragment Assembly using an Ant Colony System Algorithm. In *Proceedings of the IEEE Congress on Evolutionary Computation*, volume 3, pages 1756–1763, December 2003.

[574] C. Melhuish, O. Holland, and S. Hoddell. Collective Sorting and Segregation in Robots with Minimal Sensing. In *Robotics and Autonomous Systems*, volume 28, pages 207–216, 1998.

[575] R. Mendes, P. Cortez, M. Rocha, and J. Neves. Particle Swarms for Feedforward Neural Network Training. In *Proceedings of the International Joint Conference on Neural Networks*, pages 1895–1899, 2002.

[576] R. Mendes, J. Kennedy, and J. Neves. Watch thy Neighbor or How the Swarm can Learn from its Environment. In *Proceedings of the IEEE Swarm Intelligence Symposium*, pages 88–94, 2003.

[577] R. Mendes and A.S. Mohais. DynDE: A Differential Evolution for Dynamic Optimization Problems. In *Proceedings of the IEEE Congress on Evolutionary Computation*, volume 3, pages 2808–2815, 2005.

[578] O.J. Mengshoel and D.E. Goldberg. Probabilistic Crowding: Deterministic Crowding with Probabilistic Replacement. In W. Banzhaf *et al.*, editor, *Proceedings of the Genetic and Evolutionary Computation Conference 1999*, pages 409–416, San Fransisco, USA, Morgan Kaufmann, 1999.

[579] D. Merkle, M. Middendorf, and H. Schmeck. Ant Colony Optimization for Resource-Constrained Project Scheduling. *IEEE Transactions on Evolutionary Computation*, 6(4):333–346, August 2002.

[580] L. Messerschmidt and A.P. Engelbrecht. Learning to Play Games using a PSO-Based Competitive Learning Approach. *IEEE Transactions on Evolutionary Computation*, 8(3):280–288, 2004.

[581] N. Metropolis, A.W. Rosenbluth, M.N. Rosenbluth, A.H. Teller, and E. Teller. Equations of State Calculations by Fast Computing Machines. *Journal of Chemical Physics*, 21:1087–1092, 1958.

[582] B.J. Meyer, H.S. Meij, S.V. Grey, and A.C. Meyer. *Fisiologie van die mens - Biochemiese, fisiese en fisiologiese begrippe*. Kagiso Tersier - Cape Town, first edition, 1996.

[583] Z. Michalewicz. *Genetic Algorithms + Data Structures = Evolutionary Programs*. Springer, Berlin, 1992.

[584] Z. Michalewicz. A Survey of Constraint Handling Techniques in Evolutionary Computation Methods. In *Proceedings of the Fourth Annual Conference on Evolutionary Programming*, pages 135–155, 1995.

[585] Z. Michalewicz. Genetic Algorithms, Numerical Optimization, and Constraints. In *Proceedings of the 6th International Conference on Genetic Algorithms*, pages 151–158, 1995.

[586] Z. Michalewicz. *Genetic Algorithms + Data Structures = Evolutionary Programs*. Springer, Berlin, third edition, 1996.

[587] Z. Michalewicz and N. Attia. Evolutionary Optimization of Constrained Problems. In A.V. Sebald and L.J. Fogel, editors, *Proceedings of the Third Annual Conference on Evolutionary Programming*, pages 98–108, 1994.

[588] Z. Michalewicz and C. Janikow. Handling Constraints in Genetic Algorithms. In *Proceedings of the Fourth International Conference on Genetic Algorithms*, pages 151–157, 1991.

[589] Z. Michalewicz and G. Nazhiyath. Genocop III: A Co-Evolutionary Algorithm for Numerical Optimization Problems with Nonlinear Constraints. In *Proceedings of the IEEE International Conference on Evolutionary Computation*, volume 2, pages 647–651, 1995.

[590] Z. Michalewicz, G. Nazhiyath, and M. Michalewicz. A Note on the Usefulness of Geometrical Crossover for Numerical Optimization Problems. In L.J. Fogel, P.J. Angeline, and T. Bäck, editors, *Proceedings of the Fifth Annual Conference on Evolutionary Programming*, pages 305–312, Cambridge, M.A., 1996. MIT Press.

[591] R. Michel and M. Middendorf. An ACO Algorithm for the Shortest Common Supersequence Problem. In M. Dorigo D. Corne and F. Glover, editors, *New Ideas in Optimization*, pages 51–61. McGraw-Hill, 1999.

[592] M. Middendorf, F. Reischle, and H. Schmeck. Information Exchange in Multi Colony Ant Algorithms. In *Proceedings of the Workshop on Bio-Inspired Solutions to Parallel Processing Problems, Lecture Notes in Computer Science*, volume 1800, pages 645–652. Springer-Verlag, 2000.

[593] B.L. Miller and M.J. Shaw. Genetic Algorithms with Dynamic Niche Sharing for Multimodal Function Optimization. In *International Conference on Evolutionary Computation*, pages 786–791, 1996.

[594] J.H. Miller. *The Evolution of Automata in the Repeated Prisoner's Dilemma*. PhD thesis, Department of Economics, University of Michigan, 1988.

[595] J.H. Miller. The Co-Evolution of Automata in the Repeated Prisoner's Dilemma. Technical report, Sante Fe Institute Report 89-003, 1989.

[596] V. Miranda and N. Fonseca. EPSO – Best-of-two-worlds Meta-heuristic Applied to Power System Problems. In *Proceedings of the IEEE Congress on Evolutionary Computation*, volume 2, pages 1080–1085, 2002.

[597] W. Mo and S-U. Guan. Particle Swarm Assisted Incremental Evolution Strategy for Function Optimization. In *Proceedings of the IEEE Conference on Cybernetics and Intelligent Systems*, pages 1–6, 2006.

[598] S. Moalla, A.M. Alimi, and N. Derbel. Design of Beta Neural Systems Using Differential Evolution. In *Proceedings of the IEEE International Conference on Systems, Man, and Cybernetics*, volume 3, 2002.

[599] C.G. Moles, J.R. Banga, and K. Keller. Solving Nonconvex Climate Control Problems: Pitfalls and Algorithm Performances. *Applied Soft Computing*, 5(1):35–44, 2004.

[600] T. Mollestad. *A Rough Set Approach to Data Mining: Extracting a Logic of Default Rules from Data*. PhD thesis, Department of Computer Science, The Norwegian University of Science and Technology, 1997.

[601] N. Monmarché, M. Slimane, and G. Venturini. AntClass: Discovery of Clusters in Numeric Data by an Hybridization of an Ant Colony with the K-Means Algorithm. Technical report, Laboratoire d'Informatique, University of Tours, 1999.

[602] J. Moody and J. Utans. Architecture Selection Strategies for Neural Networks: Application to Corporate Bond Rating Prediction. In A.N. Refenes, editor, *Neural Networks in the Capital Markets*, pages 277–300. John Wiley & Sons, 1995.

[603] J.E. Moody. The Effective Number of Parameters: An Analysis of Generalization and Regularization in Nonlinear Learning Systems. In J. Moody, S.J. Hanson, and R. Lippmann, editors, *Advances in Neural Information Processing Systems*, volume 4, pages 847–854, 1992.

[604] J.E. Moody. Prediction Risk and Architecture Selection for Neural Networks. In V. Cherkassky, J.H. Friedman, and H. Wechsler, editors, *From Statistics to Neural Networks: Theory and Pattern Recognition Applications*, pages 147–165. Springer, 1994.

[605] J.E. Moody and C. Darken. Learning with Localized Receptive Fields. In D. Touretzky, G. Hinton, and T. Sejnowski, editors, *Proceedings of the Connectionist Models Summer School*, pages 133–143, San Mateo, C.A., 1989. Morgan Kaufmann.

[606] K. Mori, M. Tsukiyama, and T. Fukada. Immune Algorithm with Searching Diversity and Its Application to Resource Allocation Problems. *Transactions of the Institute of Electrical Engineers of Japan*, 113(10):872–878, 1993.

[607] N. Mori, S. Imanishi, H. Kita, and Y. Nishikawa. Adaptation to Changing Environments by Means of the Memory Based Thermodynamical Genetic Algorithm. In *Proceedings of the Seventh International Conference on Genetic Algorithms*, pages 299–306, 1997.

[608] P. Morillo, M. Fernández, and J.M. Orduña. An ACS-Based Partitioning Method for Distributed Virtual Environment Systems. In *Proceedings of the International Parallel and Distributed Processing Symposium*, page 148, 2003.

[609] S. Mostaghim and J. Teich. Strategies for Finding Local Guides in Multi-Objective Particle Swarm Optimization (MOPSO). In *Proceedings of the IEEE Swarm Intelligence Symposium*, pages 26–33, 2003.

[610] S. Mostaghim and J. Teich. The Role of $\varepsilon$-dominance in Multi-objective Particle Swarm Optimization Methods. In *Proceedings of the IEEE Congress on Evolutionary Computation*, pages 1764–1771, 2003.

[611] M.C. Mozer and P. Smolensky. Skeletonization: A Technique for Trimming the Fat from a Network via Relevance Assessment. In D.S. Touretzky, editor, *Advances in Neural Information Processing Systems*, volume 1, pages 107–115, 1989.

[612] K-R. Müller, M. Finke, N. Murata, K. Schulten, and S. Amari. A Numerical Study on Learning Curves in Stochastic Multi-Layer Feed-Forward Networks. *Neural Computation*, 8(5):1085–1106, 1995.

[613] S.D. Müller, I.F. Sbalzarini, J.H. Walther, and P.D. Koumoutsakos. Evolution Strategies for the Optimization of Microdevices. In *Proceedings of the IEEE Congress on Evolutionary Computation*, volume 1, pages 302–309, 2001.

[614] S.D. Müller, N.N. Schraudolph, and P.D. Koumoutsakos. Step Size Adaptation in Evolution Strategies using Reinforcement Learning. In *Proceedings of the IEEE Congress on Evolutionary Computation*, volume 1, pages 151–156, 2002.

[615] Y. Murakami, H. Sato, and A. Namatame. Co-evolution in Negotiation Games. In *Proceedings of the Fourth International Conference on Computational Intelligence and Multimedia Applications*, pages 241–245, 2001.

[616] N. Murata, S. Yoshizawa, and S. Amari. A Criterion for Determining the Number of Parameters in an Artificial Neural Network Model. In T. Kohonen, K. Mäkisara, O. Simula, and J. Kangas, editors, *Artificial Neural Networks*, pages 9–14. Elsevier Science Publishers, 1991.

[617] N. Murata, S. Yoshizawa, and S. Amari. Learning Curves, Model Selection and Complexity of Neural Networks. In C. Lee Giles, S.J. Hanson, and J.D. Cowan, editors, *Advances in Neural Information Processing Systems*, volume 5, pages 607–614, 1994.

[618] N. Murata, S. Yoshizawa, and S. Amari. Network Information Criterion – Determining the Number of Hidden Units for an Artificial Neural Network Model. *IEEE Transactions on Neural Networks*, 5(6):865–872, 1994.

[619] S. Naka, T. Genji, T. Yura, and Y. Fukuyama. Practical Distribution State Estimation using Hybrid Particle Swarm Optimization. In *IEEE Power Engineering Society Winter Meeting*, volume 2, pages 815–820, 2001.

[620] D. Nam, Y.D. Seo, L-J. Park, C.H. Park, and B. Kim. Parameter Optimization of an On-Chip Voltage Reference Circuit using Evolutionary Programming. *IEEE Transactions on Evolutionary Computation*, 5(4):414–421, 2001.

[621] H. Narihisa, T. Taniguchi, M. Thuda, and K. Katayama. Efficiency of Parallel Exponential Evolutionary Programming. In *Proceedings of the International Conference Workshop on Parallel Processing*, pages 588–595, 2005.

[622] O. Nasraoui, D. Dasgupta, and F. Gonzalez. The Promise and Challenges of Artificial Immune System Based Web Usage Mining: Preliminary Results. In *Proceedings of the Second SIAM International Conference on Data Mining*, pages 29–39, 2002.

[623] O. Nasraoui, F. Gonzalez, C. Cardona, C. Rojas, and D. Dasgupta. A Scalable Artificial Immune System Model for Dynamic Unsupervised Learning. In *Proceedings of the Genetic and Evolutionary Computation Conference, Lecture Notes in Computer Science*, volume 2723, pages 219–230. Springer-Verlag, 2003.

[624] D. Naug and R. Gadagkar. The Role of Age in Temporal Polyethism in a Primitively Eusocial Wasp. *Behavioral Ecology and Sociobiology*, 42:37–47, 1998.

[625] D. Naug and R. Gadagkar. Flexible Division of Labor Mediated by Social Interactions in an Insect Colony – A Simulation Model. *Journal of Theoretical Biology*, 197:123–133, 1999.

[626] M. Neal. An Artificial Immune System for Continuous Analysis of Time-varying Data. In *Proceedings of the First International Conference on Artificial Immune Systems*, volume 1, pages 76–85, 2002.

[627] M. Neethling and A.P. Engelbrecht. Determining RNA Secondary Structure using Set-Based Particle Swarm Optimization. In *Proceedings of the IEEE Congress on Evolutionary Computation*, pages 1670–1677, 2006.

[628] L. Nemes and T. Roska. A CNN Model of Oscillation and Chaos in Ant Colonies: A Case Study. *IEEE Transactions on Circuits and Systems I: Fundamental Theory and Applications*, 42(10):741–745, 1995.

[629] N.J. Nilsson. *Artificial Intelligence: A New Synthesis*. Morgan Kaufmann, 1998.

[630] M. Niranjan and F. Fallside. Neural Networks and Radial Basis Functions in Classifying Static Speech Patterns. Technical Report CUEDIF-INFENG17R22, Engineering Department, Cambridge University, 1988.

[631] G. Nitschke. Co-Evolution of Cooperation in a Pursuit Evasion Game. In *Proceedings of the IEEE/RSJ International Conference on Intelligent Robots and Systems*, volume 2, pages 2037–2042, 2003.

[632] S.J. Nowlan. Maximum Likelihood Competitive Learning. In *Advances in Information Processing Systems*, volume 2, pages 574–582, San Mateo, C.A., 1990. Morgan Kaufmann.

[633] S.J. Nowlan and G.E. Hinton. Simplifying Neural Networks By Soft Weight-Sharing. *Neural Computation*, 4:473–493, 1992.

[634] N. Ohnishi, A. Okamoto, and N. Sugiem. Selective Presentation of Learning Samples for Efficient Learning in Multi-Layer Perceptron. In *Proceedings of the IEEE International Joint Conference on Neural Networks*, volume 1, pages 688–691, 1990.

[635] E. Oja. A Simplified Neuron Model as a Principal Component Analyzer. *Journal of Mathematical Biology*, 15:267–273, 1982.

[636] E. Oja and J. Karhuner. On Stochastic Approximation of the Eigenvectors and Eigenvalues of the Expectation of a Random Matrix. *Journal of Mathematical Analysis and Applications*, 104:69–84, 1985.

[637] M. Oltean, C. Grosan, A. Abraham, and M. Köppen. Multiobjective Optimization using Adaptive Pareto Archived Evolution Strategy. In *Proceedings of the Fifth International Conference on Intelligent System Design and Applications*, pages 558–563, 2005.

[638] M. Omran, A. Salman, and A.P. Engelbrecht. Image Classification using Particle Swarm Optimization. In *Proceedings of the Fourth Asia-Pacific Conference on Simulated Evolution and Learning*, pages 370–374, 2002.

[639] M.G. Omran, A.P Engelbrecht, and A. Salman. Image Classification using Particle Swarm Optimization. In K.C. Tan, M.H. Lim, X. Yao, and L. Wang, editors, *Recent Advances in Simulated Evolution and Learning, Advances in Natural Computation*, volume 2, pages 347–365. World Scientific, 2004.

[640] M.G.H. Omran, A.P. Engelbrecht, and A. Salman. Differential Evolution Methods for Unsupervised Image Classification. In *Proceedings of the IEEE Congress on Evolutionary Computation*, volume 2, pages 966–973, 2005.

[641] M.G.H. Omran, A.P. Engelbrecht, and A. Salman. Empirical Analysis of Self-Adaptive Differential Evolution. *European Journal of Operational Research (in press)*, 2007.

[642] I. Ono and S. Kobayashi. A Real-Coded Genetic Algorithm for Function Optimization using Unimodal Normal Distribution Crossover. In *Proceedings of the Seventh International Conference on Genetic Algorithms*, pages 246–253, 1997.

[643] M. Opper. Learning and Generalization in a Two-Layer Neural Network: The Role of the Vapnik-Chervonenkis Dimension. *Physical Review Letters*, 72(13):2133–2166, 1994.

[644] G.B. Orr and T.K. Leen. Momentum and Optimal Stochastic Search. In M.C. Mozer, P. Smolensky, D.S. Touretzky, J.L. Elman, and A.S. Weigend, editors, *Proceedings of the 1993 Connectionist Models Summer School*, 1993.

[645] A. Ostermeier and N. Hansen. An Evolution Strategy with Coordinate System Invariant Adaptation of Arbitrary Normal Mutation Distributions within The Concept of Mutative Strategy Parameter Control. In *Proceedings of the Genetic and Evolutionary Computation Conference*, pages 902–909, San Francisco, 1999. Morgan Kaufmann.

[646] M. Ostertag, E. Nock, and U. Kiencke. Optimization of Airbag Release Algorithms using Evolutionary Strategies. In *Proceedings of the Fourth IEEE Conference on Control Applications*, pages 275–280, 1995.

[647] D.A. Ostrowski and R.G. Reynolds. Knowledge-Based Software Testing Agent using Evolutionary Learning with Cultural Algorithms. In *Proceedings of the IEEE Congress on Evolutionary Computation*, volume 3, pages 1657–1663, 1999.

[648] F.E.B. Otero, M.M.S. Silva, A.A. Freitas, and J.C. Nievola. Genetic Programming for Attribute Construction in Data Mining. In C. Ryan, M. Keijzer, R. Poli, T. Soule, E. Tsang, and E. Costa, editors, *Proceedings of the Sixth European Conference on Genetic Programming*, volume 2610 of *Lecture Notes in Computer Science*, pages 384–393. Springer-Verlag, 2003.

[649] R.H.J.M. Otten and L.P.P.P. van Ginneken. *The Annealing Algorithm*. Kluwer, 1989.

[650] P.S. Ow and T.E. Morton. Filtered Beam Search in Scheduling. *International Journal of Production Research*, 26:297–307, 1988.

[651] E. Ozcan and C.K. Mohan. Analysis of a Simple Particle Swarm Optimization System. In *Intelligent Engineering Systems through Artificial Neural Networks*, pages 253–258, 1998.

[652] E. Ozcan and C.K. Mohan. Particle Swarm Optimization: Surfing the Waves. In *Proceedings of the IEEE Congress on Evolutionary Computation*, volume 3, July 1999.

[653] G. Pampará, A.P. Engelbrecht, and N. Franken. Binary Differential Evolution. In *IEEE World Congress on Computational Intelligence, Proceedings of the Congress on Evolutionary Computation*, pages 1873–1879, 2006.

[654] E. Papacostantis, AP. Engelbrecht, and N. Franken. Coevolving Probabilistic Game Playing Agents using Particle Swarm Optimization Algorithms. In *Proceedings of the IEEE Evolutionary Computation in Games Symposium*, pages 195–202, 2005.

[655] U. Paquet. Training Support Vector Machines with Particle Swarms. Master's thesis, Department of Computer Science, University of Pretoria, 2003.

[656] U. Paquet and A.P. Engelbrecht. Training Support Vector Machines with Particle Swarms. In *Proceedings of the IEEE International Joint Conference on Neural Networks*, volume 2, pages 1593–1598, July 2003.

[657] J. Paredis. Coevolutionary Computation. *Artificial Life*, 2(4):355–375, 1995.

[658] G.B. Parker. The Co-Evolution of Model Parameters and Control Programs in Evolutionary Robotics. In *Proceedings of the IEEE International Symposium on Computational Intelligence in Robotics and Automation*, pages 162–167, 1999.

[659] K.E. Parsopoulos, D.K. Tasoulis, N.G. Pavlidis, V.P. Plagianakos, and M.N. Vrahatis. Vector Evaluated Differential Evolution for Multiobjective Optimization. In *Proceedings of the IEEE Congress on Evolutionary Computation*, pages 204–211, 2004.

[660] K.E. Parsopoulos, D.K. Tasoulis, and M.N. Vrahatis. Multiobjective Optimization using Parallel Vector Evaluated Particle Swarm Optimization. In *Proceedings of the IASTED International Conference on Artificial Intelligence and Applications*, volume 2, pages 823–828, 2004.

[661] K.E. Parsopoulos and M.N. Vrahatis. Particle Swarm Optimizer in Noisy and Continuously Changing Environments. In *Proceedings of the IASTED International Conference on Artificial Intelligence and Soft Computing*, pages 289–294, 2001.

[662] K.E. Parsopoulos and M.N. Vrahatis. Initializing the Particle Swarm Optimizer using the Nonlinear Simplex Method. In N. Mastorakis A. Grmela, editor, *Advances in Intelligent Systems, Fuzzy Systems, Evolutionary Computation*, pages 216–221, 2002.

[663] K.E. Parsopoulos and M.N. Vrahatis. Particle Swarm Optimization Method for Constrained Optimization Problems. In P. Sincak, J.Vascak, V. Kvasnicka, and J. Pospichal, editors, *Intelligent Technologies -- Theory and Applications: New Trends in Intelligent Technologies*, pages 214–220. IOS Press, 2002.

[664] K.E. Parsopoulos and M.N. Vrahatis. Particle Swarm Optimization Method in Multiobjective Problems. In *Proceedings of the ACM Symposium on Applied Computing*, pages 603–607, 2002.

[665] K.E. Parsopoulos and M.N. Vrahatis. Recent Approaches to Global Optimization Problems through Particle Swarm Optimization. *Natural Computing*, 1(2-3):235–306, 2002.

[666] J.M. Pasteels, J-L. Deneubourg, and S.Goss. Self-Organization Mechanisms in Ant Societies (I): Trail Recruitment to Newly Discovered Food Sources. *Experientia Suppl.*, 76:579–581, 1989.

[667] S. Paterlini and T. Krink. High Performance Clustering with Differential Evolution. In *Proceedings of the IEEE Congress on Evolutionary Computation*, volume 2, pages 2004–2011, 2004.

[668] Z. Pawlak. Rough Sets. *International Journal of Computer and Information Sciences*, 11:341–356, 1982.

[669] M.W. Pedersen, L.K. Hansen, and J. Larsen. Pruning with Generalization Based Weight Saliencies: $\gamma$ OBD, $\gamma$ OBS. In D.S. Touretzky, M.C. Mozer, and M.E. Hasselmo, editors, *Advances in Neural Information Processing Systems*, volume 8, pages 521–528, 1996.

[670] E.S. Peer, F. van den Bergh, and A.P. Engelbrecht. Using Neighborhoods with the Guaranteed Convergence PSO. In *Proceedings of the IEEE Swarm Intelligence Symposium*, pages 235–242, 2003.

[671] C.A. Pena-Reyes and M. Sipper. Applying Fuzzy CoCo to Breast Cancer Diagnosis. In *Proceedings of the IEEE Congress on Evolutionary Computation*, volume 2, pages 1168–1175, 2000.

[672] B. Peng, R.G Reynolds, and J. Brewster. Cultural Swarms. In *Proceedings of the IEEE Congress on Evolutionary Computation*, volume 3, pages 1965–1971, 2003.

[673] J. Peng, Y. Chen, and R.C. Eberhart. Battery Pack State of Charge Estimator Design using Computational Intelligence Approaches. In *Proceedings of the Annual Battery Conference on Applications and Advances*, pages 173–177, 2000.

[674] J. Peng and R.J. Williams. Incremental Multi-step Q-learning. In W. Cohen and H. Hirsh, editors, *Proceedings of the Eleventh International Machine Learning Conference*, pages 226–232, New Brunswick, N.J., 1994. Morgan Kaufmann.

[675] T. Peram, K. Veeramachaneni, and C.K. Mohan. Fitness-Distance-Ratio based Particle Swarm Optimization. In *Proceedings of the IEEE Swarm Intelligence Symposium*, pages 174–181, 2003.

[676] W. Perdrycz. *Computational Intelligence: An Introduction*. CRC Press, 1998.

[677] A.S. Perelson. Immune Network Theory. *Immunological Review*, 110:5–36, 1989.

[678] A.S. Perelson and G. Weisbuch. Immunology for Physicists. *Reviews of Modern Physics*, 69(4), 1997.

[679] T.K. Peters, H-E. Koralewski, and E.W. Zerbst. The Evolution Strategy – A Search Strategy Used in Individual Optimization of Electrical Parameters for Therapeutic Carotid Sinus Nerve Stimulation. *IEEE Transactions on Biomedical Engineering*, 36(7):668–675, 1989.

[680] D.A. Plaut, S. Nowlan, and G. Hinton. Experiments on Learning by Back Propagation. Technical Report CMU-CS-86-126, Department of Computer Science, Carnegie-Mellon University, 1986.

[681] M. Plutowski and H. White. Selecting Concise Training Sets from Clean Data. *IEEE Transactions on Neural Networks*, 4(2):305–318, 1993.

[682] T. Poggio and F. Girosi. A Theory of Networks for Approximation and Learning. Technical Report A.I. Memo 1140, MIT, Cambridge, M.A., 1989.

[683] L. Polkowski and A. Skowron. *Rough Sets in Knowledge Discovery 2: Applications, Case Studies, and Software Systems*. Springer Verlag, 1998.

[684] J.B. Pollack and A.D. Blair. Co-Evolution in the Successful Learning of Backgammon Strategy. *Machine Learning*, 32(1):225–240, 1998.

[685] G. Potgieter. Mining Continuous Classes using Evolutionary Computing. Master's thesis, Department of Computer Science, University of Pretoria, 2003.

[686] M.A. Potter. *The Design and Analysis of a Computational Model of Cooperative Coevolution*. PhD thesis, George Mason University, Fairfax, V.A., USA, 1997.

[687] M.A. Potter and K. de Jong. A Cooperative Coevolutionary Approach to Function Optimization. In Y. Davidor, H-P. Schwefel, and R. Männer, editors, *Proceedings of the Parallel Problem Solving from Nature*, pages 249–257, Berlin, 1994. Springer.

[688] M.A. Potter and K. de Jong. Evolving Neural Networks with Collaborative Species. In *Proceedings of the Summer Computer Simulation Conference*, pages 340–345, 1995.

[689] M.A. Potter and K.A. de Jong. The Coevolution of Antibodies for Concept Learning. In *Proceedings of the Fifth International Conference on Parallel Problem Solving from Nature*, pages 530–539, 1998.

[690] M.A. Potter, K.A. de Jong, and J.J. Grefenstette. A Coevolutionary Approach to Learning Sequential Decision Rules. In L. Eshelman, editor, *Proceedings of the Sixth International Conference on Genetic Algorithms*, pages 366–372, San Francisco, CA, 1995. Morgan Kaufmann.

[691] D. Powell and M.M. Skolnick. Using Genetic Algorithms in Engineering Design and Optimization with Nonlinear Constraints. In *Proceedings of the Fifth International Conference on Genetic Algorithms*, pages 424–430, 1993.

[692] S. Pramanik, R. Kozma, and D. Dasgupta. Dynamical Neuro-Representation of an Immune Model and its Application for Data Classification. In *IEEE World Congress on Computational Intelligence, Proceedings of the International Joint Conference on Neural Networks*, volume 1, pages 130–135, 2002.

[693] S.C. Pratt, E.B. Mallon, D.J.T. Sumpter, and N.R. Franks. Quorum Sensing, Recruitement, and Collective Decision-Making during Colony Emigration by the Ant *Leptothorax albipennis*. *Behavioral Ecology and Sociobiology*, 52:117–127, 2002.

[694] L. Prechelt. Adaptive Parameter Pruning in Neural Networks. Technical Report TR-95-009, International Computer Science Institute, Berkeley, California, 1995.

[695] K.V. Price. Differential Evolution vs. The Functions of The $2^{nd}$ ICEO. In *Proceedings of the IEEE International Conference on Evolutionary Computation*, pages 153–157, 1997.

[696] K.V. Price, R.M. Storn, and J.A. Lampinen. *Differential Evolution: A Practical Approach to Global Optimization*. Springer, 2005.

[697] J.G. Proakis and M. Salehi. *Communication System Engineering*. Prentice Hall Publishers, second edition, 2002.

[698] A.K. Qin and P.N. Suganthan. Self-Adaptive Differential Evolution Algorithm for Numerical Optimization. In *Proceedings of the IEEE Congress on Evolutionary Computation*, volume 2, pages 1785–1791, 2005.

[699] A. Rae and S. Parameswaran. Application-Specific Heterogeneous Multiprocessor Synthesis using Differential Evolution. In *Proceedings of the Eleventh International Symposium On System Synthesis*, pages 83–88, 1998.

[700] Y. Rahmat-Samii, D. Gies, and J. Robinson. Particle Swarm Optimization (PSO): A Novel Paradigm for Antenna Designs. *The Radio Science Bulletin*, 304:14–22, 2003.

[701] J. Rajesh, K. Gupta, H.S. Kusumakar, V.K. Jayaraman, and B.D. Kulkarni. Dynamic Optimization of Chemical Processes using Ant Colony Framework. *Computers and Chemistry*, 25:583–595, 2001.

[702] V. Ramos, C. Fernandes, and A.C. Rosa. Social Cognitive Maps, Swarm Collective Perception and Distributed Search on Dynamic Landscapes. Technical report, Instituto Superior Técnico, Lisboa, Portugal, http://alfa.ist.utl.pt/~cvrm/staff/vramos/ref_58.html, 2005.

[703] V. Ramos and J.J. Merelo. Self-Organized Stigmergic Document Maps: Environments as A Mechanism for Context Learning. In *Proceedings of the First Spanish Conference on Evolutionary and Bio-Inspired Algorithms*, pages 284–293, 2002.

[704] M. Randall. Heuristics for Ant Colony Optimisation using the Generalised Assignment Problem. In *Proceedings of the IEEE Congress on Evolutionary Computation*, volume 2, pages 1916–1923, 2004.

[705] T.K. Rasmussen and T. Krink. Improved Hidden Markov Model Training for Multiple Sequence Alignment by a Particle Swarm Optimization-Evolutionary Algorithm Hybrid. *Biosystems*, 72(1-2):5–17, 2003.

[706] A. Ratnaweera, S. Halgamuge, and H. Watson. Particle Swarm Optimization with Self-Adaptive Acceleration Coefficients. In *Proceedings of the First International Conference on Fuzzy Systems and Knowledge Discovery*, pages 264–268, 2003.

[707] A. Ratnaweera, H. Watson, and S.K. Halgamuge. Optimisation of Valve Timing Events of Internal Combustion Engines. In *Proceedings of the IEEE Congress on Evolutionary Computation*, volume 4, pages 2411–2418, 2003.

[708] I. Rechenberg. Cybernetic Solution Path of an Experimental Problem. Technical report, Ministery of Aviation, 1965.

[709] I. Rechenberg. *Evolutionsstrategie: Optimierung technischer Systeme nach Prinzipien der Biologischen Evolution.* Frammann-Holzboog Verlag, Stuttgart, 1973.

[710] I. Rechenberg. Evolution Strategy. In J.M. Zurada, R. Marks II, and C. Robinson, editors, *Computational Intelligence: Imitating Life*, pages 147–159, 1994.

[711] J. Reed, R. Toombs, and N.A. Barricelli. Simulation of Biological Evolution and Machine Learning. *Journal of Theoretical Biology*, 17:319–342, 1967.

[712] R. Reed. Pruning Algorithms – A Survey. *IEEE Transactions on Neural Networks*, 4(5):740–747, 1993.

[713] R. Reed, R.J. Marks II, and S. Oh. Similarities of Error Regularization, Sigmoid Gain Scaling, Target Smoothing, and Training with Jitter. *IEEE Transactions on Neural Networks*, 6:529–538, 1995.

[714] J.-M. Renders and H. Bersini. Hybridizing Genetic Algorithms with Hill-Climbing Methods for Global Optimization: Two Possible Ways. In *Proceedings of the First IEEE Conference on Evolutionary Computation*, pages 312–317, 1994.

[715] R.G. Reynolds. *An Adaptive Computer Model of the Evolution of Agriculture.* PhD thesis, University of Michigan, Michigan, 1979.

[716] R.G. Reynolds. Version Space Controlled Genetic Algorithms. In *Proceedings of the Second Annual Conference on Artificial Intelligence Simulation and Planning in High Autonomy Systems*, pages 6–14, April 1991.

[717] R.G. Reynolds. Cultural Algorithms: Theory and Application. In D. Corne, M. Dorigo, and F. Glover, editors, *New Ideas in Optimization*, pages 367–378. McGraw-Hill, 1999.

[718] R.G. Reynolds and H. Al-Shehri. The Use of Cultural Algorithms with Evolutionary Programming to Guide Decision Tree Induction in Large Databases. In *IEEE World Congress on Computational Intelligence, Proceedings of the International Conference on Evolutionary Computation*, pages 541–546, 1998.

[719] R.G. Reynolds and C. Chung. Fuzzy Approaches to Acquiring Experimental Knowledge in Cultural Algorithms. In *Proceedings of the Nineth IEEE Internaitonal Conference on Tools with Artificial Intelligence*, pages 260–267, 1997.

[720] R.G. Reynolds and C. Chung. Knowledge-based Self-Adaptation in Evolutionary Programming using Cultural Algorithms. In *Proceedings of the IEEE Congress on Evolutionary Computation*, pages 71–76, 1997.

[721] R.G. Reynolds and B. Peng. Cultural Algorithms: Modeling of How Cultures Learn to Solve Problems. In *Proceedings of the Sixteenth IEEE International Conference on Tools with Artificial Intelligence*, pages 166–172, 2004.

[722] R.G. Reynolds, B. Peng, and R.S. Alomari. Cultural Evolution of Ensemble Learning for Problem Solving. In *Proceedings of the IEEE Congress on Evolutionary Computation*, pages 1119–1126, 2006.

[723] R.G. Reynolds and S.R. Rolnick. Learning the Parameters for a Gradient-Based Approach to Image Segmentation from The Results of a Region Growing Approach using Cultural Algorithms. In *Proceedings of the First International Symposium on Intelligence in Neural and Biological Systems*, page 240, 1995.

[724] R.G. Reynolds and W. Sverdlik. Problem Solving using Cultural Algorithms. In *IEEE World Congress on Computational Intelligence, Proceedings of the International Conference on Evolutionary Computation*, pages 1004–1008, 1994.

[725] R.G. Reynolds and S. Zhu. Knowledge-Based Function Optimization using Fuzzy Cultural Algorithms with Evolutionary Programming. *IEEE Transactions on Systems, Man, and Cybernetics*, 31(1):1–18, 2001.

[726] J.T. Richardson, M.R. Palmer, G. Liepins, and M. Hilliard. Some Guidelines for Genetic Algorithms with Penalty Functions. In *Proceedings of the Third International Conference on Genetic Algorithms*, pages 191–197, 1989.

[727] M. Riedmiller and H. Braun. RPROP – A Fast Adaptive Learning Algorithm. In *Proceedings of the Seventh International Symposium on Computer and Information Sciences*, pages 279–285, 1992.

[728] M. Riedmiller and H. Braun. A Direct Adaptive Method for Faster Backpropagation Learning: The RPROP Algorithm. In *Proceedings of the IEEE International Conference on Neural Networks*, pages 586–591, 1993.

[729] J. Riget and J.S. Vesterstrøm. A Diversity-Guided Particle Swarm Optimizer – The ARPSO. Technical Report 2002-02, Department of Computer Science, University of Aarhus, 2002.

[730] J. Riget and J.S. Vesterstrøm. Controlling Diversity in Particle Swarm Optimization. Master's thesis, University of Aarhus, Denmark, 2002.

[731] B.D. Ripley. *Pattern Recognition and Neural Networks*. Cambridge University Press, 1996.

[732] A. Röbel. The Dynamic Pattern Selection Algorithm: Effective Training and Controlled Generalization of Backpropagation Neural Networks. Technical report, Institut für Angewandte Informatik, Technische Universität, Berlin, 1994.

[733] G.E. Robinson. Modulation of Alarm Pheromone Perception in the Honey Bee: Evidence for Division of Labour Based on Hormonally Regulated Response Thresholds. *Journal of Computational Physiology A*, 160:619, 1987.

[734] A. Rogers and A. Prügel-Bennett. Modelling the Dynamics of a Steady State Genetic Algorithm. In W. Banzhaf and C. Reeves, editors, *Foundations of Genetic Algorithms*, volume 5, pages 57–68, San Francisco, C.A., 1999. Morgan Kaufmann.

[735] J. Rönkkönen, S. Kukkonen, and K.V. Price. Real-Parameter Optimization with Differential Evolution. In *Proceedings of the IEEE Congress on Evolutionary Computation*, volume 1, pages 506–513, 2005.

[736] B.E. Rosen and J.M. Goodwin. Optimizing Neural Networks Using Very Fast Simulated Annealing. *Neural, Parallel & Scientific Computations*, 5(3):383–392, 1997.

[737] D.J. Rosenkrantz, R.E. Stearns, and P.M. Lewis. An Analysis of Several Heuristics for the Traveling Salesman Problem. *SIAM Journal on Computing*, 6:563–581, 1977.

[738] C.D. Rosin. *Coevolutionary Search Among Adversaries.* PhD thesis, University of California at San Diego, 1997.

[739] C.D. Rosin and R.K. Belew. Methods for Competitive Co-evolution: Finding Opponents Worth Beating. In Larry Eshelman, editor, *Proceedings of the Sixth International Conference on Genetic Algorithms*, pages 373–380, San Francisco, CA, 1995. Morgan Kaufmann.

[740] C.D. Rosin and R.K. Belew. New Methods for Competitive Coevolution. *Evolutionary Computation*, 5(1):1–29, 1997.

[741] S. Rouwhorst and A.P. Engelbrecht. Searching the Forest: Using Decision Trees as Building Blocks for Evolutionary Search in Classification Databases. In *International Congress on Evolutionary Computing*, pages 633–638, 2000.

[742] S.E. Rouwhorst and A.P. Engelbrecht. Searching the Forest: Using Decision Trees as Building Blocks for Evolutionary Search in Classification Databases. In *Proceedings of the IEEE International Conference on Evolutionary Computation*, 2000.

[743] O. Roux, C. Fonlupt, and E-G. Talbi. ANTabu. Technical Report LIL-98-04, Laboratoire d'Informatique du Littoral, Université du Littoral, Calais, France, 1998.

[744] O. Roux, C. Fonlupt, E-G. Talbi, and D. Robilliard. ANTabu – Enhanced Version. Technical Report LIL-99-01, Laboratoire d'Informatique du Littoral, Université du Littoral, Calais, France, 1999.

[745] G.A. Rummery and M. Niranjan. On-Line Q-Learning using Connectionist Systems. Technical Report CUED/F-INFENG/TR166, Cambridge University, 1994.

[746] T.P. Runarsson and X. Yao. Continuous Selection and Self-Adaptation Evolution Strategies. In *Proceedings of the IEEE Congress on Evolutionary Computation*, pages 279–284, 2002.

[747] N. Rychtyckyj and R.G. Reynolds. Using Cultural Algorithms to Improve Performance in Semantic Networks. In *Proceedings of the IEEE International Congress on Evolutionary Computation*, volume 3, pages 1651–1656, July 1999.

[748] K. Rzadca and F. Seredynski. Heterogeneous Multiprocessor Scheduling with Differential Evolution. In *Proceedings of the IEEE Congress on Evolutionary Computation*, volume 3, pages 2840–2847, 2005.

[749] K. Saastamoinen, J. Ketola, and E. Turunen. Defining Athlete's Anaerobic and Aerobic Thresholds by Using Similarity Measures and Differential Evolution. In *Proceedings of the IEEE International Conference on Systems, Man, and Cybernetics*, volume 2, pages 1331–1335, 2004.

[750] H. Sakanashi and Y. Kakazu. Co-Evolving Genetic Algorithm with Filtered Evaluation Function. In *Proceedings of the IEEE Symposium on Emerging Technologies and Factory Automation*, pages 454–457, 1994.

[751] A. Sakurai. n-h-1 Networks Use No Less than nh+1 Examples, but Sometimes More. *Proceedings of the IEEE*, 3:936–941, 1992.

[752] S. Saleem and R.G. Reynolds. Cultural Algorithms in Dynamic Environments. In *Proceedings of the IEEE Congress on Evolutionary Computation*, volume 2, pages 1513–1520, 2000.

[753] J. Salerno. Using the Particle Swarm Optimization Technique to Train a Recurrent Neural Model. In *Proceedings of the IEEE International Conference on Tools with Artificial Intelligence*, pages 45–49, November 1997.

[754] A. Salman, I. Ahmad, and S. Al-Madani. Particle Swarm Optimization for Task Assignment Problem. *Microprocessors and Microsystems*, 26(8):363–371, 2002.

[755] R. Salomon and J.L. van Hemmen. Accelerating Backpropagation through Dynamic Self-Adaptation. *Neural Networks*, 9(4):589–601, 1996.

[756] T. Sanger. Optimal Unsupervised Learning in a Single-Layer Linear Feedforward Neural Network. *Neural Networks*, 2:459–473, 1989.

[757] S. Sarafijanovic and J. Le Boudec. An Artificial Immune System for Misbehavior Detection in Mobile Ad-Hoc Networks with Virtual Thymus, Clustering, Danger Signal and Memory Detectors. In *Proceedings of Third International Conference on Artificial Immune Systems*, pages 342–356, 2004.

[758] H. Sarimveis and A. Nikolakopoulos. A Line Up Evolutionary Algorithm for Solving Nonlinear Constrained Optimization Problems. *Computers & Operations Research*, 32(6):1499–1514, 2005.

[759] M.A. Sartori and P.J. Antsaklis. A Simple Method to Derive Bounds on the Size and to Train Multilayer Neural Networks. *IEEE Transactions on Neural Networks*, 2(4):467–471, 1991.

[760] J.D. Schaffer. *Some Experiments in Machine Learning using Vector Evaluated Genetic Algorithms*. PhD thesis, Vanderbilt University, 1984.

[761] J.D. Schaffer. Multiple Objective Optimization with Vector Evaluated Genetic Algorithms. In *Proceedings of the First International Conference on Genetic Algorithms*, pages 93–100, 1985.

[762] L. Schindler, D. Kerrigan, and J. Kelly. Understanding the Immune System. *Science Behind the News - National Cancer Institute, http://newscenter.cancer.gov/cancertopics/understandingcancer/immunesystem*, 2002.

[763] C. Schittenkopf, G. Deco, and W. Brauer. Two Strategies to Avoid Overfitting in Feedforward Neural Networks. *Neural Networks*, 10(30):505–516, 1997.

[764] H. Schmidt and G. Thierauf. A Combined Heuristic Optimization Technique. *Advances in Engineering Software*, 36(1):11–19, 2005.

[765] J.F. Schutte and A.A. Groenwold. A Study of Global Optimization Using Particle Swarms. *Journal of Global Optimization*, 31(1):93–108, 2001.

[766] J.F. Schutte and A.A. Groenwold. Sizing Design of Truss Structures using Particle Swarms. *Structural and Multidisciplinary Optimization*, 25(4):261–269, 2003.

[767] D.B. Schwartz, V.K. Samalam, S.A. Solla, and J.S. Denker. Exhaustive Learning. *Neural Computation*, 2:374–385, 1990.

[768] H.-P. Schwefel. *Evolutionsstrategie und numerische Optimierung*. PhD thesis, Technical University Berlin, 1975.

[769] H.-P. Schwefel. *Numerical Optimization of Computer Models*. John Wiley, Chichester, U.K., 1981.

[770] H.-P Schwefel. *Evolution and Optimum Seeking*. Wiley, New York, 1995.

[771] B.R. Secrest. Travelling Salesman Problem for Surveillance Mission Planning using Particle Swarm Optimization. Master's thesis, School of Engineering and Management of the Air Force Institute of Technology, Air University, 2001.

[772] B.R. Secrest and G.B. Lamont. Visualizing Particle Swarm Optimization – Gaussian Particle Swarm Optimization. In *Proceedings of the IEEE Swarm Intelligence Symposium*, pages 198–204, 2003.

[773] T. Sejnowski. Storing Covariance with Nonlinearly Interacting Neurons. *Journal of Mathematical Biology*, 4:303–321, 1997.

[774] A.B. Sendova-Franks and N.R. Franks. Self-Assembly, Self-Organization and Division of Labour. *Philosophical Transactions of the Royal Society of London*, 354:1395–1405, 1999.

[775] P.S. Sensarma, M. Rahmani, and A. Carvalho. A Comprehensive Method for Optimal Expansion Planning using Particle Swarm Optimization. In *Proceedings of the IEEE Power Engineering Society Transmission and Distribution Conference*, volume 2, pages 1317–1322, 2002.

[776] M. Settles and B. Rylander. Neural Network Learning using Particle Swarm Optimizers. *Advances in Information Science and Soft Computing*, pages 224–226, 2002.

[777] H.S. Seung, M. Opper, and H. Sompolinsky. Query by Committee. In *Proceedings of the Fifth Annual ACM Workshop on Computational Learning Theory*, pages 287–299, 1992.

[778] S. Sevenster and A.P. Engelbrecht. GARTNet: A Genetic Algorithm for Routing in Telecommunications Networks. In *Proceedings of CESA96 IMACS Multiconference on Computational Engineering in Systems Applications, Symposium on Control, Optimization and Supervision*, volume 2, pages 1106–1111, 1996.

[779] L. Shi, G. Xu, and Z. Hua. A New Heuristic Evolutionary Programming and its Application in Solution of the Optimal Power Flow. I. Primary Principle of Heuristic Evolutionary Programming. In *Proceedings of the International Conference on Power System Technology*, volume 1, pages 762–770, 1998.

[780] Y. Shi and R.C. Eberhart. A Modified Particle Swarm Optimizer. In *Proceedings of the IEEE Congress on Evolutionary Computation*, pages 69–73, 1998.

[781] Y. Shi and R.C. Eberhart. Parameter Selection in Particle Swarm Optimization. In *Proceedings of the Seventh Annual Conference on Evolutionary Programming*, pages 591–600, 1998.

[782] Y. Shi and R.C. Eberhart. Empirical Study of Particle Swarm Optimization. In *Proceedings of the IEEE Congress on Evolutionary Computation*, volume 3, pages 1945–1950, 1999.

[783] Y. Shi and R.C. Eberhart. Fuzzy Adaptive Particle Swarm Optimization. In *Proceedings of the IEEE Congress on Evolutionary Computation*, volume 1, pages 101–106, 2001.

[784] Y. Shi and R.A. Krohling. Co-Evolutionary Particle Swarm Optimization to Solve Min-Max Problems. In *Proceedings of the IEEE Congress on Evolutionary Computation*, volume 2, pages 1682–1687, 2002.

[785] O.M. Shir and T Bäck. Dynamic Niching in Evolution Strategies with Covariance Matrix Adaptation. In *Proceedings of the IEEE Congress on Evolutionary Computation*, volume 3, pages 2584–2591, 2005.

[786] G.M. Shiraz, R.E. Marks, D.F. Midgley, and L.G. Cooper. Using Genetic Algorithms to Breed Competitive Marketing Strategies. In *Proceedings of the IEEE International Conference on Systems, Man, and Cybernetics*, volume 3, pages 2367–2372, 1998.

[787] A. Shmygelska and H.H. Hoos. An Improved Ant Colony Optimisation Algorithm for the 2D HP Protein Folding Problem. In *Proceedings of the Canadian Conference on Artificial Intelligence*, pages 400–407, 2004.

[788] A. Sierra and A. Echeverría. The Polar Evolution Strategy. In *Proceedings of the IEEE Congress on Evolutionary Computation*, pages 2301–2306, 2006.

[789] J. Sietsma and R.J.F. Dow. Creating Artificial Neural Networks that Generalize. *Neural Networks*, 4:67–79, 1991.

[790] A. Silva, A. Neves, and E. Costa. An Empirical Comparison of Particle Swarm and Predator Prey Optimisation. In *Proceedings of the Thirteenth Irish Conference on Artificial Intelligence and Cognitive Science, Lecture Notes in Artificial Intelligence*, volume 2464, pages 103–110. Springer-Verlag, 2002.

[791] K.M. Sim and W.H. Sun. Multiple Ant-Colony Optimization for Network Routing. In *Proceedings of the First International Symposium on Cyber Worlds*, pages 277–281, 2002.

[792] K. Sims. Artificial Evolution for Computer Graphics. *Computer Graphics*, 25(4):319–328, 1991.

[793] K. Sims. Evolving Virtual Creatures. In *Computer Graphics, Annual Conference Series, ACM SIGGRAPH*, pages 15–22, 1994.

[794] N. Sinha and B. Purkayastha. PSO Embedded Evolutionary Programming Technique for Nonconvex economic Load Dispatch. In *Proceedings of the IEEE Power Systems Conference and Exposition*, volume 1, pages 66–71, 2004.

[795] N. Sinha, R. Shakrabarti, and P.K. Chattopadhyay. Fast Evolutionary Programming Techniques for Short-Term Hydrothermal Scheduling. *IEEE Transactions on Power Systems*, 18(1):214–220, 2003.

[796] P. Slade and T.D. Gedeon. Bimodal Distribution Removal. In J. Mira, J. Cabestany, and A. Prieto, editors, *Proceedings of the International Workshop on Artificial Neural Networks*, pages 249–254, Berlin, 1993. Springer-Verlag.

[797] J. Smith and T.C. Fogarty. Self Adaptation of Mutation Rates in A Steady State Genetic Algorithm. In *Proceedings of the IEEE International Conference on Evolutionary Computation*, pages 318–323, 1996.

[798] J. Smith and F. Vavak. Replacement Strategies in Steady State Genetic Algorithms: Static Environments. Technical report, Intelligent Computer System Centre, University of the West of England, Bristol, England, 1998.

[799] J.A. Snyman. A New and Dynamic Method for Unconstrained Minimization. *Applied Mathematical Modelling*, 6:449–462, 1882.

[800] J.A. Snyman. An Improved Version of the Original LeapFrog Dynamic Method for Unconstrained Minimization: LFOP1(b). *Applied Mathematical Modelling*, 7:216–218, 1983.

[801] K. Socha. The Influence of Run-Time Limits on Choosing Ant System Parameters. In *Proceedings of GECCO 2003 -- Genetic and Evolutionary Computation Conference*, volume 2723, pages 49–60, 2003.

[802] F.J. Solis and R.J.-B. Wets. Minimization by Random Search Techniques. *Mathematical Operations Research*, 6:19–30, 1981.

[803] A. Somayaji and S. Forrest. Automated Response Using System-Call Delays. In *Proceedings of the Nineth Unisex Security Symposium*, pages 185–197, 2000.

[804] A. Somayaji, S. Hofmeyr, and S. Forrest. Principles of a Computer Immune System. In *Proceedings of the ACM New Security Paradigms Workshop*, pages 75–82, 1997.

[805] T. Sousa, A. Neves, and A. Silva. Swarm Optimisation as a New Tool for Data Mining. In *Proceedings of the Parallel and Distributed Processing Symposium*, pages 144–149, 2003.

[806] J.C. Spall. *Introduction to Stochastic Search and Optimization*. Wiley Inter-Science, 2003.

[807] N. Srinivas and K. Deb. Multiobjective Optimization using Nondominated Sorting in Genetic Algorithms. *Evolutionary Computation*, 2(3):221–248, 1991.

[808] A. Stacey, M. Jancic, and I. Grundy. Particle Swarm Optimization with Mutation. In *Proceedings of the IEEE Congress on Evolutionary Computation*, volume 2, pages 1425–1430, 2003.

[809] J.M. Steppe, K.W. Bauer, and S.K. Rogers. Integrated Feature and Architecture Selection. *IEEE Transactions on Neural Networks*, 7(4):1007–1014, 1996.

[810] R. Storn. Differential Evolution Design of an IIR-Filter. In *Proceedings of the IEEE International Conference on Evolutionary Computation*, pages 268–273, 1996.

[811] R. Storn. On the Usage of Differential Evolution for Function Optimization. In *Proceedings of the Biennial Conference of the North American Fuzzy Information Processing Society*, pages 519–523, 1996.

[812] R. Storn. System Design by Constraint Adaptation and Differential Evolution. *IEEE Transactions on Evolutionary Computation*, 3(1):22–34, 1999.

[813] R. Storn and K. Price. Differential Evolution – A Simple and Efficient Heuristic for global Optimization over Continuous Spaces. *Journal of Global Optimization*, 11(4):431–359, 1997.

[814] T. Stützle. Parallelization Strategies for Ant Colony Optimization. In A.E. Eiben, T. Bäck, M. Schoenauer, and H-P. Schwefel, editors, *Proceedings of the Parallel Problem Solving from Nature Conference, Lecture Notes in Computer Science*, volume 1498, pages 722–731. Springer-Verlag, 1998.

[815] T. Stützle and H. Hoos. MAX-MIN Ant System and Local Search for The Traveling Salesman Problem. In *Proceedings of the IEEE International Conference on Evolutionary Computation*, pages 309–314, 1997.

[816] T. Stützle and H.H. Hoos. MAX-MIN Ant System. *Future Generation Computer Systems*, 16(8):889–914, 2000.

[817] C-T. Su and C-S. Lee. Network Reconfiguration of Distribution Systems using Improved Mixed-Integer Hybrid Differential Evolution. *IEEE Transactions on Power Delivery*, 18(3):1022–1027, 2002.

[818] M.C. Su, T.A. Liu, and H.T. Chang. An Efficient Initialization Scheme for the Self-Organizing Feature Map Algorithm. In *Proceedings of the IEEE International Joint Conference in Neural Networks*, volume 3, pages 1906–1910, 1999.

[819] G.A. Suer. Evolutionary Programming for Designing Manufacturing Cells. In *Proceedings of the IEEE International Conference on Evolutionary Computation*, pages 379–384, 1997.

[820] P.N. Suganthan. Particle Swarm Optimiser with Neighborhood Operator. In *Proceedings of the IEEE Congress on Evolutionary Computation*, pages 1958–1962, 1999.

[821] R. Sun, S. Tatsumo, and Z. Gang. Multiagent Reinforcement Learning Method with an Improved Ant Colony System. In *Proceedings of the IEEE International Conference on Systems, Man, and Cybernetics*, volume 3, pages 1612–1617, 2001.

[822] K.K. Sung and P. Niyogi. A Formulation for Active Learning with Applications to Object Detection. Technical Report 1438, Artificial Intelligence Laboratory, Massachusetts Institute of Technology, 1996.

[823] R.S. Sutton. *Temporal Credit Assignment in Reinforcement Learning*. PhD thesis, University of Massachusetts, Amherst, 1984.

[824] R.S. Sutton. Learning to Predict by the Method of Temporal Differences. *Machine Learning*, 3(1):9–44, 1988.

[825] R.S. Sutton. Implementation Details of the TD($\lambda$) Procedure for the Case of Vector Predictions and Backpropagation. Technical Report TN87-509.1, GTE Laboratories, 1989.

[826] W. Sverdlik, R.G. Reynolds, and E. Zannoni. HYBAL: A Self-Tuning Algorithm for Concept Learning in Highly Autonomous Systems. In *Proceedings of the Third Annual Conference on Artificial Intelligence, Simulation, and Planning in High Autonomy Systems*, pages 15–22, 1992.

[827] A.K. Swain and A.S. Morris. A Novel Hybrid Evolutionary Programming Method for Function Optimization. In *Proceedings of the IEEE Congress on Evolutionary Computation*, volume 1, pages 699–705, 2000.

[828] G. Syswerda. Uniform Crossover in Genetic Algorithms. In *Proceedings of the Third International Conference on Genetic Algorithms*, pages 2–9, 1989.

[829] G. Syswerda. Schedule Optimization using Genetic Algorithms. In L. Davis, editor, *Handbook of Genetic Algorithms*, pages 331–349. International Thomson Computer Press, 1991.

[830] M-J. Tahk and B-C. Sun. Coevolutionary Augmented Lagrangian Methods for Constrained Optimization. *IEEE Transactions on Evolutionary Computation*, 4(2):114–124, 2000.

[831] É.D. Taillard. FANT: Fast Ant System. Technical Report IDSIA 46-98, IDSIA, Lugano, Switzerland, 1998.

[832] É.D. Taillard and L.M. Gambardella. Adaptive Memories for the Quadratic Assignment Problem. Technical Report IDSIA-87-97, IDSIA, Lugano, Switzerland, 1997.

[833] T. Takagi and M. Sugeno. Fuzzy Identification of Systems and its Application to Modeling and Control. *IEEE Transactions on Systems, Man, and Cybernetics*, 15(1):116–132, 1985.

[834] T. Takahashi. Principal Component Analysis is a Group Action of SO(N) which Minimizes an Entropy Function. In *Proceedings of the International Joint Conference on Neural Networks*, volume 1, pages 355–358, 1993.

[835] E-G. Talbi, O. Rouz, C. Fonlupt, and D. Robillard. Parallel Ant Colonies for the quadratic assignment problem. *Future Generation Computer Systems*, 14(4):441–449, 2001.

[836] H. Talbi and M. Batouche. Hybrid Particle Swarm with Differential Evolution for Multimodal Image Registration. In *Proceedings of the IEEE International Conference on Industrial Technology*, volume 3, pages 1567–1573, 2004.

[837] S. Tamura, M. Tateishi, M. Matumoto, and S. Akita. Determination of the Number of Redundant Hidden Units in a Three-layer Feed-Forward Neural Network. In *Proceedings of the International Joint Conference on Neural Networks*, volume 1, pages 335–338, 1993.

[838] V. Tandon, H. El-Mounayri, and H. Kishawy. NC End Milling Optimization using Evolutionary Computation. *International Journal of Machine Tools and Manufacture*, 42(5):595–605, 2002.

[839] O. Tezak, D. Dolinar, and M. Milanovic. Snubber Design Approach for dc-dc Converter Based on Differential Evolution Method. In *Proceedings of the Eight IEEE International Workshop on Advanced Motion Control*, pages 87–91, 2004.

[840] G. Théraulaz and E. Bonabeau. A Brief History of Stigmergy. *Artificial Life*, 5:97–116, 1999.

[841] G. Théraulaz, E. Bonabeau, and J-L. Deneubourg. Response Threshold Reinforcement and Division of Labour in Insect Societies. *Proceedings of the Royal Society of London, Series B*, 265:327–332, 1998.

[842] G. Théraulaz, S. Goss, J. Gervet, and J-L. Deneubourg. Task Differentation in Polists Wasp Colonies: A Model for Self-Organizing of Robots. In J-A. Meyer and S.W. Wilson, editors, *Proceedings of the First International Conference on Simulation of Adaptive Behavior: From Animals to Animats*, pages 346–355. MIT Press, 1991.

[843] H. Thiele. Fuzzy Rough Sets versus Rough Fuzzy Sets – An Interpretation and Comparative Study using Concepts of Modal Logics. Technical Report CI-30/98, University of Dortmund, 1998.

[844] H.H. Thodberg. Improving Generalization of Neural Networks through Pruning. *International Journal of Neural Systems*, 1(4):317–326, 1991.

[845] J. Timmis. *Artificial Immune Systems: A Novel Data Analysis Technique Inspired by the Immune Network Theory*. PhD thesis, University of Wales, Aberystwyth, August 2000.

[846] J. Timmis and M. Neal. A Resource Limited Artificial Immune System for Data Analysis. In *Research and Development in Intelligent Systems*, volume 14, pages 19–32, Cambridge, UK, 2000. Springer.

[847] J. Timmis, M. Neal, and J. Hunt. Data Analysis using Artificial Immune Systems, Cluster Analysis and Kohonen Networks: Some Comparisons. In *Proceedings of IEEE International Conference on Systems, Man and Cybernetics*, volume 3, pages 922–927, 1999.

[848] F. Tin-Loi and N.S. Que. Identification of Cohesive Crack Fracture Parameters by Evolutionary Search. *Computer Methods in Applied Mechanics and Engineering*, 191(49-50):5741–5760, 2002.

[849] S.J.P. Todd and W. Latham. Mutator, A Subjective Human Interface for Evolution of Computer Sculptures. Technical report, IBM United Kingdom Scientific Centre Report 248, 1991.

[850] J.F.A. Traniello and R.B. Rosengaus. Ecology, Evolution and Division of Labour in Social Insects. *Animal Behaviour*, 53:209–213, 1997.

[851] I.C. Trelea. The Particle Swarm Optimization Algorithm: Convergence Analysis and Parameter Selection. *Information Processing Letters*, 85(6):317–325, 2003.

[852] C-F. Tsai, C-W. Tsai, and C-C. Tseng. A Novel and Efficient Ant-Based Algorithm for Solving Traveling Salesman Problem. In *Proceedings of the IEEE International Conference on Systems, Man and Cybernetics*, volume 2, pages 320–325, 2002.

[853] C-F. Tsai, C-W. Tsai, H-C. Wu, and T. Yang. ACODF: A Novel Data Clustering Approach for Data Mining in Large Databases. *Journal of Systems and Software*, 73:133–145, 2004.

[854] D. Tsou and C. MacNish. Adaptive Particle Swarm Optimisation for High-Dimensional Highly Convex Search Spaces. In *Proceedings of the IEEE Congress on Evolutionary Computation*, volume 2, pages 783–789, 2003.

[855] Y. Tsujimura, Y. Mafune, and M. Gen. Introducing Co-Evolution and Sub-Evolution Processes into Genetic Algorithm for Job-Shop Scheduling. In *Proceedings of the Twenty-Sixth Annual Conference of the IEEE Industrial Electronics Society*, volume 4, pages 2827–2830, 2000.

[856] T. Tsukada, T. Tamura, S. Kitagawa, and Y. Fukuyama. Optimal Operational Planning for Cogeneration System using Particle Swarm Optimization. In *Proceedings of the IEEE Swarm Intelligence Symposium*, pages 138–143, 2003.

[857] S. Tsutsui and D.E. Goldberg. Simplex Crossover and Linkage Identification: Single-Stage Evolution vs. Multi-Stage Evolution. In *Proceedings of the IEEE Congress on Evolutionary Computation*, volume 1, pages 974–979, 2002.

[858] A.M. Turing. Computing Machinery and Intelligence. *Mind*, 59:433–460, 1950.

[859] E. Uchibe, M. Nakamura, and M. Asada. Co-Evolution for Cooperative Behavior Acquisition in a Multiple Mobile Robot Environment. In *Proceedings of the IEEE/RSJ International Conference on Intelligent Robots and Systems*, volume 1, pages 425–430, 1998.

[860] H. Ulmer, F. Streichert, and A. Zell. Evolution Strategies Assisted by Gaussian Processes with Improved Pre-Selection Criterion. In *Proceedings of the IEEE Congress on Evolutionary Computation*, volume 1, pages 692–699, 2003.

[861] R.K. Ursem. Diversity-Guided Evolutionary Algorithms. In *Proceedings of the Parallel Problem Solving from Nature Conference*, pages 462–471, 2002.

[862] F van den Bergh. Particle Swarm Weight Initialization in Multi-Layer Perceptron Artificial Neural Networks. In *Proceedings of the International Conference on Artificial Intelligence*, pages 42–45, 1999.

[863] F. van den Bergh. *An Analysis of Particle Swarm Optimizers*. PhD thesis, Department of Computer Science, University of Pretoria, Pretoria, South Africa, 2002.

[864] F. van den Bergh and A.P. Engelbrecht. Cooperative Learning in Neural Networks using Particle Swarm Optimizers. *South African Computer Journal*, 26:84–90, 2000.

[865] F. van den Bergh and A.P Engelbrecht. Effects of Swarm Size on Cooperative Particle Swarm Optimisers. In *Proceedings of the Genetic and Evolutionary Computation Conference*, pages 892–899, 2001.

[866] F. van den Bergh and A.P. Engelbrecht. Training Product Unit Networks using Cooperative Particle Swarm Optimisers. In *Proceedings of the IEEE International Joint Conference on Neural Networks*, volume 1, pages 126–131, 2001.

[867] F. van den Bergh and A.P Engelbrecht. Training Product Unit Networks using Cooperative Particle Swarm Optimisers. In *Proceedings of the IEEE International Joint Conference on Neural Networks*, volume 1, pages 126–131, July 2001.

[868] F. van den Bergh and A.P. Engelbrecht. A New Locally Convergent Particle Swarm Optimizer. In *Proceedings of the IEEE International Conference on Systems, Man, and Cybernetics*, pages 96–101, 2002.

[869] F. van den Bergh and A.P. Engelbrecht. A Cooperative Approach to Particle Swarm Optimization. *IEEE Transactions on Evolutionary Computation*, 8(3):225–239, 2004.

[870] F. van den Bergh and A.P. Engelbrecht. A Study of Particle Swarm Optimization Particle Trajectories. *Information Sciences*, 176(8):937–971, 2006.

[871] G.N. Varela and M.C. Sinclair. Ant Colony Optimisation for Virtual-Wavelength-Path Routing and Wavelength Allocation. In *Proceedings of the IEEE Congress on Evolutionary Computation*, volume 3, page 1816, July 1999.

[872] F. Vavak and T.C. Fogarty. Comparison of Steady State and Generational Genetic Algorithms for Use in Nonstationary Environments. In *Proceedings of the IEEE International Conference on Evolutionary Computation*, pages 192–195, 1996.

[873] F. Vavak, K. Jukes, and T.C. Fogarty. Learning the Local Search Range for Genetic Optimisation in Nonstationary Environments. In *Proceedings of the IEEE International Conference on Evolutionary Computation*, pages 355–360, 1997.

[874] G. Venter and J. Sobieszczanski-Sobieski. Multidisciplinary Optimization of a Transport Aircraft Wing using Particle Swarm Optimization. *Structural and Multidisciplinary Optimization*, 26(1-2):121–131, 2003.

[875] G. Venter and J. Sobieszczanski-Sobieski. Particle Swarm Optimization. *Journal for the American Institute of Aeronautics and Astronautics*, 41(8):1583–1589, 2003.

[876] J. Vesterstrøm and R. Thomsen. A Comparative Study of Differential Evolution, Particle Swarm Optimization, and Evolutionary Algorithms on Numerical Benchmark Problems. In *Proceedings of the IEEE Congress on Evolutionary Computation*, volume 2, pages 1980–1987, 2004.

[877] J.S. Vesterstrøm and J. Riget. Particle Swarms: Extensions for Improved Local, Multi-Modal, and Dynamic Search in Numerical Optimization. Master's thesis, Department of Computer Science, University of Aarhus, 2002.

[878] J.S. Vesterstrøm, J. Riget, and T. Krink. Division of Labor in Particle Swarm Optimization. In *Proceedings of the IEEE Congress on Evolutionary Computation*, pages 1570–1575, 2002.

[879] H.L. Viktor, A.P. Engelbrecht, and I. Cloete. Reduction of Symbolic Rules from Artificial Neural Networks using Sensitivity Analysis. In *Proceedings of the IEEE International Conference on Neural Networks*, pages 1788–1793, 1995.

[880] T.P. Vogl, J.K. Mangis, A.K. Rigler, W.T. Zink, and D.L. Alken. Accelerating the Convergence of the Back-Propagation Method. *Biological Cybernetics*, 59:257–263, 1988.

[881] M. Vogt. Combination of Radial Basis Function Neural Networks with Optimized Learning Vector Quantization. In *Proceedings of the IEEE International Conference on Neural Networks*, volume 3, pages 1841–1846, New York, 1993.

[882] V.P. Volny and D.M. Gordon. Genetic Basis for Queen-Worker Dimorphism in a Social Insect. *Proceedings of the National Academy of Sciences of the United States of America*, 99(9):6108–6111, 2002.

[883] J. Vondras and P. Martinek. Multi-Criterion Filter Design via Differential Evolution Method for Function Minimization. In *Proceedings of the First IEEE International Conference on Circuits and Systems for Communications*, pages 106–109, 2002.

[884] F-S. Wang and J-P. Chiou. Differential Evolution for Dynamic Optimization of Differential-Algebraic Systems. In *Proceedings of the IEEE International Conference on Evolutionary Computation*, pages 531–536, 1997.

[885] F-S. Wang and H-J. Jang. Parameter Estimation of a Bioreaction Model by Hybrid Differential Evolution. In *Proceedings of the IEEE Congress on Evolutionary Computation*, volume 1, pages 410–417, 2000.

[886] K-P. Wang, L. Huang, C-G. Zhou, and W. Pang. Particle Swarm Optimization for Travelling Salesman Problem. In *Proceedings of International Conference on Machine Learning and Cybernetics*, pages 1583–1585, 2003.

[887] L. Wang, X-P. Wang, and Q-D. Wu. Ant System Algorithm Based Rosenbrock Function Optimization in Multi-Dimension Space. In *Proceedings of the First International Conference on Machine Learning and Cybernetics*, pages 710–714, 2002.

[888] L.X. Wang and J.M. Mendel. Generating Fuzzy Rules by Learning from Examples. *IEEE Transactions on Systems, Man, and Cybernetics*, 22(6):1413–1426, 1992.

[889] X. Wang, H. Dong, and D. Chen. PID self-tuning control based on evolutionary programming. In *Proceedings of the Fourth World Congress on Intelligent Control and Automation*, volume 4, pages 3132–3135, 2002.

[890] A Watkins and J Timmis. Artificial Immune Recognition System (AIRS): Revisions and Refinements. In *Proceedings of the First International Conference on Artificial Immune Systems*, volume 1, pages 173–181, 2002.

[891] C.J.H.C. Watkins. *Learning from Delayed Rewards*. PhD thesis, King's College, Cambridge University, U.K., 1989.

[892] D.J. Watts and S.H. Strogatz. Collective Dynamics of 'Small-World' Networks. *Nature*, 393(6684):440–442, 1998.

[893] C. Wei, Z. He, Y. Zheng, and W. Pi. Swarm Directions Embedded in Fast Evolutionary Programming. In *Proceedings of the IEEE Congress on Evolutionary Computation*, volume 2, pages 1278–1283, May 2002.

[894] C. Wei, S. Yao, and Z. He. A Modified Evolutionary Programming. In *Proceedings of the IEEE International Conference on Evolutionary Computation*, pages 135–138, 1996.

[895] A.S. Weigend, D.E. Rumelhart, and B.A. Huberman. Generalization by Weight-Elimination with Application to Forecasting. In R. Lippmann, J. Moody, and D.S. Touretzky, editors, *Advances in Neural Information Processing Systems*, volume 3, pages 875–882, 1991.

[896] J.Y. Wen, Q.H. Wu, L. Jiang, and S.J. Cheng. Pseudo-Gradient Based Evolutionary Programming. *Electronic Letters*, 39(7):631–632, 2003.

[897] P.J. Werbos. *Beyond Regression: New Tools for Prediction and Analysis in the Behavioural Sciences*. PhD thesis, Harvard University, Boston, USA, 1974.

[898] L.F.A. Wessels and E. Barnard. Avoiding False Local Minima by Proper Initialization of Connections. *IEEE Transactions on Neural Networks*, 3(6):899–905, 1992.

[899] D. Wettschereck and T. Dietterich. Improving the Performance of Radial Basis Function Networks by Learning Center Locations. In J.E. Moody, S.J. Hanson, and R.P. Lippmann, editors, *Advances in Neural Information Processing Systems*, volume 4, pages 1133–1140, San Mateo, C.A., 1992. Morgan Kaufmann.

[900] C.M. White and G.G. Yen. A Hybrid Evolutionary Algorithm for TSP. In *Proceedings of the IEEE Congress on Evolutionary Computation*, volume 2, pages 1473–1478, 2004.

[901] D. White and P. Ligomenides. GANNet: A Genetic Algorithm for Optimizing Topology and Weights in Neural Network Design. In J. Mira, J. Cabestany, and A. Prieto, editors, *International Workshop on Artificial Neural Networks, Lecture Notes in Computer Science*, volume 686, pages 332–327, Berlin, 1993. Springer-Verlag.

[902] T. White and B. Pagurek. Towards Multi-Swarm Problem Solving in Networks. In *Proceedings of the International Conference on Multi-Agent Systems*, pages 333–340, 1998.

[903] D. Whitley. A Review of Models for Simple Genetic Algorithms and Cellular Genetic Algorithms. In V. Rayward-Smith, editor, *Applications of Modern Heuristics Methods*, pages 55–67, 1995.

[904] D. Whitley and C. Bogart. The Evolution of Connectivity: Pruning Neural Networks using Genetic Algorithms. In *Proceedings of the IEEE International Joint Conference on Neural Networks*, volume 1, pages 134–137, 1990.

[905] D. Whitley, T. Starkweather, and D. Fuquay. Scheduling Problems and the Travelings Salesmen: The Genetic Edge Recombination Operator. In *Proceedings of the Third International Conference on Genetic Algorithms*, pages 116–121. Morgan Kaufmann, 1989.

[906] D. Whitley and N.-W. Yu. Modeling Simple Genetic Algorithms for Permutation Problems. In D. Whitley and M. Vose, editors, *Foundations of Genetic Algorithms*, volume 3, pages 163–184, San Mateo, 1995. Morgan Kaufmann.

[907] B. Widrow. ADALINE and MADALINE – 1963, Plenary Speech. In *Proceedings of the First IEEE International Joint Conference on Neural Networks*, volume 1, pages 148–158, 1987.

[908] B. Widrow and M.A. Lehr. 30 Years of Neural Networks: Perceptron, Madaline and Backpropagation. *Proceedings of the IEEE*, 78:1415–1442, 1990.

[909] S.T. Wierchoń and U. Kużelewska. Stable Clusters Formation in an Artificial Immune System. In *Proceedings of the First International Conference on Artificial Immune Systems*, volume 1, pages 68–75, 2002.

[910] P.M. Williams. Bayesian Regularization and Pruning Using a Laplace Prior. *Neural Computation*, 7:117–143, 1995.

[911] R.J. Williams. A Class of Gradient-Estimating Algorithms for Reinforcement Learning in Neural Networks. In *Proceedings of the IEEE First International Conference on Neural Networks*, pages 601–608, 1987.

[912] M. Wilson, C. Melhuish, and A. Sendova-Franks. Creating Annular Structures Inspired by Ant Colony Behaviour using Minimalist Robots. In *Proceedings of the IEEE International Conference on Systems, Man and Cybernetics*, volume 2, pages 53–58, October 2002.

[913] M. Wilson, C. Melhuish, and A. Sendova-Franks. Multi-Object Segregation: Ant-Like Brood Sorting using Minimalism Robots. In *Proceedings of the Seventh International Conference on Simulation of Adaptive Behaviour*, pages 369–370, 2002.

[914] A. Wloszek and P.D. Domanski. Application of the Coevolutionary System to the Fuzzy Model Design. In *IEEE International Conference on Fuzzy Systems*, volume 1, pages 391–395, 1997.

[915] M. Wodrich and C. Bilchev. Cooperative Distributed Search: The Ant's Way. *Control Cybernetics*, 26:413, 1997.

[916] K.P. Wong and J. Yuryevich. Evolutionary-Programming-Based Algorithms for Environmentally-Constrained Economic Dispatch. *IEEE Transactions on Power Systems*, 13(2):301–306, 1997.

[917] M.L. Wong and K.S. Leung. *Data Mining Using Grammar Based Genetic Programming and Applications*. Springer, 2000.

[918] A. Wright. Genetic Algorithms for Real Parameter Optimization. In G.J.E. Rawlins, editor, *Foundations of Genetic Algorithms*, pages 205–220, San Mateo, C.A., 1991. Morgan Kaufmann.

[919] B. Wu and Z. Shi. A Clustering Algorithm Based on Swarm Intelligence. In *Proceedings of the International Conference on Info-tech and Info-net*, volume 3, pages 58–66, 2001.

[920] B.L. Wu and X.H. Yu. Enhanced Evolutionary Programming for Function Optimization. In *IEEE World Congress on Computational Intelligence, Proceedings of the IEEE Congress on Evolutionary Computation*, pages 695–698, 1998.

[921] X. Xiao, E.R. Dow, R.C. Eberhart, Z. Ben Miled, and R.J. Oppelt. Gene Clustering using Self-Organizing Maps and Particle Swarm Optimization. In *Proceedings of the Second IEEE International Workshop on High Performance Computational Biology*, page 10, 2003.

[922] X. Xie, W. Zhang, and Z. Yang. A Dissipative Particle Swarm Optimization. In *Proceedings of the IEEE Congress on Evolutionary Computation*, volume 2, pages 1456–1461, 2002.

[923] X. Xie, W. Zhang, and Z. Yang. Adaptive Particle Swarm Optimization on Individual Level. In *Proceedings of the Sixth International Conference on Signal Processing*, volume 2, pages 1215–1218, 2002.

[924] X. Xie, W. Zhang, and Z. Yang. Hybrid Particle Swarm Optimizer with Mass Extinction. In *Proceedings of the International Conference on Communication, Circuits and Systems*, volume 2, pages 1170–1173, 2002.

[925] L. Xu, T. Reinikainen, W. Ren, B.P. Wang, Z. Han, and D. Agonafer. A simulation-based multi-objective design optimization of electronic packages under thermal cycling and bending. *Microelectronics Reliability*, 44(12):1977–1983, 2004.

[926] X. Xu and R.D. Dony. Differential Evolution with Powell's Direction Set Method in Medical Image Registration. In *Proceedings of the IEEE International Symposium on Biomedical Imaging*, volume 1, pages 732–735, 2004.

[927] Z. Xu, X. Hou, and J. Sun. Ant Algorithm-Based Task Scheduling in Grid Computing. In *Canadian Conference on Electrical and Computer Engineering*, volume 2, pages 1107–1110, May 2003.

[928] F. Xue, A.C. Sanderson, and R.J. Graves. Pareto-Based Multi-Objective Differential Evolution. In *Proceedings of the IEEE Congress on Evolutionary Computation*, volume 2, pages 862–869, 2003.

[929] Q. Xue, Y. Hu, and W.J. Tompkins. Analyses of the Hidden Units of Back-Propagation Model. In *Proceedings of the IEEE International Joint Conference on Neural Networks*, volume 1, pages 739–742, 1990.

[930] R. Yager. Multiple objective decision-making using fuzzy sets. *International Journal of Man-Machine Studies*, 9:375–382, 1977.

[931] Y. Yang and M. Kamel. Clustering Ensemble using Swarm Intelligence. In *Proceedings of the IEEE Swarm Intelligence Symposium*, pages 65–71, 2003.

[932] X. Yao, G. Lin, and Y. Liu. An Analysis of Evolutionary Algorithms Based on Neighbourhood and Step Sizes. In P.J. Angeline, R.G. Reynolds, J.R. McDonnell, and R. Eberhart, editors, *Proceedings of the Sixth Annual Conference on Evolutionary Programming*, pages 297–307, 1997.

[933] X. Yao and Y. Liu. Evolving Artificial Neural Networks through Evolutionary Programming. In *Proceedings of the Fifth Annual Conference on Evolutionary Programming*, pages 257–266, 1996.

[934] X. Yao and Y. Liu. Fast Evolutionary Programming. In L.J. Fogel, P.J. Angeline, and T.B. Bäck, editors, *Proceedings of the Fifth Annual Conference on Evolutionary Programming*, pages 451–460. MIT Press, 1996.

[935] X. Yao and Y. Liu. Fast Evolution Strategies. In P.J. Angeline, R.G. Reynolds, J.R. McDonnell, and R. Eberhart, editors, *Evolutionary Programming VI*, pages 151–161, Berlin, 1997. Springer.

[936] X. Yao, Y. Liu, and G. Liu. Evolutionary Programming Made Faster. *IEEE Transactions on Evolutionary Computation*, 3(2):82–102, 1999.

[937] K. Yasuda, A. Ide, and N. Iwasaki. Adaptive Particle Swarm Optimization. In *Proceedings of the IEEE International Conference on Systems, Man, and Cybernetics*, volume 2, pages 1554–1559, 2003.

[938] S. Yasui. Convergence Suppression and Divergence Facilitation: Minimum and Joint Use of Hidden Units by Multiple Outputs. *Neural Networks*, 10(2):353–367, 1997.

[939] Z-W. Ye and Z-B. Zheng. Research in the Configuration of Parameter $\alpha, \beta, \rho$ in Ant Algorithm Exemplified by TSP. In *Proceedings of the International Conference on Machine Learning and Cybernetics*, volume 4, pages 2106–2111, 2003.

[940] G.G. Yen and H. Lu. Dynamic Population Strategy Assisted Particle Swarm Optimization. In *Proceedings of the IEEE International Symposium on Intelligent Control*, pages 697–702, 2003.

[941] H. Yoshida, Y. Fukuyama, S. Takayama, and Y Nakanishi. A Particle Swarm Optimization for Reactive Power and Voltage Control in Electric Power Systems Considering Voltage Security Assessment. In *Proceedings of the IEEE International Conference on Systems, Man, and Cybernetics*, volume 6, pages 497–502, October 1999.

[942] X-H. Yu and G-A. Chen. Efficient Backpropagation Learning using Optimal Learning Rate and Momentum. *Neural Networks*, 10(3):517–527, 1997.

[943] J. Yuryevich and K.P. Wong. Evolutionary Programming Based Optimal Power Flow Algorithm. *IEEE Transactions on Power Systems*, 14(4):1245–1250, 1999.

[944] L.A. Zadeh. Fuzzy Sets. *Information and Control*, 8:338–353, 1965.

[945] L.A. Zadeh. Fuzzy Algorithms. *Information and Control*, 12:94–102, 1968.

[946] L.A. Zadeh. The Concept of a Linguistic Variable and its Application to Approximate Reasoning, Parts 1 and 2. *Information Sciences*, pages 338–353, 1975.

[947] L.A. Zadeh. Soft Computing and Fuzzy Logic. *IEEE Software*, 11(6):48–56, 1994.

[948] B-T. Zhang. Accelerated Learning by Active Example Selection. *International Journal of Neural Systems*, 5(1):67–75, 1994.

[949] C. Zhang and H. Shao. An ANN's Evolved by a New Evolutionary System and Its Application. In *Proceedings of the Thirty-Ninth IEEE Conference on Decision and Control*, volume 4, pages 3563–3563, 2000.

[950] C. Zhang, H. Shao, and Y. Li. Particle Swarm Optimization for Evolving Artificial Neural Network. In *Proceedings of the IEEE International Conference on Systems, Man, and Cybernetics*, pages 2487–2490, 2000.

[951] F. Zhang and D. Xue. An Optimal Concurrent Design Model using Distributed Product Development Life-Cycle Databases. In *Sixth International Conference on Computer Supported Cooperative Work in Design*, pages 273–278, 2001.

[952] G. Zhang and Y. Zhang. Optimal Design of High Voltage Bushing Electrode in Transformer with Evolution Strategy. *IEEE Transactions on Magnetics*, 35(3):1690–1693, 1999.

[953] J. Zhang and J. Xu. Evolutionary Programming Based on Uniform Design with Application to Multiobjective Optimization. In *Proceedings of the Fifth World Congress on Intelligent Control and Automation*, volume 3, pages 2298–2302, 2004.

[954] W-J. Zhang and X-F. Xie. DEPSO: Hybrid Particle Swarm with Differential Evolution Operator. In *Proceedings of the IEEE International Conference on System, Man, and Cybernetics*, volume 4, pages 3816–3821, 2003.

[955] X. Zhang, L. Yu, Y. Zheng, Y. Shen, G. Zhou, L. Chen, L. Xi, T. Yuan, J. Zhang, and B. Yang. Two-Stage Adaptive PMD Compensation in a 10 Gbit/s Optical Communication System using Particle Swarm Optimization Algorithm. *Optics Communications*, 231(1-6):233–242, 2004.

[956] Y. Zhang and S. Huang. Multiobjective Optimization using Distance-Based Particle Swarm Optimization. In *Proceedings of the International Conference on Computational Intelligence, Robotics and Autonomous Systems*, 2003.

[957] Y. Zhang and A. Kandel. *Compensatory Genetic Fuzzy Neural Networks and Their Applications*. World Scientific, 1998.

[958] Y. Zheng, L. Ma, L. Zhang, and J. Qian. On the Convergence Analysis and Parameter Selection in Particle Swarm Optimization. In *Proceedings of the International Conference on Machine Learning and Cybernetics*, volume 3, pages 1802–1807, 2003.

[959] Y. Zheng, L. Ma, L. Zhang, and J. Qian. Robust PID Controller Design using Particle Swarm Optimization. In *Proceedings of the IEEE International Symposium on Intelligence Control*, pages 974–979, 2003.

[960] Q. Zhou and Y. Li. Directed Variation in Evolution Strategies. *IEEE Transactions on Evolutionary Computation*, 7(4):356–366, 2006.

[961] J. Zurada. Lambda Learning Rule for Feedforward Neural Networks. In *Proceedings of the IEEE International Joint Conference on Neural Networks*, volume 3, pages 1808–1811, 1992.

[962] J.M. Zurada, A. Malinowski, and I. Cloete. Sensitivity Analysis for Minimization of Input Data Dimension for Feedforward Neural Network. In *Proceedings of the IEEE International Symposium on Circuits and Systems*, volume 6, pages 447–450, 1994.

# Appendix A

# Optimization Theory

Optimization algorithms are search methods, where the goal is to find a solution to an optimization problem, such that a given quantity is optimized, possibly subject to a set of constraints. Although this definition is simple, it hides a number of complex issues. For example, the solution may consist of a combination of different data types, nonlinear constraints may restrict the search area, the search space can be convoluted with many candidate solutions, the characteristics of the problem may change over time, or the quantity being optimized may have conflicting objectives. This is just a short list of issues, given to illustrate some of the complexities an optimization algorithm may have to face. This chapter provides a crisp summary of these issues, and characterizes different problem types.

Section A.1 summarizes the main ingredients of optimization problems. A classification of optimization problems is given in Section A.2. Section A.3 discusses optima types. A list of optimization method classes is provided in Section A.4. Unconstrained optimization problems are defined and discussed in Section A.5, constrained optimization in Section A.6, multi-solution problems in Section A.7, multi-objective problems in Section A.8, and dynamic environments in Section A.9.

## A.1 Basic Ingredients of Optimization Problems

Each optimization problem consists of the following basic ingredients:

- An **objective function**, which represents the quantity to be optimized, that is, the quantity to be minimized or maximized. Let $f$ denote the objective function. Then a maximum of $f$ is a minimum of $-f$. Some problems, specifically constraint-satisfaction problems (CSP), do not define an explicit objective function. Instead, the objective is to find a solution that satisfies all of a set of constraints.

- A **set of unknowns or variables**, which affects the value of the objective function. If $\mathbf{x}$ represents the unknowns, also referred to as the independent variables, then $f(\mathbf{x})$ quantifies the quality of the candidate solution, $\mathbf{x}$.

- A **set of constraints**, which restricts the values that can be assigned to the unknowns. Most problems define at least a set of boundary constraints, which

*Computational Intelligence: An Introduction*, Second Edition A.P. Engelbrecht
©2007 John Wiley & Sons, Ltd

define the domain of values for each variable. Constraints can, however, be more complex, excluding certain candidate solutions from being considered as solutions.

The goal of an optimization method is then to assign values, from the allowed domain, to the unknowns such that the objective function is optimized and all constraints are satisfied. To achieve this goal, the optimization algorithm searches for a solution in a search space, $\mathcal{S}$, of candidate solutions. In the case of constrained problems, a solution is found in the feasible space, $\mathcal{F} \subseteq \mathcal{S}$.

# A.2    Optimization Problem Classifications

Optimization problems are classified based on a number of characteristics:

- The **number of variables** that influences the objective function: A problem with only one variable to be optimized is referred to as a univariate problem. If more than one variable is considered, the problem is referred to as a multivariate problem.

- The **type of variables**: A continuous problem has continuous-valued variables, i.e. $x_j \in \mathbb{R}$, for each $j = 1, \ldots, n_x$. If $x_i \in \mathbb{Z}$, the problem is referred to as an integer or discrete optimization problem. A mixed integer problem has both continuous-valued and integer-valued variables. Problems where solutions are permutations of integer-valued variables are referred to as combinatorial optimization problems.

- The **degree of nonlinearity of the objective function**: Linear problems have an objective function that is linear in the variables. Quadratic problems use quadratic functions. When other nonlinear objective functions are used, the problem is referred to as a nonlinear problem.

- The **constraints used**: A problem that uses only boundary constraints is referred to as an unconstrained problem. Constrained problems have additional equality and/or inequality constraints.

- The **number of optima**: If there exists only one clear solution, the problem is unimodal. If more than one optimum exists, the problem is multimodal. Some problems may have false optima, in which case the problem is referred to as being deceptive.

- The **number of optimization criteria**: If the quantity to be optimized is expressed using only one objective function, the problem is referred to as a uni-objective (or single-objective) problem. A multi-objective problem specifies more than one sub-objective, which need to be simultaneously optimized.

The optimization methods used to solve the above problem types differ significantly, as will be illustrated in the parts that follow.

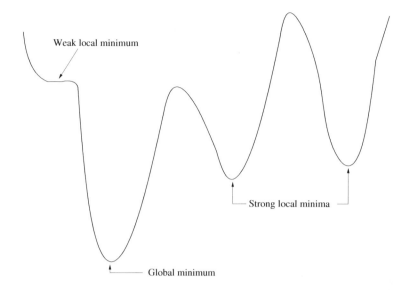

**Figure A.1** Types of Optima for Unconstrained Problems

## A.3 Optima Types

Solutions found by optimization algorithms are classified by the quality of the solution. The main types of solutions are referred to as local optima or global optima. For the purpose of this section, a minimization problem is assumed.

A global optimum (considering a minimization problem) is formally defined as follows:

**Definition A.1 Global minimum:** *The solution* $\mathbf{x}^* \in \mathcal{F}$, *is a global optimum of the objective function, $f$, if*

$$f(\mathbf{x}^*) < f(\mathbf{x}), \ \forall \mathbf{x} \in \mathcal{F} \tag{A.1}$$

*where* $\mathcal{F} \subseteq \mathcal{S}$.

The global optimum is therefore the best of a set of candidate solutions, as illustrated in Figure A.1 for a minimization problem. As illustrated in Figure A.2, a problem may have more than one global optimum.

A local minimum is defined as follows (as illustrated in Figure A.1):

**Definition A.2 Strong local minimum:** *The solution,* $\mathbf{x}_{\mathcal{N}}^* \in \mathcal{N} \subseteq \mathcal{F}$, *is a strong local minimum of $f$, if*

$$f(\mathbf{x}_{\mathcal{N}}^*) < f(\mathbf{x}), \ \forall \mathbf{x} \in \mathcal{N} \tag{A.2}$$

*where* $\mathcal{N} \subseteq \mathcal{F}$ *is a set of feasible points in the neighborhood of* $\mathbf{x}_{\mathcal{N}}^*$.

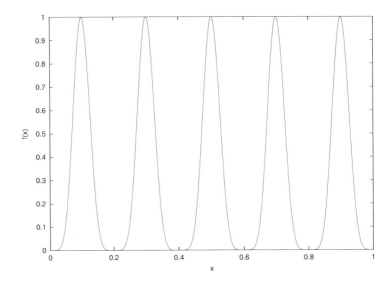

**Figure A.2** Problem with Multiple Global Optima, with $f(x) = \sin^6(5\pi x)$

**Definition A.3 Weak local minimum:** *The solution, $\mathbf{x}_{\mathcal{N}}^* \in \mathcal{N} \subseteq \mathcal{F}$, is a weak local minimum of $f$, if*

$$f(\mathbf{x}_{\mathcal{N}}^*) \leq f(\mathbf{x}), \ \forall \mathbf{x} \in \mathcal{N} \tag{A.3}$$

*where $\mathcal{N} \subseteq \mathcal{F}$ is a set of feasible points in the neighborhood of $\mathbf{x}_{\mathcal{N}}^*$.*

## A.4    Optimization Method Classes

An optimization algorithm searches for an optimum solution by iteratively transforming a current candidate solution into a new, hopefully better, solution. Optimization methods can be divided into two main classes, based on the type of solution that is located. Local search algorithms use only local information of the search space surrounding the current solution to produce a new solution. Since only local information is used, local search algorithms locate local optima (which may be a global minimum). A global search algorithm uses more information about the search space to locate a global optimum. It is said that global search algorithms explore the entire search space, while local search algorithms exploit neighborhoods. Optimization algorithms are further classified into deterministic and stochastic methods. Stochastic methods use random elements to transform one candidate solution into a new solution. The new point can therefore not be predicted. Deterministic methods, on the other hand, do not make use of random elements.

Based on the problem characteristics, optimization methods are grouped in the following classes (within each of these classes further subdivision occurs based on whether local or global optima are located, and based on whether random elements are used to investigate new points in the search space):

- **unconstrained methods**, used to optimize unconstrained problems;
- **constrained methods**, used to find solutions in constrained search spaces;
- **multi-objective optimization methods** for problems with more than one objective to optimize;
- **multi-solution (niching) methods** with the ability to locate more than one solution; and
- **dynamic methods** with the ability to locate and track changing optima.

Subsequent sections discuss each of these optimization method classes.

# A.5   Unconstrained Optimization

Except for boundary constraints, unconstrained optimization problems place no restrictions on the values that can be assigned to variables of the problem. The feasible space is simply the entire search space. This chapter provides a formal definition of unconstrained optimization problems in Section A.5.1. Section A.5.2 summarizes algorithms used in later chapters, and Section A.5.3 gives a list of classical benchmark problems.

## A.5.1   Problem Definition

The general unconstrained optimization problem is defined as

**Definition A.4 Unconstrained optimization problem:**

$$
\begin{aligned}
minimize \quad & f(\mathbf{x}), \quad \mathbf{x} = (x_1, x_2, \dots, x_{n_x}) \\
subject\ to \quad & x_j \in dom(x_j)
\end{aligned}
\tag{A.4}
$$

*where $\mathbf{x} \in \mathcal{F} = \mathcal{S}$, and $dom(x_j)$ is the domain of variable $x_j$.*

For a continuous problem, the domain of each variable is $\mathbb{R}$, i.e. $x_j \in \mathbb{R}$. For an integer problem, $x_j \in \mathbb{Z}$, while $dom(x_i)$ for a general discrete problem is a finite set of values. Note that an integer problem is simply a special case of a discrete problem.

## A.5.2   Optimization Algorithms

Many optimization algorithms have been developed to solve unconstrained problems. This section summarizes only a few that support the material presented in this book. The reader is referred to [201, 262, 263, 317, 806] for more detail.

### General Local Search Procedure

Local search methods follow the same basic structure as given in Algorithm A.1. A starting point, $\mathbf{x}(0)$, is selected, and its quality evaluated. Then, iteratively a search direction is determined and a move is made in that direction.

---

**Algorithm A.1** General Local Search Algorithm

---

Find starting point $\mathbf{x}(0) \in \mathcal{S}$;
$t = 0$;
**repeat**
    Evaluate $f(\mathbf{x}(t))$;
    Calculate a search direction, $\mathbf{q}(t)$;
    Calculate step length $\eta(t)$;
    Set $\mathbf{x}(t+1)$ to $\mathbf{x}(t) + \eta(t)\mathbf{q}(t)$;
    $t = t + 1$;
**until** *stopping condition is true*;
Return $\mathbf{x}(t)$ as the solution;

---

Search directions and step lengths can be determined using steepest gradient descent, conjugate gradients, or Newton methods (amongst many others).

### Beam Search

Beam search (BS) is a classical tree local search method [650]. The search method allows the extension of partial, or approximate, solutions in a number of ways as determined by tree branches. Each step of the algorithm extends a partial solution from the set $\mathcal{B}$ (referred to as the beam) in at most $n_c$ possible ways. A newly constructed solution is stored in the set of complete solutions, $\mathcal{B}_c$, if that solution represents a complete solution. If not, the new partial solution is added to the set $\mathcal{B}_p$ of partial solutions. At the end of each iteration, a new beam, $\mathcal{B}$, is created by selecting $n_b$ solutions from $\mathcal{B}_p$, where $n_b$ is the beam width. BS uses a lower bound on the objective function value to select partial solutions from $\mathcal{B}_p$. The lower bound specifies the minimum objective function value for any complete solution that can be constructed from a partial solution. Partial solutions are usually extended using a deterministic greedy policy, which is based on a weighting function that assigns weights to all the possible extensions.

### Tabu Search

Tabu search (TS) is an iterative neighborhood search algorithm [319, 362], where the neighborhood changes dynamically. TS enhances local search by actively avoiding points in the search space already visited. By avoiding already visited points, loops in search trajectories are avoided and local optima can be escaped. The main feature of TS is the use of an explicit memory. A simple TS usually implements two forms of memory:

- A **frequency-based memory**, which maintains information about how often a search point has been visited (or how often a move has been made) during a specified time interval.

- A **recency-based memory**, which maintains information about how recently a search point has been visited (or how recently a move has been made). Recency is based on the iteration at which the event occurred.

If, for example, the frequency count of a search point exceeds a given threshold, then that point is classified as being tabu for the next cycle of iterations. Positions specified in the tabu list are excluded from the neighborhood of candidate positions that can be visited from the current position. Positions remain in the tabu list for a specified time period.

The following may be used to terminate TS:

- the neighborhood is empty, i.e. all possible neighboring points have already been visited, or

- when the number of iterations since the last improvement is larger than a specified threshold.

**Simulated Annealing**

Annealing refers to the cooling process of a liquid or solid, and the analysis of the behavior of substances as they cool [581]. As temperature reduces, the mobility of molecules reduces, with the tendency that molecules may align themselves in a crystalline structure. The aligned structure is the minimum energy state for the system. To ensure that this alignment is obtained, cooling must occur at a sufficiently slow rate. If the substance is cooled at a too rapid rate, an amorphous state may be reached.

Simulated annealing is an optimization process based on the physical process described above. In the context of mathematical optimization, the minimum of an objective function represents the minimum energy of the system. Simulated annealing is an algorithmic implementation of the cooling process to find the optimum of an objective function [466, 467, 649, 806].

Simulated annealing (SA) uses a random search strategy, which not only accepts new positions that decrease the objective function (assuming a minimization problem), but also accepts positions that increase objective function values. The latter is accepted probabilistically based on the Boltzmann–Gibbs distribution. If $P_{ij}$ is the probability of moving from point $\mathbf{x}_i$ to $\mathbf{x}_j$, then $P_{ij}$ is calculated using

$$P_{ij} = \begin{cases} 1 & \text{if } f(\mathbf{x}_j) < f(\mathbf{x}_i) \\ e^{-\frac{f(\mathbf{x}_j)-f(\mathbf{x}_i)}{c_b T}} & \text{otherwise} \end{cases} \tag{A.5}$$

where $c_b > 0$ is the Boltzmann constant and $T$ is the temperature of the system.

Algorithm A.2 summarizes the SA algorithm. The algorithm requires specification of the following components:

- A **representation of possible solutions**, which is usually a vector of floating-point values.

- A **mechanism to generate new solutions** by adding small random changes to current solutions. For example, for continuous-valued vectors,

$$\mathbf{x}(t+1) = \mathbf{x}(t) + \mathbf{D}(t)\mathbf{r}(t) \tag{A.6}$$

  where $\mathbf{r}(t) \sim U(-1, 1)^{n_x}$, and $\mathbf{D}$ is a diagonal matrix that defines the maximum change allowed in each variable. When an improved solution is found,

$$\mathbf{D}(t+1) = (1 - \alpha)\mathbf{D}(t) + \alpha\omega\mathbf{R}(t) \tag{A.7}$$

  where $\mathbf{R}(t)$ is a diagonal matrix whose elements are the magnitudes of the successful changes made to each variable, and $\alpha$ and $\omega$ are constants.

  For integer problems,
$$\mathbf{x}(t+1) = \mathbf{x}(t) + \mathbf{r}(t) \tag{A.8}$$

  where each element of $\mathbf{r}(t)$ is randomly selected from the set $\{-1, 0, 1\}$.

- A **method to evaluate solutions**, which is usually just the objective function in the case of unconstrained problems.

- An **annealing schedule**, which consists of an initial temperature and rules for lowering the temperature with increase in number of iterations. The annealing schedule determines the degree of uphill movement (objective function increase) allowed during the search. An initial high temperature is selected, which is then incrementally reduced using, for example,

  - **Exponential cooling**: $T(t+1) = \alpha T(t)$, where $\alpha \in (0, 1)$.
  - **Linear cooling**: $T(t+1) = T(t) - \Delta T$, where, e.g. $\Delta T = (T(0) - T(n_t))/n_t$; $T(0)$ is the initial large temperature, and $T(n_t)$ is the final temperature at the last iteration, $n_t$.

---

**Algorithm A.2** Simulated Annealing Algorithm

---

Create initial solution, $\mathbf{x}(0)$;
Set initial temperature, $T(0)$;
$t = 0$;
**repeat**
    Generate new solution, $\mathbf{x}$;
    Determine quality, $f(\mathbf{x})$;
    Calculate acceptance probability using equation (A.5);
    **if** $U(0, 1) \leq$ *acceptance probability* **then**
        $\mathbf{x}(t) = \mathbf{x}$;
    **end**
**until** *stopping condition is true*;
Return $\mathbf{x}(t)$ as the solution;

---

## LeapFrog Algorithm

LeapFrog is an optimization approach based on the physical problem of the motion of a particle of unit mass in an $n_x$-dimensional conservative force field [799, 800]. For more detail refer to Section 3.2.4.

## A.5.3 Example Benchmark Problems

This section lists a number of the classical benchmark functions used to evaluate the performance of optimization algorithms for unconstrained optimization. The purpose of the section is not to provide an extensive list of example problems, but to provide a list that can be used as a good starting point when analyzing the performance of optimization methods.

**Spherical:**

$$f(\mathbf{x}) = \sum_{j=1}^{n_x} x_j^2 \tag{A.9}$$

with $x_j \in [-100, 100]$ and $f^*(\mathbf{x}) = 0.0$.

**Quadric:**

$$f(\mathbf{x}) = \sum_{j=1}^{n_x} \left( \sum_{k=1}^{j} x_j \right)^2 \tag{A.10}$$

with $x_j \in [-100, 100]$ and $f^*(\mathbf{x}) = 0.0$.

**Ackley:**

$$f(\mathbf{x}) = -20e^{-0.2\sqrt{\frac{1}{n_x}\sum_{j=1}^{n_x} x_j^2}} - e^{\frac{1}{n_x}\sum_{j=1}^{n_x}\cos(2\pi x_j)} + 20 + e \tag{A.11}$$

with $x_j \in [-30, 30]$ and $f^*(\mathbf{x}) = 0.0$.

**Bohachevsky 1:**

$$f(x_1, x_2) = x_1^2 + 2x_2^2 - 0.3\cos(3\pi x_1) - 0.4\cos(4\pi x_2) + 0.7 \tag{A.12}$$

with $x_1, x_2 \in [-50, 50]$ and $f^*(x_1, x_2) = 0.0$.

**Colville:**

$$
\begin{aligned}
f(x_1, x_2, x_3, x_4) =\ & 100(x_2 - x_1^2)^2 + (1 - x_1)^2 + 90(x_4 - x_3^2)^2 \\
& + (1 - x_3)^2 + 10.1((x_2 - 1)^2 + (x_4 - 1)^2) \\
& + 19.8(x_2 - 1)(x_4 - 1)
\end{aligned} \tag{A.13}
$$

with $x_1, x_2, x_3, x_4 \in [-10, 10]$ and $f^*(x_1, x_2, x_3, x_4) = 0.0$.

**Easom:**

$$f(x_1, x_2) = -\cos(x_1)\cos(x_2)e^{-(x_1-\pi)^2-(x_2-\pi)^2} \tag{A.14}$$

with $x_1, x_2 \in [-100, 100]$ and $f^*(x_1, x_2) = -1.0$.

**Griewank:**

$$f(\mathbf{x}) = 1 + \frac{1}{4000} \sum_{j=1}^{n_x} x_j^2 - \prod_{j=1}^{n_x} \cos\left(\frac{x_j}{\sqrt{j}}\right) \tag{A.15}$$

with $x_j \in [-600, 600]$ and $f^*(\mathbf{x}) = 0.0$.

**Hyperellipsoid:**

$$f(\mathbf{x}) = \sum_{j=1}^{n_x} j^2 x_j^2 \tag{A.16}$$

with $x_j \in [-1, 1]$ and $f^*(\mathbf{x}) = 0.0$.

**Rastrigin:**

$$f(\mathbf{x}) = \sum_{j=1}^{n_x} (x_j^2 - 10\cos(2\pi x_j) + 10) \tag{A.17}$$

with $x_j \in [-5.12, 5.12]$ and $f^*(\mathbf{x}) = 0.0$.

**Rosenbrock:**

$$f(\mathbf{x}) = \sum_{j=1}^{n_x/2} [100(x_{2j} - x_{2j-1}^2)^2 + (1 - x_{2j-1})^2] \tag{A.18}$$

with $x_j \in [-2.048, 2.048]$ and $f^*(\mathbf{x}) = 0.0$.

**Schwefel:**

$$f(\mathbf{x}) = \sum_{j=1}^{n_x} x_j \sin\left(\sqrt{|x_j|}\right) + 418.9829 n_x \tag{A.19}$$

with $x_j \in [-500, 500]$ and $f^*(\mathbf{x}) = 0.0$.

## A.6  Constrained Optimization

Many real-world optimization problems are solved subject to sets of constraints. Constraints place restrictions on the search space, specifying regions of the space that are infeasible. Optimization algorithms have to find solutions that do not lie in infeasible regions. That is, solutions have to satisfy all specified constraints. This chapter provides a mathematical definition of the constrained optimization problem in Section A.6.1. An overview of constraint handling methods is given in Section A.6.2. Section A.6.3 provides a list of benchmark functions.

### A.6.1  Problem Definition

Assuming a minimization problem, the general constrained problem is defined as:

**Definition A.5 Constrained optimization problem:**

$$minimize \quad f(\mathbf{x}), \quad \mathbf{x} = (x_1, \ldots, x_{n_x})$$
$$subject\ to \quad g_m(\mathbf{x}) \le 0, \ m = 1, \ldots, n_g$$
$$h_m(\mathbf{x}) = 0, \ m = n_g + 1, \ldots, n_g + n_h$$
$$x_j \in dom(x_j) \quad\quad\quad (A.20)$$

*where $n_g$ and $n_h$ are the number of inequality and equality constraints respectively, and $dom(x_j)$ is as defined in Section A.5.1.*

A special instance of the constrained optimization problem is defined below, where only linear equality constraints of the form $\mathbf{Ax} = \mathbf{b}$ are defined:

**Definition A.6 Constrained optimization with linear equality constraints:**

$$minimize \quad f(\mathbf{x}), \quad \mathbf{x} = (x_1, \ldots, x_{n_x})$$
$$subject\ to \quad\quad \mathbf{Ax} = \mathbf{b} \quad\quad\quad (A.21)$$

*where $\mathbf{A} \in \mathbb{R}^{n_h \times n_x}$ and $\mathbf{b} \in \mathbb{R}^{n_h}$.*

Figure A.3 illustrates the effect of constraints. The shaded area indicates the infeasible region of the search space. Note how the global optimum for the unconstrained function is no longer the global optimum for the constrained problem. Instead, the best solution is an extremum (not called an optimum, since the derivative of the best solution is not zero).

## A.6.2   Constraint Handling Methods

The following types of constraints can be found:

- **Boundary constraints**, which basically define the borders of the search space. Upper and lower bounds on each dimension of the search space define the hypercube in which solutions must be found. While boundaries are usually defined by specifying upper and lower bounds on variables, such box constraints are not the only way in which boundaries are specified. The boundary of a search space can, for example, be on the circumference of a hypersphere. It is also the case that a problem can be unbounded.

- **Equality constraints** specify that a function of the variables of the problem must be equal to a constant.

- **Inequality constraints** specify that a function of the variables must be less than or equal to (or, greater than or equal to) a constant.

Constraints can be linear or nonlinear.

Constraint handling methods have to consider a number of important questions, which relate mainly to the trade-off between feasible and infeasible solutions:

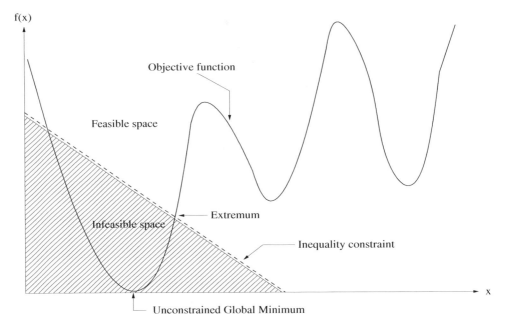

**Figure A.3** Constrained Problem Illustration

- How should two feasible solutions be compared? The answer to this question is somewhat obvious: the solution with the better objective function value should be preferred.

- How should two infeasible solutions be compared? In this case the answer is not at all obvious, and is usually problem-dependent. The issues to consider are:

  - should the infeasible solution with the best objective function value be preferred?
  - should the solution with the least number of constraint violations or lowest degree of violation be preferred?
  - should a balance be found between best objective function value and degree of violation?

- Should it be assumed that any feasible solution is better than any unfeasible solution? Alternatively, can objective function value and degree of violation be optimally balanced? Again, the answer to this problem may be problem-dependent. In financial-critical, or life-critical problems, the first strategy should be preferred to ensure no financial loss, or loss of life. Less critical problems, such as time-tabling, may consider solutions where less severe constraints are violated.

Research in constraint handling methods are numerous in the evolutionary computation (EC) and swarm intelligence (SI) paradigms. Based on these research efforts, constraint handling methods have been categorized in a number of classes [140, 487, 584]:

- **Reject infeasible solutions**, where solutions are not constrained to the feasible space. Solutions that find themselves in infeasible space are simply rejected or ignored.

- **Penalty function methods**, which add a penalty to the objective function to discourage search in infeasible areas of the search space.

- **Convert the constrained problem to an unconstrained problem**, then solve the unconstrained problem.

- **Preserving feasibility methods**, which assumes that solutions are initialized in feasible space, and applies specialized operators to transform feasible solutions to new, feasible solutions. These methods constrict solutions to move only in feasible space, where all constraints are satisfied at all times.

- **Pareto ranking methods**, which use concepts from multi-objective optimization, such as non-dominance (refer to Section A.8), to rank solutions based on degree of violation.

- **Repair methods**, which apply special operators or actions to infeasible solutions to facilitate changing infeasible solutions to feasible solutions.

The rest of this section provides a short definition and discussion of two of these approaches.

## Penalty Methods

Penalty methods add a function to the objective function to penalize vectors that represent infeasible solutions. Assuming a constrained minimization problem as defined in Definition A.5,

**Definition A.7 Penalty method:**

$$minimize \quad F(\mathbf{x}, t) = f(\mathbf{x}, t) + \lambda p(\mathbf{x}, t) \tag{A.22}$$

*where $\lambda$ is the penalty coefficient and $p(\mathbf{x}, t)$ is the (possibly) time-dependent penalty function.*

A major problem of penalty-based methods is due to the added penalty function, which changes the shape of the objective function. While the change in objective function is necessary to increase the slope of the function in the infeasible areas (assuming minimization), the penalty function may introduce false minima into the original search space as defined by the unpenalized function. These false minima may trap search algorithms. Figure A.4 illustrates the effect of penalty functions. Figure A.4(a) illustrates the function

$$f(x_1, x_2) = \frac{x_1 \cos(x_1)}{20} + 2e^{-x_1^2 - (x_2 - 1)^2} + 0.01 x_1 x_2 \tag{A.23}$$

with one clear maximum at $\mathbf{x}^* = (0.0, 0.0)$. Figure A.4(b) illustrates the penalized objective,

$$F(x_1, x_2) = f(x_1, x_2) + \lambda p(x_1, x_2) \tag{A.24}$$

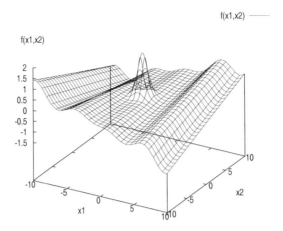

(a) $f(x_1, x_2) = \frac{x_1 \cos(x_1)}{20} + 2e^{-x_1^2 - (x_2-1)^2} + 0.01x_1x_2$

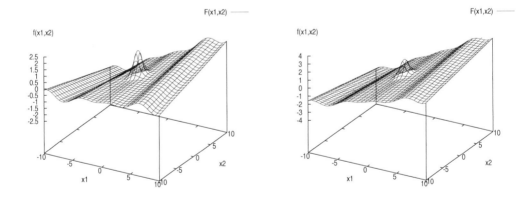

(b) With penalty $p(x_1, x_2) = 3x_1$ and $\lambda = 0.05$        (c) With penalty $p(x_1, x_2) = 3x_1$ and $\lambda = 0.1$

**Figure A.4** Illustration of the Effect of Penalty Functions

where $\lambda = 0.05$ and $p(x_1, x_2) = 3x_1$. Notice how the penalty causes the slope of the function to increase for large $x_1$ and $x_2$ values, with function values exceeding that of the maximum of the original function $f(x_1, x_2)$. This effect is for a low penalty coefficient of $\lambda = 0.05$. For larger values, the slope increases even more as illustrated in Figure A.4(c) for $\lambda = 0.1$.

The penalty function is usually constructed from a set of functions, one for each of the constraints, quantifying the degree to which a solution violates the constraint. That is,

$$p(\mathbf{x}_i, t) = \sum_{m=1}^{n_g+n_h} \lambda_m(t) p_m(\mathbf{x}_i) \tag{A.25}$$

where

$$p_m(\mathbf{x}_i) = \begin{cases} \max\{0, g_m(\mathbf{x}_i)^\alpha\} & \text{if } m \in [1, \ldots, n_g] \text{ (inequality)} \\ |h_m(\mathbf{x}_i)|^\alpha & \text{if } m \in [n_g + 1, \ldots, n_g + n_h] \text{ (equality)} \end{cases} \tag{A.26}$$

with $\alpha$ a positive constant, representing the power of the penalty. In equation (A.25), $\lambda_m(t)$ represents a time-varying degree to which violation of the $m$-th constraint contributes to the overall penalty. More emphasis can therefore be given to crucial constraints. The constraint penalty coefficient can of course be static, i.e. $\lambda_m(t) = \lambda_m$.

## Convert Constrained to Unconstrained Problem

A constrained problem can be converted to an unconstrained problem by defining the Lagrangian for the constrained problem, and then by maximizing the Lagrangian. Consider the standard constrained optimization problem as defined in Definition A.5, referred to as the *primal problem*. The constraints in equation (A.20) can be introduced into the objective function, $f$, by augmenting it with a weighted sum of the constraint functions. Let $\lambda_g \in \mathbb{R}^{n_g}$ be the weights associated with the $n_g$ inequality constraints, and $\lambda_h \in \mathbb{R}^{n_h}$ be the weights associated with the $n_h$ equality constraints.

These vectors, referred to as the Lagrange multiplier vectors, define the Lagrangian, $L : \mathbb{R}^{n_x} \times \mathbb{R}^{n_g} \times \mathbb{R}^{n_h}$,

$$L(\mathbf{x}, \lambda_g, \lambda_h) = f(\mathbf{x}) + \sum_{m=1}^{n_g} \lambda_{gm} g_m(\mathbf{x}) + \sum_{m=n_g+1}^{n_g+n_h} \lambda_{hm} h_m(\mathbf{x}) \tag{A.27}$$

The dual problem associated with the primal problem in equation (A.20) is then defined as

## Definition A.8 Dual problem:

$$\begin{align} maximize_{\lambda_g, \lambda_h} \quad & L(\mathbf{x}, \lambda_g, \lambda_h) \\ subject\ to \quad & \lambda_{gm} \geq 0, \quad m = 1, \ldots, n_g + n_h \end{align} \tag{A.28}$$

If the primal problem is convex over the search space $\mathcal{S}$, then the solution to the primal problem is the vector $\mathbf{x}^*$ of the saddle point, $(\mathbf{x}^*, \lambda_g^*, \lambda_h^*)$, of the Lagrangian in equation (A.27), such that

$$L(\mathbf{x}^*, \lambda_g, \lambda_h) \leq L(\mathbf{x}^*, \lambda_g^*, \lambda_h^*) \leq L(\mathbf{x}, \lambda_g^*, \lambda_h^*) \tag{A.29}$$

The vector $\mathbf{x}^*$ that solves the primal problem, as well as the Lagrange multiplier vectors, $\lambda_g^*$ and $\lambda_h^*$, can be found by solving the min-max problem,

$$\min_{\mathbf{x}} \max_{\lambda_g, \lambda_h} L(\mathbf{x}, \lambda_g, \lambda_h) \tag{A.30}$$

For non-convex problems, the solution of the dual problem does not coincide with the solution of the primal problem. For non-convex problems, the Lagrangian is augmented by adding a penalty term, i.e.

$$L'(\mathbf{x}, \lambda_g, \lambda_h) = L(\mathbf{x}, \lambda_g, \lambda_h) + \lambda p(\mathbf{x}, t) \tag{A.31}$$

where $\lambda > 0$, $L(\mathbf{x}, \lambda_g, \lambda_h)$ is as defined in equation (A.27), and the penalty $p(\mathbf{x}, t)$ is as defined in equations (A.25) and (A.26) with $\lambda_m(t) = 1$ and $\alpha = 2$.

## A.6.3   Example Benchmark Problems

A number of benchmark functions for constrained optimization are listed in this section. Again, the list is not intended to be complete. The objective is to provide a list of constrained problems as a starting point in evaluating algorithms for constrained optimization.

**Constrained problem 1:** Minimize the function

$$f(\mathbf{x}) = 100(x_2 - x_1^2)^2 + (1 - x_1)^2 \tag{A.32}$$

subject to the nonlinear constraints,

$$x_1 + x_2^2 \geq 0$$
$$x_1^2 + x_2 \geq 0$$

with $x_1 \in [-0.5, 0.5]$ and $x_2 \leq 1.0$. The global optimum is $\mathbf{x}^* = (0.5, 0.25)$, with $f(\mathbf{x}^*) = 0.25$.

**Constrained problem 2:** Minimize the function

$$f(\mathbf{x}) = (x_1 - 2)^2 - (x_2 - 1)^2 \tag{A.33}$$

subject to the nonlinear constraint

$$-x_1^2 + x_2 \geq 0$$

and the linear constraint

$$x_1 + x_2 \leq 2$$

with $\mathbf{x}^* = (1, 1)$ and $f(\mathbf{x}^*) = 1$.

**Constrained problem 3:** Minimize the function

$$f(\mathbf{x}) = 5x_1 + 5x_2 + 5x_3 + 5x_4 - 5\sum_{j=1}^{4} x_j^2 - \sum_{j=5}^{13} x_j \tag{A.34}$$

subject to the constraints

$$
\begin{array}{ll}
2x_1 + 2x_2 + x_{10} + x_{11} \leq 10 & \quad 2x_1 + 2x_3 + x_{10} + x_{12} \leq 10 \\
2x_2 + 2x_3 + x_{11} + x_{12} \leq 10 & \quad -8x1 + x_{10} \leq 0 \\
-8x_2 + x_{11} \leq 0 & \quad -8x_3 + x_{12} \leq 0 \\
-2x_4 - x_5 + x_10 \leq 0 & \quad -2x_6 - x_7 + x_{11} \leq 0 \\
-2x_8 - x_9 + x_{12} \leq 0 &
\end{array}
$$

with $x_j \in [0, 1]$ for $j = 1, \ldots, 9$, $x_j \in [0, 100]$ for $j = 10, 11, 12$, and $x_{13} \in [0, 1]$. The solution is $\mathbf{x}^* = (1, 1, 1, 1, 1, 1, 1, 1, 1, 3, 3, 3, 1)$, with $f(\mathbf{x}^*) = -15$.

**Constrained problem 4:** Maximize the function

$$f(\mathbf{x}) = (\sqrt{n_x})^{n_x} \prod_{j=1}^{n_x} x_j \qquad (A.35)$$

subject to the equality constraint,

$$\sum_{j=1}^{n_x} x_j^2 = 1$$

with $x_j \in [0, 1]$. The solution is $\mathbf{x}^* = (\frac{1}{\sqrt{n_x}}, \ldots, \frac{1}{\sqrt{n_x}})$, with $f(\mathbf{x}^*) = 1$.

**Constrained problem 5:** Minimize the function

$$-10.5x_1 - 7.5x_2 - 3.5x_3 - 2.5x_4 - 1.5x_5 - 10x_6 - 0.5 \sum_{j=1}^{5} x_j^2 \qquad (A.36)$$

subject to the constraints

$$6x_1 + 3x_2 + 3x_3 + 2x_4 + x_5 - 6.5 \;\leq\; 0$$
$$10x_1 + 10x_3 + x_6 \leq 20$$

with $x_j \in [0, 1]$ for $j = 1, \ldots, 5$, and $x_6 \geq 0$. The best known solution is $f(\mathbf{x}) = -213.0$.

## A.7 Multi-Solution Problems

Multi-solution problems are multi-modal, containing many optima. These optima may include more than one global optimum and a number of local minima, or just one global optimum together with more than one local optimum. The objective of multi-solution optimization methods is to locate as many as possible of these optima. A formal definition is given in Section A.7.1, with different algorithm categories listed in Section A.7.2. Example benchmark problems are given in Section A.7.3.

### A.7.1 Problem Definition

A multi-solution problem is formally defined as follows (assuming minimization):

**Definition A.9 Multi-solution problem:** *Find a set of solutions,* $\mathcal{X} = \{\mathbf{x}_1^*, \mathbf{x}_2^*, \ldots, \mathbf{x}_{n_x}^*\}$, *such that each* $\mathbf{x}^* \in \mathcal{X}$ *is a minimum of the general optimization problem as defined in Definition A.5. That is, for each* $\mathbf{x}^* \in \mathcal{X}$,

$$||f'(\mathbf{x}^*)|| \leq \epsilon(1 + |f(\mathbf{x}^*)|) \qquad (A.37)$$

*where* $\epsilon$ *is, for example, the square root of machine precision.*

In the evolutionary computation literature, multi-solution algorithms are referred to as *niching* or *speciation* algorithms.

## A.7.2 Niching Algorithm Categories

Niching algorithms can be categorized based on the way that niches are located. Three categories are identified:

- **Sequential niching** (or temporal niching) develops niches over time. The process iteratively locates a niche (or optimum), and removes any references to it from the search space. Removal of references to niches usually involves modification of the search space. The search for, and removal of niches continues sequentially until a convergence criterion is met, for example, no more niches can be obtained over a number of generations.

- **Parallel niching** locates all niches in parallel. Individuals dynamically self-organize, or speciate, on the locations of optima. In addition to locating niches, parallel niching algorithms need to organize individuals such that they maintain their positions around optimal locations over time. That is, once a niche is found, individuals should keep grouping around the niche.

- **Quasi-sequential niching** locates niches sequentially, but does not change the search space to remove the niche. Instead, the search for a new niche continues, while the found niches are refined and maintained in parallel.

Regardless of the way in which niches are located, a further categorization of niching algorithms can be made according to speciation behavior [553]:

- **Sympatric speciation**, where individuals form species that coexist in the same search space, but evolve to exploit different resources. For example, different kinds of fish feed on different food sources in the same environment.

- **Allopatric speciation**, where differentiation between individuals is based on spatial isolation in the search space. There is no interspecies communication, and subspecies can develop only through deviation from the available genetic information (triggered by mutation). As an example, consider different fish species that live and play around their food sources, with no concern about the existence of other species living in different areas.

- **Parapatric speciation**, where development of new species is evolved as a result of segregated species sharing a common border. Communication between the original species may not have been encouraged or intended. For example, new fish species may evolve based on the interaction with a small percentage of different schools of fish. The new species may have different food requirements and may eventually upset the environment's stability.

## A.7.3   Example Benchmark Problems

This section lists five easy functions to test niching algorithms. In addition to these, any of the multi-modal functions listed in the previous chapters can be used.

**Niching problem 1:** Maximize

$$f(x) = \sin^6(5\pi x) \tag{A.38}$$

for $x \in [0, 1]$. The solutions are located at $x = 0.1, x = 0.3, x = 0.5, x = 0.7$ and $x = 0.9$.

**Niching problem 2:** Maximize

$$f(x) = \left(e^{-2\log(2)\times\left(\frac{x-0.1}{0.8}\right)^2}\right) \times \sin^6(5\pi x) \tag{A.39}$$

for $x \in [0, 1]$. The solutions are located at $x = 0.08, x = 0.25, x = 0.45, x = 0.68$ and $x = 0.93$.

**Niching problem 3:** Maximize

$$f(x) = \sin^6(5\pi(x^{3/4} - 0.05)) \tag{A.40}$$

for $x \in [0, 1]$. The solutions are located at $x = 0.1, x = 0.3, x = 0.5, x = 0.7$ and $x = 0.9$.

**Niching problem 4:** Maximize

$$f(x) = \left(e^{-2\log(2)\times\left(\frac{x-0.08}{0.854}\right)^2}\right) \times \sin^6(5\pi(x^{3/4} - 0.05)) \tag{A.41}$$

for $x \in [0, 1]$. The solutions are located at $x = 0.08, x = 0.25, x = 0.45, x = 0.68$ and $x = 0.93$.

**Niching problem 5 (modified Himmelblau function):** Maximize

$$f(x_1, x_2) = 200 - (x_1^2 + x_2 - 11)^2 - (x_1 + x_2^2 - 7)^2 \tag{A.42}$$

with $x_1, x_2 \in [-5, 5]$. Maxima are located at $(-2.81, 3.13)$, $(3.0, 2.0)$, $(3.58, -1.85)$ and $(-3.78, -3.28)$.

# A.8   Multi-Objective Optimization

Many real-world problems require the simultaneous optimization of a number of objective functions. Some of these objectives may be in conflict with one another. For example, consider finding optimal routes in data communications networks, where the objectives may include to minimize routing cost, to minimize route length, to minimize congestion, and to maximize utilization of physical infrastructure. There is an important trade-off between the last two objectives: minimization of congestion is achieved by reducing the utilization of links. A reduction in utilization, on the other hand, means that infrastructure, for which high installation and maintenance costs are incurred, is under-utilized.

This chapter provides a theoretical overview of multi-objective optimization (MOO), focusing on definitions that are needed in later chapters. The objective of this chapter is by no means to give a complete treatment of MOO. The reader can find more in-depth treatments in [150, 195]. Section A.8.1 defines the multi-objective problem (MOP), and discusses the meaning of an optimum in terms of MOO. Section A.8.2 summarizes weight aggregation approaches to solve MOPs. Section A.8.3 provides definitions of Pareto-optimality and dominance, and a lists a few example problems.

## A.8.1   Multi-objective Problem

Let $\mathcal{S} \subseteq \mathbb{R}^{n_x}$ denote the $n_x$-dimensional search space, and $\mathcal{F} \subseteq \mathcal{S}$ the feasible space. With no constraints, the feasible space is the same as the search space. Let $\mathbf{x} = (x_1, x_2, \ldots, x_{n_x}) \in \mathcal{S}$, referred to as a *decision vector*. A single objective function, $f_k(\mathbf{x})$, is defined as $f_k : \mathbb{R}^{n_x} \to \mathbb{R}$. Let $\mathbf{f}(\mathbf{x}) = (f_1(\mathbf{x}), f_2(\mathbf{x}), \ldots, f_{n_k}(\mathbf{x})) \in \mathcal{O} \subseteq \mathbb{R}^{n_k}$ be an *objective vector* containing $n_k$ objective function evaluations; $\mathcal{O}$ is referred to as the *objective space*. The search space, $\mathcal{S}$ is also referred to as the *decision space*.

Using the notation above, the multi-objective optimization problem is defined as:

**Definition A.10  Multi-objective problem:**

$$
\begin{aligned}
minimize \quad & \mathbf{f}(\mathbf{x}) \\
subject\ to \quad & g_m(\mathbf{x}) \leq 0, \quad m = 1, \ldots, n_g \\
& h_m(\mathbf{x}) = 0, \quad m = n_g + 1, \ldots, n_g + n_h \\
& \mathbf{x} \in [\mathbf{x}_{min}, \mathbf{x}_{max}]^{n_x}
\end{aligned}
\tag{A.43}
$$

In equation (A.43), $g_m$ and $h_m$ are respectively the inequality and equality constraints, while $\mathbf{x} \in [\mathbf{x}_{min}, \mathbf{x}_{max}]$ represents the boundary constraints. Solutions, $\mathbf{x}^*$, to the MOP are in the feasible space, i.e. all $\mathbf{x}^* \in \mathcal{F}$.

The meaning of an "optimum" has to be redefined for MOO. In terms of uni-objective optimization (UOO) where only one objective is optimized, a local optimum and global optimum is as defined in Section A.3. In terms of MOO, the definition of optimality is not that simple. The main problem is the presence of conflicting objectives, where improvement in one objective may cause a deterioration in another objective. For example, maximization of the structural stability of a mechanical structure may cause an increase in costs, working against the additional objective to minimize costs. Trade-offs exist between such conflicting objectives, and the task is to find solutions that balance these trade-offs. Such a balance is achieved when a solution cannot improve any objective without degrading one or more of the other objectives. These solutions are referred to as *non-dominated solutions*, of which many may exist.

The objective when solving a MOP is therefore to produce a set of good compromises, instead of a single solution. This set of solutions is referred to as the *non-dominated set*, or the *Pareto-optimal set*. The corresponding objective vectors in objective space are referred to as the *Pareto front*. The concepts of dominance and Pareto-optimality are defined in the next section.

## A.8.2   Weighted Aggregation Methods

One of the simplest approaches to deal with MOPs, is to define an aggregate objective function as a weighted sum of the objectives. Uni-objective optimization algorithms can then be applied, without any changes to the algorithm, to find optimum solutions. For aggregation methods, the MOP is redefined as

$$
\begin{aligned}
\text{minimize} \quad & \sum_{k=1}^{n_k} \omega_k f_k(\mathbf{x}) \\
\text{subject to} \quad & g_m(\mathbf{x}) \leq 0, \quad m = 1, \ldots, n_g \\
& h_m(\mathbf{x}) = 0, \quad m = n_g + 1, \ldots, n_g + n_h \\
& \mathbf{x} \in [\mathbf{x}_{min}, \mathbf{x}_{max}]^{n_x} \\
& \omega_k \geq 0, k = 1, \ldots, n_k
\end{aligned}
\tag{A.44}
$$

It is also usually assumed that $\sum_{k=1}^{n_k} \omega_k = 1$.

The aggregation approach does, however, have a number of problems:

- The algorithm has to be applied repeatedly to find different solutions if a single-solution PSO is used. However, even for repeated applications, there is no guarantee that different solutions will be found. Alternatively, a niching strategy can be used to find multiple solutions.

- It is difficult to get the best weight values, $\omega_k$, since these are problem-dependent.

- Aggregation methods can only be applied to generate members of the Pareto-optimal set when the Pareto front is concave, regardless of the values of $\omega_k$ [174, 422].

The second problem can be addressed by not using fixed values, but to have these weights change dynamically. The following two schemes can be used to dynamically adjust the weights for two-objective problems [664, 665]:

- **Bang-bang weighted aggregation:**

$$
\begin{aligned}
\omega_1(t) &= \text{sign}(\sin(2\pi t/\tau)) \\
\omega_2(t) &= 1 - \omega_1(t)
\end{aligned}
\tag{A.45}
$$

  where $\tau$ is the weights' change frequency. Weights are changed abruptly due to the use of sign in equation (A.45).

- **Dynamic weighted aggregation:**

$$
\begin{aligned}
\omega_1(t) &= |\sin(2\pi t/\tau)| \\
\omega_2(t) &= 1 - \omega_1(t)
\end{aligned}
\tag{A.46}
$$

  With this approach, weights change more gradually.

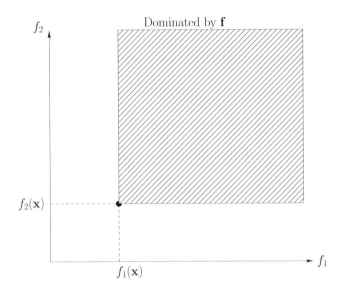

**Figure A.5** Illustration of Dominance

## A.8.3 Pareto-Optimality

This section provides a number of definitions that are needed when talking about MOO. Definitions include dominance, Pareto-optimal, Pareto-optimal front, and others. These definitions assume minimization.

**Definition A.11 Domination:** *A decision vector,* $\mathbf{x}_1$ *dominates a decision vector,* $\mathbf{x}_2$ *(denoted by* $\mathbf{x}_1 \prec \mathbf{x}_2$*), if and only if*

- $\mathbf{x}_1$ *is not worse than* $\mathbf{x}_2$ *in all objectives, i.e.* $f_k(\mathbf{x}_1) \leq f_k(\mathbf{x}_2), \forall k = 1, \dots, n_k,$ *and*

- $\mathbf{x}_1$ *is strictly better than* $\mathbf{x}_2$ *in at least one objective, i.e.* $\exists k = 1, \dots, n_k : f_k(\mathbf{x}_1) < f_k(\mathbf{x}_2).$

Similarly, an objective vector, $\mathbf{f}_1$, dominates another objective vector, $\mathbf{f}_2$, if $\mathbf{f}_1$ is not worse than $\mathbf{f}_2$ in all objective values, and $\mathbf{f}_1$ is better than $\mathbf{f}_2$ in at least one of the objective values. Objective vector dominance is denoted by $\mathbf{f}_1 \prec \mathbf{f}_2$.

From Definition A.11, solution $\mathbf{x}_1$ is better than solution $\mathbf{x}_2$ if $\mathbf{x}_1 \prec \mathbf{x}_2$ (i.e. $\mathbf{x}_1$ dominates $\mathbf{x}_2$), which happens when $\mathbf{f}_1 \prec \mathbf{f}_2$.

The concept of dominance is illustrated in Figure A.5, for a two-objective function, $\mathbf{f}(\mathbf{x}) = (f_1(\mathbf{x}), f_2(\mathbf{x}))$. The striped area denotes the area of objective vectors dominated by $\mathbf{f}$.

**Definition A.12 Weak domination:** *A decision vector,* $\mathbf{x}_1$*, weakly dominates a decision vector,* $\mathbf{x}_2$ *(denoted by* $\mathbf{x}_1 \preceq \mathbf{x}_2$*), if and only if*

- $\mathbf{x}_1$ *is not worse that* $\mathbf{x}_2$ *in all objectives, i.e.* $f_k(\mathbf{x}_1) \leq f_k(\mathbf{x}_2), \forall k = 1, \ldots, n_k.$

**Definition A.13 Pareto-optimal:** *A decision vector,* $\mathbf{x}^* \in \mathcal{F}$ *is Pareto-optimal if there does not exist a decision vector,* $\mathbf{x} \neq \mathbf{x}^* \in \mathcal{F}$ *that dominates it. That is,* $\nexists k : f_k(\mathbf{x}) < f_k(\mathbf{x}^*).$ *An objective vector,* $\mathbf{f}^*(\mathbf{x})$*, is Pareto-optimal if* $\mathbf{x}$ *is Pareto-optimal.*

The concept of Pareto-optimality is named after the mathematician Vilfredo Pareto, who generalized this concept, first introduced by Francis Ysidro Edgeworth.

**Definition A.14 Pareto-optimal set:** *The set of all Pareto-optimal decision vectors form the Pareto-optimal set,* $\mathcal{P}^*$*. That is,*

$$\mathcal{P}^* = \{\mathbf{x}^* \in \mathcal{F} | \nexists \mathbf{x} \in \mathcal{F} : \mathbf{x} \prec \mathbf{x}^*\} \tag{A.47}$$

The Pareto-optimal set therefore contains the set of solutions, or balanced trade-offs, for the MOP. The corresponding objective vectors are referred to as the Pareto-optimal front:

**Definition A.15 Pareto-optimal front:** *Given the objective vector,* $\mathbf{f}(\mathbf{x})$*, and the Pareto-optimal solution set,* $\mathcal{P}^*$*, then the Pareto-optimal front,* $\mathcal{PF}^* \subseteq \mathcal{O}$*, is defined as*

$$\mathcal{PF}^* = \{\mathbf{f} = (f_1(\mathbf{x}^*), f_2(\mathbf{x}^*), \ldots, f_k(\mathbf{x}^*)) | \mathbf{x}^* \in \mathcal{P}\} \tag{A.48}$$

The Pareto front therefore contains all the objective vectors corresponding to decision vectors that are not dominated by any other decision vector. An example Pareto front is illustrated for the following functions in Figure A.6 [664, 665]:

- The convex, uniform Pareto front in Figure A.6(a) for MOP

$$\mathbf{f}(\mathbf{x}) = (f_1(\mathbf{x}), f_2(\mathbf{x})) \tag{A.49}$$

$$f_1(\mathbf{x}) = \frac{1}{n_x} \sum_{j=1}^{n_x} x_j^2$$

$$f_2(\mathbf{x}) = \frac{1}{n_x} \sum_{j=1}^{n_x} (x_j - 2)^2$$

- The convex, non-uniform Pareto front in Figure A.6(b) for MOP

$$\mathbf{f}(\mathbf{x}) = (f_1(\mathbf{x}), f_2(\mathbf{x})) \tag{A.50}$$
$$f_1(\mathbf{x}) = x_1$$
$$f_2(\mathbf{x}) = g(\mathbf{x})(1 - \sqrt{f_1(\mathbf{x})/g(\mathbf{x})})$$

where

$$g(\mathbf{x}) = 1 + \frac{9}{n_x - 1} \sum_{j=2}^{n_x} x_j$$

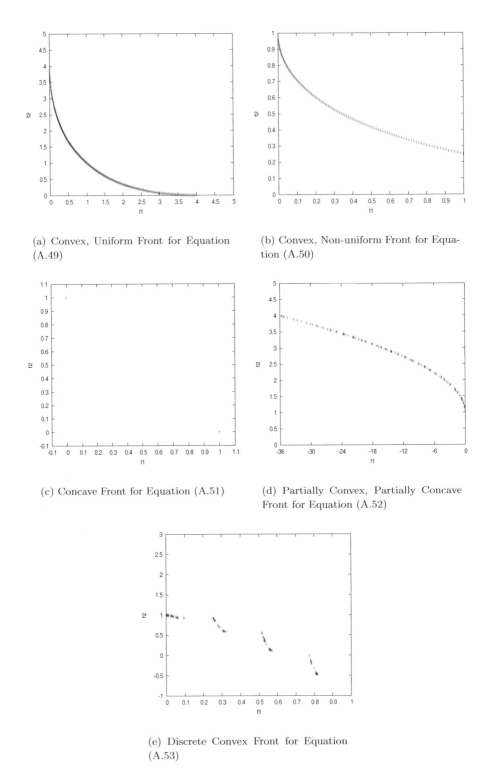

(a) Convex, Uniform Front for Equation (A.49)

(b) Convex, Non-uniform Front for Equation (A.50)

(c) Concave Front for Equation (A.51)

(d) Partially Convex, Partially Concave Front for Equation (A.52)

(e) Discrete Convex Front for Equation (A.53)

**Figure A.6** Example Pareto-Optimal Fronts

- The concave Pareto front in Figure A.6(c) for the MOP of Figure A.6(b), but with

$$f_2(\mathbf{x}) = g(\mathbf{x})(1 - (f_1(\mathbf{x})/g(\mathbf{x}))^2) \tag{A.51}$$

- The partially concave and partially convex Pareto front in Figure A.6(d) for the MOP of Figure A.6(b), but with

$$f_2(\mathbf{x}) = g(\mathbf{x})(1 - \sqrt[4]{f_1(\mathbf{x})/g(\mathbf{x})} - (f_1(\mathbf{x})/g(\mathbf{x}))^4) \tag{A.52}$$

- The discrete, convex Pareto front in Figure A.6(e) for the MOP of Figure A.6(b), but with

$$f_2(\mathbf{x}) = g(\mathbf{x})(1 - \sqrt{f_1(\mathbf{x})/g(\mathbf{x})} - (f_1(\mathbf{x})/g(\mathbf{x}))\sin(10\pi f_1(\mathbf{x}))) \tag{A.53}$$

The objective when solving a MOP is to approximate the true Pareto-optimal front, and then to select the solution that represents the best trade-off (for problems which, in the end, require only one solution). To find the exact true Pareto-optimal front (i.e. to find all the Pareto-optimal solutions in $\mathcal{F}$) is usually computationally prohibitive. The task is therefore reduced to finding an approximation to the true Pareto front such that

- the distance to the Pareto front is minimized,
- the set of non-dominated solutions, i.e. the Pareto-optimal set, is as diverse as possible, and
- already found non-dominated solutions are maintained.

The task of finding an approximation to the true Pareto front is therefore in itself a MOP, where the first objective ensures an accurate approximation, and the second objective ensures that the entire Pareto front is covered.

## A.9 Dynamic Optimization Problems

Dynamic optimization problems have objective functions that change over time. Such changes in objective function cause changes in the position of optima, and the characteristics of the search space. Existing optima may disappear while new optima may appear. This chapter provides a formal definition of a dynamic problem in Section A.9.1, and lists different types of dynamic problems in Section A.9.2. Example benchmark problems are given in Section A.9.3.

### A.9.1 Definition

A dynamic optimization problem is formally defined as

**Definition A.16 Dynamic optimization problem:**

$$minimize \quad f(\mathbf{x}, \varpi(t)), \quad \mathbf{x} = (x_1, \ldots, x_{n_x}), \varpi(t) = (\varpi_1(t), \ldots, \varpi_{n_\varpi})$$
$$subject \ to \quad g_m(\mathbf{x}) \le 0, \ m = 1, \ldots, n_g$$
$$h_m(\mathbf{x}) = 0, \ m = n_g + 1, \ldots, n_g + n_h$$
$$x_j \in dom(x_j) \quad \text{(A.54)}$$

*where $\varpi(t)$ is a vector of time-dependent objective function control parameters. The objective is to find*

$$\mathbf{x}^*(t) = \min_{\mathbf{x}} f(\mathbf{x}, \varpi(t)) \quad \text{(A.55)}$$

*where $\mathbf{x}^*(t)$ is the optimum found at time step $t$.*

The goal of an optimization algorithm for dynamic environments is then to locate an optimum and to track its trajectory as closely as possible.

## A.9.2 Dynamic Environment Types

In order to track the optimum over time, the optimization algorithm needs to detect and track changes. The environment may change on any timescale, referred to as *temporal severity*. Changes can be continuously spread over time, at irregular time intervals or periodically. Due to these changes, the position of an optimum may change by any amount, referred to as *spatial severity*.

Eberhart *et al.* defines three types of dynamic environments [228, 385]:

- **Type I environments**, where the location of the optimum in problem space is subject to change. The change in the optimum, $\mathbf{x}^*(t)$ is quantified by the severity parameter, $\zeta$, which measures the jump in location of the optimum.

- **Type II environments**, where the location of the optimum remains the same, but the value, $f(\mathbf{x}^*(t))$, of the optimum changes.

- **Type III environments**, where both the location of the optimum and its value changes.

The changes in the environment, as caused by the control parameters, can be in one or more of the dimensions of the problem. If the change is in all the dimensions, then for type I environments, the change in optimum is quantified by $\zeta \mathbf{I}$, where $\mathbf{I}$ is the unit vector.

Examples of these types of dynamic environments are illustrated in Figures A.7 and A.8. Figure A.7 illustrates the dynamic function,

$$f(\mathbf{x}, \varpi(t)) = \sum_{j=1}^{n_x} (x_j - \varpi_1(t))^2 + \varpi_2(t) \quad \text{(A.56)}$$

for $n_x = 2$. Figure A.7(a) illustrates the static function with both control parameters set to zero. A type I environment is illustrated in A.7(b) with $\varpi_1 = 3$ and $\varpi_2 = 0$.

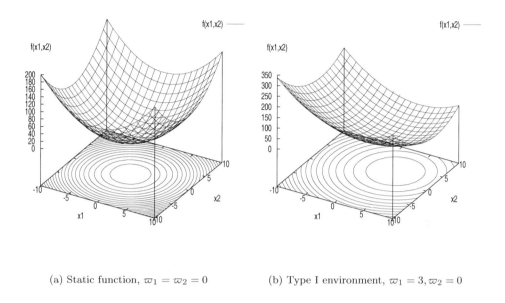

(a) Static function, $\varpi_1 = \varpi_2 = 0$        (b) Type I environment, $\varpi_1 = 3, \varpi_2 = 0$

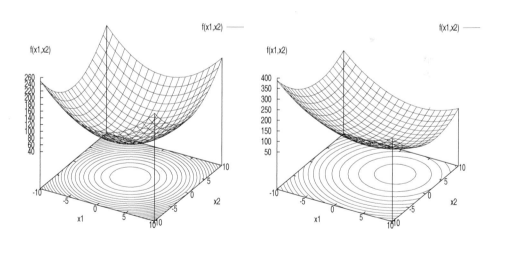

(c) Type II environment, $\varpi_1 = 0, \varpi_2 = 50$      (d) Type III environment, $\varpi_1 = 3, \varpi_2 = 50$

**Figure A.7** Dynamic Parabola Objective Function

With control parameter $\varpi_1 \neq 0$, the location of the optimum, $\mathbf{x}^* = (0, 0)$, moves with severity, $\zeta = \varpi_1^2$. Figure A.7(c) illustrates a type II environment with $\varpi_1 = 0$ and $\varpi_2 = 50$. The position of the optimum remains at $\mathbf{x}^* = (0, 0)$, but its value changes form 0.0 to $\varpi_2$. A type III environment is illustrated in Figure A.7(d), with $\varpi_1 = 3$ and $\varpi_2 = 50$.

Figure A.8 illustrates the static version, type I, II and III environments respectively in subfigures (a), (b), (c) and (d), for the following function:

$$f(\mathbf{x}, \varpi) = |f_1(\mathbf{x}, \varpi) + f_2(\mathbf{x}, \varpi) + f_3(\mathbf{x}, \varpi)| \tag{A.57}$$

with

$$
\begin{aligned}
f_1(\mathbf{x}, \varpi) &= \varpi_1(1 - x_1)^2 e^{(-x_1^2 - (x_2 - 1)^2)} \\
f_2(\mathbf{x}, \varpi) &= -0.1 \left( \frac{x_1}{5} - \varpi_2 x_1^3 - x_2^5 \right)^2 e^{(-x_1^2 - x_2^2)} \\
f_3(\mathbf{x}, \varpi) &= 0.5 e^{(-(x_1 + 1)^2 - x_2^2)}
\end{aligned}
$$

with

| | |
|---|---|
| Static | $\varpi_1 = 1, \varpi_2 = 1$ |
| Type I | $\varpi_1 = 1, \varpi_2 = 2.5$ |
| Type II | $\varpi_1 = 3, \varpi_2 = 1$ |
| Type III | $\varpi_1 = 0.5, \varpi_2 = 5$ |

Note, for the type II environment, the appearance of more maxima, which becomes the global optimum for the type III environment.

Noisy environments (a special form of dynamic environment) can easily be visualized and tested by adding Gaussian noise to the objective function:

$$f(\mathbf{x}, \varpi(t)) = f(\mathbf{x})(1 + N(0, \varpi(t))) \tag{A.58}$$

## A.9.3   Example Benchmark Problems

In addition to the functions given in equations (A.56) and (A.57), the following dynamic test function generator [193] can be used to generate functions to test optimization algorithms for dynamic environments:

$$f(x_1, x_2) = \max_{l=1, \ldots, n_\chi} \left[ H_l - R_l \sqrt{(x_1 - x_{1l})^2 + (x_2 - x_{2l})^2} \right] \tag{A.59}$$

where $n_\chi$ is the number of optima, and the $l$-th optimum is specified by its position $(x_{1l}, x_{2l})$, height $H_l$, and slope $R_l$. Dynamic environments can be simulated by changing these parameters over time.

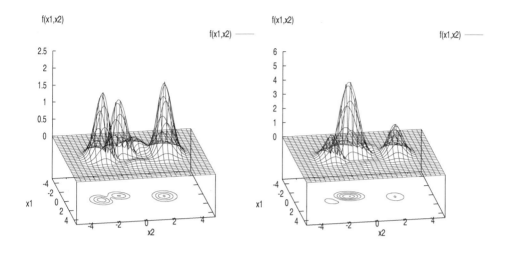

(a) Static function, $\varpi_1 = \varpi_2 = 1$      (b) Type I environment, $\varpi_1 = 1, \varpi_2 = 2.5$

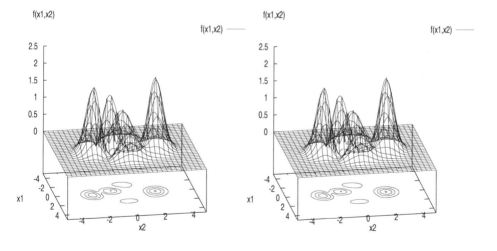

(c) Type II environment, $\varpi_1 = 3, \varpi_2 = 1$      (d) Type III environment, $\varpi_1 = 0.5, \varpi_2 = 5$

**Figure A.8** Dynamic Objective Function

# Index

Printed in the USA/Agawam, MA
August 6, 2013